ISBN 978-1-330-16592-8
PIBN 10042384

FB &c Ltd, Dalton House, 60 Windsor Avenue, London, SW19 2RR.
Company number 08720141. Registered in England and Wales.

For support please visit www.forgottenbooks.com

COMMENTARY

ON

NEWTON'S PRINCIPIA.

WITH

A SUPPLEMENTARY VOLUME.

DESIGNED FOR THE USE OF STUDENTS AT THE UNIVERSITIES.

BY

J. M. F. WRIGHT, A. B.

LATE SCHOLAR OF TRINITY COLLEGE, CAMBRIDGE, AUTHOR OF SOLUTIONS
OF THE CAMBRIDGE PROBLEMS, &c. &c.

IN TWO VOLUMES.

VOL. II.

LONDON:
PRINTED FOR T. T. & J. TEGG, 73, CHEAPSIDE;
AND RICHARD GRIFFIN & CO., GLASGOW.

MDCCCXXXIII.

Gift of J. Wangenheim
to Math. Dept.

GLASGOW:
GEORGE BROOKMAN, PRINTER, VILLAFIELD.

INTRODUCTION

TO

VOLUME II.

AND TO THE

MECANIQUE CELESTE.

ANALYTICAL GEOMETRY

1. *To determine the position of a point in fixed space.*

Assume any point A in fixed space as known and immoveable, and let

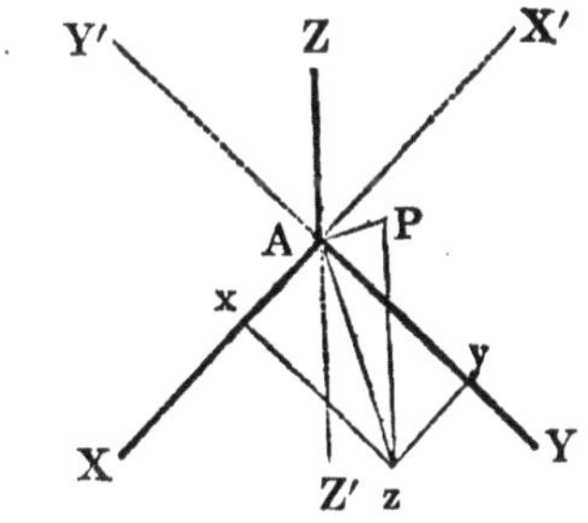

three fixed planes of indefinite extent, be taken at right angles to one another and passing through A. Then shall their intersections A X′, A Y′, A Z′ pass through A and be at right angles to one another.

This being premised, let P be any point in fixed space; from P draw P z parallel to A Z, and from z where it meets the plane X A Y, draw z x, z y parallel to A Y, A X respectively. Make

$$A x = x, \quad A y = y, \quad P z = z.$$

Then it is evident that if x, y, z are given, the point P can be found *practically* by taking A x = x, A y = y, drawing x z, y z parallel to A Y, A X; lastly, from their intersection, making z P parallel to A Z and equal to z. Hence x, y, z determine the position of the point P.

The lines x, y, z are called the rectangular coordinates of the point P; the point A the origin of coordinates; the lines A X, A Y, A Z the axes of coordinates, A X being further designated the axis of x, A Y the axis of y, and A Z the axis of z; and the planes X A Y, Z A X, Z A Y coordinate planes.

These coordinate planes are respectively denoted by

plane (x, y), plane (x, z), plane (y, z);

and in like manner, any point whose coordinates are x, y, z is denoted briefly by

point (x, y, z).

If the coordinates x, y, z when measured along A X, A Y, A Z be always considered positive; when measured in the opposite directions, viz. along A X′ A Y′, A Z′, they must be taken negatively. Thus accordingly as P is in the spaces

Z A X Y, Z A Y X′, Z A X′ Y′, Z A Y′ X;
Z′ A X Y, Z′ A Y X′, Z′ A X′ Y′, Z′ A Y′ X,

the point P will be denoted by
point (x, y, z), point (— x, y, z), point (— x, — y, z), point (x, — y, z); point (x, y, – z), point (– x, y, – z), point (– x, – y, – z), point (x, – y, – z) respectively.

2. *Given the position of two points* (α, β, γ), $(\alpha', \beta', \gamma')$ *in fixed space, to find the distance between them.*

The distance P P′ is evidently the diagonal of a rectangular parallelopiped whose three edges are parallel to A X, A Y, A Z and equal to

$$\alpha \backsim \alpha', \quad \beta \backsim \beta', \quad \gamma \backsim \gamma'.$$

Hence

$$P P' = \sqrt{(\alpha - \alpha')^2 + (\beta - \beta')^2 + (\gamma - \gamma')^2} \quad \ldots \quad (1)$$

the distance required.

Hence if P′ coincides with A or α', β', γ' equal zero,

$$P A = \sqrt{\alpha^2 + \beta^2 + \gamma^2} \quad \ldots \quad (2)$$

3. Calling the distance of any point P (x, y, z) from the origin A of coordinates the *radius-vector*, and denoting it by ϱ, suppose it inclined to the axes A X, A Y, A Z or to the planes (y, z), (x, z), (x, y), by the angles X, Y, Z.

Then it is easily seen that

$$x = \varrho \cos. X, \; y = \varrho \cos. Y, \; z = \varrho \cos. Z \; \ldots\ldots \quad (3)$$

Hence (see 2)

$$\cos. X = \frac{x}{\sqrt{(x^2 + y^2 + z^2)}}, \; \cos. Y = \frac{y}{\sqrt{(x^2 + y^2 + z^2)}},$$

$$\cos. Z = \frac{z}{\sqrt{(x^2 + y^2 + z^2)}} \; \ldots\ldots \quad (4)$$

so that *when the coordinates of a point are given, the angles which the radius-vector makes with each of the axes may hence be found.*

Again, adding together the squares of equations (3), we have

$$(x^2 + y^2 + z^2) = \varrho^2 (\cos.^2 X + \cos.^2 Y + \cos.^2 Z).$$

But

$$\varrho^2 = x^2 + y^2 + z^2 \text{ (see 2)},$$

$$\therefore \cos.^2 X + \cos.^2 Y + \cos.^2 Z = 1 \; \ldots\ldots \quad (5)$$

which shows that when two of these angles are given the other may be found.

4. *Given two points in space, viz.* (α, β, γ), $(\alpha', \beta', \gamma')$, *and one of the coordinates of any other point* (x, y, z) *in the straight line that passes through them, to determine this other point*; that is, *required the equations to a straight line given in space.*

The distances of the point (α, β, γ) from the points $(\alpha', \beta', \gamma')$, and (x, y, z) are respectively, (see 2)

$$P P' = \sqrt{(\alpha - \alpha')^2 + (\beta - \beta')^2 + (\gamma - \gamma')^2},$$

and

$$P Q = \sqrt{(\alpha - x)^2 + (\beta - y)^2 + (\gamma - z)^2}.$$

But from similar triangles, we get

$$(\gamma - z)^2 : (P Q)^2 :: (\gamma - \gamma')^2 : (P P')^2$$

whence

$$(\gamma - z)^2 = \frac{(\gamma - \gamma')^2}{(\alpha - \alpha')^2 + (\beta - \beta')^2 + (\gamma - \gamma')^2} \cdot \{(\alpha - x)^2 + (\beta - y)^2 + (\gamma - z)^2\}$$

which gives

$$\{(\alpha - \alpha')^2 + (\beta - \beta')^2\} (\gamma - z)^2 = (\gamma - \gamma')^2 . \{(\alpha - x)^2 + (\beta - y)^2\}$$

But since α, α' are independent of β, β' and vice versa, the two first terms of the equation,

$$(\alpha - \alpha')^2 . (\gamma - z)^2 - (\gamma - \gamma')^2 (\alpha - x)^2 - (\gamma - \gamma')^2 (\beta - y)^2 + (\beta - \beta')^2 (\gamma - z)^2 = 0$$

are essentially different from the last. Consequently by (6 vol. 1.)

$$(\alpha - \alpha')^2 (\gamma - z)^2 = (\gamma - \gamma')^2 (\alpha - x)^2$$
$$(\beta - \beta')^2 (\gamma - z)^2 = (\gamma - \gamma')^2 (\beta - y)^2$$

which give

$$\left.\begin{array}{l} z - \gamma = \pm \dfrac{\gamma - \gamma'}{\alpha - \alpha} (\alpha - x') \\ z - \gamma = + \dfrac{\gamma - \gamma'}{\beta - \beta'} (\beta - y) \end{array}\right\} \quad \ldots\ldots\ldots \quad (6)$$

These results may be otherwise obtained; thus, p g p′, is the projection of the given line on the plane (x, y) &c. as in fig.

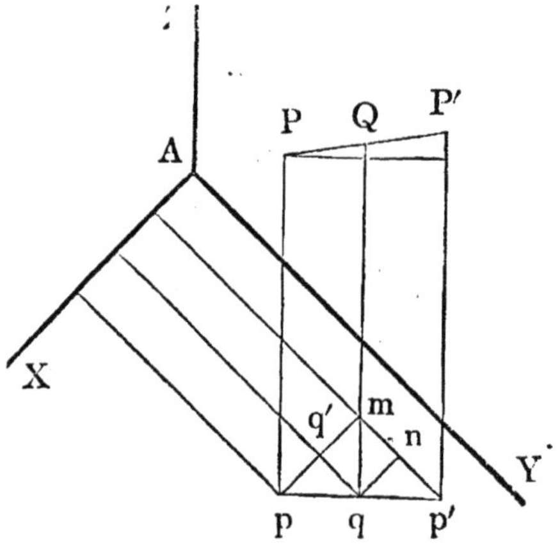

Hence

$$z - \gamma : \gamma' - \gamma :: p\,q : p\,p'$$
$$:: m\,n : m\,p'$$
$$:: y - \beta : \beta' - \beta$$

Also

$$z - \gamma : \gamma' - \gamma :: p\,q : p\,p' :: p\,r : p\,m$$
$$:: \alpha - x : \alpha - \alpha'.$$

Hence the general forms of the equations to a straight line given in space, not considering signs, are

$$\left.\begin{array}{l} z = \alpha x + b \\ z = \alpha' y + b' \end{array}\right\} \quad \ldots\ldots\ldots\ldots \quad (7)$$

To find where the straight line meets the planes, (x, y), (x, z), (y, z), we make

$$z = 0, \; y = 0, \; x = 0,$$

which give

$$x = - \frac{b}{a}$$

$$y = -\frac{b}{a'}$$
$$z = b'$$
$$x = \frac{b' - b}{a}$$
$$z = b$$
$$y = \frac{b - b'}{a'}$$

which determine the points required.

To find when the straight line is parallel to the planes, (x, y), (x, z), (y, z), we must make z, y, x, respectively constant, and the equations become of the form

$$\left.\begin{aligned} z &= c \\ a'y &= ax + b - b' \end{aligned}\right\} \quad \ldots\ldots\ldots \quad (8)$$

To find when the straight line is perpendicular to the planes, (x, y), (x, z) (y, z), or parallel to the axes of z, y, x, we must assume x, y; x, z; y, z; respectively constant, and z, y, x, will be any whatever.

To find the equations to a straight line passing through the origin of coordinates; we have, since $x = 0$, and $y = 0$, when $z = 0$,

$$\left.\begin{aligned} z &= ax \\ z &= a'y \end{aligned}\right\} \quad \ldots\ldots\ldots\ldots \quad (9)$$

5. *To find the conditions that two straight lines in fixed space may intersect one another; and also their point of intersection.*

Let their equations be

$$\left.\begin{aligned} z &= ax + A \\ z &= by + B \end{aligned}\right\}$$
$$\left.\begin{aligned} z &= a'x + A' \\ z &= b'y + B' \end{aligned}\right\}$$

from which eliminating x, y, z, we get the equation of condition

$$\frac{a'A - aA'}{a' - a} = \frac{b'B - bB'}{b' - b}.$$

Also when this condition is fulfilled, the point is found from

$$x = \frac{A - A'}{a' - a}, \quad y = \frac{B - B'}{b' - b}, \quad z = \frac{a'A - aA'}{a' - a}. \quad \ldots \quad (10)$$

6. *To find the angle I, at which these lines intersect.*

Take an isosceles triangle, whose equal sides measured along these lines equal 1, and let the side opposite the angle required be called i; then it is evident that

$$\cos. I = 1 - \tfrac{1}{2}\, i^2$$

But if at the extremities of the line i, the points in the intersecting lines be (x', y', z'), (x'', y'', z''), then (see 2)

$$i^2 = (x' - x'')^2 + (y' - y')^2 + (z' - z'')^2$$

$$\therefore\ 2 \cos. I = 2 - \{(x' - x'')^2 + (y' - y'')^2 + (z' - z'')^2\}$$

But by the equations to the straight lines, we have (5)

$$\left.\begin{array}{l} z' = a x' + A \\ z' = b y' + B \end{array}\right\}$$

$$\left.\begin{array}{l} z'' = a' x'' + A' \\ z'' = b' y'' + B' \end{array}\right\}$$

and by the construction, and Art. 2, if (x, y, z) be the point of intersection,

$$\left.\begin{array}{l} (x - x')^2 + (y - y')^2 + (z - z')^2 = 1 \\ (x - x'')^2 + (y - y'')^2 + (z - z'')^2 = 1 \end{array}\right\}$$

Also at the point of intersection,

$$\left.\begin{array}{l} z = a x + A = a' x + A' \\ z = b y + B = b' y + B' \end{array}\right\}$$

From these several equations we easify get

$$z - z' = a (x - a')$$

$$y - y' = \frac{a}{b} (x - x')$$

$$z - z'' = a' (x - x'')$$

$$y - y'' = \frac{a'}{b'} (x - x'')$$

whence by substitution,

$$(x - x')^2 + a^2 (x - x')^2 + \frac{a^2}{b^2} (x - x')^2 = 1$$

$$(x - x'')^2 + a'^2 (x - x'')^2 + \frac{a'^2}{b'^2} (x - x'')^2 = 1$$

which give

$$x - x' = \frac{1}{\sqrt{\left(1 + a^2 + \frac{a^2}{b^2}\right)}}$$

$$x - x'' = \frac{1}{\sqrt{\left(1 + a'^2 + \frac{a'^2}{b'^2}\right)}}$$

Hence

$$(x' - x'')^2 = \frac{1}{1 + a^2 + \frac{a^2}{b^2}} + \frac{1}{1 + a'^2 + \frac{a'^2}{b'^2}} - \frac{2}{\sqrt{\left(1 + a^2 + \frac{a^2}{b^2}\right)} \sqrt{\left(1 + a'^2 + \frac{a'^2}{b'^2}\right)}}$$

Also, since

$$y - y' = \frac{a}{b}(x - x')$$

$$y - y'' = \frac{a'}{b'}(x - x'')$$

and

$$z - z' = a\ (x - x')$$
$$z - z'' = a'\ (x - x'')$$

we have

$$(y'-y'')^2 = \frac{a^2}{b^2} \cdot \frac{1}{1+a^2+\frac{a^2}{b^2}} + \frac{a'^2}{b'^2} \cdot \frac{1}{1+a'^2+\frac{a'^2}{b'^2}} - \frac{aa'}{bb'} \cdot \frac{2}{\sqrt{\left(1+a^2+\frac{a^2}{b^2}\right)}\sqrt{\left(1+a'^2+\frac{a'^2}{b'^2}\right)}}$$

$$(z'-z'')^2 = \frac{a^2}{1+a^2+\frac{a^2}{b^2}} + \frac{a'^2}{1+a'^2+\frac{a'^2}{b'^2}} - aa' \frac{2}{\sqrt{\left(1+a^2+\frac{a^2}{b^2}\right)}\sqrt{\left(1+a'^2+\frac{a'^2}{b'^2}\right)}}$$

Hence by adding these squares together we get

$$2 \cos. I = 2 - \left\{ 1+1 - \frac{2\left(1 + aa' + \frac{aa'}{bb'}\right)}{\sqrt{\left(1 + a^2 + \frac{a^2}{b^2}\right)}\sqrt{\left(1 + a'^2 + \frac{a'^2}{b'^2}\right)}} \right\}$$

which gives

$$\cos. I = \frac{1 + aa' + \frac{aa'}{bb'}}{\sqrt{\left(1 + a^2 + \frac{a^2}{b^2}\right)}\sqrt{\left(1 + a'^2 + \frac{a'^2}{b'^2}\right)}} \quad \ldots\ldots \quad (11)$$

This result may be obtained with less trouble by drawing straight lines from the origin of coordinates, parallel to the intersecting lines; and then finding the cosine of the angle formed by these new lines. For the new angle is equal to the one sought, and the equations simplify into

$$\left.\begin{array}{l} z' = ax' = by', \quad z'' = a'x'' = b'y'' \\ z = ax = by, \quad z = a'x = b'y \\ x'^2 + y'^2 + z'^2 = 1 \\ x''^2 + y''^2 + z''^2 = 1 \end{array}\right\}$$

From the above general expression for the angle formed by two intersecting lines, many particular consequences may be deduced.

For instance, *required the conditions requisite that two straight lines given in space may intersect at right angles.*

That they intersect at all, this equation must be fulfilled. (see 5)

$$\frac{a'A - aA'}{a' - a} = \frac{b'B - bB'}{b' - b};$$

and that being the case, in order for them to intersect at right angles, we have

$$I = \frac{\pi}{2}, \cos. I = 0$$

and therefore

$$1 + a a' + \frac{a a'}{b b'} = 0. \quad . \quad . \quad . \quad . \quad . \quad . \quad . \quad (12)$$

7. In the preceding No. the angle between two intersecting lines is expressed in a function of the rectangular coordinates, which determine the positions of those lines. But since the lines themselves would be given in *parallel* position, if their inclinations to the planes, (x, y), (x, z), (y, z), were given, it may be required, from other *data*, to find the same angle.

Hence denoting generally the complements of the inclinations of a straight line to the planes, (x, y), (x, z), (y, z), by Z, Y, X, the problem may be stated and resolved, as follows:

Required the angle made by the two straight lines, whose angles of inclination are Z, Y, X; Z′, Y,′, X′.

Let two lines be drawn, from the origin of the coordinates, parallel to given lines. These make the same angles with the coordinate planes, and with one another, as the given lines. Moreover, making an isosceles triangle, whose vertex is the origin, and equal sides equal unity, we have, as in (6),

$$\cos. I = 1 - \tfrac{1}{2} i^2 = 1 - \tfrac{1}{2}\{(x - x')^2 + (y - y')^2 + (z - z')^2\}$$

the points in the straight lines equally distant from the origin being (x, y, z), (x′, y′, z′).

But in this case,

$$x^2 + y^2 + z^2 = 1$$
$$x'^2 + y'^2 + z'^2 = 1$$

and

$$x = \cos. X, \; y = \cos. Y, \; z = \cos. Z$$
$$x' = \cos. X', \; y' = \cos. Y', \; z' = \cos. Z'$$

$$\therefore \cos. I = x x' + y y' + z z'$$
$$= \cos. X. \cos. X' + \cos. Y. \cos. Y' + \cos. Z. \cos. Z'. \quad . \quad (13)$$

Hence when the lines pass through the origin of coordinates, the same expression for their mutual inclination holds good; but at the same time X, Y, Z; X′, Y′, Z′, not only mean the *complements of the inclinations* to the planes as above described, but also the *inclinations* of the lines to the axes of coordinates of x, y, z, respectively.

8. *Given the length* (L) *of a straight line and the complements of its inclinations to the planes* (x, y), (x, z), (y z), viz. Z, Y, X, *to find the lengths of its projections upon those planes.*

By the figure in (4) it is easily seen that

$$\left.\begin{array}{l} \text{L projected on the plane } (x, y) = L.\sin. Z \\ \text{————————} (x, z) = L.\sin. Y \\ \text{————————} (y, z) = L.\sin. X \end{array}\right\} \quad . \quad . \quad . \quad (14)$$

9. Instead of determining the parallelism or direction of a straight line in space by the angles Z, Y, X, it is more concise to do it by means of Z (for instance) and the angle θ which its projection on the plane (x, y) makes with the axis of x.

For, drawing a line parallel to the given line from the origin of the coordinates, the projection of this line is parallel to that of the given line, and letting fall from any point (x, y, z) of the new line, perpendiculars upon the plane (x, y) and upon the axes of x and of y, it easily appears, that

$$x = L \cos. X = (L \sin. Z) \cos. \theta \text{ (see No. 8)}$$
$$y = L.\cos. Y = (L.\sin. Z) \sin. \theta$$

which give

$$\left.\begin{array}{l} \cos. X = \sin. Z.\cos. \theta \\ \cos. Y = \sin. Z.\sin. \theta \end{array}\right\} \quad . \quad . \quad . \quad . \quad . \quad . \quad . \quad (15)$$

Hence the expression (13) assumes this form,

$$\cos. I = \sin. Z.\sin. Z' (\cos. \theta \cos. \theta' + \sin. \theta \sin. \theta') + \cos. Z \cos. Z'$$
$$= \sin. Z.\sin. Z' \cos. (\theta - \theta') + \cos. Z.\cos. Z' \quad . \quad . \quad . \quad . \quad (16)$$

which may easily be adapted to logarithmic computation.

The expression (5) is merely verified by the substitution.

10. *Given the angle of intersection* (I) *between two lines in space and their inclinations to the plane* (x, y), *to find the angle at which their projections upon that plane intersect one another.*

If, as above, Z, Z′ be the complements of the inclinations of the lines upon the plane, and θ, θ' the inclinations of the projections to the axis of x, we have from (16)

$$\cos. (\theta - \theta') = \frac{\cos. I - \cos. Z.\cos. Z'}{\sin. Z.\sin. Z'} \quad . \quad . \quad . \quad . \quad . \quad (17)$$

This result indicates that I, Z, Z′ are sides of a spherical triangle (radius = I), $\theta - \theta'$ being the angle subtended by I. The form may at once indeed be obtained by taking the origin of coordinates as the center of the sphere, and radii to pass through the angles of the spherical triangle, measured along the axis of z, and along lines parallel to the given lines.

Having considered at some length the mode of determining the position and properties of points and straight lines in fixed space, we proceed to treat, in like manner, of planes.

It is evident that the position of a plane is fixed or determinate in position when three of its points are known. Hence is suggested the following problem.

11. *Having given the three points* (α, β, γ), $(\alpha', \beta', \gamma')$, $(\alpha'', \beta'', \gamma'')$ *in space, to find the equation to the plane passing through them;* that is, *to find the relation between the coordinates of any other point in the plane.*

Suppose the plane to make with the planes (z, y), (z, x) the intersec-

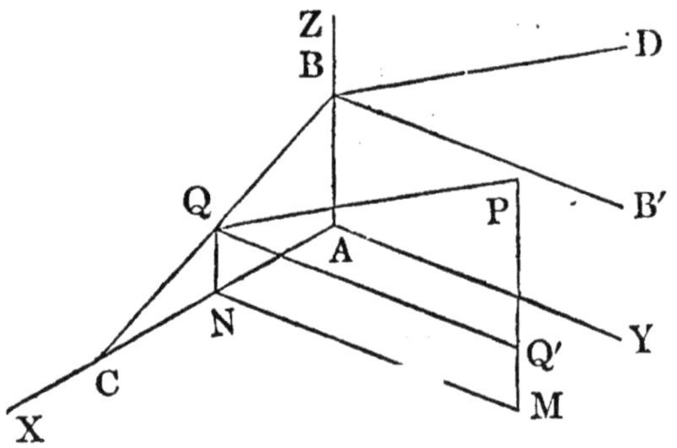

tions or traces B D, B C respectively, and let P be any point whatever in the plane; then through P draw P Q in that plane parallel to B D, &c. as above. Then

$$z - QN = PQ' = QQ' \cot. DBZ$$
$$= y \cot. DBZ.$$

But

$$QN = AB - NA . \cot. CBA$$
$$= AB + x \cot. CBZ,$$
$$\therefore z = AB + x \cot. CBZ + y \cot. DBZ.$$

Consequently if we put A B = g, and denote by (X, Z), (Y, Z) the inclinations to A Z of the traces in the planes of (x, z), (y, z) respectively, we have

$$z = g + x \cot. (X, Z) + y \cot. (Y, Z) \quad . \quad . \quad . \quad . \quad . \quad (18)$$

Hence the form of the equation to the plane is generally

$$z = Ax + By + C \quad . \quad . \quad . \quad . \quad . \quad . \quad . \quad . \quad . \quad . \quad . \quad (19)$$

Now to find these constants there are given the coordinates of three points of the plane; that is

$$\gamma = A\,\alpha + B\,\beta + C$$
$$\gamma' = A\,\alpha' + B\,\beta' + C$$
$$\gamma'' = A\,\alpha'' + B\,\beta'' + C$$

from which we get

$$A = \frac{\gamma\,\beta' - \gamma'\,\beta + \gamma'\,\beta'' - \gamma''\,\beta' + \gamma''\,\beta - \gamma\,\beta''}{\alpha\,\beta' - \alpha'\,\beta + \alpha'\,\beta'' - \alpha''\,\beta' + \alpha''\,\beta - \alpha\,\beta''} = \text{cot.}\,(X, Z) \ldots (20)$$

$$B = \frac{\gamma\,\alpha' - \gamma'\,\alpha + \gamma'\,\alpha'' - \gamma''\,\alpha' + \gamma''\,\alpha - \gamma\,\alpha''}{\alpha\,\beta' - \alpha'\,\beta + \alpha'\,\beta'' - \alpha''\,\beta' + \alpha''\,\beta - \alpha\,\beta''} = \text{cot.}\,(Y, Z) \ldots (21)$$

$$C = \frac{\beta''\,(\gamma\,\alpha' - \gamma'\,\alpha) + \beta\,(\gamma'\,\alpha'' - \gamma''\,\alpha') + \beta'\,(\gamma''\,\alpha - \gamma\,\alpha'')}{\alpha\,\beta' - \alpha'\,\beta + \alpha'\,\beta'' - \alpha''\,\beta' + \alpha''\,\beta - \alpha\,\beta''} \ldots (22)$$

Hence when the trace coincides with the axis of x, we have

$$C = 0,$$

and

$$A = \text{cot.}\,\frac{\pi}{2} = 0$$

$$\left.\begin{array}{r}\beta''\,(\gamma\,\alpha_{,} - \gamma'\,\alpha) + \beta\,(\gamma'\,\alpha'' - \gamma''\,\alpha') + \beta'\,(\gamma''\,\alpha - \gamma\,\alpha'') = 0\\ \gamma\,\beta' - \gamma'\,\beta + \gamma'\,\beta'' - \gamma''\,\beta' + \gamma''\,\beta - \gamma\,\beta'' = 0\end{array}\right\} \ldots (23)$$

$$B = \frac{1}{\beta''} \cdot \frac{(\beta - \beta_{II}) \,.\, (\gamma'\,\alpha'' - \gamma''\,\alpha') + (\beta' - \beta'') \,.\, (\gamma''\,\alpha - \gamma\,\alpha'')}{\alpha\,\beta' - \alpha'\,\beta + \alpha'\,\beta'' - \alpha''\,\beta' + \alpha''\,\beta - \alpha\,\beta''} \ldots\ldots (24)$$

and the equation to the plane becomes

$$z = B\,y \quad . \; . \; . \; . \; . \; . \; . \; . \; . \; . \quad (25)$$

When the plane is parallel to the plane (x, y),

$$A = 0, \quad B = 0,$$

and

$$z = C \; . \; . \; . \; . \; . \; . \; . \; . \; . \; . \; . \; . \; . \quad (26)$$

from which, by means of A = 0, B = 0, any two of the quantities $\gamma, \gamma', \gamma''$ being eliminated, the value of C will be somewhat simplified.

Hence also will easily be deduced a number of other particular results connected with the theory of the plane, the point, and the straight line, of which the following are some.

To find the projections on the planes (x, y), (x, z), (y, z) *of the intersection of the planes,*

$$z = A\,x + B\,y + C,$$
$$z = A'\,x + B'\,y + C'.$$

Eliminating z, we have

$$(A - A')\,x + (B - B')\,y + C - C' = 0 \quad . \; . \; . \; . \; . \quad (27)$$

which is the equation to the projection on (x, y).

Eliminating x, we get

$$(A'-A)z+(AB'-A'B)y+AC'-A'C=0 \quad . \quad . \quad . \quad . \quad (28)$$

which is the equation to the projection on the plane (y, z).
And in like manner, we obtain

$$(B'-B)z+(A'B-AB')x+BC'-B'C=0 \quad . \quad . \quad . \quad . \quad (29)$$

for the projection on the plane (x, z).

To find the conditions requisite that a plane and straight line shall be parallel or coincide.

Let the equations to the straight line and plane be

$$\left.\begin{array}{l} x = a z + A \\ y = b z + B \end{array}\right\}$$

$$z = A' x + B' y + C'.$$

Then by substitution in the latter, we get

$$z(A'a+B'b-1)+A'A+B'B+C'=0.$$

Now if the straight line and plane have only one point common, we should thus at once have the coordinates to that point.

Also if the straight line coincide with the plane in the above equation, z is indeterminate, and (Art. 6. vol. I.)

$$A'a+B'b-1=0,\ A'A+B'B+C'=0 \quad . \quad . \quad . \quad (27)$$

But finally if the straight line is merely to be parallel to the plane, the above conditions ought to be fulfilled even when the plane and line are moved parallelly up to the origin or when A, B, C′ are zero. The only condition in this case is

$$A'a+B'b=1 \quad . \quad . \quad . \quad . \quad . \quad . \quad . \quad . \quad . \quad . \quad (28)$$

To find the conditions that a straight line be perpendicular to a plane $z = Ax + By + C$.

Since the straight line is to be perpendicular to the given plane, the plane which projects it upon (x, y) is at right angles both to the plane (x, y) and to the given plane. The intersection, therefore, of the plane (x, y) and the given plane is perpendicular to the projecting plane. Hence the trace of the given plane upon (x, y) is perpendicular to the projection on (x, y) of the given straight line. But the equations of the traces of the plane on (x, z), (y, z), are

$$\left.\begin{array}{l} z = Ax + C,\ z = By + C \\ \text{or} \\ x = \frac{1}{A}z - \frac{C}{A},\ y = \frac{1}{B}z - \frac{C}{B} \end{array}\right\} \quad . \quad . \quad . \quad . \quad . \quad . \quad . \quad (29)$$

and if those of the perpendicular be

$$\left.\begin{array}{l} x = a z + A_{,} \\ y = b z + B_{,} \end{array}\right\}$$

it is easily seen from (11) or at once, that the condition of these traces being at right angles to the projections, are

$$A + a = 0, \quad A + b = 0.$$

To draw a straight line passing through a given point (α, β, γ) *at right angles to a given plane.*

The equations to the straight line, are clearly

$$x - \alpha + A(z - \gamma) = 0, \quad y - \beta + B(z - \gamma) = 0. \quad \ldots \quad (30)$$

To find the distance of a given point (α, β, γ) *from a given plane.*

The distance is (30) evidently, when (x, y, z) is the common point in the plane and perpendicular

$$\sqrt{(z-\gamma)^2 + (y-\beta)^2 + (x-\alpha)^2} = (z-\gamma)\sqrt{1 + A^2 + B^2}.$$

But the equation to the plane then also subsists, viz.

$$z = Ax + By + C$$

from which, and the equations to the perpendicular, we have

$$z - \gamma = C - \gamma + A\alpha + B\beta,$$

therefore the distance required is

$$\frac{C - \gamma + A\alpha + B\beta}{A^2 + B^2} \quad \ldots\ldots\ldots \quad (31)$$

To find the angle I *formed by two planes*

$$z = Ax + By + C,$$
$$z = A'x + B'y + C'.$$

If from the origin perpendiculars be let fall upon the planes, the angle which they make is equal to that of the planes themselves. Hence, if generally, the equations to a line passing through the origin be

$$\left.\begin{array}{l} x = az \\ y = bz \end{array}\right\}$$

the conditions that it shall be perpendicular to the first plane are

$$A + a = 0,$$
$$B + b = 0,$$

and for the other plane

$$A' + a = 0,$$
$$B' + b = 0.$$

Hence the equations to these perpendiculars are

$$\left.\begin{array}{l} x + Az = 0 \\ y + Bz = 0 \end{array}\right\}$$
$$\left.\begin{array}{l} x + A'z = 0 \\ y + B'z = 0, \end{array}\right\}$$

which may also be deduced from the forms (30).

Hence from (11) we get

$$\cos. I = \frac{1 + A A' + B B'}{\surd (1 + A^2 + B^2) \surd (1 + A'^2 + B'^2)} \quad . \quad . \quad . \quad (32)$$

Hence *to find the inclination* (ε) *of a plane with the plane* (x, y).

We make the second plane coincident with (x, y), which gives

$$A' = 0, \quad B' = 0,$$

and therefore

$$\cos. \varepsilon = \frac{1}{\surd (1 + A^2 + B^2)} \quad . \quad . \quad . \quad . \quad . \quad . \quad . \quad (33)$$

In like manner may the inclinations (ζ), (η) of a plane

$$z = A x + B y + C$$

to the planes (x, z), (y, z) be expressed by

$$\left.\begin{aligned} \cos. \zeta &= \frac{B}{\surd (1 + A^2 + B^2)} \\ \cos. \eta &= \frac{A}{\surd (1 + A^2 + B^2)} \end{aligned}\right\} \quad . \quad . \quad . \quad . \quad . \quad . \quad . \quad (34)$$

Hence

$$\cos.^2 \varepsilon + \cos.^2 \zeta + \cos.^2 \eta = 1 \quad . \quad . \quad . \quad . \quad . \quad . \quad . \quad (35)$$

Hence also, if ε', ζ', η' be the inclinations of another plane to (x, y), (x, z), (y, z).

$$\cos. I = \cos. \varepsilon \cos. \varepsilon' + \cos. \zeta \cos. \zeta' + \cos. \eta \cos. \eta' \quad . \quad . \quad . \quad (36)$$

To find the inclination υ *of a straight line* $x = a z + A'$, $y = b z + B'$, *to the plane* $z = A x + B y + C$.

The angle required is that which it makes with its projection upon the plane. If we let fall from any part of the straight line a perpendicular upon the plane, the angle of these two lines will be the complement of υ. From the origin, draw any straight line whatever, viz. $x = a' z$, $y = b' z$. Then in order that it may be perpendicular to the plane, we must have

$$a' = - A, \quad b' = - B.$$

The angle which this makes with the given line can be found from (11); consequently by that expression

$$\sin. \upsilon = \frac{1 - A a - B b}{\surd (1 + a^2 + b^2) \surd (1 + A^2 + B^2)} \quad . \quad . \quad . \quad . \quad (37)$$

Hence we easily find that the angles made by this line and the coordinate planes (x, y), (x, z), (y, z), viz. Z, Y, X are found from

$$\cos. Z = \frac{1}{\surd (1 + a^2 + b^2)},$$

$$\cos. Y = \frac{b}{\sqrt{(1 + a^2 + b^2)}},$$

$$\cos. X = \frac{a}{\sqrt{(1 + a^2 + b^2)}} \quad . \quad . \quad . \quad . \quad . \quad . \quad . \quad . \quad . \quad . \quad (38)$$

which agrees with what is done in (3).

TRANSFORMATION OF COORDINATES.

12. *To transfer the origin of coordinates to the point (α, β, γ) without changing their direction.*

Let it be premised that instead of supposing the coordinate planes at right angles to one another, we shall here suppose them to make any angles whatever with each other. In this case the axes cease to be rectangular, but the coordinates x, y, z are still drawn parallel to the axes.

This being understood, assume

$$x = x' + \alpha, \quad y = y' + \beta, \quad z = z' + \gamma \quad . \quad . \quad . \quad . \quad . \quad (39)$$

and substitute in the expression involving x, y, z. The result will contain x′, y′, z′ the coordinates referred to the origin (α, β, γ).

When the substitution is made, the signs of α, β, γ as explained in (1), must be attended to.

13. *To change the direction of the axes from rectangular, without affecting the origin.*

Conceive three new axes A x′, A y′, A z′, the first axes being supposed rectangular, and these having any given arbitrary direction whatever. Take any point; draw the coordinates x′, y′, z′ of this point, and project them upon the axis A X. The abscissa x will equal the sum, taken with their proper signs, of these three projections, (as is easily seen by drawing the figure); but if (x x′), (y, y′), (z, z′) denote the angles between the axes A x, A x′; A y, A y′; A z, A z′ respectively; these projections are

$$x' \cos. (x' x), \quad y' \cos. (y' x), \quad z' \cos. (z' x).$$

In like manner we proceed with the other axes, and therefore get

$$\left.\begin{array}{l} x = x' \cos. (x' x) + y' \cos. (y' x) + z' \cos. (z' x) \\ y = y' \cos. (y' y) + z' \cos. (z' y) + x' \cos. (x' y) \\ z = z' \cos. (z' z) + y' \cos. (y' z) + x' \cos. (x' z) \end{array}\right\} \quad . \quad . \quad . \quad . \quad (40)$$

Since (x′ x), (x′ y), (x′ z) are the angles which the staight line A x′, makes with the rectangular axes of x, y, z, we have (5)

$$\left.\begin{aligned} \cos.^2 (x' x) + \cos.^2 (x' y) + \cos.^2 x' z &= 1 \\ \cos.^2 (y' x) + \cos.^2 (y' y) + \cos.^2 (y' z) &= 1 \\ \cos.^2 (z' x) + \cos.^2 (z' y) + \cos.^2 (z' x) &= 1 \end{aligned}\right\} \quad . \; . \; . \; . \quad (41)$$

We also have from (13) p.

$$\left.\begin{aligned} \cos.(x'y') &= \cos.(x'x)\cos.(y'x) + \cos.(x'y)\cos.(y'y) + \cos.(x'z)\cos.(y'z) \\ \cos.(x'z') &= \cos.(x'x)\cos.(z'x) + \cos.(x'y)\cos.(z'y) + \cos.(x'z)\cos.(z'z) \\ \cos.(y'z') &= \cos.(y'x)\cos.(z'x) + \cos.)y'y)\cos.(z'y) + \cos.(y'z)\cos.(z'z) \end{aligned}\right\} . \; (42)$$

The equations (40) and (41), contain the nine angles which the axes of x′, y′, z′ make with the axes of x, y, z.

Since the equations (41) determine three of these angles only, six of them remain arbitrary. Also when the new system is likewise rectangular, or cos. (x′ y′) = cos. (x′ z′) = cos. (y′ z′) = 1, three others of the arbitraries are determined by equations (42). Hence in that case there remain but three arbitrary angles.

14. *Required to transform the rectangular axe of coordinates to other rectangular axes, having the same origin, but two of which shall be situated in a given plane.*

Let the given plane be Y′ A C, of which the trace in the plane (z, x) is

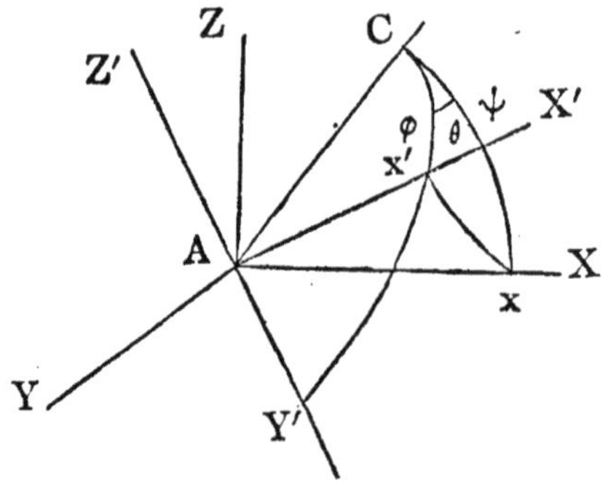

A C. At the distance A C describe the arcs C Y′, C x, x x′ in the planes C A Y′, (z, x), and X′ A X. Then if one of the new axes of the coordinates be A X′, its position and that of the other two, A Y′, A Z′, will be determined by C x′ = φ, C x = ψ, and the spherical angle x C x′ = θ = inclination of the given plane to the plane (z, x).

Hence to transform the axes, it only remains to express the angles (x′x), (y′x), &c. which enter the equations (40) in terms of θ ψ and φ.

By spherics

$$\cos.(x'x) = \cos.\psi\cos.\varphi + \sin.\psi\sin.\varphi\cos.\theta.$$

In like manner forming other spherical triangles, we get

$$\cos.(y'x) = \cos.(90^\circ + \varphi)\cos.\psi + \sin.\psi\sin.(90^\circ + \varphi)\cos.\theta$$
$$\cos.(x'y) = \cos.(90^\circ + \psi)\cos.\varphi + \sin.(90^\circ + \psi)\sin.\varphi\cos.\theta$$
$$\cos.(y'y) = \cos.(90^\circ + \psi)\cos.(90 + \varphi) + \sin.(90^\circ + \psi)\sin.(90^\circ + \varphi)\cos.\theta$$

So that we obtain these four equations

$$\left.\begin{array}{l} \cos.(x'x) = \cos.\psi\cos.\varphi + \sin.\psi\sin.\varphi\cos.\theta \\ \cos.(y'x) = -\sin.\psi\sin.\varphi + \sin.\psi\cos.\varphi\cos.\theta \\ \cos.(x'y) = -\sin.\psi\cos.\varphi + \cos.\psi\sin.\varphi\cos.\theta \\ \cos.(y'y) = \sin.\psi\sin.\varphi + \cos.\psi\cos.\varphi\cos.\theta \end{array}\right\} \quad . \quad . \quad (43)$$

Again by spherics, (since A Z′ is perpendicular to A C, and the inclination of the planes Z′A C, (x, y) is $90^\circ - \theta$) we have

$$\left.\begin{array}{l} \cos(z'x) = \sin.\psi\sin.\theta \\ \cos.(z'y) = \cos.\psi\sin.\theta \end{array}\right\} \quad . \quad . \quad . \quad . \quad . \quad . \quad . \quad . \quad . \quad (44)$$

And by considering that the angle between the planes Z A C, Z A X′, = $90^\circ + \theta$, by spherics, we also get

$$\left.\begin{array}{l} \cos.(x'z) = -\sin.\varphi\sin.\theta \\ \cos.(y'z) = -\cos.\varphi\sin.\theta \\ \cos.(z'z) = \cos.\theta \end{array}\right\} \quad . \quad . \quad . \quad . \quad . \quad . \quad . \quad . \quad (45)$$

which give the nine coefficients of equations (40).

Equations (41), (42) will also hereby be satisfied when the systems are rectangular.

15. *To find the section of a surface made by a plane.*

The last transformation of axes is of great use in determining the nature of the section of a surface, made by a plane, or of the section made by any two surfaces, plane or not, provided the section lies in one plane; for having transformed the axes to others, A Z′, A X′, A Y′, the two latter lying in the plane of the section, by the equations (40), and the determinations of the last article, by putting $z' = 0$ in the equation to the surface, we have that of the section at once. It is better, however, to make $z = 0$ in the equations (40), and to seek directly the values of cos. (x′x), cos. (y′x), &c. The equations (40) thus become

$$\left.\begin{array}{l} x = x'\cos.\psi + y'\sin.\psi\cos.\theta \\ y = x'\sin.\psi - y'\cos.\psi\cos.\theta \\ z = y'\sin.\theta \end{array}\right\} \quad . \quad . \quad . \quad . \quad . \quad . \quad (46)$$

16. *To determine the nature and position of all surfaces of the second order*, or *to discuss the general equation of the second order*, viz.

$$ax^2 + by^2 + cz^2 + 2dxy + 2exz + 2fyz + gx + hy + iz = k \quad . \quad . \quad . \quad (a)$$

First simplify it by such a transformation of coordinates as shall destroy the terms in x y, x z, y z; the axes from rectangular will become oblique, by substituting the values (40), and the nine angles which enter these, being subjected to the conditions (41), there will remain six of them arbitrary, which we may dispose of in an infinity of ways. Equate to zero the coefficients of the terms in $x'y'$, $x'z'$, $y'z'$.

But if it be required that the new axes shall be also rectangular, as this condition will be expressed by putting each of the equations (42) equal zero, the *six* arbitrary angles will be reduced to *three*, which the three coefficients to be destroyed will make known, and the problem will thus be determined.

This investigation will be rendered easier by the following process:

Let $x = \alpha z$, $y = \beta z$ be the equations of the axis of x'; then for brevity making

$$l = \frac{1}{\sqrt{(1 + \alpha^2 + \beta^2)}}$$

we find that (3)

$$\cos. (x'x) = \alpha l, \ \cos. (x'y) = \beta l, \ \cos. x'z = l.$$

Reasoning thus also as to the equations $x = \alpha' z$, $y = \beta' z$ of the axis of y', and the same for the axis of z', we get

$$\cos. (y'x) = \alpha' l', \ \cos. (y'y) = \beta' l', \ \cos. (y'z) = l'$$
$$\cos. (z'x) = \alpha'' l'', \ \cos. (z'y) = \beta'' l'', \ \cos. (z'z) = l''.$$

Hence by substitution the equations (40) become

$$\left.\begin{aligned} x &= l\alpha x' + l'\alpha' y' + l''\alpha'' z' \\ y &= l\beta x' + l'\beta' y' + l''\beta'' z' \\ z &= l x' + l' y' + l'' z'. \end{aligned}\right\}$$

The nine angles of the problem are replaced by the six unknowns α, α', α'', β, β', β'', provided the equations (41) are thereby also satisfied.

Substitute therefore these values of x, y, z, in the general equation of the 2d degree, and equate to zero the coefficients of $x'y'$, $x'z'$, $y'z'$, and we get

$$\left.\begin{aligned} (a\alpha + d\beta + e)\alpha' + (d\alpha + b\beta + f)\beta' + e\alpha + f\beta + c = 0 \\ (a\alpha + d\beta + e)\alpha'' + (d\alpha + b\beta + f)\beta'' + e\alpha + f\beta + c = 0 \\ (a\alpha'' + d\beta'' + e)\alpha' + (d\alpha'' + b\beta'' + f)\beta' + e\alpha'' + f\beta'' + c = 0 \end{aligned}\right\}$$

One of these equations may be found without the others, and by making the substitution only in part. Moreover from the symmetry of the process the other two equations may be found from this one. Eliminate α', β' from the first of them, and the equations $x = \alpha' z$, $y = \beta' z$, of the axis of y'; the resulting equation

$$(a\alpha + d\beta + e)x + (d\alpha + b\beta + f)y + (e\alpha + f\beta + c)z = 0 \quad . \quad . \quad (b)$$

is that of a plane (19).

But the first equation is the condition which destroys the term $x' y'$: since if we only consider it, α, β, α', β', may be any whatever that will satisfy it; it suffices therefore that the axis of y' be traced in the plane above alluded to, in order that the transformed equations may not contain any term in $x' y'$.

In the same manner eliminating α'', β'', from the 2d equation by means of the equations of the axis of z', viz. $x = \alpha'' z$, $y = \beta'' z$, we shall have a plane such, that if we take for the axis of z' every straight line which it will there trace out, the transformed equation will not contain the term in $x' z'$. But, from the form of the two first equations, it is evident that this second plane is the same as the first: therefore, if we there trace the axes of y' and z' at pleasure, this plane will be that of y' and z', and the transformed equation will have no terms involving $x' y'$ or $x z'$. The direction of these axes in the plane being any whatever, we have an infinity of systems which will serve this purpose; the equation (b) will be that of a plane parallel to the plane which bisects all the parallels to x, and which is therefore called the *Diametrical Plane.*

Again, if we wish to make the term in $y' z'$ disappear, the third equation will give α', β', and there are an infinity of oblique axes which will answer the three required conditions. But make x', y', z', rectangular. The axis of x' must be perpendicular to the plane $(y' z')$ whose equation we have just found; and that $x = \alpha z$, $y = \beta z$, may be the equations (see equations b) we must have

$$a\alpha + d\beta + e = (e\alpha + f\beta + c)\alpha \quad \ldots \quad (c)$$

$$d\alpha + b\beta + f = (e\alpha + f\beta + c)\beta \quad \ldots \quad (d)$$

Substituting in (c) the value of α found from (d) we get

$$\{(a-b)fe + (f^2-e^2)d\}\beta^3$$
$$+ \{(a-b)(c-b)e + (2d^2-f^2-e^2)e + (2c-a-b)fd\}\beta^2$$
$$+ \{(c-a)(c-b)d + (2e^2-f^2-d^2)d + (2b-a-c)fe\}\beta$$
$$+ (a-c)fd + (f^2-d^2)e = 0.$$

This equation of the 3d degree gives for β at least one real root; hence the equation (d) gives one for α; so that the axis of x' is determined so as to be perpendicular to the plane $(y', z',)$ and to be free from terms in $x' z'$, and $y' z'$. It remains to make in this plane $(y', z',)$ the axes at right angles and such that the term $x' y'$ may also disappear. But it is evident that we shall find at the same time a plane $(x', z',)$ such that the axis of y' is perpendicular to it, and also that the terms in $x' y'$, $z' y'$ are not involved. But it happens that the conditions for making the axis of y' perpendicular to this plane are both (c) and (d) so that the same equation of the 3d degree must give also β''. The same holds good for the axis of z. Consequently the three roots of the equation β are all real, and are the values of β, β', β''. Therefore α, α', α'', are given by the equation (d). Hence, *There is only one system of rectangular axes which eliminates* $x'y'$, $x'z'$, $x'y'$; *and there exists one in all cases.* These axes are called the *Principal axes of the Surface.*

Let us analyze the case which the cubic in β presents.

I. If we make

$$(a - b)\,fe + (f^2 - e^2)\,d = 0$$

the equation is deprived of its first term. This shows that then one of the roots of β is infinite, as well as that α derived from equation (d) becomes $e\alpha + f\beta = 0$. The corresponding angles are right angles. One of the axes, that of z' for instance, falls upon the plane (x, y), and we obtain its equation by eliminating α and β from the equations $x = \alpha z$, $y = \beta z$, which equation is

$$ex + fy = 0$$

The directions of y', z' are given by the equation in β reduced to a quadrature.

2ndly. If besides this first coefficient the second is also $= 0$, by substituting b, found from the above equation, in the factor of β^2, it reduces to the last term of the equation in β, viz.

$$(a - c)\,fd + (f^2 - d^2)\,e = 0.$$

These two equations express the condition required. But the coefficient of β is deduced from that of β^2 by changing b into c and d into e, and the same holds for the first and last term of the equation in β. Therefore the cubic equation is also thus satisfied. There exists therefore an infinity of rectangular systems, which destroy the terms in $x'y'$, $x'z'$, $y'z'$. Eliminating a and b from equations (c) and (d) by aid of the above two equations of condition, we find that they are the product of $f\alpha - d$ and $e\beta - d$ by the common factor $ed\alpha + fd\beta + fe$. These factors are therefore nothing; and eliminating α and β, we find

$$fx = dz,\ ey = dz,\ edx + fdy + fez = 0.$$

The two first are the equations of one of the axes. The third that of a plane which is perpendicular to it, and in which are traced the two other axes under arbitrary directions. This plane will cut the surface in a curve wherein all the rectangular axes are principal, which curve is therefore a circle, the only one of curves of the second order which has that property. The surface is one then of revolution round the axis whose equations we have just given.

The equation once freed from the three rectangles, becomes of the form

$$kz^2 + my^2 + nx^2 + qx + q'y + q''z = h \quad \ldots\ldots \quad (e)$$

The terms of the first dimension are evidently destroyed by removing the origin (39). It is clear this can be effected, except in the case where one of the squares x^2, y^2, z^2 is deficient. We shall examine these cases separately. First, let us discuss the equation

$$kz^2 + my^2 + nx^2 = h \quad \ldots\ldots\ldots \quad (f)$$

Every straight line passing through the origin, cuts the surface in two points at equal distances on both sides, since the equation remains the same after having changed the signs of x, y, z. The origin being in the middle of all the chords drawn through this point is a *center*. *The surface therefore has the property of possessing a center whenever the transformed equation has the squares of all the variables.*

We shall always take *n* positive: it remains to examine the cases where k and m are both positive, both negative, or of different signs.

If in the equation (f) k, m, and n, are all positive, h is also positive; and if h is nothing, we have $x = 0$, $y = 0$, $z = 0$, and the surface has but one point.

But when h is positive by making x, y, or z, separately equal zero, we find the equations to an ellipse, curves which result from the section of the surface in question by the three coordinate planes. Every plane parallel to them gives also an ellipse, and it will be easy to show the same of all plane sections. Hence the surface is termed an *Ellipsoid.*

The lengths A, B, C, of the *three principal axes* are obtained by finding the sections of the surface through the axes of x, y, and z. These give

$$kC^2 = h, \; mB^2 = h, \; nA^2 = h.$$

from which eliminating k, m and n, and substituting in equation (f) we get

$$\left.\begin{aligned} \frac{z^2}{C^2} + \frac{y^2}{B^2} + \frac{x^2}{A^2} = 1 \\ A^2B^2z^2 + A^2C^2y^2 + B^2C^2x^2 = A^2B^2C^2 \end{aligned}\right\} \quad (47)$$

which is the equation to an *Ellipsoid* referred to its *center and principal axes.*

We may conceive this surface to be generated by an ellipse, traced in the plane (x, y), moving parallel to itself, whilst its two axes vary, the curve sliding along another ellipse, traced in the plane (x, z) as a directrix. If two of the quantities A, B, C, are equal, we have an ellipsoid of revolution. If all three are equal, we have a sphere.

Now suppose k negative, and m and h positive or

$$k z^2 - m y^2 - a x^2 = - h.$$

Making x or y equal zero, we perceive that the sections by the planes (y z) and (x z), are hyperbolas, whose axis of z is the second axis. All planes passing through the axis of z, give this same curve. Hence the surface is called an *hyperboloid*. Sections parallel to the plane (x y) are always real ellipses, A, B, C $\sqrt{-1}$ designating the lengths upon the axes from the origin, the equation is the same as the above equation excepting the first term becoming negative.

Lastly, when k and h are negative

$$k z^2 + m y^2 + n x^2 = - h;$$

all the planes which pass through the axis of z cut the surface in hyperbolas, whose axis of z is the first axis. The plane (x y) does not meet the surface and its parallels passing through the opposite limits, give ellipses. This is a *hyperboloid* also, but having two vertexes about the axis of z. The equation in A, B, C is still the same as above, excepting that the term in z^2 is the only positive one.

When h = 0, we have, in these two cases,

$$k z^2 = m y^2 + n x^2 \quad . \quad . \quad . \quad . \quad . \quad . \quad . \quad . \quad . \quad (48)$$

the equation to a cone, which cone is the same to these hyperboloids that an asymptote is to hyperbolas.

It remains to consider the case of k and m being negative. But by a simple inversion in the axes, this is referred to the two preceding ones. The hyperboloid in this case has one or two vertexes about the axis of x according as h is negative or positive.

When the equation (e) is deprived of one of the squares, of x^2 for instance, by transferring the origin, we may disengage that equation from the constant term and from those in y and z; thus it becomes

$$k z^2 + m y^2 = h x \quad . \quad . \quad . \quad . \quad . \quad . \quad . \quad . \quad (49)$$

The sections due to the planes (x z), (x y) are parabolas in the same or opposite directions according to the signs of k, m, h; the planes parallel to these give also parabolas. The planes parallel to that of (y z) give ellipses or parabolas according to the sign of m. The surface is an *elliptic paraboloid* in the one case, and a *hyperbolic paraboloid* in the other case. When k = m, it is a *paraboloid of revolution.*

When h = 0, the equation takes the form

$$a^2 z^2 \pm b^2 y^2 = 0$$

according to the signs of k and m. In the one case we have

$$z = 0, \quad y = 0$$

whatever may be the value of x, and the surface reduces to the axis of x. In the other case.

$$(a z + b y)(a z - b y) = 0$$

shows that we make another factor equal zero; thus we have the system of two planes which increase along the axis of x.

When the equation (e) is deprived of two squares, for instance of x^2, y^2, by transferring the origin parallelly to z, we reduce the equation to

$$k z^2 + p y + q x = 0 \quad . \quad . \quad . \quad . \quad . \quad . \quad . \quad . \quad (50)$$

The sections made by the planes drawn according to the axis of z, are parabolas. The plane (x y) and its parallels give straight lines parallel to them. The surface is, therefore, a cylinder whose base is a parabola, or a *parabolic cylinder*.

If the three squares in (e) are wanting, it will be that of a plane.

It is easy to recognise the case where the proposed equation is decomposable into rational factors; the case where it is formed of positive squares, which resolve into two equations representing the sector of two planes.

PARTIAL DIFFERENCES.

17. If $u = f(x, y, z, \&c.)$ denote any function of the variable x, y, z, &c. d u be taken on the supposition that y, z, &c. are constant, then is the result termed the partial difference of u relative to x, and is thus written

$$\left(\frac{d u}{d x}\right) d x.$$

Similarly

$$\left(\frac{d u}{d y}\right) d y, \quad \left(\frac{d u}{d z}\right) d z, \ \&c.$$

denote the partial differences of u relatively to y, z, &c. respectively.

Now since the total difference of u arises from the increase or decrease of its variables, it is evident that

$$d u = \left(\frac{d u}{d x}\right) d x + \left(\frac{d u}{d y}\right) d y + \left(\frac{d u}{d z}\right) d z + \&c. \quad . \quad . \quad . \quad (51)$$

But, by the general principle laid down in (6) Vol. I, this result may be demonstrated as follows; Let

$$\begin{aligned} u + du = A &+ A\,dx + B\,dy + C\,dz + \&c. \\ &A'dx^2 + B'dy^2 + C'dz^2 + \&c. \\ &+ M\,dx.dy + N\,dx.dz + \&c. \\ &+ A''dx^3 + \&c. \end{aligned}$$

Then equating quantities of the same nature, we have

$$du = A\,dx + B\,dy + C\,dz + \&c.$$

Hence when dy, dz, &c. $= 0$, or when y, z, &c. are considered constant

$$du = A\,dx$$

or according to the above notation

$$A = \left(\frac{du}{dx}\right).$$

In the same manner it is shown, that

$$B = \left(\frac{du}{dy}\right)$$

$$C = \left(\frac{du}{dz}\right)$$

&c.

Hence

$$du = \left(\frac{du}{dx}\right)dx + \left(\frac{du}{dy}\right)dy + \left(\frac{du}{dz}\right)dz + \&c. \text{ as before.}$$

Ex. 1. $u = xyz$, &c.

$$\left(\frac{du}{dx}\right) = yz,\ \left(\frac{du}{dy}\right) = xz,\ \left(\frac{du}{dz}\right) = xy,$$

$$\therefore du = yz\,dx + xz\,dy + xy\,dz$$

$$\text{or } \frac{du}{u} = \frac{dx}{x} + \frac{dy}{y} + \frac{dz}{z}.$$

Ex. 2. $u = xyz$, &c. Here as above

$$\frac{du}{u} = \frac{dx}{x} + \frac{dy}{y} + \frac{dz}{z} + \&c.$$

And in like manner the total difference of any function of any number of variables may be found, viz. by first taking the partial differences, as in the rules laid down in the Comments upon the first section of the first book of the Principia.

But this is not the only use of partial differences. They are frequently used to abbreviate expressions. Thus, in p. 13, and 14, Vol. II. we learn that the actions of M, μ, μ'', &c. upon μ resolved parallel to x, amount to

$$\frac{d^2(\zeta+x)}{dt^2} = \frac{\mu'(x'-x)}{[(x'-x)^2+(y'-y)^2+(z'-z)^2]^{\frac{3}{2}}} + \frac{\mu''(x''-x)}{[(x''-x)^2+(y''-y)^2+(z''-z)^2]^{\frac{3}{2}}}$$

$$+\frac{\mu'''(x'''-x)}{[(x'''-x)^2+(y'''-y)^2+(z'''-z)^2]^{\frac{3}{2}}} + \&c. - \frac{Mx}{[(x^2+y^2+z^2)]^{\frac{3}{2}}}$$

retaining the notation there adopted.

But if we make

$$\sqrt{(x'-x)^2+(y'-y)^2+(z'-z)^2} = \rho_{0,1}$$

and generally

$$\sqrt{(x'^{\cdots n}-x'^{\cdots m})^2+(y'^{\cdots n}-y'^{\cdots m})^2+(z'^{\cdots n}-z'^{\cdots m})^2} = \rho_{n,m},$$

and then assume

$$\lambda = \frac{\mu\mu'}{\rho_{0,1}} + \frac{\mu\mu''}{\rho_{0,2}} + \&c. \quad \ldots\ldots\ldots \quad (A)$$

$$+\frac{\mu'\mu''}{\rho_{1,2}} + \frac{\mu'\mu'''}{\rho_{1,3}} + \&c. \quad \ldots\ldots\ldots \quad (B)$$

$$+\frac{\mu''\mu'''}{\rho_{2,3}} + \frac{\mu''\mu''''}{\rho_{2,4}} + \&c. \quad \ldots\ldots\ldots \quad (C)$$

&c.

we get

$$\left.\begin{aligned}
\left(\frac{d\lambda}{dx}\right) &= \left(\frac{dA}{dx}\right) = \frac{\mu\mu'(x'-x)}{\rho^3_{0,1}} + \frac{\mu\mu''(x''-x)}{\rho^3_{0,2}} + \&c. \\
\left(\frac{d\lambda}{dy}\right) &= \left(\frac{dA}{dy}\right) = \frac{\mu\mu'(y'-y)}{\rho^3_{0,1}} + \frac{\mu\mu''(y''-y)}{\rho^3_{0,2}} + \&c. \\
\left(\frac{d\lambda}{dz}\right) &= \left(\frac{dA}{dz}\right) = \frac{\mu\mu'(z'-z)}{\rho^3_{0,1}} + \frac{\mu\mu''(z''-z)}{\rho^3_{0,2}} + \&c.
\end{aligned}\right\}$$

We also get

$$\left(\frac{d\lambda}{dx'}\right) = -\frac{\mu\mu'(x'-x)}{\rho^3_{0,1}} + \left(\frac{dB}{dx'}\right)$$

$$\left(\frac{d\lambda}{dx''}\right) = -\frac{\mu\mu''(x''-x)}{\rho^3_{0,2}} - \frac{\mu'\mu''(x''-x')}{\rho^3_{1,2}} + \left(\frac{dC}{dx''}\right)$$

$$\left(\frac{d\lambda}{dx'''}\right) = -\frac{\mu\mu'''(x'''-x)}{\rho^3_{0,3}} - \frac{\mu'\mu'''(x'''-x')}{\rho^3_{1,3}} - \frac{\mu''\mu'''(x'''-x'')}{\rho^3_{2,3}} + \left(\frac{dD}{dx'''}\right)$$

Hence since (B) has one term less than (A); (C) one term less than (B); and so on; it is evident that

$$\left(\frac{d\lambda}{dx'}\right) + \left(\frac{d\lambda}{dx''}\right) + \left(\frac{d\lambda}{dx'''}\right) + \&c. = -\left(\frac{d\lambda}{dx}\right) - \left(\frac{dB}{dx'}\right) - \left(\frac{dC}{dx}\right) - \&c.$$
$$+\left(\frac{dB}{dx'}\right) + \left(\frac{dC}{dx''}\right) + \&c.$$

and therefore that

$$\Sigma.\left(\frac{d\lambda}{dx}\right) = \left(\frac{d\lambda}{dx}\right) + \left(\frac{d\lambda}{dx'}\right) + \left(\frac{d\lambda}{dx''}\right) + \&c. = 0$$

See p. 15, Vol. II.

Hence then λ is so assumed that the sum of its partial differences relative to x, x', x'' &c. shall equal zero, and at the same time abbreviate the expression for the forces upon μ along x from the above complex formula into

$$\frac{d^2(\xi + x)}{dt^2} = \frac{1}{\mu}\left(\frac{d\lambda}{dx}\right) - \frac{Mx}{\rho^3};$$

and the same may be said relatively to the forces resolved parallel to y, z, &c. &c.

Another consequence of this assumption is

$$x\left(\frac{d\lambda}{dy}\right) + x'\left(\frac{d\lambda}{dy'}\right) + \&c. = y\left(\frac{d\lambda}{dx}\right) + y'\left(\frac{d\lambda}{dx'}\right) + \&c.$$

$$\text{or } \Sigma.x\left(\frac{d\lambda}{dy}\right) = \Sigma.y\left(\frac{d\lambda}{dx}\right).$$

For

$$y\left(\frac{d\lambda}{dx}\right) = \frac{\mu\mu'(x'-x)y}{\rho^3_{0,1}} + \frac{\mu\mu''(x''-x)y}{\rho^3_{0,2}} + \frac{\mu\mu'''(x'''-x)y}{\rho^3_{0,3}} + \&c.$$

$$y'\left(\frac{d\lambda}{dx'}\right) = \frac{\mu'\mu''(x''-x')y'}{\rho^3_{1,2}} + \frac{\mu'\mu'''(x'''-x')y'}{\rho^3_{1,3}} + \&c. - \frac{\mu\mu'(x'-x)y'}{\rho^3_{0,1}}$$

$$y''\left(\frac{d\lambda}{dx''}\right) = \frac{\mu''\mu'''(x'''-x'')y''}{\rho^3_{2,3}} + \frac{\mu''\mu''''(x''''-x'')y''}{\rho^3_{2,4}} + \&c. \frac{\mu\mu''(x''-x)y''}{\rho^3_{0,2}} \quad \frac{(\mu'\mu''(x''-x')y''}{\rho^3_{1,2}}$$

&c.

Hence it is evident that

$$\Sigma.y\left(\frac{d\lambda}{dx}\right) = \frac{\mu\mu'(x'-x)(y-y')}{\rho^3_{0,1}} + \frac{\mu\mu''(x''-x)(y-y'')}{\rho^3_{0,2}} + \&c.$$
$$+ \frac{\mu'\mu''(x''-x')(y'-y'')}{\rho^3_{12}} + \frac{\mu'\mu''(x'''-x')(y-y''')}{\rho^3_{12}} + \&c.$$
$$+ \frac{\mu''\mu'''(x'''-x'')(y''-y''')}{\rho^3_{2,3}} + \frac{\mu''\mu''''(x''''-x'')(y''-y'''')}{\rho^3_{2,4}} + \&c.$$

&c.

In like manner it is found that

$$\Sigma . x \left(\frac{d\lambda}{dy}\right) = \frac{\mu\mu'(y'-y)(x-x')}{\varrho^3_{0,1}} + \frac{\mu\mu''(y''-y)(x-x'')}{\varrho^3_{0,2}} + \&c.$$

$$+ \frac{\mu'\mu''(y''-y')(x'-x'')}{\varrho^3_{1,2}} + \frac{\mu'\mu'''(y'''-y')(x'-x''')}{\varrho^3_{1,3}} + \&c.$$

&c.

which is also perceptible from the substitution in the above equation of y for x, x for y; y′ for x′, x′ for y′; and so on.

But

$$(y'-y)(x-x') = (x'-x)(y-y')$$
$$(y''-y)(x-x'') = (x''-x)(y-y'')$$

&c.

consequently

$$\Sigma . x \left(\frac{d\lambda}{dy}\right) = \Sigma . y \left(\frac{d\lambda}{dx}\right)$$

See p. 16. For similar uses of partial differences, see also pp. 22, and 105.

CHANGE OF THE INDEPENDENT VARIABLE.

When an expression is given containing differential coefficients, such as

$$\frac{dy}{dx}, \frac{d^2y}{dx^2} \&c.$$

in which the first differential only of x and its powers are to be found, it shows that the differential had been taken on the supposition that d x is constant, or that $d^2x = 0$, $d^3x = 0$, and so on. But it may be required to transform this expression to another in which d^2x, d^3x shall appear, and in which d y shall be constant, or from which d^2y, &c. shall be excluded. This is performed as follows:

For instance if we have the expression

$$\frac{1 + \frac{dy^2}{dx^2}}{\frac{d^2y}{dx^2}} \frac{dy}{dx}$$

the differential coefficients

$$\frac{dy}{dx}, \frac{d^2y}{dx}$$

may be eliminated by means of the equation of the curve to which we mean to apply that expression. For instance, from the equation to a parabola $y = a x^2$, we derive the values of

$$\frac{dy}{dx} \text{ and } \frac{d^2 y}{dx^2}$$

which being substituted in the above formula, these differential coefficients will disappear. If we consider

$$\frac{dy}{dx}, \frac{d^2 y}{dx^2}$$

unknown, we must in general have two equations to eliminate them from one formula, and these equations will be given by twice differentiating the equation to the curve.

When by algebriacal operations, dx ceases to be placed underneath dy, as in this form

$$\frac{y(dx^2 + dy^2)}{dx^2 + dy^2 - y d^2 y} \quad \ldots\ldots\ldots \quad (52)$$

the substitution is effected by regarding dx, dy, $d^2 y$ as unknown; but then in order to eliminate them, there must be in general the same number of equations as of unknowns, and consequently it would seem the elimination cannot be accomplished, because by means of the equation to the curve, only two of the equations between dx, dy, $d^2 y$ can be obtained. It must be remarked, however, that when by means of these two equations we shall have eliminated dy and $d^2 y$, there will remain a common factor dx^2, which will also vanish. For example, if the curve is always a parabola represented by the equation $y = a x$, by differentiating twice we obtain

$$dy = 2axdx \; 0 \; d^2 y = 2a\,dx^2$$

and these being substituted in the formula immediately above, we shall obtain, after suppressing the common factor dx^2,

$$\frac{y(1 + 4a^2 x^2)}{4a^2 x^2 - 2ay}.$$

The reason why dx^2 becomes a common factor is perceptible at once, for when from a formula which primitively contained

$$\frac{d^2 y}{dx^2}, \frac{dy}{dx},$$

we have taken away the denominator of $\frac{d^2 y}{dx^2}$ all the terms independent of $\frac{d^2 y}{dx^2}$ and $\frac{dy}{dx}$ must acquire the factor dx^2; then the terms which were affected by $\frac{d^2 y}{dx^2}$, do not contain dx, whilst those affected by $\frac{dy}{dx}$

contain d x. When we afterwards differentiate the equation of the curve, and obtain results of the form $dy = M\,dx$, $d^2y = N\,dx^2$, these values being substituted in the terms in d^2y, and in $dy\,dx$, will change them, as likewise the other terms, into products of dx^2.

What has been said of a formula containing differentials of the two first orders applying equally to those in which these differentials rise to superior orders, it thence follows that by differentiating the equation of the curve as often as is necessary, we can always make disappear from the expression proposed, the differentials therein contained.

The same will also hold good if, beside these differentials which we have just been considering, the formula contain terms in d^2x, in d^3x, &c.; for suppose that there enter the formula these differentials dx, dy, d^2x, d^2y and that by twice differentiating the equation represented by $y = fx$, we obtain these equations

$$F(x, y, dy, dx) = 0$$
$$F(x, y, dx, dy, d^2x, d^2y) = 0,$$

we can only find two of the three differentials dy, d^2x, d^2y, and we see it will be impossible to eliminate all the differentials of the formula; there is therefore a condition tacitly expressed by the differential d^2x; it is that the variable x is itself considered a function of a third variable which does not enter the formula, and which we call the *independent variable.* This will become manifest if we observe, that the equation $y = fx$ may be derived from the system of two equations

$$x = Ft, \quad y = \varphi t$$

from which we may eliminate t. Thus the equation

$$y = a \cdot \frac{(x - c)^2}{b^2}$$

is derived from the system of two equations

$$x = bt + c, \quad y = at^2,$$

and we see that x and y must vary by virtue of the variation which t may undergo. But this hypothesis that x and y vary as t alters, supposes that there are relations between x and t, and between y and t. One of these relations is arbitrary, for the equation which we represent generally by $y = fx$, for example

$$y = \frac{a}{b}(x - c)^2,$$

if we substitute between x and t, the arbitrary relation,

$$x = \frac{t^3}{c^2},$$

this value being put in the equation

$$y = \frac{a}{b^2}(x - c)^2$$

will change it to

$$y = a\,\frac{(t^3 - c^3)^2}{b^2 c^4},$$

an equation which, being combined with this,

$$x = \frac{t^3}{c^2},$$

ought to reproduce by elimination,

$$y = a\,\frac{(x - c)^2}{b^2},$$

the only condition which we ought to regard in the selection of the variable t.

We may therefore determine the independent variable t at pleasure. For example, we may take the chord, the arc, the abscissa or ordinate for this independent variable; if t represent the arc of the curve, we have

$$t = \sqrt{(d x^2 + d y^2)};$$

if t denote the chord and the origin be at the vertex of the curve, we have

$$t = \sqrt{(x^2 + y^2)};$$

lastly, if t be the abscissa or ordinate of the curve, we shall have

$$t = x, \text{ or } t = y.$$

The choice of one of the three hypotheses or of any other, becoming indispensible in order that the formula which contains the differentials, may be delivered from them, if we do not always adopt it, it is even then tacitly supposed that the independent variable has been determined. For example, in the usual case where a formula contains only the differentials $d\,x$, $d\,y$, $d^2 y$, $d^3 y$, &c. the hypothesis is that the independent variable t has been taken for the abscissa, for then it results that

$$t = x, \quad \frac{dx}{dt} = 1,$$

$$\frac{d^2x}{dt^2} = 0,$$

$$\frac{d^3x}{dt^2} = 0, \text{ \&c.}$$

and we see that the formula does not contain the second, third, &c. differentials.

To establish this formula, in all its generality, we must, as above, suppose x and y to be functions of a third variable t, and then we have

$$\frac{dy}{dt} = \frac{dy}{dx} \cdot \frac{dx}{dt};$$

from which we get

$$\frac{dy}{dx} = \frac{\frac{dy}{dt}}{\frac{dx}{dt}} \quad \ldots\ldots\ldots\ldots \quad (53)$$

taking the second differential of y and operating upon the second membei as if a fraction, we shall get

$$\frac{d^2 y}{dx} = \frac{\frac{dx}{dt} \cdot \frac{d^2 y}{dt} - \frac{dy}{dt} \cdot \frac{d^2 x}{dt}}{\frac{dx^2}{dt^2}}.$$

In this expression, d t has two uses; the one is to indicate that it is the independent variable, and the other to enter as a sign of algebra. In the second relation only will it be considered, if we keep in view that t is the independent variable. Then supposing dt^2 the common factor, the above expression simplifies into

$$\frac{d^2 y}{dx} = \frac{dx\,d^2 y - dy\,d^2 x}{dx^2},$$

and dividing by d x, it will become

$$\frac{d^2 y}{dx^2} = \frac{dx\,d^2 y - dy\,d^2 x}{dx^3}.$$

Operating in the same way upon the equation (53), we see that in taking t as the independent variable, the second member of the equation ought to become identical with the first; consequently we have only one change to make in the formula which contains the differential coefficients $\frac{dy}{dx}$, $\frac{d^2 y}{dx^2}$, viz. to replace $\frac{d^2 y}{dx^2}$ by

$$\frac{dx\,d^2 y - dy\,d^2 x}{dx^3} \quad \ldots\ldots\ldots \quad (54)$$

To apply these considerations to the radius of curvature which is given by the equation See p. 61. vol. I.)

$$R = \frac{\left(1 + \frac{dy^2}{dx^2}\right)^{\frac{3}{2}}}{\frac{d^2 y}{dx^2}},$$

if we wish to have the value of R, in the case where t shall be the independent variable, we must change the equation to

$$R = \frac{\left(1 + \frac{dy^2}{dx^2}\right)^{\frac{3}{2}}}{\frac{dx\,d^2y - dy\,d^2x}{dx^3}};$$

and observing that the numerator amounts to

$$\frac{(dx^2 + dy^2)^{\frac{3}{2}}}{dx^3}$$

we shall have

$$R = \frac{(dx^2 + dy^2)^{\frac{3}{2}}}{dx\,d^2y - dy^2\,d^2x} \quad . \quad . \quad . \quad . \quad . \quad (55)$$

This value of R supposes therefore that x and y are functions of a third independent variable. But if x be considered this variable, that is to say, if $t = x$, we shall have $d^2x = 0$, and the expression again reverts to the common one

$$R = \frac{(dx^2 + dy^2)^{\frac{3}{2}}}{dx\,d^2y} = \frac{\left(1 + \frac{dy^2}{dx^2}\right)^{\frac{3}{2}}}{\frac{d^2y}{dx^2}}.$$

But if, instead of x for the independent variable, we wish to have the ordinate y, this condition is expressed by $y = t$; and differentiating this equation twice, we have

$$\frac{dy}{dt} = 1, \quad \frac{d^2y}{dt^2} = 0.$$

The first of these equations merely indicates that y is the independent variable, which effects no change in the formula; but the second shows us that d^2y ought to be zero, and then the equation (55) becomes

$$R = -\frac{(dx^2 + dy^2)^{\frac{3}{2}}}{dy\,d^2x} \quad . \quad . \quad . \quad . \quad . \quad . \quad . \quad . \quad (56)$$

We next remark, that when x is the independent variable, and consequently $d^2x = 0$, this equation indicates that dx is constant. Whence it follows, that generally the independent variable has always a constant differential.

Lastly, if we take the arc for the independent variable, we shall have

$$dt = \sqrt{(dx^2 + dy^2)};$$

Hence, we easily deduce

$$\frac{dx^2}{dt^2} + \frac{dy^2}{dt^2} = 1;$$

differentiating this equation, we shall regard d t as constant, since t is the independent variable; we get

$$\frac{2\,dx\,d^2x}{dt^2} + \frac{2\,dy\,d^2y}{dt^2} = 0;$$

which gives

$$dx\,d^2x = -dy\,d^2y.$$

Consequently, if we substitute the value of d^2x, or that of d^2y, in the equation (55), we shall have in the first case

$$R = \frac{(dx^2 + dy^2)^{\frac{3}{2}}}{(dx\ \ + dy^2)\,d^2y}\,dx = \frac{\sqrt{(dx^2 + dy\ )}}{d^2y}\,dx \quad . \quad . \quad (57)$$

and in the second case,

$$R = -\frac{(dx^2 + dy^2)^{\frac{3}{2}}}{(dx\ \ + dy^2)\,d^2x}\,dy = -\frac{\sqrt{(dx^2 + dy^2)}}{d^2x}\,dy \quad . \quad (58)$$

In what precedes, we have only considered the two differential coefficients

$$\frac{dy}{dx}, \frac{d^2y}{dx^2};$$

but if the formula contain coefficients of a higher order, we must, by means analogous to those here used, determine the values of

$$\frac{d^3y}{dx^3} \text{ of } \frac{d^4y}{dx^4}, \text{ \&c.}$$

which will belong to the case where x and y are functions of a third independent variable.

PROPERTIES OF HOMOGENEOUS FUNCTIONS.

If $M\,dx + N\,dy + P\,dt +$ &c. $= dz$, *be a homogeneous function of any number of variables*, x, y, t, &c. *in which the dimension of each term is* n, *then is*

$$Mx + Ny + Pt + \text{\&c.} = nz.$$

For let $M\,dx + N\,dy$ be the differential of a homogeneous function z between two variables x and y; if we represent by n the sum of the exponents of the variables, in one of the terms which compose this function, we shall have therefore the equation

$$M\,dx + N\,dy = dz.$$

Making $\frac{y}{x} = q$, we shall find (vol. I.)

$$F(q) \times x^n = z;$$

and replacing, in the above equation, y by its value q x, and calling M′ N′, what M and N then become, that equation transforms to

$$M' dx + N' d. qx = dz;$$

and substituting the value of z, we shall have

$$M' dx + N' d(qz) = d(x^n F. q.)$$

But $d(qz) = q dx + x dq$. Therefore

$$(M' + N'q) dx + N' x dq = d(x^n F. q).$$

But, $(M' + N' q) dx$ being the differential of $x^n F q$ relatively to x, we have (Art. 6. vol. 1.)

$$M' + N'q = n x^{n-1} \times F. q.$$

If in this equation y be put for q x, it will become

$$M + N\frac{y}{x} = n x^{n-1} F. q,$$

or

$$Mx + Ny = nz.$$

This theorem is applicable to homogeneous functions of any number of variables; for if we have, for example, the equation

$$M dx + N dy + P dt = dz,$$

in which the dimension is n in every term, it will suffice to make

$$\frac{y}{x} = q, \frac{t}{x} = r$$

to prove, by reasoning analogous to the above, that we get $z = x^n F(q, r)$, and, consequently, that

$$Mx + Ny + Pt = nz \quad . \quad . \quad . \quad . \quad . \quad . \quad . \quad . \quad (59)$$

and so on for more variables.

THEORY OF ARBITRARY CONSTANTS.

An equation V = 0 between x, y, and constants, may be considered as the complete integral of a certain differential equation, of which the order depends on the number of constants contained in V = 0. These constants are named *arbitrary constants*, because if one of them is represented by *a*, and V or one of its differentials is put under the form f (x, y) = *a*, we see that *a* will be nothing else than the arbitrary constant given by the integration of d f (x, y). Hence, if the differential equation in question is of the order *n*, each integration introducing an arbitrary constant, we have V = 0, which is given by the last of three integrations, and contains, at least, n arbitrary constants more than the given differential equation. Let therefore

$$F(x, y) = 0, F\left(x, y, \frac{dy}{dx}\right) = 0, F\left(x, y, \frac{dy}{dx}, \frac{d^2y}{dx^2}\right) = 0 \text{ \&c.} \quad (a)$$

be the primitive equation of a differential equation of the second order and its immediate differentials.

Hence we may eliminate from the two first of these three equations, the constants a and b, and obtain

$$\varphi\left(x, y, \frac{dy}{dx}, b\right) = 0, \varphi\left(x, y, \frac{dy}{dx}, a\right) = 0 \quad . \quad . \quad . \quad . \quad (b)$$

If, without knowing $F(x, y) = 0$, we find these equations, it will be sufficient to eliminate from them $\frac{dy}{dx}$, to obtain $F(x, y) = 0$, which will be the complete integral, sinee it will contain the arbitrary constants a, b.

If, on the contrary, we eliminate these two constants between the above three equations, we shall arrive at an equation which, containing the same differential coefficients, may be denoted by

$$\varphi\left(x, y, \frac{dy}{dx}, \frac{d^2y}{dx^2}\right) = 0 \quad . \quad . \quad . \quad . \quad . \quad . \quad . \quad . \quad (c)$$

But each of the equations (b) will give the same. In fact, by eliminating the constant contained in one of these equations and its immediate differential, we shall obtain separately two equations of the second order, which do not differ from equation (c) otherwise than the values of x and of y are not the same in both. Hence it follows, that a differential equation of the second order may result from two equations of the first order which are necessarily different, since the arbitrary constant of the one is different from that of the other. The equations (b) are what we call the first integrals of the equation (c), which is independent, and the equation $F(x, y) = 0$ is the second integral of it.

Take, for example, the equation $y = ax + b$, which, because of its two constants, may be regarded as the primitive equation of an equation of the second order. Hence, by differentiation, and then by elimination of a, we get

$$\frac{dy}{dx} = a, \; y = x\frac{dy}{dx} + b.$$

These two first integrals of the equation of the second order which we are seeking, being differentiated each in particular, conduct equally, by the elimination of a, b, to the independent equation $\frac{d^2y}{dx^2} = 0$. In the case where the number of constants exceeds that of the required arbitrary constants, the surplus constants, being connected with the same equations, do not acquire any new relation. Required, for instance, the equation of the second order, whose primitive is

$$y = \tfrac{1}{2} a x^2 + b x + c = 0;$$

differentiating we get

$$\frac{dy}{dx} = a x + b.$$

The elimination of a, and then that of b, from these equations, give separately these two first integrals

$$\frac{dy}{dx} = a x + b, \quad y = x\frac{dy}{dx} - \tfrac{1}{2} a x^2 + c \quad . \quad . \quad . \quad (d)$$

Combining them each with their immediate differentials, we arrive, by two different ways, at $\frac{d^2 y}{d x^2} = a$. If, on the contrary, we had climinated the third constant *a* between the primitive equation and its immediate differential, that would not have produced a different result; for we should have arrived at the same result as that which would lead to the elimination of a from the equations (d), and we should then have fallen upon the equation $x\frac{d^2 y}{d x^2} = \frac{dy}{dx} - b$, an equation which reduces to $\frac{d^2 y}{d x^2} = a$ by combining it with the first of the equations (d).

Let us apply these considerations to a differential equation of the third order: differentiating three times successively the equation F (x, y) = 0, we shall have

$$F\left(x, y, \frac{dy}{dx}\right) = 0, \; F\left(x, y, \frac{dy}{dx}, \frac{d^2 y}{d x^2}\right) = 0, \; F\left(x, y, \frac{dy}{dx}, \frac{d^2 y}{d x^2}, \frac{d^3 y}{d x^3}\right) = 0$$

These equations admitting the same values for each of the arbitrary constants contained by F (x, y) = 0, we may generally eliminate these constants between this latter equation and the three preceding ones, and obtain a result which we shall denote by

$$f\left(x, y, \frac{dy}{dx}, \frac{d^2 y}{d x^2}, \frac{d^3 y}{d x^3}\right) = 0 \quad . \quad . \quad . \quad . \quad . \quad . \quad . \quad (e)$$

This will be the differential equation of the third order of F (x, y) = 0, and whose three arbitrary constants are eliminated; reciprocally, F (x, y) = 0, will be the third integral of the equation (e).

If we eliminate successively each of the arbitrary constants from the equation F (x, y) = 0, and that which we have derived by differentiation, we shall obtain three equations of the first order, which will be the second integrals of the equation (e).

Finally, if we eliminate two of the three arbitrary constants by means of the equation F (x, y) = 0, and the equations which we deduce by two successive differentiations, that is to say, if we eliminate these constants from the equations

$$F\left(x, y\right) = 0,\ F\left(x, y, \frac{dy}{dx}\right) = 0,\ F\left(x, y, \frac{dy}{dx}, \frac{d^2y}{dx^2}\right) = 0\ .\ .\ (f)$$

we shall get, successively, in the equation which arises from the elimination, one of the three arbitrary constants; consequently, we shall have as many equations as arbitrary constants. Let a, b, c, be these arbitrary constants. Then the equations in question, considered only with regard to the arbitrary constants which they contain, may be represented by

$$\varphi c = 0,\ \varphi b = 0,\ \varphi a = 0\ .\ .\ .\ .\ .\ .\ .\ (g)$$

Since the equations (f) all aid in the elimination which gives us one of these last equations, it thence follows that the equations (g) will each be of the second order; we call them the first integrals of the equation (e).

Generally, a differential equation of an order *n* will have a number n of first integrals, which will contain therefore the differential coefficients from $\frac{dy}{dx}$ to $\frac{d^{n-1}y}{dx^{n-1}}$ inclusively; that is to say, a number $_{n-1}$ of differential coefficients; and we see that then, when these equations are all known, to obtain the primitive equation it will suffice to eliminate from these equations the several differential coefficients.

PARTICULAR SOLUTIONS OF DIFFERENTIAL EQUATIONS.

It is easily seen that a particular integral may always be deduced from the complete integral, by giving a suitable value to the arbitrary constant.

For example, if we have given the equation

$$x\,dx + y\,dy = dy\sqrt{x^2 + y^2 - a^2},$$

whose complete integral is

$$y + c = \sqrt{(x^2 + y^2 - a^2)},$$

whence (for convenience, by rationalizing,) we get

$$(a^2 - x^2)\frac{dy^2}{dx^2} + 2xy\frac{dy}{dx} + x^2 = 0\ .\ .\ .\ .\ .\ (h)$$

and the complete integral becomes

$$2 c y + c^2 - x^2 + a^2 = 0 \quad . \quad . \quad . \quad . \quad \text{(i)}$$

Hence, in taking for c an arbitrary constant value $c = 2 a$, we shall obtain this particular integral

$$2 c y + 5 a^2 - x^2 = 0,$$

which will have the property of satisfying the proposed equation (h) as well also as the complete integral. In fact, we shall derive from this particular integral

$$y = \frac{x^2 - 5 a^2}{2 c}, \ \frac{d y}{d x} = \frac{x}{c};$$

these values reduce the proposed to

$$(x^2 - a^2) \frac{x^2}{c^2} = \frac{x^2}{c^2} (x^2 + c^2 - 5 a^2),$$

an equation which becomes identical, by substituting in the second member, the value of c^2, which gives the relation $c = 2 a$. Let

$$M d x + N d y = 0,$$

be a differential equation of the first order of a function of two variables x and y; we may conceive this equation as derived by the elimination of a constant c from a certain equation of the same order, which we shall represent by

$$m d x + n d y = 0,$$

and the complete integral

$$F (x, y, c) = 0,$$

which we shall designate by u. But, since every thing is reduced to taking the constant c such, that the equation

$$M d x + N d y = 0,$$

may be the result of elimination, we perceive that is at the same time permitted to vary the constant c, provided the equation

$$M d x + N d y = 0,$$

holds good; in this case, the complete integral

$$F (x, y, c) = 0$$

will take a greater generality, and will represent an infinity of curves of the same kind, differing from one another by a parameter, that is, by a constant.

Suppose therefore that the complete integral being differentiated, by considering c as the variable, we have obtained

$$d y = \left(\frac{d y}{d x}\right) d x + \left(\frac{d y}{d c}\right) d c$$

an equation which, for brevity, we shall write

$$d y = p d x + q d c. \quad . \quad . \quad . \quad . \quad . \quad . \quad . \quad . \quad . \quad (k)$$

Hence it is clear, that if p remaining finite, q d c is nothing, the result of the elimination of c as a variable from

$$F (x, y, c) = 0,$$

and the equation (k), will be the same as that arising from c considered constant, from

$$F (x, y, c) = 0,$$

and the equation

$$d y = p d x$$

(this result is on the hypothesis

$$M d x + N d y = 0),$$

for the equation (k), since

$$q d c = 0,$$

does not differ from

$$d y = p d x;$$

but in order to have

$$q d c = 0,$$

one of the factors of this equation = constant, that is to say, that we have

$$d c = 0, \text{ or } q = 0.$$

In the first case, d c = 0, gives c = *constant*, since that takes place for particular integrals; the second case, only therefore conducts to a particular solution. But, q being the coefficient of d c of the equation (k), we see that q = 0, gives

$$\frac{d y}{d x} = 0.$$

This equation will contain c or be independent of it. If it contain c, there will be two cases; either the equation q = 0, will contain only c and constants, or this equation will contain c with variables. In the first case, the equation q = 0, will still give c = *constant*, and in the second case, it will give c = f (x, y); this value being substituted in the equation F (x, y, c) = 0, will change it into another function of x, y, which will satisfy the proposed, without being comprised in its complete integral, and consequently will be a singular solution; but we shall have a particular integral if the equation c = f (x, y), by means of the complete integral, is reduced to a constant.

When the factor $q = 0$ from the equation $q\,d\,c = 0$ not containing the arbitrary constant c, we shall perceive whether the equation $q = 0$ gives rise to a particular solution, by combining this equation with the complete integral. For example, if from $q = 0$, we get $x = M$, and put this value in the complete integral $F(x, y, c) = 0$, we shall obtain

$$c = \text{constant} = B \text{ or } c = f y;$$

In the first case, $q = 0$, gives a particular integral; for by changing c into B in the complete integral, we only give a particular value to the constant, which is the same as when we pass from the complete integral to a particular integral. In the second case, on the contrary, the value $f y$ introduced instead of c in the complete integral, will establish between x and y a relation different from that which was found by merely replacing c by an arbitrary constant. In this case, therefore, we shall have a particular solution. What has been said of y, applies equally to x.

It happens sometimes that the value of c presents itself under the form $\frac{0}{0}$: this indicates a factor common to the equations u and U which is extraneous to them, and which must be made to disappear.

Let us apply this theory to the research of particular solutions, when the complete integral is given.

Let the equation be

$$y\,d\,x - x\,d\,y = a\,\sqrt{(d\,x^2 + d\,y^2)}$$

of which the complete integral is thus found.

Dividing the equation by $d\,x$, and making

$$\frac{d\,y}{d\,x} = p$$

we obtain

$$y - p\,x = a\,\sqrt{(1 + p^2)}.$$

Then differentiating relatively to x and to p, we get

$$d\,y - p\,d\,x - x\,d\,p = \frac{a\,p\,d\,p}{\sqrt{(1 + p^2)}};$$

observing that

$$d\,y = p\,d\,x,$$

this equation reduces to

$$x\,d\,p + \frac{a\,p\,d\,p}{\sqrt{(1 + p^2)}} = 0$$

and this is satisfied by making $d\,p = 0$. This hypothesis gives $p =$ constant $= c$, a value which being put in the above equation gives

$$y - cx = a\sqrt{(1 + c^2)} \quad . \quad . \quad . \quad . \quad . \quad . \quad . \quad . \quad . \quad (l)$$

This equation containing an arbitrary constant c, which is not to be found in the proposed equation, is the complete integral of it.

This being accomplished, the part $q\,d\,c$ of the equation $d\,y = p\,d\,x + q\,d\,c$ will be obtained by differentiating the last equation relatively to c regarded as the only variable. Operating thus we shall have

$$x\,d\,c + \frac{a\,c\,d\,c}{\sqrt{(1 + c^2)}} = 0;$$

consequently the coefficients of $d\,c$, equated to zero, will give us

$$x = -\frac{a\,c}{\sqrt{(1 + c^2)}} \quad . \quad . \quad . \quad . \quad . \quad . \quad . \quad . \quad . \quad . \quad (m)$$

To find the value of c, we have

$$(1 + c^2)\,x^2 = a^2 c^2,$$

which gives

$$c^2 = \frac{x^2}{a^2 - x^2},\quad 1 + c^2 = \frac{a^2}{a^2 - x^2},$$

and

$$\sqrt{(1 + c^2)} = \frac{a}{\sqrt{(a^2 - x^2)}};$$

by means of this last equation, eliminating the radical of the equation (m) we shall thus obtain

$$c = -\frac{x}{\sqrt{(a^2 - x^2)}} \quad . \quad . \quad . \quad . \quad . \quad . \quad . \quad . \quad . \quad . \quad (n)$$

This value and that of $\sqrt{(1 + c^2)}$ being substituted in the equation (l) will give us

$$y + \frac{x^2}{\sqrt{(a^2 - x^2)}} = \frac{a^2}{\sqrt{(a^2 - x^2)}}$$

whence is derived

$$y = \sqrt{(a^2 - x^2)},$$

an equation which, being squared, will give us

$$y^2 = a^2 - x^2;$$

and we see that this equation is a particular solution, for by differentiating it we obtain

$$d\,y = -\frac{x\,d\,x}{y};$$

this value and that of $\sqrt{(x^2 + y^2)}$, being substituted in the equation originally proposed, reduce it to

$$a^2 = a^2.$$

In the application which we have just given, we have determined the value of c by equating to zero the differential coefficient $\left(\frac{dy}{dc}\right)$. This process may sometimes prove insufficient. In fact, the equation

$$dy = p\,dx + q\,dc$$

being put under this form

$$A\,dx + B\,dy + C\,dc = 0$$

where A, B, C, are functions of x and y, we shall thence obtain

$$dy = -\frac{A}{B}dx - \frac{C}{B}dc \quad . \quad . \quad . \quad . \quad . \quad . \quad . \quad . \quad . \quad (o)$$

$$dx = -\frac{B}{A}dy - \frac{C}{A}dc \quad . \quad . \quad . \quad . \quad . \quad . \quad . \quad . \quad . \quad (p)$$

and we perceive that if all that has been said of y considered a function of x, is applied to x considered a function of y, the value of the coefficient of d c will not be the same, and that it will suffice merely that any factor of B destroys in C another factor than that which may destroy a factor of A, in order that the value of the coefficient of d c, on both hypotheses, may appear entirely different. Thus although very often the equations

$$\frac{C}{B} = 0, \; \frac{C}{A} = 0$$

give for c the same value, that will not always happen; the reason of which is, that when we shall have determined c by means of the equation

$$\frac{dy}{dc} = 0,$$

it will not be useless to see whether the hypothesis of $\frac{dx}{dc}$ gives the same result.

Clairaut was the first to remark a general class of equations susceptible of a particular solution; these equations are contained in the form

$$y = \frac{dy}{dx}x + F.\frac{dy}{dx}$$

an equation which we shall represent by

$$y = px + Fp \quad . \quad . \quad . \quad . \quad . \quad . \quad . \quad . \quad . \quad . \quad (r)$$

By differentiating it, we shall find

$$dy = p\,dx + x\,dp + \left(\frac{dFp}{dp}\right)dp;$$

this equation, since $dy = p\,dx$, becomes

$$x\,dp + \left(\frac{dFp}{dp}\right)dp = 0;$$

and since d p is a common factor, it may be thus written:

$$\left\{x + \left(\frac{d\,F\,p}{d\,p}\right)\right\} d\,p = 0$$

We satisfy this equation by making $d\,p = 0$, which gives $p =$ const. $= c$; consequently, by substituting this value in the equation (r) we shall find

$$y = c\,x + F\,c.$$

This equation is the complete integral of the equation proposed, since an arbitrary constant c has been introduced by integration. If we differentiate relatively to c we shall get

$$\left\{x + \left(\frac{d\,F\,c}{d\,c}\right)\right\} d\,c.$$

Consequently, by equating to zero the coefficients of d c, we have

$$x + \frac{d\,F\,c}{d\,c} = 0,$$

which being substituted in the complete integral, will give the particular solution.

THE INTEGRATION OF EQUATIONS OF PARTIAL DIFFERENCES.

An equation which subsists between the differential coefficients, combined with variables and constants, is, in general, a partial differential equation, or an equation of partial differences. These equations are thus named, because the notation of the differential coefficients which they contain indicates that the differentiation can only be effected partially; that is to say, by regarding certain variables as constant. This supposes, therefore, that the function proposed contains only one variable.

The first equation which we shall integrate is this; viz.

$$\left(\frac{d\,z}{d\,x}\right) = a.$$

If contrary to the hypothesis, z instead of being a function of two variables x, y, contains only x, we shall have an ordinary differential equation, which, being integrated, will give

$$z = a\,x + c$$

but, in the present case, z being a function of x and of y, the *y*s contained in z have been made to disappear by differentiation, since differentiating relatively to x, we have considered y as constant. We ought, therefore, when integrating, to preserve the same hypothesis, and suppose that the arbitrary constant is in general a function of y; consequently, we shall have for the integral of the proposed equation

$$z = a x + \varphi y.$$

Required to integrate the equation

$$\left(\frac{dz}{dx}\right) = X,$$

in which X is any function of x. Multiplying by d x, and integrating, we get

$$z = \int X \, dx + \varphi y.$$

For example, if the function X were $x^2 + a^2$, the integral would be

$$z = \frac{x^3}{3} + a^2 x + \varphi y.$$

In like manner, it is found that the integral of

$$\left(\frac{dz}{dx}\right) = Y$$

is

$$z = x Y + \varphi y.$$

Similarly, we shall integrate every equation in which $\left(\frac{dz}{dx}\right)$ is equal to a function of two variables x, y. If, for example,

$$\left(\frac{dz}{dx}\right) = \frac{x}{\sqrt{a y + x^2}},$$

considering y as constant, we integrate by the ordinary rules, making the arbitrary constant a function of y. This gives

$$z = \sqrt{(a y + x^2)} + \varphi y.$$

Finally, if we wish to integrate the equation

$$\left(\frac{dz}{dx}\right) = \frac{1}{\sqrt{(y^2 - x^2)}}$$

regarding y as constant, we get

$$z = \sin.^{-1} \frac{x}{y} + \varphi y.$$

Generally to integrate the equation

$$\left(\frac{dz}{dx}\right) dx = F(x, y) \, dx,$$

we shall take the integral relatively to x, and adding to it an arbitrary function of y, as the constant, to complete it, we shall find

$$z = \int F(x, y) \, dx + \varphi y.$$

Now let us consider the equations of partial differences which contain two differential coefficients of the first order; and let the equation be

$$M.\left(\frac{dz}{dx}\right) + N\left(\frac{dz}{dy}\right) = 0$$

in which M and N represent given functions of x, y. Hence

$$\left(\frac{dz}{dy}\right) = -\frac{M}{N}\left(\frac{dz}{dx}\right);$$

substituting this value in the formula

$$dz = \left(\frac{dz}{dx}\right) dx + \left(\frac{dz}{dy}\right) dy,$$

which has no other meaning than to express the condition that z is a function of x and of y, we obtain

$$dz = \left(\frac{dz}{dx}\right)\left\{dx - \frac{M}{N} dy\right\}$$

or

$$dz = \left(\frac{dz}{dx}\right)\frac{N\,dx - M\,dy}{N}.$$

Let λ be the factor proper to make N d x — M d y a complete differential d s; we shall have

$$\lambda\,(N\,dx - M\,dy) = ds.$$

By means of this equation, we shall eliminate N d x — M d y from the preceding equation, and we shall obtain

$$dz = \frac{1}{\lambda N}.\left(\frac{dz}{dx}\right).\,ds.$$

Finally, if we remark that the value of $\left(\frac{dz}{dx}\right)$ is indeterminate, we may take it such that $\frac{1}{\lambda N}.\left(\frac{dz}{dx}\right)$ d s may be integrable, which would make it a function of s; for we know that the differential of every given function of s must be of the form F s . d s. It therefore follows, that we may assume

$$\frac{1}{\lambda N}.\left(\frac{dz}{dx}\right) = F s,$$

an equation which will change the preceding one into

$$dz = F s.ds$$

which gives

$$z = \varphi\, s.$$

Integrating by this method the equation

$$x\left(\frac{dz}{dy}\right) - y\left(\frac{dz}{dx}\right) = 0$$

we have in this case

$$M = -y,$$

and

$$N = x;$$

consequently

$$ds = \lambda (x\,dx + y\,dy).$$

It is evident that the factor necessary to make this integrable is z. Substituting this for λ and integrating, we get

$$s = x^2 + y^2.$$

Hence the integral of the proposed equation is

$$z = \varphi (x^2 + y^2).$$

Now let us consider the equation

$$P\left(\frac{dz}{dx}\right) + Q\left(\frac{dz}{dy}\right) + R = 0;$$

in which P, Q, R are functions of the variables x, y, z; dividing it by P and making

$$\frac{Q}{P} = M, \frac{R}{P} = N,$$

we shall put it under this form:

$$\left(\frac{dz}{dx}\right) + M\left(\frac{dz}{dy}\right) + N = 0;$$

and again making

$$\left(\frac{dz}{dx}\right) = p,$$

and

$$\left(\frac{dz}{dy}\right) = q,$$

it becomes

$$p + Mq + N = 0. \quad \ldots\ldots\ldots\ldots \quad (a)$$

This equation establishes a relation between the coefficients p and q of the general formula

$$dz = \left(\frac{dz}{dx}\right) dx + \left(\frac{dz}{dy}\right) dy$$
$$= p\,dx + q\,dy;$$

without which relation p and q would be perfectly arbitrary, for as it has been already observed, this formula has no other meaning than to indicate that z is a function of two variables x, y, and that function may be any whatever; so that we ought to regard p and q as indeterminate in this last equation. Eliminating p from it, we shall obtain

$$dz + N\,dx = q\,(dy - M\,dx)$$

and q will remain always indeterminate. Hence the two members of this equation are heterogeneous (See Art. 6. vol. 1), and consequently

$$dz + N\,dx = 0, \quad dy - M\,dx = 0. \quad \ldots \quad (b)$$

If P, Q, R do not contain the variable z, it will be the same of M and N; so that the second of these equations will be an equation of two variables x and y, and may become a complete differential by means of a factor λ. This gives

$$\lambda\,(dy - M\,dx) = 0.$$

The integral of this equation will be a function of x and of y, to which we must add an arbitrary constant s; so that we shall have

$$F\,(x, y) = s;$$

whence we derive

$$y = f\,(x, s).$$

Such will be the value of y given us by the second of the above equations; and to show that they subsist simultaneously we must substitute this value in the first of them. But although the variable y is not shown, it is contained in N. This substitution of the value of y just found, amounts to considering y in the first equation as a function of x and of the arbitrary constant s. Integrating therefore this first equation on that hypothesis we find

$$z = -\int N\,dx + \varphi\,s.$$

To give an example of this integration, take the equation

$$x\left(\frac{dz}{dx}\right) + y\left(\frac{dz}{dy}\right) = a\sqrt{(x^2 + y^2)};$$

and comparing it with the general equation (a), we have

$$M = \frac{y}{x}, \quad N = -\frac{a}{x}\sqrt{(x^2 + y^2)}.$$

These values being substituted in the equations (b) will change them to

$$dz - \frac{a}{x}\sqrt{(x^2 + y^2)}\,dx = 0, \quad dy - \frac{y}{x}\,dx = 0. \quad \ldots \quad (c)$$

Let λ be the factor necessary to make the last of these integrable, and we have

$$\lambda\left(dy - \frac{y}{x}\,dx\right) = 0,$$

or rather

$$\lambda.\,\frac{x\,dy - y\,dx}{x} = 0;$$

which is integrable when $\lambda = \frac{1}{x}$; for then the integral is $\frac{y}{x}$ = constant. Put therefore

$$\frac{y}{x} = s$$

and consequently

$$y = s\,x.$$

By means of this value of y, we change the first of the equations (c) into

$$d\,z - a\,\frac{\sqrt{x^2 - s^2 x^2}}{x}\,.\,d\,x = 0,$$

or rather into

$$d\,z = a\,d\,x\,\sqrt{(1 + s^2)}.$$

Integrating on the supposition that s is constant, we shall obtain

$$z = a \int d\,x\,\sqrt{(1 + s^2)} + \varphi\,s$$

and consequently

$$z = a\,x\,\sqrt{(1 + s^2)} + \varphi\,s.$$

Substituting for s its value we get

$$z = a\,x\,\sqrt{\left(1 + \frac{y^2}{x^2}\right)} + \varphi\left(\frac{y}{x}\right)$$

$$= a\,\sqrt{(x^2 + y^2)} + \varphi\left(\frac{y}{x}\right).$$

In the more general case where the coefficients P, Q, R of the equation contain the three variables x, y, z it may happen that the equations (b) contain only the variables which are visible, and which consequently we may put under the forms

$$d\,z = f\,(x, z)\,d\,x = 0, \quad d\,y = F\,(x, y)\,d\,x.$$

These equations may be treated distinctly, by writing them as above,

$$z = \int f\,(x, z)\,d\,x + \quad z, \quad y = \int F\,(x, y)\,d\,x + \Phi\,y$$

for then we see we may make z constant in the first equation and y in the second; contradictory hypotheses, since one of three coordinates x, y, z cannot be supposed constant in the first equation without its being not constant in the second.

Let us now see in what way the equations (b) may be integrated in the case where they only contain the variables which are seen in them.

Let μ and λ be the factors which make the equations (b) integrable. If their integrals thus obtained be denoted by U and by V, we have

$$\lambda\,(d\,z + N\,d\,x) = d\,U, \quad \mu\,(d\,y - M\,d\,x) = d\,V.$$

By means of these values the above equation will become

$$dU = q\frac{\lambda}{\mu}dV \quad . \; . \; . \; . \; . \; . \; . \; . \; . \; . \quad (d).$$

Since the first member of this equation is a complete differential the second is also a complete differential, which requires $q\frac{\lambda}{\mu}$ to be a function of V. Represent this function by φV. Then the equation (d) will become

$$dU = \varphi V . dV$$

which gives, by integrating,

$$U = \Phi V.$$

Take, for example, the equation

$$xy\left(\frac{dz}{dx}\right) + x^2\left(\frac{dz}{dy}\right) = yz;$$

which being written thus, viz.

$$\left(\frac{dz}{dx}\right) + \frac{x}{y}\left(\frac{dz}{dy}\right) - \frac{z}{x} = 0$$

we compare it with the equation

$$\left(\frac{dz}{dx}\right) + M\left(\frac{dz}{dy}\right) + N = 0$$

and obtain

$$M = \frac{x}{y}, \; N = -\frac{z}{x}$$

By means of these values the equations (b) becomes

$$dz - \frac{z}{x} . dx = 0, \; dy - \frac{x}{y} dx = 0;$$

which reduce to

$$xdz - zdx = 0, \; ydy - xdx = 0.$$

The factors necessary to make these integrable are evidently $\frac{1}{x^2}$ and 2. Substituting which and integrating, we find $\frac{z}{x}$ and $y^2 - x^2$ for the integrals. Putting, therefore, these values for U and V in the equation $U = \Phi V$, we shall obtain, for the integral of the proposed equation,

$$\frac{z}{x} = \Phi(y^2 - x^2)$$

It must be remarked, that, if we had eliminated q instead of p, the equations (b) would have been replaced by these

$$Mdz + Ndy = 0, \; dy - Mdx = 0 \quad . \; . \; . \; . \; . \; . \quad (e)$$

and since all that has been said of equations (b) applies equally to these, it follows that, in the case where the first of equations (b) was not integrable, we may replace those equations by the system of equations (e), which amounts to employing the first of the equations (e) instead of the first of the equations (b).

For instance, if we had

$$a z \left(\frac{d z}{d x}\right) - z x \left(\frac{d z}{d y}\right) + x y = 0;$$

this equation being divided by a z and compared with

$$\left(\frac{d z}{d x}\right) + M \left(\frac{d z}{d y}\right) + N = 0$$

will give us

$$M = -\frac{x}{a}, \; N = \frac{x y}{a z}$$

and the equations (b) will become

$$d z + \frac{x y}{a z} d x = 0, \; d y + \frac{x}{a} d x = 0;$$

which reduce to

$$a z d z + x y d x = 0, \; a d y + x d x = 0 \quad . \; . \; . \quad (f)$$

The first of these equations, which, containing three variables, is not immediately integrable, we replace by the first of the equations (e), and we shall have, instead of the equations (f), these

$$-\frac{x}{a} d z + \frac{x y}{a z} d y = 0, \; a d y + x d x = 0;$$

which reduce to

$$2 y d y - 2 z d z = 0, \; 2 a d y + 2 x d x = 0;$$

equations, whose integrals are

$$y^2 - z^2, \; 2 a y + x^2.$$

These values being substituted for U and V, will give us

$$y^2 - z^2 = \varphi (2 a y + x^2).$$

It may be remarked, that the first of equations (e) is nothing else than the result of the elimination of d x from the equations (b).

Generally we may eliminate every variable contained in the coefficients M, N, and in a word, combine these equations after any manner whatever; if after having performed these operations, and we obtain two integrals, represented by U = a, V = b, a and b being arbitrary constants, we can always conclude that the integral is U = Φ V. In fact, since a and b are two arbitrary constants, having taken b at pleasure, we may compose a in terms of b in any way whatsoever; which is tantamount to saying that we may take for *a* an arbitrary function of b. This condition will be expressed by the equations a = φ (b). Consequently, we shall have the equations $U = \varphi b$, $V = b$, in which x, y, z represent the same coordinates. If we eliminate (b) from these equations, we shall obtain $U = \varphi V$.

This equation also shows us that in making $V = b$, we ought to have $U = \varphi b =$ *constant;* that is to say, that U and V are at the same time constant; without which a and b would depend upon one another, whereas the function φ is arbitrary. But this is precisely the condition expressed by the equations $U = a$, $V = b$.

To give an application of this theorem, let

$$z x \left(\frac{dz}{dx}\right) - z y \left(\frac{dz}{dy}\right) - y^2 = 0.$$

Dividing by $z x$ and comparing it with the general equation we have

$$M = -\frac{y}{x},\ N = -\frac{y^2}{zx};$$

and the equations (b) give us

$$dz - \frac{y^2}{zx} dx = 0,\ dy + \frac{y}{x} dx = 0$$

or

$$z x dz - y^2 dx = 0,\ x dy + y dz = 0.$$

The first of these equations containing three variables we shall not attempt its integration in that state; but if we substitute in it for $y dx$ its value derived from the second equation, it will acquire a common factor x, which being suppressed, the equation becomes

$$z dz + y dy = 0,$$

and we perceive that by multiplying by 2 it becomes integrable. The other equation is already integrable, and by integrating we find

$$z^2 + y^2 = a,\ x y = b,$$

whence we conclude that

$$z^2 + y^2 = \varphi x y.$$

We shall conclude what we have to say upon equations of partial differences of the first order, by the solution of this problem.

Given an equation which contains an arbitrary function of one or more variables, to find the equation of partial differences which produced it.

Suppose we have

$$z = F (x^2 + y^2).$$

Make

$$x^2 + y^2 = u \quad \ldots\ldots\ldots\ldots \quad (f)$$

and the equation becomes

$$z = F u.$$

The differential of F u must be of the form φ u . d u. Consequently

$$dz = du . \varphi u$$

If we take the differential of z relatively to x only, that is to say, in regarding y as constant, we ought to take also d u on the same hypothesis. Consequently, dividing the preceding equation by d x, we get

$$\left(\frac{dz}{dx}\right) = \left(\frac{du}{dx}\right) \varphi u.$$

Again, considering x as constant and y as variable, we shall similarly find

$$\left(\frac{dz}{dy}\right) = \left(\frac{du}{dy}\right) \varphi u.$$

But the values of these coefficients are found from the equation (f), which gives

$$\left(\frac{du}{dx}\right) = 2x, \left(\frac{du}{dy}\right) = 2y.$$

Hence our equations become

$$\left(\frac{dz}{dx}\right) = 2x \varphi u, \left(\frac{dz}{dy}\right) = 2y \varphi u;$$

and eliminating φ u from these, we get the equation required; viz.

$$y \left(\frac{dz}{dx}\right) = x \left(\frac{dz}{dy}\right).$$

As another example, take this equation

$$z^2 + 2ax = F(x - y).$$

Making

$$x - y = u,$$

it becomes

$$z^2 + 2ax = Fu$$

and differentiating, we get

$$d(z^2 + 2ax) = du \varphi u.$$

Then taking the differential relatively to x, we have

$$2z \left(\frac{dz}{dx}\right) + 2a = \left(\frac{du}{dx}\right) \varphi u.$$

and similarly, with regard to y, we get

$$2z \left(\frac{dz}{dy}\right) = \left(\frac{du}{dy}\right) \varphi u.$$

But since

$$x - y = u$$

$$\left(\frac{d u}{d x}\right) = 1, \left(\frac{d u}{d y}\right) = -1,$$

which, being substituted in the above equation, gives us

$$2 z \left(\frac{d z}{d x}\right) + 2 a = \varphi u, 2 z \left(\frac{d z}{d y}\right) = -\varphi u$$

and eliminating φ u from these, we have the equation required; viz.

$$\left(\frac{d z}{d x}\right) + \left(\frac{d z}{d y}\right) + \frac{a}{z} = 0.$$

We now come to

EQUATIONS OF PARTIAL DIFFERENCES OF THE SECOND ORDER.

An equation of Partial Differences of the second order in which z is a function of two variables x, y ought always to contain one or more of the differential coefficients

$$\left(\frac{d^2 z}{d x^2}\right), \left(\frac{d^2 z}{d y^2}\right), \left(\frac{d^2 z}{d x . d y}\right)$$

independently of the differential coefficients which enter equations of the first order.

We shall merely integrate the simplest equations of this kind, and shall begin with this, viz.

$$\left(\frac{d^2 z}{d x^2}\right) = 0.$$

Multiplying by d x and integrating relatively to x we add to the integral an arbitrary function of y; and we shall thus get,

$$\left(\frac{d z}{d x}\right) = \varphi y$$

Again multiplying by d x and integrating, the integral will be completed when we add another arbitrary function of y, viz. ψ y. We thus obtain

$$z = x \varphi y + \psi y.$$

Now let us integrate the equation.

$$\left(\frac{d^2 z}{d x^2}\right) = P$$

in which P is any function of x, y. Operating as before we first obtain

$$\left(\frac{d z}{d x}\right) = \int P d x + \varphi y;$$

and the second integration gives us

$$z = \int \{\int P d x + \varphi y\} d x + \psi y.$$

In the same manner we integrate

$$\left(\frac{d^2 z}{d y^2}\right) = P$$

and find

$$z = \int \{\varphi x + \int P d y\} d y + \psi x.$$

The equation

$$\left(\frac{d^2 z}{d y d x}\right) = P$$

must be integrated first relatively to one of the variables, and then relatively to the other, which will give

$$z = \int \{\varphi y + \int P d x\} d y + \varphi x.$$

In general, similarly may be treated the several equations

$$\left(\frac{d^n z}{d y^n}\right) = P, \left(\frac{{}^n z}{d x d y^{n-1}}\right) = Q, \left(\frac{d^n z}{d x^2 d y^{n-2}}\right) = R, \&c.$$

in which P, Q, R, &c. are functions of x, y, which gives place to a series of integrations, introducing for each of them an arbitrary function.

One of the next easiest equations to integrate is this:

$$\left(\frac{d^2 z}{d y^2}\right) + P \left(\frac{d z}{d y}\right) = Q;$$

in which P and Q will always denote two functions of x and y.

Make

$$\left(\frac{d z}{d y}\right) = u,$$

and the proposed will transform to

$$\left(\frac{d u}{d y}\right) + P u = Q.$$

To integrate this, we consider x constant, and then it contains only two variables y and u, and it will be of the same form as the equation

$$d y + P y d x = Q d x$$

whose integral (see Vol. I. p. 109) is

$$y = e^{-\int P d x} \{\int Q e^{\int P d x} d x + C\}.$$

Hence our equation gives

$$u = e^{-\int P d y} \{\int Q e^{\int P d y} d y + \varphi x\}.$$

But

$$u = \left(\frac{d z}{d y}\right).$$

Hence by integration we get

$$z = \int \{ e^{-\int P d y} \left(\int Q e^{\int P d y} d y\right) + \varphi x \} d y + \psi x.$$

By the same method we may integrate

$$\left(\frac{d^2 z}{d x d y}\right) + P \left(\frac{d z}{d x}\right) = Q, \quad \frac{d^2 z}{d x d y} + P \left(\frac{d z}{d y}\right) = Q,$$

in which P, Q represent functions of x, and because of the divisor d x d y, we perceive that the value of z will not contain arbitrary functions of the same variable.

THE DETERMINATION OF THE ARBITRARY FUNCTIONS WHICH ENTER THE INTEGRALS OF EQUATIONS OF PARTIAL DIFFERENCES OF THE FIRST ORDER.

The arbitrary functions which complete the integrals of equations of partial differences, ought to be given by the conditions arising from the nature of the problems from which originated these equations; problems generally belonging to the physical branches of the Mathematics.

But in order to keep in view the subject we are discussing, we shall limit ourselves to considerations purely analytical, and we shall first seek what are the conditions contained in the equation

$$\left(\frac{d z}{d x}\right) = a.$$

Since z is a function of x, y, this equation may be considered as that of a surface. This surface, from the nature of its equation, has the following property, that $\left(\frac{d z}{d x}\right)$ must always be constant. Hence it follows that every section of this surface made by a plane parallel to that of x, y is a straight line. In fact, whatever may be the nature of this section, if we divide it into an infinity of parts, these, to a small extent, may be considered straight lines, and will represent the elements of the section, one of these elements making with a parallel to the axis of abscissæ, an angle whose tangent is $\left(\frac{d z}{d x}\right)$. Since this angle is constant, it follows that all the angles formed in like manner by the elements of the curve, with parallels to the axis of abscissæ will be equal. Which proves that the section in question is a straight line.

We might arrive at the same result by considering the integral of the equation

$$\left(\frac{dz}{dx}\right) = a$$

which we know to be

$$z = ax + \varphi y,$$

since for all the points of the surface which in the cutting plane, the ordinate is equal to a constant c. Replacing therefore φy by φc, and making $\varphi c = C$, the above equation becomes

$$z = ax + C;$$

this equation being that of a straight line, shows that the section is a straight line.

The same holding good relatively to other cutting planes which may be drawn parallel to that of x, z, we conclude that all these planes will cut the surface in straight lines, which will be parallel, since they will each form with a parallel to the axis of x, an angle whose tangent is a.

If, however, we make $x = 0$, the equation $z = ax + \varphi y$ reduces to $z = \varphi y$, and will be that of a curve traced upon the plane of y, z; this curve containing all the points of the surface whose coordinates are $x = 0$, will meet the plane in a point whose coordinate is $x = 0$; and since we have also $y = c$, the third coordinate by means of the equation

$$z = ax + C$$

will be

$$z = C.$$

What has been said of this one plane, applies equally to all others which are parallel to it, and it thence results that through all the points of the curve whose equation is $z = \varphi y$, and which is traced in the plane of y, z, will pass straight lines parallel to the axis of x. This is expressed by the equations

$$\left(\frac{dz}{dx}\right) = a$$

and

$$z = ax + \varphi y;$$

and since this condition is always fulfilled, whatever may be the figure of the curve whose equation is $z = \varphi y$, we see that this curve is arbitrary.

From what precedes, it follows that the curve whose equation is $z = \varphi y$, may be composed of arcs of different curves, which unite at their extremities, as in this diagram

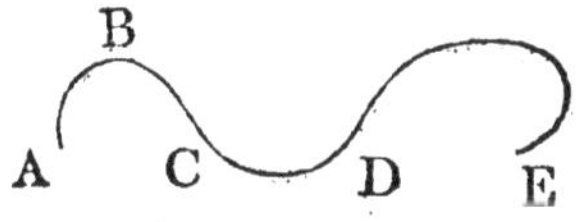

or which have a break off in their course, as in this figure.

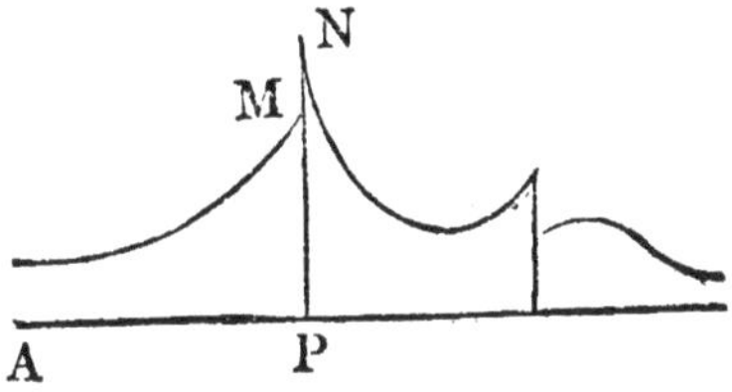

In the first case the curve is *discontinuous*, and in the second it is *discontiguous*. We may remark that in this last case, two different ordinates P M, P N corresponding to the same abscissa A P; finally, it is possible, that without being discontiguous, the curve may be composed of an infinite series of arcs indefinitely small, which belong each of them to different curves; in this case, the curve is irregular, as will be, for instance, the flourishes of the pen made at random; but in whatever way it is formed, the curve whose equation is $z = \varphi y$, it will suffice, to construct the surface, to make a straight line move parallelly with this condition, that its general point shall trace out the curve whose equation is

$$z = \varphi y,$$

and which is traced at random upon the plane of y, z.

If instead of the equation

$$\left(\frac{dz}{dx}\right) = a,$$

we had

$$\left(\frac{dz}{dx}\right) = X,$$

in which X was a function of x, then in drawing a plane parallel to the plane (x, z), the surface will be cut by it no longer in a straight line, as in the preceding case. In fact, for every point taken in this section, the tangent of the angle formed by the element produced of the section, with a parallel to the axis of x, will be equal to a function X of the abscissa x of this point; and since the abscissa x is different for every point, it follows that this angle will be different at each point of the section, which section, therefore, is no longer, as before, a straight line. The surface will be constructed, as before, by moving the section parallelly, so that its point may ride continually in the curve whose equation is $z = \varphi y$.

Suppose now that in the preceding equation, instead of X we have a function, P of x, and of y. The equation

$$\left(\frac{dz}{dx}\right) = P,$$

containing three variables will belong still to a curve surface. If we cut this surface by a plane parallel to that of x, z, we shall have a section in which y will be constant; and since in all its points $\left(\frac{dz}{dx}\right)$ will be equal to a function of the variable x, this section must be a curve, as in the preceding case. The equation

$$\left(\frac{dz}{dx}\right) = P$$

being integrated, we shall have for that of the surface

$$z = \int P\,dx + \varphi y;$$

if in this equation we give successively to y the increasing values y', y'', y''', &c. and make P', P'', P''', &c. what the function P becomes in these cases, we shall have the equations

$$\left.\begin{array}{ll} z = \int P'dx + \varphi y', & z = \int P''dx + \varphi y'' \\ z = \int P'''dx + \varphi y''', & z = \int P''''dx + \varphi y'''' \text{ \&c.} \end{array}\right\}$$

and we see that these equations will belong to curves of the same nature, but different in form, since the values of the constant y will not be the same. These curves are nothing else than the sections of the surface made by planes parallel to the plane (x, z); and in meeting the plane (y, z) they will form a curve whose equation will be obtained by equating to zero, the value of x in that of the surface. Call the value of $\int P\,dx$, in this case, Y, and we shall have

$$z = Y + \varphi y;$$

and we perceive that by reason of φy, the curve determined by this equation must be arbitrary. Thus, having traced at pleasure a curve, Q R S, upon the plane (y, z), if we represent by R L the section whose equation

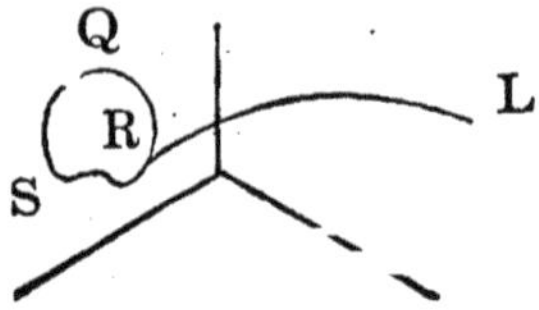

is $z = \int P'dx + \varphi y'$, we shall move this section, always keeping the extremity R applied to the curve Q R S; but so that this section as it moves, may assume the successive forms determined by the above group of equations, and we shall thus construct the surface to which will belong the equation

$$\left(\frac{d\,z}{d\,x}\right) = P.$$

Finally let us consider the general equation

$$\left(\frac{d\,z}{d\,x}\right) + M\left(\frac{d\,z}{d\,y}\right) + N = 0,$$

whose integral is $U = \varphi V$. Since $U = a$, $V = b$, each of these equations subsisting between three coordinates, we may regard them as belonging to two surfaces; and since the coordinates are common, they ought to belong to the curve of intersection of the two surfaces. This being shown, a and b being arbitrary constants, if in $U = a$, we give to x and y the values x', y' we shall obtain for z, a function of x', of y' and of a, which will determine a point of the surface whose equation is $U = a$. This point, which is any whatever, will vary in position if we give successively different values to the arbitrary constant a, which amounts to saying that by making *a* vary, we shall pass the surface whose equation is $U = a$, through a new system of points. This applies equally to $V = b$, and we conclude that the curve of intersection of the two surfaces will change continually in position, and consequently will describe a curved surface in which a, b may be considered as two coordinates; and since the relation $a = \varphi b$ which connects these two coordinates, is arbitrary we perceive that the determination of the function φ amounts to making a surface pass through a curve traced arbitrarily.

To show how this sort of problems may conduct to analytical conditions, let us examine what is the surface whose equation is

$$y\left(\frac{d\,z}{d\,x}\right) = x\left(\frac{d\,z}{d\,y}\right).$$

We have seen that this equation being integrated gives

$$z = \varphi\,(x^2 + y^2).$$

Reciprocally we hence derive

$$x^2 + y^2 = \Phi\,z.$$

If we cut the surface by a plane parallel to the plane (x, y) the equation of the section will be

$$x^2 + y^2 = \Phi\,c;$$

and representing by a^2 the constant $\Phi\,c$, we shall have

$$x^2 + y^2 = a^2.$$

This equation belongs to the circle. Consequently the surface will have this property, viz. that every section made by a plane parallel to the plane (x, y) will be a circle.

This property is also indicated by the equation

$$y\left(\frac{dz}{dx}\right) = x\left(\frac{dz}{dy}\right)$$

for this equation gives

$$x = y\frac{dy}{dx}.$$

This equation shows us that the subnormal ought to be always equal to the abscissa which is the property of the circle.

The equation $z = \varphi(x^2 + y^2)$ showing merely that all the sections parallel to the plane (x, y) are circles, it follows thence that the law according to which the radii of these sections ought to increase, is not comprised in this equation, and that consequently, every surface of revolution will satisfy the problem; for we know that in this sort of surfaces, the sections parallel to the plane (x, y) are always circles, and it is needless to say that the generatrix which, during a revolution, describes the surface, may be a curve discontinued, discontiguous, regular or irregular.

Let us therefore investigate the surface for which this generatrix will be a parabola A N, and suppose that, in this hypothesis, the surface is cut by a plane A B, which shall pass through the axis of z; the trace of

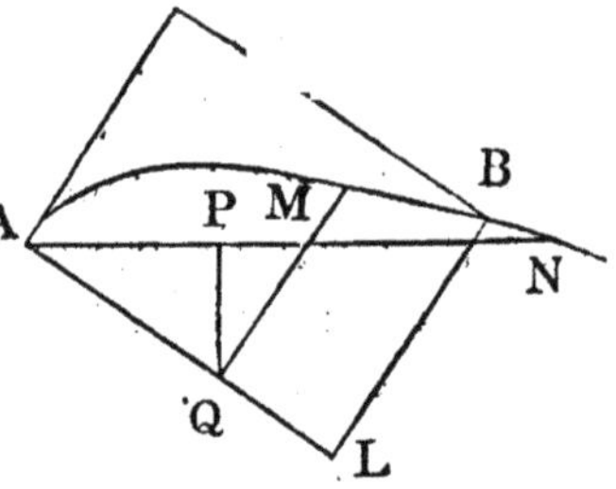

this plane upon the plane (x, y) will be a straight line A L, which, being drawn through the origin, will have the equation $y = ax$; if we represent by t the hypothenuse of the right angled triangle A P Q, constructed upon the plane (x, y) we shall have

$$t^2 = x^2 + y^2;$$

but t being the abscissa of the parabola A M, of which Q M = z is the ordinate, we have, by the nature of the curve,

$$t^2 = bz.$$

Putting for t^2 its value $x^2 + y^2$, we get

$$z = \frac{1}{b}(y^2 + x^2), \text{ or } z = \frac{1}{b}(a^2x^2 + x^2) = \frac{1}{b}x^2(1 + a^2);$$

and making

$$\frac{1}{b}(a^2 + 1) = m,$$

we shall obtain

$$z = m x^2;$$

so that the condition prescribed in the hypothesis, where the generatrix is a parabola, is that we ought to have

$$z = m x^2, \text{ when } y = a x.$$

Let us now investigate, by means of these conditions, the arbitrary function which enters the equation $z = \varphi (x^2 + y^2)$. For that purpose, we shall represent by U the quantity $x^2 + y^2$, which is effected by the symbol φ, and the equation then becomes

$$z = \varphi U;$$

and we shall have the three equations

$$x^2 + y^2 = U, \; y = a x, \; z = m x^2.$$

By means of the two first we eliminate y and obtain the value of x^2 which being put into the third, will give

$$Z = m \cdot \frac{U}{1 + a^2},$$

an equation which reduces to

$$Z = \frac{1}{b} U;$$

the value of z being substituted in the equation $z = \varphi U$, will change it to

$$\varphi U = \frac{1}{b} U;$$

and putting the value of U in this equation, we shall find that

$$\varphi (x^2 + y^2) = \frac{1}{b}(x^2 + y^2),$$

and we see that the function is determined. Substituting this value of $\varphi (x^2 + y^2)$ in the equation $z = \varphi (x^2 + y^2)$, we get

$$z = \frac{1}{b}(x^2 + y^2),$$

for the integral sought, an equation which has the property required, since the hypothesis of $y = a x$ gives

$$z = m x^2.$$

This process is general; for, supposing the conditions which determine the arbitrary constant to be that the integral gives $F (x, y, z) = 0$, when we have $f (x, y, z) = 0$, we shall obtain a third equation by equating to

U the quantity which follows φ, and then by eliminating, successively, two of the variables x, y, z, we shall obtain each of these variables in a function of U; putting these values in the integral, we shall get an equation whose first member is φ U, and whose second member is a compound expression in terms of U; restoring the value of U in terms of the varibles, the arbitrary function will be determined.

THE ARBITRARY FUNCTIONS WHICH ENTER THE INTEGRALS OF THE EQUATIONS OF PARTIAL DIFFERENCES OF THE SECOND ORDER.

Equations of partial differences of the second order conduct to integrals which contain two arbitrary functions; the determination of these functions amounts to making the surface pass through two curves which may be discontinuous or discontiguous. For example, take the equation

$$\left(\frac{d^2 z}{d x^2}\right) = 0,$$

whose integral has been found to be

$$z = x \varphi y + \psi y$$

Let A x, A y, A z, be the axis of coordinates; if we draw a plane

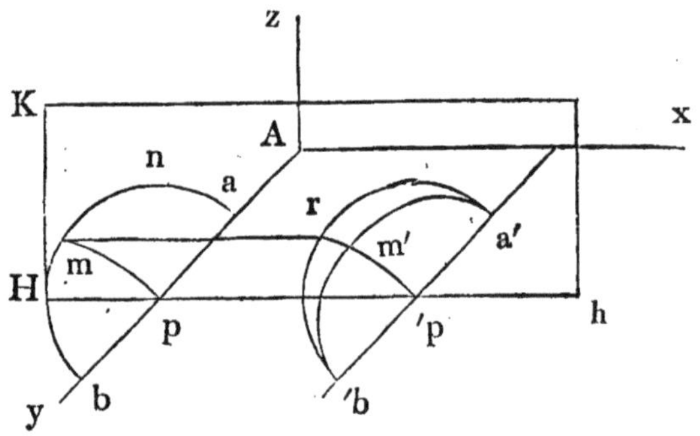

K L parallel to the plane (x, z), the section of the surface by this plane will be a straight line; since, for all the points of this section, y being equal to A p, if we represent A p by a constant c, the quantities φ y, ψ y will become φ c, ψ c, and, consequently, may be replaced by two constants, a, b, so that the equation

$$z = x \varphi y + \psi y$$

will become

$$z = a x + b,$$

and this is the equation to the section made by the plane K L.

To find the point where this section meets the plane (y, z) make $x = 0$, and the equation above gives $z = \psi y$, which indicates a curve a m b, traced upon the plane (y, z). It will be easy to show that the section meets the curve a m b in a point m; and since this section is a straight line, it is only requisite, to find the position of it, to find a second point. For that purpose, observe that when $x = 0$, the first equation reduces to

$$z = \psi y,$$

whilst, when $x = 1$, the same equation reduces to

$$z = \varphi y + \psi y.$$

Making, as above, $y = A p = c$, these two values of z will become

$$z = b, \; z = a + b,$$

and determining two points m and r, taken upon the same section, m r we know to be in a straight line. To construct these points we thus proceed: we shall arbitrarily trace upon the plane (y, z) the curve a m b, and through the point p, where the cutting plane K L meets the axis of y, raise the perpendicular $p m = b$, which will be an ordinate to the curve; we shall then take at the intersection H L of the cutting plane, and the plane (x, y), the part p p′ equal to unity, and through the point p′, we shall draw a plane parallel to the plane (y, z), and in this plane construct the curve a′ m′ b′, after the modulus of the curve a m b, and so as to be similarly disposed; then the ordinate m′ p′ will be equal to m p; and if we produce m′ p′ by m′ r, which will represent a, we shall determine the point r of the section.

If, by a second process, we then produce all the ordinates of the curve a′ m′ b′, we shall construct a new curve a′ r′ b′, which will be such, that drawing through this curve and through a m b, a plane parallel to the plane (x, z), the two points where the curves meet, will belong to the same section of the surface.

From what precedes, it follows that the surface may be constructed, by moving the straight line m r so as continually to touch the two curves, a m b, a′ m′ b′.

This example suffices to show that the determination of the arbitrary functions which complete the integrals of equations of partial differences of the second order, is the same as making the surface pass through two curves, which, as well as the functions themselves, may be discontinuous, discontiguous, regular or irregular.

CALCULUS OF VARIATIONS.

If we have given a function $Z = F, (x, y, y', y'')$, wherein y', y'' mean

$$\left(\frac{d\,y}{d\,x}\right), \left(\frac{d^2 y}{d\,x^2}\right),$$

y itself being a function of x, it may be required to make L have certain properties, (such as that of being a maximum, for instance) whether by assigning to x, y numerical values, or by establishing relations between these variables, and connecting them by equations. When the equation $y = \varphi x$ is given, we may then deduce $y, y', y'' \ldots$ in terms of x and substituting, we have the form

$$Z = f\,x.$$

By the known rules of the differential calculus, we may assign the values of x, when we make of x a *maximum* or *minimum*. Thus we determine what are the points of a given curve, for which the proposed function Z, is greater or less than for every other point of the same curve.

But if the equation $y = \varphi\,x$ is not given, then taking successively for $\varphi\,x$ different forms, the function $Z = f\,x$ will, at the same time, assume different functions of x. It may be proposed to assign to $\varphi\,x$ such a form as shall make Z greater or less than every other form of $\varphi\,x$, *for the same numerical value of x whatever it may be in other respects.* This latter species of problem belongs to the calculus of variations. This theory relates not to maxima and minima only; but we shall confine ourselves to these considerations, because it will suffice to make known all the rules of the calculus. We must always bear in mind, that the variables x, y are not *independent*, but that the equation $y = \varphi\,x$ is unknown, and that we only suppose it given to facilitate the resolution of the problem. We must consider x as any quantity whatever which remains the same for all the differential forms of $\varphi\,x$; the forms of $\varphi, \varphi', \varphi'' \ldots$ are therefore variable, whilst x is constant.

In $Z = F(x, y, y', y'' \ldots)$ put $y + k$ for y, $y' + k'$, for $y' \ldots$, k being an arbitrary function of x, and $k', k'' \ldots$ the quantities

$$\frac{d\,k}{d\,x}, \frac{d^2 k}{d\,x^2} \ldots$$

But, Z will become

$$Z_{,} = F(x, y + k, y' + k', y'' + k'' \ldots)$$

Taylor's theorem holds good whether the quantities x, y, k be dependent or independent. Hence we have

$$Z_{,} = Z + \left\{k\left(\frac{dZ}{dy}\right) + k'\left(\frac{dZ}{dy'}\right) + k''\left(\frac{dZ}{dy''}\right) + \ldots\right\} + \&c.$$

so that we may consider x, y, y′, y″ . . . as so many independent variables.

The nature of the question requires that the equation $y = \varphi x$ should be determined, so that for the same value of x, we may have always $Z_{,} > Z$, or $Z_{,} < Z$: reasoning as in the ordinary *maxima and minima*, we perceive that the terms of the first order must equal zero, or that we have

$$k\left(\frac{dZ}{dy}\right) + k'\left(\frac{dZ}{dy'}\right) + k''\left(\frac{dZ}{dy''}\right) + \&c. = 0.$$

Since k is arbitrary for every value of x, and it is not necessary that its value or its form should remain the same, when x varies or is constant, k′, k″ . . . is as well arbitrary as k. For we may suppose for any value $x = X$ that $k = a + b(x - X) + \frac{1}{2}c(x - X)^2 + \&c.$, X, a, b, c . . . being taken at pleasure; and since this equation, and its differentials, ought to hold good, whatever is x, they ought also to subsist when $x = X$, which gives $k = a$, $k' = b$, $k'' = c$, &c. Hence the equation $Z_{,} = Z + \ldots$ cannot be satisfied when a, b, c . . . are considered independent, unless (see 6, vol. I.)

$$\left(\frac{dZ}{dy}\right) = 0,\ \left(\frac{dZ}{dy'}\right) = 0,\ \left(\frac{dZ}{dy''}\right) = 0 \ldots \left(\frac{dZ}{dy^{(n)}}\right) = 0,$$

n being the highest order of y in Z. These different equations subsist simultaneously, whatever may be the value of x; and if so, there ought to be a maximum or minimum; and the relation which then subsists between x, y will be the equation sought, viz. $y = \varphi x$, which will have the property of making Z greater or less than every other relation between x and y can make it. We can distinguish the maximum from the minimum from the signs of the terms of the second order, as in vol. I. p. (31.)

But if all these equations give different relations between x, y, the problem will be impossible in the state of generality which we have ascribed to it; and if it happen that some only of these equations subsist mutually, then the function Z will have *maxima and minima*, relative to some of the quantities y, y′, y″ . . . without their being common to them all. The equations which thus subsist, will give the relative *maxima and minima*. And if we wish to make X a maximum or minimum only relatively to one of the quantities y, y′, y″ . . ., since then we have only one equation to satisfy, the problem will be always possible.

From the preceding considerations it follows, that first, the quantities x, y depend upon one another, and that, nevertheless, we ought to make them vary, as if they were independent, for this is but an artifice to get the more readily at the result.

Secondly, that these variations are not indefinitely small; and if we employ the differential calculus to obtain them, it is only an expeditious means of getting the second term of the developement, the only one which is here necessary.

Let us apply these general notions to some examples.

Ex. I. Take, upon the axis of x of a curve, two abscissas m, n; and draw indefinite parallels to the axis of y. Let $y = \varphi x$ be the equation of this curve: if through any point whatever, we draw a tangent, it will cut the parallels in points whose ordinates are

$$l = y + y' (m - x), \ h = y + y' (n - x).$$

If the form of φ is given, every thing else is known; but if it is not given, it may be asked, what is the curve which has the property of having for each point of tangency, the product of these two ordinates less than for every other curve.

Here we have $l \times h$; or

$$Z = \{ y \times (m - x) y' \} + \{ y + (n - x) y' \}.$$

From the enunciation of the problem, the curves *which pass through the same point* (x, y) have tangents taking different directions, and that which is required, ought to have a tangent, such that the condition Z = *maximum* is fulfilled. We may consider x and y constant; whence

$$\left(\frac{dZ}{dy'}\right) = 0, \ \frac{2y'}{y} = \frac{2x - m - n}{(x-m)(x-n)} = \frac{1}{x-m} + \frac{1}{x-n}.$$

Then integrating we get

$$y^2 = C (x - m) (x - n).$$

The curve is an ellipse or a hyperbola, according as C is positive or negative; the vertexes are given by $x = m$, $x = n$; in the first case, the product $h \times l$ or Z is a *maximum*, because y″ is negative; in the second, Z is a *minimum* or rather a negative *maximum*; this product is moreover constant, and $l h = -\frac{1}{4} C (m - n)^2$, the square of the semi-axis.

Ex. 2. *What is the curve for which, in each of its points, the square of the subnormal added to the abscissa is a minimum?*

We have in this case

$$Z = (y y' + x)^2$$

whence we get two equations subsisting mutually by making

$$y y' + x = 0$$

and thence

$$x^2 + y^2 = r^2.$$

Therefore all the circles described from the origin as a center will alone satisfy the equation.

The theory just expounded has not been greatly extended; but it serves as a preliminary developement of great use for the comprehension of a far more interesting problem which remains to be considered. This requires all the preceding reasonings to be applied to a function of the form $\int Z$: the sign $\int$ indicates the function Z to be a differential and that after having integrated it between prescribed limits it is required to endow it with the preceding properties. The difficulty here to be overcome is that of resolving the problem without integrating.

When a body is in motion, we may compare together either the different points of the body in one of its positions or the plane occupied successively by a given point. In the first case, the body is considered fixed, and the symbol d will relate to the change of the coordinates of its surface; in the second, we must express by a convenient symbol, *variations* altogether independent of the first, which shall be denoted by δ. When we consider a curve immoveable, or even variable, but taken in one of its positions, $d x$, $d y$... announce a comparison between its coordinates; but to consider the different planes which the same point of a curve occupies, the curve varying in form according to any law whatever, we shall write δx, δy... which denote the increments considered under this point of view, and are functions of x, y... In like manner, $d x$ becoming $d (x + \delta x)$ will increase by $d \delta x$; $d^2 x$ will increase by $d^2 \delta x$, &c.

Observe that the variations indicated by the symbol δ are finite, and wholly independent of those which d represents; the operations to which these symbols relate being equally independent, the order in which they are used must be equally a matter of indifference as to the result. So that we have

$$\delta . d x = d . \delta x$$

$$d^2 . \delta x = \delta . d^2 x$$

&c.

$$\int \delta U = \delta \int U.$$

and so on.

It remains to establish relations between x, y, z... such that $\int Z$ may be a *maximum or a minimum between given limits.* That the calculus may be rendered the more symmetrical, we shall not suppose any differential constant; moreover we shall only introduce three variables because it will be easy to generalise the result. To abridge the labour of the process, make

$$d\,x = x_{,},\quad d^2\,x = x_{,,}\ \&c.$$

so that

$$z = F\,(x,\ x_{,},\ x_{,,},\ldots y,\ y_{,},\ y_{,,},\ldots z,\ z_{,},\ z_{,,},\ldots).$$

Now x, y and z receiving the arbitrary and finite increments $\delta\,x$, $\delta\,y$, $\delta\,z$, $d\,x$ or $x_{,}$ becomes

$$d\,(x + \delta\,x) = d\,x + \delta\,d\,x \text{ or } x_{,} + \delta\,x_{,}.$$

In the same manner, $x_{,,}$ increases by $\delta\,x_{,,}$ and so on; so that developing $Z_{,}$ by Taylor's theorem, and integrating $\int Z$ becomes

$$\int Z_{,} = \int Z + \int \left\{ \left(\frac{d\,Z}{d\,x}\right) \delta\,x + \left(\frac{d\,Z}{d\,y}\right) \delta\,y + \left(\frac{d\,Z}{d\,z}\right) \delta\,z + \left(\frac{d\,Z}{d\,x_{,}}\right) \delta\,x_{,} \right.$$
$$\left. + \left(\frac{d\,Z}{d\,y_{,}}\right) \delta\,y_{,} + \left(\frac{d\,Z}{d\,z_{,}}\right) \delta\,z_{,} + \left(\frac{d\,Z}{d\,x_{,,}}\right) \delta\,x_{,,} + \&c. \right\} + \int \&c.$$

The condition of a *maximum* or *minimum* requires the integral of the terms of the first order to be zero between given limits *whatever may be* $\delta\,x$, $\delta\,y$, $\delta\,z$ as we have already seen. Take the differential of the known function Z considering $x,\ x_{,},\ x_{,,}\ldots y,\ y_{,},\ y_{,,}\ldots$ as so many independent variables; we shall have

$$d\,Z = m\,d\,x + n\,d\,x_{,} + p\,d\,x_{,,} + \ldots M\,d\,y + N\,d\,y_{,}\ldots + \mu\,d\,z + \nu\,d\,z_{,}\ldots$$

$m,\ n\ldots M,\ N\ldots \mu,\ \nu\ldots$ being the coefficients of the partial differences of Z relatively to $x,\ x_{,}\ldots y,\ y_{,}\ldots z,\ z_{,},\ldots$ considered as so many variables; these are therefore known functions for each proposed value of Z. Performing this differentiation exactly in the same manner by the symbol δ, we have

$$\left.\begin{aligned} \delta\,Z = m\,\delta\,x + n\,\delta\,d\,x + p\,\delta\,d^2 x + q\,\delta\,d^3 x + \ldots \\ + M\,\delta\,y + N\,\delta\,d\,y + P\,\delta\,d^2 y + q\,\delta\,d^3 y + \ldots \\ + \mu\,\delta\,z + \nu\,\delta\,d\,z + \pi\,\delta\,d^2 z + \chi\,\delta\,d^3 y + \ldots \end{aligned}\right\}(A)$$

But this known quantity, whose number of terms is limited, is precisely that which is under the sign $\int$, in the terms of the first order of the developement: so that the required condition of *max.* or *min.* is that

$$\int \delta\,Z = 0,$$

between given limits, whatever may be the variations $\delta\,x$, $\delta\,y$, $\delta\,z$. Observe, that here, as before, the differential calculus is only employed as a means of obtaining easily the assemblage of terms to be equated to zero; so that the variations are still any whatever and finite.

We have said that d . δ x may be put for d . δ x; thus the first line is equivalent to

$$m\,\delta x + n\,.\,d\,\delta x + p\,.\,d^2\,\delta x + q\,.\,d^3\,\delta x + \&c.$$

m, n . . . contains differentials, so that the defect of homogeneity is here only apparent. To integrate this, we shall see that it is necessary to disengage from the symbol $\int$ as often as possible, the terms which contain d δ. To effect this, we integrate *by parts* which gives

$$\begin{aligned}
&\int n\, d\,\delta x = n\,.\,\delta x - \int d\, n\,.\,\delta x \\
&\int p\,.\,d^2\delta x = p\, d\,\delta x - d\, p\ \delta x + \int d^2 p\,\delta x \\
&\int q\, d^3\,\delta x = q\, d^2\delta x - d\, q\,.\,d\,\delta x + d^2 q\,.\,d\,x - \int d^3 q\,.\,\delta x \\
&\qquad \&c.
\end{aligned}$$

Collecting these results, we have this series, the law of which is easily recognised; viz.

$$\begin{aligned}
&\int (m - d\, n + d^2 p - d^3 q + d^4 r - \ldots)\,\delta x \\
&+ (n - d\, p + d^2 q - d^3 r + d^4 s - \ldots)\,\delta x \\
&+ (p - d\, q + d^2 r - d^3 s + d^4 t - \ldots)\, d\,\delta x \\
&+ (q - d\, r + \ldots)\, d^2\,\delta x \\
&+ \&c.
\end{aligned}$$

The integral of (A) or $\int .\ \delta z = 0$, becomes therefore

$$(B) \ldots \int \{(m - d\, n + d^2 p - \ldots)\delta x + (M - d\, N + d^2 P - \ldots)\delta y + (\mu - d\, \nu - \ldots)\delta z\} = 0$$

$$(C) \ldots \left\{ \begin{aligned}
&(n - d\, p + d^2 q \ldots)\,\delta x + (N - d\, P + d^2 Q - \ldots)\,\delta y + (\nu - d\,\pi \ldots)\, d\, z \\
&+ (p - d\, q + d^2 r \ldots)\, d\,\delta x + (P - d\, Q + \ldots) d\,\delta y + (\pi - d\,\chi \ldots) d\,\delta z \\
&+ (q - d\, r \ldots)\, d^2\delta x \ldots + K = 0
\end{aligned} \right.$$

K being the arbitrary constant. The equation has been split into two, because the terms which remain under the sign $\int$ cannot be integrated, at least whilst δ x, δ y, δ z are arbitrary. In the same manner, if the nature of the question does not establish some relation between δ x, δ y, δ z, the independence of these variations requires also that equation (B) shall again make three others; viz.

$$\left. \begin{aligned}
0 &= m - d\, n + d^2 p - d^3 q + d^4 r - \ldots \\
0 &= M - d\, N + d^2 P - d^3 Q + d^4 R - \ldots \\
0 &= \mu - d\,\nu + d^2 \pi - d^3 \chi + d^4 \varrho - \ldots
\end{aligned} \right\} \ldots \quad (D)$$

Consequently, to find the relations between x, y, z, which make $\int$ Z a *maximum*, we must take the differential of the given function Z by considering x, y, z, d x, d y, d z, d^2 x, . . . as so many independent variables, and use the letter δ to signify their increase; this is what is termed *taking the variation* of Z. Comparing the result with the equation (A), we shall observe the values of m, M, μ, n, N . . . in terms of x, y, z, and their differences expressed by d. We must then substitute these in the equations (C), (D); the first refers to the limits between which the *maximum* should subsist; the equations (D) constitute the relations required; they are the differentials of x, y, z, and, excepting a case of absurdity, may form distinct conditions, since they will determine numerical values for the variables. If the question proposed relate to Geometry, these equations are those of a curve or of a surface, to which belongs the required property.

As the integration is effected and should be taken between given limits, the terms which remain and compose the equation (C) belong to these limits: it is become of the form $K + L = 0$, L being a function of $x, y, z, \delta x, \delta y, \delta z \ldots$ Mark with one and two accents the numerical values of these variables at the first and second limit. Then, since the integral is to be taken between these limits, we must mark the different terms of L which compose the equation C, first with one, and then with two accents; take the first result from the second and equate the difference to zero; so that the equation

$$L_{,,} - L_{,} = 0$$

contains no variables, because x, d x . . . will have taken the values $x_{,}, \delta x_{,} \ldots x_{,,}, \delta x_{,,} \ldots$ assigned by the limits of the integration. We must remember that these accents merely belong to the limits of the integral.

There are to be considered four separate cases.

1. *If the limits are given and fixed*, that is to say, if the extreme values of x, y, z are constant, since $\delta x_{,}, d\,\delta x_{,} \ldots d\,x_{,,}, d\,\delta x_{,,}$ &c. are zero, all the terms of $L_{,}$ and $L_{,,}$ are zero, and the equation (C) is satisfied. Thus we determine the constants which integration introduces into the equations (D), by the conditions conferred by the limits.

2. *If the limits are arbitrary and independent*, then each of the coefficients $\delta x_{,}, \delta x_{,,} \ldots$ in the equation (C) is zero in particular.

3. *If there exist equations of condition*, (which signifies geometrically that the curve required is terminated at points which are not fixed, but which are situated upon two given curves or surfaces,) for the limits, that is to say, if the nature of the question connects together by equations, some of the quantities $x_{,}, y_{,}, z_{,}, x_{,,}, y_{,,}, z_{,,}$ we use the differentials of these equations to obtain more variations $\delta x_{,}, \delta y_{,}, \delta z_{,}, d\,x_{,,}$, &c. in functions of the others; substituting in $L_{,,} - L_{,} = 0$, these variations will be reduced to the least number possible: the last being absolutely independent, the equation will split again into many others by equating separately their coefficients to zero.

Instead of this process, we may adopt the following one, which is more elegant. Let

$$u = 0,\ v = 0,\ \&c.$$

be the given equations of condition; we shall multiply their variations δu, δv . . . by the indeterminates λ, λ'. . . This will give $\lambda\ \delta\ u + \lambda' \delta\ v + \ldots$ a known function of $\delta\ x_{,}$, $\delta\ x_{,,}$, $\delta\ y_{,}$ . . . Adding this sum to $L_{,,} - L_{,}$, we shall get

$$L_{,,} - L_{,} + \lambda\ \delta\ u + \lambda'\ d\ v + \ldots = 0\ .\ .\ .\ .\ (E).$$

Consider all the variations $\delta\ x_{,}$, $\delta\ x_{,,}$, . . . as independent, and equate their coefficients separately to zero. Then we shall eliminate the indeterminates λ, λ'. . . from these equations. By this process, we shall arrive at the same result as by the former one; for we have only made legitimate operations, and we shall obtain the same number of final equations.

It must be observed, that we are not to conclude from $u = 0$, $v = 0$, that at the limits we have $d\ u = 0$, $d\ v = 0$; these conditions are independent, and may easily not coexist. In the contrary case, we must consider $d\ u = 0$, $d\ v = 0$, as new conditions, and besides $\lambda\ \delta\ u$, we must also take $\lambda'\ \delta\ d\ u$ . . .

4. Nothing need be said as to the case where one of the limits is fixed and the other subject to certain conditions, or even altogether arbitrary, because it is included in the three preceding ones.

It may happen also that the nature of the question subjects the variations $\delta\ x$, $\delta\ y$, $\delta\ z$, to certain conditions, given by the equations

$$\epsilon = 0,\ \theta = 0,$$

and independently of limits; thus, for example, when the required curve is to be traced upon a given curve surface. Then the equation (B) will not split into three equations, and the equations (D) will not subsist. We must first reduce, as follows, the variations to the smallest number possible in the formula (B), by means of the equations of condition, and equate to zero the coefficients of the variations that remain; or, which is tantamount, add to (B) the terms $\lambda\ \delta\ \epsilon + \lambda'\ \delta\ \theta + \ldots$; then split this equation into others by considering $\delta\ x$, $\delta\ y$, $\delta\ z$ as independent; and finally climinate λ, λ' . . .

It must be observed, that, in particular cases, it is often preferable to make, upon the given function Z, all the operations which have produced the equations (B), (C) instead of comparing each particular case with the general formulæ above given.

Such are the general principles of the calculus of variations: let us illustrate it with examples.

Ex. 1. *What is the curve* C M K *of which the length* M K, *comprised between the given radii-vectors* A M, A K *is the least possible.*

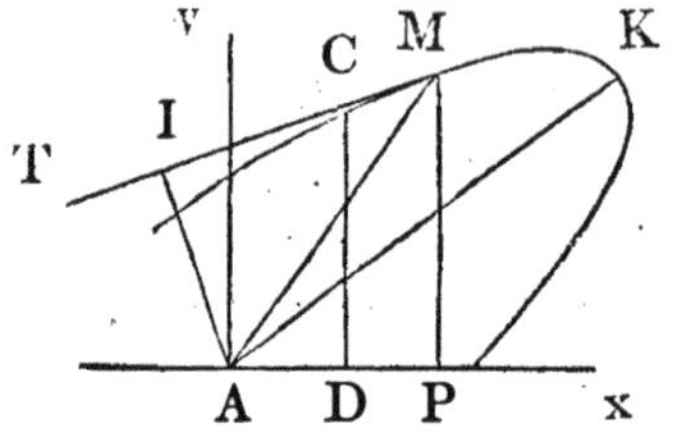

We have, (vol. I, p. 000), if r be the radius-vector,

$$s = \int (r^2 d\theta^2 + d^2) = Z$$

it is required to find the relation $r = \varphi\theta$, which renders Z a minimum the variation is

$$\delta Z = \frac{r d\theta^2 . \delta + r^2 d\theta . \delta d\theta + d r . o d r}{\sqrt{(r^2 d\theta^2 + d r^2)}};$$

Comparing with equation (A), where we suppose $x = r$, $y = \theta$, we have

$$m = \frac{r d\theta^2}{d s}, n = \frac{d r}{d s}, M = 0, N = \frac{r^2 d\theta}{d s},$$

the equations (D) are

$$\frac{r d\theta^2}{d s} = d\left(\frac{d r}{d s}\right), \frac{r^2 d\theta}{d s} = c.$$

Eliminating $d\theta$, and then $d s$, from these equations, and $d s^2 = r^2 d\theta^2 + d r^2$, we perceive that they subsist mutually or agree; so that it is sufficient to integrate one of them. But the perpendicular A I let fall from the origin A upon any tangent whatever. T M is

$$A J = A M + \sin. A M T = r \sin. \beta,$$

which is equivalent, as we easily find, to

$$\frac{r \tan. \beta}{\sqrt{(1 + \tan.^2 \beta)}}$$

which gives

$$\frac{r^2 d\theta}{\sqrt{(r^2 d\theta^2 + d r^2)}} = \frac{r^2 d\theta}{d s} = c;$$

and since this perpendicular is here constant, the required line is a straight line. The limits M and K being indeterminate, the equations (C) are unnecessary..

Ex. 2. *To find the shortest line between two given points, or two given curves.*

The length s of the line is

$$\int Z = \int \sqrt{(d x^2 + d y^2 + d z^2)}.$$

It is required to make this quantity a minimum; we have

$$\delta Z = \frac{d x}{d s} \delta d x + \frac{d y}{d s} \delta d y + \frac{d z}{d s} \delta d z,$$

and comparing with the formula (A), we find

$$m = 0, M = 0, \mu = 0, n = \frac{d x}{d s}, N = \frac{d y}{d s}, \nu = \frac{d z}{d s}:$$

the other coefficients P, p, π . . . are zero. The equations (D) become, therefore, in this case,

$$d \left(\frac{d x}{d s}\right) = 0, d \left(\frac{d y}{d s}\right) = 0, d \left(\frac{d z}{d s}\right) = 0;$$

whence, by integrating

$$d x = a d s, d y = b d s, d z = c d s.$$

Squaring and adding, we get

$$a^2 + b^2 + c^2 = 1,$$

a condition that the constants a, b, c must fulfil in order that these equations may simultaneously subsist. By division, we find

$$\frac{d y}{d x} = \frac{b}{a}, \frac{d z}{d x} = \frac{c}{a},$$

whence

$$b x = a y + a', c x = a z + b';$$

the projections of the line required are therefore straight lines—the line is therefore itself a straight line.

To find the position of it, we must know the five constants a, b, c, a′, b′. If it be required to find the shortest distance between two given fixed points $(x_{,}, y_{,}, z_{,})$, $(x_{,,}, y_{,,}, z_{,,})$, it is evident that $\delta_{,} x$, $\delta x_{,,}$, $\delta y_{,}$ . . . are zero, and that the equation (C) then holds good. Subjecting our two equations to the condition of being satisfied when we substitute therein $x_{,}$, $x_{,,}$, $y_{,}$ . . . for x, y, z, we shall obtain four equations, which, with $a^2 + b^2 + c^2 = 1$, determine the five necessary constants.

Suppose that the second limit is a fixed point $(x_{,,}, y_{,,}, z_{,,})$, in the plane (x, y), and the first a curve passing through the point $(x_{,}, y_{,} z_{,})$, and also situated in this plane; the equation

$$b x = a y + a'$$

then suffices. Let $y_{,} = f x_{,}$ be the equation of the curve; hence

$$\delta y_{,} = A \delta x_{,};$$

the equation (C) becomes

$$\text{I} = \left(\frac{d x}{d s}\right) \delta x + \left(\frac{d y}{d s}\right) \delta y;$$

and since the second limit is fixed it is sufficient to combine together the equations

$$\delta y_, = A \delta x_,$$
$$d x_, \delta x_, + d y_, \delta y_, = 0.$$

Eliminating δy, we get

$$d x_, + A d y_, = 0.$$

We might also have multiplied the equation of condition

$$\delta y_, - A \delta x_, = 0$$

by the indeterminate λ, and have added the result to $L_{,}$, which would have given

$$\left(\frac{d x_,}{d s_,}\right) \delta x_, + \left(\frac{d y'}{d s_,}\right) \delta y_, + \lambda \delta y_, - \lambda A \delta x_, = 0,$$

whence

$$\left(\frac{d x_,}{d s_,}\right) - \lambda A = 0, \quad \left(\frac{d y_,}{d s_,}\right) + \lambda = 0.$$

Eliminating λ we get

$$d x_, + A d y_, = 0.$$

But then the point $(x_{,}, y_{,})$ is upon the straight line passing through the points $(x_{,}, y_{,}, z_{,})$, $(x_{,,}, y_{,,}, z_{,,})$, and we have also

$$b d x_, = a d y_,,$$

whence

$$a = - b A$$

and

$$\frac{d y}{d x} = - \frac{I}{A} = \frac{b}{a};$$

which shows the straight line is a normal to the curve of condition. The constant a' is determined by the consideration of the second limit which is given and fixed.

It would be easy to apply the preceding reasoning to three dimensions, and we should arrive at similar conclusions; we may, therefore, infer generally that the shortest distance between two curves is the straight line which is a normal to them.

If the shortest line required were to be traced upon a curve surface whose equation is $u = 0$, then the equation (B) would not decompose into three others. We must add to it the term $\lambda \delta u$; then regarding δx, δy, δz as independent, we shall find the relations

$$d . \left(\frac{d x}{d s}\right) + \lambda \left(\frac{d u}{d x}\right) = 0,$$

$$d\left(\frac{dy}{ds}\right) + \lambda\left(\frac{du}{dy}\right) = 0,$$

$$d\left(\frac{dz}{ds}\right) + \lambda\left(\frac{du}{dz}\right) = 0.$$

From these eliminating λ, we have the two equations

$$\left(\frac{du}{dz}\right) d\left(\frac{dx}{ds}\right) = \left(\frac{du}{dx}\right) d\left(\frac{dz}{ds}\right)$$

$$\left(\frac{du}{dy}\right) d\left(\frac{dz}{ds}\right) = \left(\frac{du}{dz}\right) d\left(\frac{dy}{ds}\right)$$

which are those of the curve required.

Take for example, the least distance measured upon the surface of a

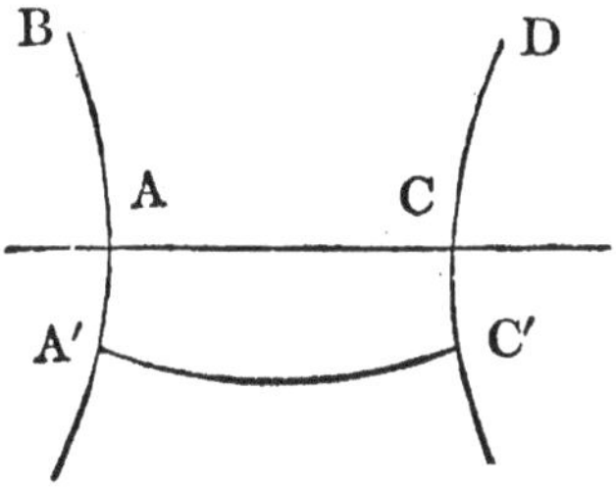

sphere, whose center is at the origin of coordinates: hence

$$u = x^2 + y^2 + z^2 - r^2 = 0,$$

$$\left(\frac{du}{dx}\right) = 2x, \quad \left(\frac{du}{dy}\right) = 2y, \quad \left(\frac{du}{dz}\right) = 2z.$$

Our equations give, making ds constant,

$$z\,d^2x = x\,d^2z, \quad z\,d^2y = y\,d^2z,$$

whence

$$y\,d^2x = x\,d^2y.$$

Integrating we have

$$z\,dx - x\,dz = a\,ds, \quad z\,dy - y\,dz = b\,ds, \quad y\,dx - x\,dy = c\,ds.$$

Multiplying the first of these equations by $-y$, the second by x, the third by z, and adding them, we get

$$a\,y = b\,x + c\,z$$

the equation of a plane passing through the origin of coordinates. Hence the curve required is a great circle which passes through the points A′ C′, or which is normal to the two curves A′ B and C′ D which are limits and are given upon the spherical surface.

When a body moves in a fluid it encounters a resistance which *ceteris paribus* depends on its form (see vol. I.): if the body be one of revolution and moves in the direction of its axis, we can show by mechanics that the resistance is the least possible when the equation of the generating curve fulfils the condition

$$\int \frac{y\,d\,dy^3}{dx^2 + dy^2} = \textit{minimum},$$

or

$$Z = \frac{y\,y'^3\,dx}{1 + y'^2}.$$

Let us determine the generating curve of the *solid of least resistance* (see Principia, vol. II.).

Taking the variation of the above expression, we get

$$m = 0,\ n = \frac{-2\,y\,dy^3\,dx}{(dx^2 + dy^2)^2} = \frac{-2\,y\,y'^3}{(1+y'^2)^2},\ p = 0, \&c.$$

$$M = \frac{dy^3}{dx^2 + dy^2} = \frac{y'^3\,dx}{1 + y'^2},\ N = \frac{y\,y'^2\,(3 + y'^2)}{(1 + y'^2)^2}, \&c.$$

the second equation (D) is

$$M - d\,N = 0;$$

and it follows from what we have done relatively to Z, that

$$d\left(\frac{y'^3\,y}{1 + y'^2}\right) = M\frac{dy}{dx} + N\,dy' = y'\,dN + N\,dy',$$

because

$$M = d\,N.$$

Thus integrating, we have

$$a + \frac{y'^3\,y}{1 + y'^2} = N\,y' = \frac{y\,y'^3\,(3 + y'^2)}{(1 + y'^2)^2}.$$

Therefore

$$a\,(1 + y'^2)^2 = 2\,y\,y'^3.$$

Observe that the first of the equations (D) or $m - d\,n = 0$, would have given the same result $-\,n = a$; so that these two equations conduct to the same result. We have

$$y = \frac{a\,(1 + y'^2)^2}{2\,y'^3},$$

$$x = \int \frac{dy}{y'} = \frac{y}{y'} + \int \frac{y\,dy'}{y'^2};$$

substituting for y its value, this integral may easily be obtained; it remains to eliminate y' from these values of x and y, and we shall obtain the equation of the required curve, containing two constants which we shall determine from the given conditions.

Ex. 3. *What is the curve* A B M *in which the area* B O D M *comprised*

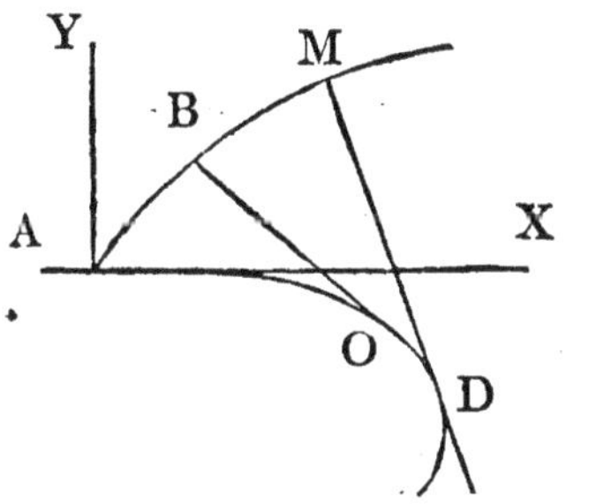

between the arc B M *the radii of curvature* B O, D M *and the arc* O D *of the evolute, is a minimum?*

The element of the arc A M is

$$ds = dx\sqrt{1 + y} ;$$

the radius of curvature M D is

$$\frac{(1 + y'^2)^{\frac{3}{2}}}{y''};$$

and their product is the element of the proposed area, or

$$Z = \frac{(1 + y'^2)\, dx}{y''} = \frac{(dx + dy^2)}{dx\ dy}.$$

It is required to find the equation $y = fx$, which makes $\int Z$, a minimum.

Take the variation δN, and consider only the second of the equations (D), which is sufficient for our object, and we get

$$M = 0,\ N - dP = 4a,$$

$$N = \frac{dx^2 + dy^2}{dx\, d^2y} . 4\, dy = \frac{1 + y'^2}{y''} 4 y',$$

$$P = - \frac{(1 + y'^2)^2}{y''^2\, dx}.$$

But

$$d\left(\frac{(1 + y'^2)^2}{y''}\right) = N\, dy' + P\, dy''\, dx$$

$$= 4a\, dy + dP\, dy' + P\, dy''\, dx,$$

putting $4a + P$ for N. Moreover $y''\, dx = dy'$, changes the last terms into

$$(y''\, dP + P\, dy'')\, dx = d\,(P y'') . dx = - d\left(\frac{(1 + y'^2)^2}{y''}\right).$$

Integrating, therefore,

$$\frac{(1+y'^2)^2}{2y''} = a y' + b,$$

$$y'' = \frac{(1+y'^2)^2}{2(a y' + b)} = \frac{d y'}{d x}, \quad d x = \frac{2(a y' + b) d y'}{(1+y'^2)^2},$$

finally,

$$x = c + \frac{b y' - a}{1 + y'^2} + b \tan.^{-1} y';$$

On the other side we have

$$y = \int y' d x = y' x - \int x d y'$$

or

$$y = y' x - c y' - \int \frac{b y' - a}{1 + y'^2} d y' - \int b d y' \tan.^{-1} y';$$

this last term integrates *by parts*, and we have

$$y = y' x - c y' - (b y' - a) \tan.^{-1} y' + f.$$

Eliminating the tangent from these values of x and y, we get

$$b y = a (x - c) + \frac{(b y' - a)^2}{1 + y'^2} + b f,$$

$$\sqrt{(b y - a x + g)} = \frac{(b y' - a) d x}{d s}, \quad d s = \frac{b d y - a d x}{\sqrt{(b y - a x + g)}};$$

finally,

$$s = 2 \sqrt{(b y - a x + g)} + h.$$

This equation shows that the curve required is a cycloid, whose four constants will be determined from the same number of conditions.

Ex. 4. *What is the curve of a given length* s, *between two fixed points, for which* $\int y\, d s$ *is a maximum?*

We easily find

$$(y + \lambda)\left(\frac{d x}{d s}\right) = c, \text{ whence } d x = \frac{c d y}{\sqrt{\{(y+\lambda)^2 - c^2\}}};$$

and it will be found that the curve required is a *catenary.*

Since $\int \frac{y\, d s}{s}$ is the vertical ordinate of the center of gravity of an arc whose length is s, we see that the center of gravity of any arc whatever of the catenary is lower than that of any other curve terminated by the same points.

Ex. 5. Reasoning in the same way for $\int y^2 d x =$ *minimum*, and $\int y\, d x =$ const. we find $y^2 + \lambda y = c$, or rather $y = c$. We have here a straight line parallel to x. Since $\frac{\int y^2 d x}{2 \int y\, d x}$ is the vertical ordinate of the center of gravity of every plane area, that of a rectangle, whose side is horizontal, is the lowest possible; so that every mass of water whose upper surface is horizontal, has its center of gravity the lowest possible.

FINITE DIFFERENCES.

If we have given a series a, b, c, d, . . . take each term of it from that which immediately follows it, and we shall form the *first differences*, viz.

$$a' = b - a,\ b' = c - b,\ c' = d - c,\ \&c.$$

In the same manner we find that this series a', b', c', d' . . . gives the *second differences*

$$a'' = b' - a',\ b'' = c' - b',\ c'' = d' - c',\ \&c.$$

which again give the third differences

$$a''' = b'' - a'',\ b''' = c'' - b'',\ c''' = d'' - c'',\ \&c.$$

These differences are indicated by Δ, and an exponent being given to it will denote the order of differences. Thus Δ^n is a term of the series of nth differences. Moreover we give to each difference the sign which belongs to it; this is —, when we take it from a decreasing series.

For example, the function

$$y = x^3 - 9x + 6$$

in making x successively equal to 0, 1, 2, 3, 4 . . . gives a series of numbers of which y is the general term, and from which we get the following differences,

for $x =$	0,	1,	2,	3,	4,	5,	6,	7 . . .
series $y =$	6,	— 2,	— 4,	6,	34,	86,	168,	286 . . .
first diff. $\Delta y =$	— 8,		— 2,	10,	28,	52,	82,	118 . . .
second diff. $\Delta^2 y =$			6,	12,	18,	24,	30,	36 . . .
third diff. $\Delta^3 y =$			6,	6,	6,	6,	6,	. . .

We perceive that the third differences are here constant, and that the second difference is an arithmetic progression: we shall always arrive at constant differences, whenever y is a rational and integer function of x; which we now demonstrate.

In the monomial $k x^m$ make $x = \alpha, \beta, \gamma, \ldots \theta, \varkappa, \lambda$ (these numbers having h for a constant difference), and we get the series

$$k\alpha^m,\ k\beta^m, \ldots k\theta^m,\ k\varkappa^m,\ k\lambda^m.$$

Since $\varkappa = \lambda - h$, by developing $k\varkappa^m = k(\lambda - h)^m$, and designating by m, A', A'' . . . the coefficients of the binomial, we find, that

$$k(\lambda^m - \varkappa^m) = k m h \lambda^{m-1} - k A' h^2 \lambda^{m-2} + k A''\ ^3 h \ldots$$

Such is the first difference of any two terms whatever of the series

$$k\,\alpha^m,\ k\,\beta^m \ldots k\,\varkappa^m,\ \&c.$$

The difference which precedes it, or $k\,(\varkappa^m - \theta^m)$ is deduced by changing λ into $\varkappa$ and $\varkappa$ into θ' and since $\varkappa = \lambda - h$, we must put $\lambda - h$ for λ in the second member:

$$k\,m\,h(\lambda-h)^{m-1} - k\,A'\,h^2\,(\lambda-h^{m-2})\ldots = k\,m\,h\,\lambda^{m-1} - \{A' + m(m-1)\}kh^2\lambda^{m-2}\ldots$$

Subtracting these differences, the two first terms will disappear, and we get for the second difference of an arbitrary rank

$$k\,m\,(m-1)\,h^2\,\lambda^{m-2} + k\,B'\,h^3\,\lambda^{m-3} + \ldots$$

In like manner, changing λ into $\lambda - h$, in this last developement, and subtracting, the two first terms disappear, and we have for the third difference

$$k\,m\,(m-1)\,(m-2)\,h^3\,\lambda^{m-3} + k\,B''\,h^4\,\lambda^{m-4}\ldots,$$

and so on continually.

Each of these differences has one term at least, in its developement, like the one above; the first has m terms; the second has m — 1 terms; third, m — 2 terms; and so on. From the form of the first term, which ends by remaining alone in the mth difference, we see this is reduced to the constant

$$1\,.\,2\,.\,3\ldots m\,k\,h^m.$$

If in the functions M and N we take for x two numbers which give the results m, n; then M + N becomes m + n. In the same manner, let m′, n′ be the results given by two other values of x; the first difference, arising from M + N, is evidently

$$(m - m') + (n - n').$$

that is, *the difference of the sum is the sum of the differences.* The same may be shown of the 3d and 4th . . . differences.

Therefore, if we make

$$x = \alpha,\ \beta,\ \gamma \ldots$$

in

$$k\,x^m + p\,x^{m-1} + \ldots,$$

the mth difference will be the same as if these were only the first term $k\,x^m$, for that of $p\,x^{m-1}$, $q\,x^{m-2}\ldots$ is nothing. Therefore *the mth difference is constant, when for* x *we substitute numbers in arithmetic progression, in a rational and integer function of* x.

We perceive, therefore, that if it be required to substitute numbers in arithmetic progression, as is the case in the resolution of numerical equations, according to Newton's Method of Divisors, it will suffice to find the (m + 1) first results, to form the first, second, &c. differences. The mth difference will have but one term; as we know it is constant and $= 1 . 2 . 3 \ldots m k h^{m}$, we can extend the series at pleasure. That of the (m — 1)th differences will then be extended to that of two known terms, since it is an arithmetic procession; that of the (m — 2)th differences will, in its turn, be extended; and so on of the rest.

This is perceptible in the preceding example, and also in this; viz.

x = 0 . 1 . 2 . 3	3d Diff.	6 .	6 .	6 .	6 .	6 .	6 .	6 . . .
Series 1 . − 1 . 1 . 13	2nd . .	4 .	10 .	16 .	22 .	28 .	34 .	40 . . .
1st. . . . − 2 . 2 . 12	1st .	— 2 .	2 .	12 .	28 .	50 .	78 .	112 . . .
2nd . 3 . 10	Results	1 . —	1 .	1 .	13 .	41 .	91 .	169 . . .
3d . . . 6	For x	0 .	1 .	2 .	3 .	4 .	5 .	6 . . .

These series are deduced from that which is constant

6 . 6 . 6 . 6 . . .

and from the initial term already found for each of them: *any term is derived by adding the two terms on the left which immediately precede it.* They may also be continued in the contrary direction, in order to obtain the results of x = — 1, — 2, — 3, &c.

In resolving an equation it is not necessary to make the series of results extend farther than the term where we ought only to meet with numbers of the same sign, which is the case when all the terms of any column are positive on the right, and alternate in the opposite direction; for the additions and subtractions by which the series are extended as required, preserve constantly the same signs in the results. We learn, therefore, by this method, the limits of the roots of an equation, whether they be positive or negative.

Let y_x denote the function of x which is the general term, viz. the x + 1th, of a proposed series

$$y_0 + y_2 + y_1 + \ldots y_x + y_{x+1} + \ldots$$

which is formed by making

$$x = 0, 1, 2, 3 \ldots$$

For example, y_5 will designate that x has been made = 5, or, with regard to the place of the terms, that there are 5 before it (in the last example this is 91). Then

$$y_1 - y_0 = \Delta y_0, \quad y_2 - y_1 = \Delta y_1, \quad y_3 - y_2 = y_2 \ldots$$
$$\Delta y^1 - \Delta y_0 = \Delta^2 y_0, \quad \Delta y^2 - \Delta^2 y_1 = \Delta^2 y_1, \quad \Delta y_3 - \Delta y_2 = \Delta^2 y_2 \ldots$$
$$\Delta^2 y_1 - \Delta^2 y_0 = \Delta^3 y_0, \quad \Delta^2 y_2 - \Delta^2 y_1 = \Delta^3 y_1, \quad \Delta^2 y_3 - \Delta^2 y_2 = \Delta^3 y_2 \ldots$$
&c.

and generally we have

$$y_x - y_{x-1} = \Delta y_{x-1}$$
$$\Delta y_x - \Delta y_{x-1} = \Delta^2 y_{x-1}$$
$$\Delta^2 y_x - \Delta^2 y_{x-1} = \Delta^3 y_{x-1}$$
&c.

Now let us form the differences of any series a, b, c, d . . . in this manner. Make

$$b = c + a'$$
$$c = b + b'$$
$$d = c + c'$$
&c.

$$b' = a' + a''$$
$$c' = b' + b''$$
$$d' = c' + c''$$
&c.

$$b'' = a'' + a'''$$
$$c'' = b'' + b'''$$
$$d'' = c'' + c'''$$
&c.

and so on continually. Then eliminating b, b′, c, c′, &c. from the first set of equations, we get

$$b = a + a'$$
$$c = a + 2\,a' + a''$$
$$d = a + 3\,a' + 3\,a'' + a'''$$
$$e = a + 4\,a' + 6\,a'' + 4\,a''' + a''''$$
$$f = a + 5\,a' + 10\,a'' + \&c.$$
&c.

Also we have

$$a' = b - a$$
$$a'' = c - 2\,b + a$$
$$a''' = d - 3\,c + 3\,b - a$$
&c.

But the letters a′, a″, a‴, &c. are nothing else than Δy_0, $\Delta^2 y_0$, $\Delta^3 y_0$ . . . a, b, c . . . being y_0, y_1, y_2 . . ., consequently

$$y_1 = y_0 + \Delta y^0$$
$$y_2 = y_0 + 2\,\Delta y_0 + \Delta^2 y$$
$$y_3 = y_0 + 3\,\Delta y_0 + 3\,\Delta^2 y_0 + \Delta^3 y_0$$
&c.

And

$$\Delta y_0 = y_1 - y_0$$
$$\Delta^2 y_0 = y_2 - 2\,y_1 + y_0$$
$$\Delta^3 y_0 = y_3 - 3\,y_2 + 3\,y_1 - y_0$$
$$\Delta^4 y_0 = y_4 - 4\,y_3 + 6\,y_2 + 4\,y_1 + y_0$$
&c.

Hence, generally, we have

$$y_x = y_0 + x\,\Delta y_0 + x\,\frac{x-1}{2}\,.\,\Delta^2 y_0 + x\,.\,\frac{x-1}{2}\,.\,\frac{x-2}{3}\,\Delta^3 y + \ldots (A)$$

$$\Delta_n y_0 = y_n - n\,y_{n-1} + n\,.\,\frac{n-1}{2}\,.\,y_{n-2} - n\,.\,\frac{n-1}{2}\,.\,\frac{n-2}{3}\,y_{n-3} + \ldots (B)$$

These equations, which are of great importance, give the general term of any series, from knowing its first term and the first term of all the orders of differences; and also the first term of the series of nth differences, from knowing all the terms of the series y_0, y_1, $y_2 \ldots$

To apply the former to the example in p. (81), we have

$$y_0 = 1$$
$$\Delta y_0 = -2$$
$$\Delta^2 y_0 = 4$$
$$\Delta^3 y_0 = 6$$
$$\Delta^4 y_0 = 0$$

whence

$$y_x = 1 - 2\,x + 2\,x\,(x-1) + x\,(x-1)\,(x-2) = x^3 - x^2 - 2\,x + 1$$

The equations (A), (B) will be better remembered by observing that

$$y_x = (1 + \Delta y_0)^x,$$
$$\Delta^n y_0 = (y - 1)^n,$$

provided that in the developements of these powers, we mean by the exponents of Δy_0, the orders of differences, and by those of y the place in the series.

It has been shown that a, b, c, d . . . may be the values of y_x, when those of x are the progressional numbers

$$m,\ m + h,\ m + 2\,h \ldots m + i\,h$$

that is

$$a = y_m,\ b = y_{m+h},\ c = \&c.$$

In the equation (A), we may, therefore, put y_{m+ih} for y_x, y_m for y_0, Δy_m for Δy_0, &c. and, finally, the coefficients of the ith power. Make $i\,h = z$, and write Δ, $\Delta^2 \ldots$ for Δy_m, $\Delta^2 y_m \ldots$ and we shall get

$$y_{m+z} = y_m + \frac{z\,\Delta\,.}{h} + \frac{z\,.\,(z-h)\,\Delta^2\,.}{2\,h^2} + \frac{z\,(z-h)\,(z-2\,h)\,\Delta^3\,.}{2\,h^3} + \ldots (C)$$

This equation will give y_x when $x = m + z$, z being either integer or fractional. We get from the proposed series the differences of all orders, and the initial terms represented by Δ, Δ^2, &c.

But in order to apply this formula, so that it may be limited, we must arrive at constant differences; or, at least, this must be the case if we would have Δ, Δ^2 . . . decreasing in value so as to form a converging series: the developement then gives an approximate value of a term corresponding to

$$x = m + z;$$

it being understood that the factors of Δ do not increase so as to destroy this convergency, a circumstance which prevents z from surpassing a certain limit.

For example, if the radius of a circle is 1000,

the arc of 60° has a chord	1000,0	$\Delta. = 74,6$	
65°	1074,6		$\Delta^2 = -2,0$
70°	1147,2	72,6	
75°	1217,5		$-2,3$

Since the difference is nearly constant from 60° to 75°, to this extent of the arc we may employ the equation (C); making $h = 5$, we get for the quantity to be added to $y_m = 1090$, this

$$\tfrac{1}{5}.\ 74,6.\ z - \tfrac{2}{50}\ z\ (z - 5) = 15,12.\ z - 0,04.\ z^2$$

So that, by taking $z = 1, 2, 3\ldots$ then adding 1000, we shall obtain the chords of 61°, 62°, 63°; in the same manner, making z the necessary *fraction*, we shall get the chord of any arc whatever, that is intermediate to those, and to the limits 60° and 75°. It will be better, however, when it is necessary thus to employ great numbers for z, to change these limits.

Let us now take

log. 3100 $= y_m$	$= 4913617$	$\Delta. = 13987$	
log. 3110	$= 4927604$		$\Delta.^2 = -45$
log. 3120	$= 4941546$	13942	
log. 3130	$= 4955443$	13897	-45

We shall here consider the decimal part of the logarithm as being an integer. By making $h = 10$, we get, for the part to be added to log. 3100, this

$$1400,95 \times z - 0,225 \times z^2.$$

To get the logarithms of 3101, 3102, 3103, &c. we make

$$z = 1, 2, 3 \ldots;$$

and in like manner, if we wish for the log. 3107, 58, we must make z — 7, 58, whence the quantity to be added to the logarithm of 3100 is 10606. Hence

$$\log. 310768 = 5,4924223.$$

The preceding methods may be usefully employed to abridge the labour of calculating tables of logarithms, tables of sines, chords, &c. Another use which we shall now consider, is that of inserting the intermediate terms in a given series, of which two distant terms are given. This is called

INTERPOLATION.

It is completely resolved by the equation (C).

When it happens that $\Delta^2 = 0$, or is very small, the series reduces to

$$\frac{z}{m} + \frac{y\,\Delta}{h}$$

whence we learn that the results have a difference which increases proportionally to z.

When Δ^2 is constant, which happens more frequently, by changing z into z + 1 in (C), and subtracting, we have the general value of the first difference of the new interpolated series; viz.

$$\text{First difference } \Delta' = \frac{\Delta.}{h} + \frac{2z - h + 1}{2h^2}\Delta.^2$$

$$\text{Second difference } \Delta'' = \frac{\Delta.^2}{h^2}.$$

If we wish to insert u terms between those of a given series, we must make

$$h = n + 1;$$

then making z = 0, we get the initial term of the differences

$$\Delta'' = \frac{\Delta.^2}{(n+1)^2}$$

$$\Delta' = \frac{\Delta.}{n+1} - \tfrac{1}{2}\,n\,\Delta'';$$

we calculate first Δ'', then Δ'; the initial term Δ' will serve to compose the series of first differences of the interpolated series, (Δ'' is the constant difference of it); and then finally the other terms are obtained by simple additions.

If we wish in the preceding example to find the log. of 3101,

3102, 3103 . . . we shall interpolate 9 numbers between those which are given: whence

$$u = 9$$
$$\Delta'' = -0,45$$
$$\Delta' = 1400,725.$$

We first form the arithmetical progression whose first term is Δ', and — 0,45 for the constant. The first differences are

1400,725; 1400,725; 1399,375; 1398,925, &c.

Successive additions, beginning with log. 3100, will give the consecutive logarithms required.

Suppose we have observed a physical phenomenon every twelve hours, and that the results ascertained by such observations have been

For 0 hours . . .	78	$\Delta = 222$	
12 . . .	300		$\Delta^2 = 144$
24 . . .	666	366	
36 . . .	1176	510	144.

&c.

If we are desirous of knowing the state corresponding to 4^h, 8^h, 12^h, &c., we must interpolate two terms; whence

$$u = z, \Delta'' = 16, \Delta' = 58$$

composing the arithmetic progression whose first term is 58, and common difference 16, we shall have the first differences of the new series, and then what follow

First differences 58, 74, 90, 106, 122, 138 . . .

Series 78, 136, 210, 300, 406, 528, 646 . . .

A 0^h, 4^h, 8^h, 16^h 20^h, 24^h.

The supposition of the second differences being constant, applies almost to all cases, because we may choose intervals of time which shall favour such an hypothesis. This method is of great use in astronomy; and even when observation or calculation gives results whose second differences are irregular, we impute the defect to errors which we correct by establishing a greater degree of regularity.

Astronomical, and geodesical tables are formed on these principles. We calculate directly different terms, which we take so near that their first or second differences may be constant; then we interpolate to obtain the intermediate numbers.

Thus, when a converging series gives the value of y by aid of that of a variable x; instead of calculating y for each known value of x, when the formula is of frequent use, we determine the results y for the continually increasing values of x, in such a manner that y shall always be nearly of the same value: we then write in the form of a table every value by the side of that of x, which we call the *argument* of this table. For the numbers x which are intermediate to them, y is given by simple propositions, and by inspection alone we then find the results required.

When the series has two variables, or arguments x and z, the values of y are disposed in a table by a sort of *double entry*; taking for coordinates x and z, the result is thus obtained. For example, having made $z = 1$, we range upon the first line all the values of y corresponding to

$$x = 1, 2, 3 \ldots;$$

we then put upon the second line which $z = z$ gives; in a third line those which $z = 3$ gives, and so on. To obtain the result which corresponds to

$$x = 3, z = 5$$

we stop at the case which, in the third column, occupies the fifth place. The intermediate values are found analogously to what has been already shown.

So far we have supposed x to increase continually by the same difference. If this is not the case and we know the results

$$y = a, b, c, d \ldots$$

which are due to any suppositions

$$x = \alpha, \beta, \gamma, \delta \ldots$$

we may either use the theory which makes a parabolic curve pass through a series of given points, or we may adopt the following:

By means of the known corresponding values

$$a, \alpha; \; b\,\beta; \; \&c.$$

we form the consecutive functions

$$A = \frac{b - a}{\beta - \alpha}$$

$$A_1 = \frac{c - b}{\gamma - \beta}$$

$$A_2 = \frac{d - c}{\delta - \gamma}$$

&c.

$$B = \frac{A_1 - A}{\gamma - \alpha}$$

$$B_1 = \frac{A_2 - A}{\delta - \beta}$$

$$B_2 = \frac{A_3 - A}{\epsilon - \gamma}$$

&c.

$$C = \frac{B_1 - B}{\delta - \alpha}$$

$$C_1 = \frac{B_2 - B_1}{\varepsilon - \beta}$$

&c.

$$D = \frac{C_1 - C}{\nu - \alpha}$$

and so on.

By elimination we easily get

$$b = a + A(\beta - \alpha)$$

$$c = a + A(\gamma - \alpha) + B(\gamma - \alpha)(\gamma - \beta)$$

$$d = a + A(\delta - \alpha) + B(\delta - \alpha)(\delta - \beta) + C(\delta - \alpha)(\delta - \beta)(\delta - \gamma)$$

&c.

and generally

$$y_x = a + A(x - \alpha) + B(x - \alpha)(x - \beta) + C(x - \alpha)(x - \beta)(x - \gamma) + \&c.$$

We must seek therefore the first differences amongst the results

$$a,\ b,\ c\ \ldots$$

and divide by the differences of

$$\alpha,\ \beta,\ \gamma\ \ldots$$

which will give

$$A,\ A_1,\ A_2,\ \&c.$$

proceeding in the same manner with these numbers, we get

$$B,\ B_1,\ B_2,\ \&c.$$

which in like manner give

$$C,\ C_1,\ C_2,\ \&c.$$

and, finally substituting, we get the general term required.

By actually multiplying, the expression assumes the form

$$a + a'x + a'x^2 + \ldots$$

of every rational and integer polynomial, which is the same as when we neglect the superior differences.

The chord of 60° = rad.	= 1000				
		35	$A = 15$		
62°.20′	= 1035				$B = -0,035$
		42	$A_1 = 14,82$	$-0,18$	
65°.10′	= 1077				$B_1 = -0,031$
		56	$A_2 = 14,61$	$-0,21$	
69°. 0′	= 1133				

We have

$$\alpha = 0,\ \beta = 2\tfrac{1}{3},\ \gamma = 5\tfrac{1}{6},\ \delta = 9.$$

We may neglect the third differences and put

$$y_x = 100 + 15,082\,x - 0,035\,x^2.$$

Considering every function of x, y_x, as being the general term of the series which gives

$$x = m,\ m + h,\ m + 2h,\ \&c.$$

if we take the differences of these results, to obtain a new series, the general term will be what is called the *first difference* of the proposed function y_x which is represented by Δy_x. Thus we obtain this difference by changing x into x + h in y_x and taking y_x from the result; the remainder will give the series of first differences by making

$$x = m, \; m + h, \; m + 2h, \; \&c.$$

Thus if

$$y_x = \frac{x^2}{a + x}$$

$$\Delta y_x = \frac{(x + h)^2}{a + x + h} - \frac{x^2}{a + x}.$$

It will remain to reduce this expression, or to develope it according to the increasing powers of h.

Taylor's theorem gives *generally* (vol. I.)

$$\Delta y_x = \frac{dy}{dx} h + \frac{d^2 y}{dx^2} \cdot \frac{h^2}{1.2} + \&c.$$

To obtain the second difference we must operate upon Δy_x as upon the proposed y_x, and so on for the third, fourth, &c. differences.

INTEGRATION OF FINITE DIFFERENCES.

Integration here means the method of finding the quantity whose difference is the proposed quantity; that is to say the general term y_x of a

$$y_m, \; y_{m+h}, \; y_{m+2h}, \; \&c.$$

from knowing that of the series of a difference of any known order. This operation is indicated by the symbol Σ.

For example

$$\Sigma (3x^2 + x - 2)$$

ought to indicate that here

$$h = 1.$$

A function y_x generates a series by making

$$x = 0, \; 1, 2, 3 \ldots$$

the first differences which here ensue, form another series of which

$$3x^2 + x - 2$$

is the general term, and it is

$$-2, \; 2, \; 12, \; 28 \ldots$$

By integrating we here propose to find y_x such, that putting x + 1 for x, and subtracting, the remainder shall be

$$3x^2 + x - 2.$$

It is easy to perceive that, first the symbols Σ and Δ destroy one another as do $\int$ and d; thus

$$\Sigma \Delta f x = f x.$$

Secondly, that

$$\Delta (a y) = a \Delta y$$

gives

$$\Sigma a y = a \Sigma y.$$

Thirdly, that as

$$\Delta (A t - B u) = A \Delta t - B \Delta u$$

so is

$$\Sigma (A t - B u) = A \Sigma t - B \Sigma u,$$

t and u being the functions of x.

The problem of determining y_x by its first difference does not contain data sufficient completely to resolve it; for in order to recompose the series derived from y_x in beginning with

$$-2, 2, 12, 28, \&c.$$

we must make the first term

$$y_0 = a$$

and by successive additions, we shall find

$$a, a - 2, a + 2, a + 12, \&c.$$

in which *a* remains arbitrary.

Every integral may be considered as comprised in the equation (A) p. 83; for by taking

$$x = 0, 1, 2, 3 \ldots$$

in the first difference given in terms of x, we shall form the series of first differences; subtracting these successively, we shall have the second differences; then in like manner, we shall get the third and fourth differences. The initial term of these series will be

$$\Delta y_0, \Delta^2 y_0 \ldots$$

and these values substituted in y_x will give y_x. Thus, in the example above, which is only that of page (81) when a = 1, we have

$$\Delta y_0 = -2, \Delta^2 y_0 = 4, \Delta^3 y_0 = 6, \Delta^4 y_0 = 0, \&c.;$$

which give

$$y_x = y_0 - 2 x - x^2 + x^3.$$

Generally, the first term y_0 of the equation (A) is an arbitrary constant, which is to be added to the integral. If the given function is a second difference, we must by a first integration reascend to the first difference and thence by another step to y_x; thus we shall have two arbitrary constants; and in fact, the equation (A) still gives y_x by finding Δ^2, Δ^3, the only difference in the matter being that y_0 and Δy_0 are arbitrary. And so on for the superior orders.

Let us now find Σx^m, the exponent m being integer and positive. Represent this developement by

$$\Sigma x^m = p x^a + q x^b + r x^c + \&c.$$

a, b, c, &c. being decreasing exponents, which as well as the coefficients p, q, &c. must be determined. Take the first difference, by suppressing Σ in the first member, then changing x into x + h in the second member and subtracting. Limiting ourselves to the two first terms, we get

$$x^m = p a h x^{a-1} + \tfrac{1}{2} p a (a-1) h^2 x^{a-2} + \ldots q b h x^{b-1} + \ldots$$

But in order that the identity may be established the exponents ought to give

$$a - 1 = m$$
$$a - 2 = b - 1$$

whence

$$a = m + 1,\ b = m.$$

Moreover the coefficients give

$$1 = p a h,\ -\tfrac{1}{2} p a (a-1) h = q b;$$

whence

$$p = \frac{1}{(m+1) h},\ q = -\tfrac{1}{2}.$$

As to the other terms, it is evident, that the exponents are all integer and positive; and we may easily perceive that they fail in the alternate terms. Make therefore

$$\Sigma x^m = p x^{m+1} - \tfrac{1}{2} x^m + \alpha x^{m-1} + \beta x^{m-3} + \gamma x^{m-5} + \ldots$$

and determine α, β, γ ... &c.

Take, as before, the first difference by putting x + h for x, and subtracting: and first transferring

$$p x^{m+1} - \tfrac{1}{2} x^m,$$

we find that the first member, by reason of

$$p h (m + 1) = 1,$$

reduces to

$$A'.\frac{h^2}{2.3} x^{m-2} + A''.\frac{m-3}{4}.\frac{3 h^4}{2.5} x^{m-4} + A''''.\frac{m-5}{6}.\frac{5 h^6}{2.7} x^{m-6} \ldots$$

To abridge the operation, we omit here the alternate terms of the developement; and we designate by

$$1,\ m,\ A',\ A'',\ \&c.$$

the coefficients of the binomial.

Making the same calculations upon

$$\alpha x^{m-1} + \beta x^{m-3} + \&c.$$

we shall have, with the same respective powers of x and of h,

$$(m-1)\alpha+(m-1).\frac{n-2}{2}.\frac{m-3}{3}\alpha+(m-1).\frac{m-2}{3}\cdots\frac{m-4}{5}\alpha+\ldots$$
$$+\qquad (m-3)\beta+(m-3).\frac{m-4}{2}\cdots\frac{m-5}{3}\beta+\ldots$$
$$+\qquad (m-4)\gamma+\ldots$$

Comparing them term by term, we easily derive

$$\alpha=\frac{m}{3.4},$$
$$\beta=\frac{-A''}{2.3.4.5},$$
$$\gamma=\frac{A''''}{6.6.7}$$
&c.

whence finally we get

$$\Sigma x^m=\frac{x^{m+1}}{(m+1)h}-\frac{x^m}{2}+m a h x^{m-1}+A'' b h^3 x^{m-3}$$
$$+A'''' c h^5 x^{m-5}+A^{vi} d h^7 x^{m-7}+\ldots(D)$$

This developement has for its coefficients those of the binomial, taken from *two to two*, multiplied by certain numerical factors a, b, c . . ., which are called the *numbers* of Bernoulli, because James Bernoulli first determined them. These factors are of great and frequent use in the theory of series; we shall give an easy method of finding them presently. These are their values

$$a=\frac{1}{12}$$
$$b=-\frac{1}{120}$$
$$c=\frac{1}{252}$$
$$d=-\frac{1}{240}$$
$$e=\frac{1}{132}$$
$$f=-\frac{691}{32780}$$
$$g=\frac{1}{12}$$
$$h=-\frac{3617}{8160}$$
$$i=\frac{43867}{14364}$$
&c.

which it will be worth the trouble fully to commit to memory.

From the above we conclude that to obtain $\Sigma\, x^m$, m being any number, integer and positive, we must besides the two first terms

$$\frac{x^{m+1}}{(m+1)\,h} - \frac{x^m}{2}$$

also take the developement of

$$(x+h)^m$$

reject the odd terms, the first, third, fifth, &c. and multiply the retained terms respectively by

$$a,\ b,\ c\ \ldots$$

Now x and h *have even exponents only when* m *is odd and reciprocally;* so that we must reject the last term h^m when it falls in a useless situation; the number of terms is $\frac{1}{2}\,m+2$ when m is even, and it is $\frac{1}{2}\,(m+3)$ when m is odd; that is to say, it is the same for two consecutive values of m.

Required the integral of x^{10}.

Besides

$$\frac{x^{11}}{11\,h} - \tfrac{1}{2}\,x^{10}$$

we must develope $(x+h)^{10}$, retaining the second, fourth, sixth, &c. terms and we shall have

$$10\,x^9\,a\,h + 120\,x^7\,b\,h^3 + 252\,x^5\,c\,h^5 + \&c.$$

Therefore

$$\Sigma\, x^{10} = \frac{x^{11}}{11\,h} - \tfrac{1}{2}\,x^{10} + \frac{5}{6}\,x^9\,h - x^7\,h^3 + x^5\,h^5 - \tfrac{1}{2}\,x^3\,h^7 + \frac{5}{66}\,x\,h^9$$

In the same manner we obtain

$$\Sigma\, x^0 = \frac{x}{h}$$

$$\Sigma\, x^1 = \frac{x^2}{2\,h} - \frac{x}{2}$$

$$\Sigma\, x^2 = \frac{x^3}{3\,h} - \frac{x^2}{2} + \frac{h\,x}{6},$$

$$\Sigma\, x^3 = \frac{x^4}{4\,h} - \frac{x^3}{2} + \frac{h\,x^2}{4}$$

$$\Sigma\, x^4 = \frac{x^5}{5\,h} - \frac{x^4}{4} + \frac{h\,x^3}{3} - \frac{h^3\,x}{30}$$

$$\Sigma\, x^5 = \frac{x^6}{6\,h} - \frac{x^5}{2} + \frac{5\,h\,x^4}{12} - \frac{h^3\,x^2}{12},$$

$$\Sigma\, x^6 = \frac{x^7}{7\,h} - \frac{x^6}{2} + \frac{h\,x^5}{2} - \frac{h^3\,x^3}{6} + \frac{h^5\,x}{42},$$

$$\Sigma\, x^7 = \frac{x^8}{8\,h} - \frac{x^7}{2} + \frac{7\,h\,x^6}{12} - \frac{7\,h^3\,x^4}{24} + \frac{h^5\,x^2}{12}$$

$$\Sigma x^8 = \frac{x^9}{9h} - \frac{x^8}{2} + \frac{2hx^7}{3} - \frac{7h^3x^5}{15} + \frac{2h^5x^3}{9} - \frac{h^7x}{30}$$

$$\Sigma x^9 = \frac{x^{10}}{10^h} - \frac{x^9}{2} + \frac{3hx^8}{4} - \frac{7h^3x^6}{10} + \frac{h^5x^4}{2} - \frac{3h^7x^2}{20}$$

$$\Sigma x^{10} = \frac{x^{11}}{11h} - \text{\&c. as before,}$$

&c.

We shall now give an easy method of determining the *Number of Bernoulli* a, b, c... In the equation (D) make

$$x = h = 1;$$

Σx^m is the general term of the series whose first difference is x^m. We shall here consider $\Sigma . x^0 = 1$, and the corresponding series which is that of the natural numbers

$$0, 1, 2, 3 \ldots$$

Take zero for the first member and transpose

$$\frac{1}{m+1} - \tfrac{1}{2}$$

which equals

$$\frac{2(m+1)}{1-m}.$$

Then we get

$$\frac{m-1}{2(m+1)} = a\,m + b\,A'' + c\,A^{iv} + d\,A^{vi} + \ldots + k\,m.$$

By making $m = 2$, the second member is reduced to am, which gives

$$a = \frac{1}{12}.$$

Making $m = 4$, we get

$$\frac{3}{10} = 4a + b\,A''$$

$$= 4a + m.\frac{m-1}{2}.\frac{m-2}{3}b$$

$$= 4a + 4b$$

$$= \tfrac{1}{3} + 4b.$$

Whence

$$b = -\frac{1}{120}.$$

Again, making $m = 6$, we get

$$\frac{5}{14} = 6a + b\,A'' + c\,A^{iv}$$

$$= 6a + 20b + 6c$$

$$= \tfrac{1}{2} - \tfrac{1}{6} + 6c$$

which gives

$$c = \frac{1}{252};$$

and proceeding thus by making

$$m = 2, 4, 6, 8, \&c.$$

we obtain at each step a new equation which has one term more than the preceding one, which last terms, viz.

$$2a, 4b, 6c, \ldots mk$$

will hence successively be found, and consequently,

$$a, b, c \ldots k.$$

Take the difference of the product

$$y_x = (x - h)\,x\,(x + h)\,(x + 2h) \ldots (x + ih),$$

by $x + h$ for x and subtracting; it gives

$$\Delta y_x = x\,(x + h)\,(x + 2h) \ldots (x + ih) \times (i + 2)\,h;$$

dividing by the last constant factor, integrating, and substituting for y_x its value, we get

$$\Sigma\, x\,(x + h)\,(x + 2h) \ldots (x + ih)$$
$$= \frac{x - h}{(i + 2)\,h} \times x.\,(x + h)\,(x + 2h) \ldots (x + ih)$$

This equation gives *the integral of a product of factors in arithmetic progression.*

Taking the difference of the second member, we verify the equation

$$\Sigma \frac{1}{x\,(x + h)\,(x + 2h) \ldots (x + ih)} = \frac{-1}{i\,h\,x\,(x + h) \ldots \{x + (i - 1)\,h\}}$$

which gives the integral of any inverse product.

Required the integral of a^x.

Let

$$y_x = a^x.$$

Then

$$\Delta y_x = a^x\,(a^h - 1)$$

whence

$$y_x = \Sigma\, a^x\,(a^h - 1) = a^x;$$

consequently

$$\Sigma\, a^x = \frac{a^x}{a^h - 1} + \text{constant}.$$

Required the integrals of sin. x, cos. x.

Since

$$\cos. B - \cos. A = 2 \sin. \tfrac{1}{2}(A + B).\ \sin. \tfrac{1}{2}(A - B)$$
$$\Delta \cos. x = \cos. (x + h) - \cos. x$$
$$= -2 \sin. \left(x + \frac{h}{2}\right) \sin. \frac{h}{2}$$

Integrating and changing $x + \frac{h}{2}$ into z, we have

$$\Sigma \sin. z = - \cos. \frac{\left(z - \frac{h}{2}\right)}{2 \sin. \frac{h}{2}} + \text{constant.}$$

In the same way we find

$$\Sigma \cos. z = \frac{\sin. \left(z - \frac{h}{2}\right)}{2 \sin. \frac{h}{2}} + \text{constant.}$$

When we wish to integrate the powers of sines and cosines, we transform them into sines and cosines of multiple arcs, and we get terms of the form

$$A \sin. q x, \; A \cos. q x.$$

Making

$$q x = x$$

the integration is performed as above.

Required the integral of a product, viz.

Assume

$$\Sigma (u z) = u \Sigma z + t$$

u, z and t being all functions of x, t being the only unknown one. By changing x into x + h in

$$u \Sigma z + t$$

u becomes $u + \Delta u$, z becomes $z + \Delta z$, &c. and we have

$$u \Sigma z + u z + \Delta u \Sigma (z + \Delta z) + t + \Delta t;$$

substituting from this the second member

$$u \Sigma z + t,$$

we obtain the difference, or u z; whence results the equation

$$0 = \Delta u \Sigma (z + \Delta z) + \Delta t$$

which gives

$$t = - \Sigma \{\Delta u \Sigma (z + \Delta z)\}.$$

Therefore

$$\Sigma (u z) = u \Sigma z - \Sigma \{\Delta u . \Sigma (z + \Delta z)\}$$

which is analogous to integrating *by parts* in differential functions.

There are but few functions of which we can find the finite integral; when we cannot integrate them exactly, we must have recourse to series.

Taylor's theorem gives us

$$\Delta y_x = \frac{d y}{d x} h + \frac{d^2 y}{d x^2} . \frac{h^2}{2} + \&c.$$

$$= y' h + \frac{y''}{2} h^2 + \&c.$$

by supposition. Hence

$$y_x = h \, \Sigma \, y' + \frac{h^2}{2} \, \Sigma \, y'' + \&c.$$

Considering y' as a given function of x, viz. z, we have

$$y' = z$$
$$y'' = z'$$
$$y''' = z''$$
$$\&c.$$

and

$$y_x = \int y' \, d\,x = \int z \, d\,x$$

whence

$$\int z \, d\,x = h \, \Sigma \, z + \frac{h^2}{2} \, \Sigma \, z' + \&c.$$

which gives

$$\Sigma \, z = h^{-1} \int z \, d\,x - \frac{h}{2} \Sigma \, z' - \tfrac{1}{6} h^2 \, \Sigma \, z'' - \&c.$$

This equation gives $\Sigma \, z$, when we know z', $\Sigma \, z''$, &c. Take the differentials of the two numbers. That of the first $\Sigma \, z$ will give, when divided by $d\,x$, $\Sigma \, z'$. Hence we get $\Sigma \, z''$, then $\Sigma \, z'''$, &c.; and even without making the calculations, it is easy to see, that the result of the substitution of these values, will be of the form

$$\Sigma \, z = h^{-1} \int z \, d\,x + A \, z + B \, h \, z' + C \, h^2 \, z'' + \&c.$$

It remains to determine the factors A, B, C, &c. But if

$$z = x^m$$

we get

$$\int z \, d\,x, \; z', \; z'', \&c.$$

and substituting, we obtain a series which should be identical with the equation (D), and consequently defective of the powers $m - 2$, $m - 4$, so that we shall have

$$\Sigma \, z = \frac{\int z \, d\,x}{h} \cdot \frac{z}{2} + \frac{a \, h \, z'}{1} + \frac{b \, h^3 \, z'''}{1.2} + \frac{c h^5 z'''''}{2.3.4} + \frac{d h^7 z'''''''}{2 \ldots 6} + \&c.$$

a, b, c, &c. being the numbers of Bernoulli.

For example, if

$$z = l \, x$$
$$\int l \, x \cdot d\,x = x \, l \, x - x$$
$$z' = x^{-1}$$
$$z'' = \&c.$$

consequently

$$\Sigma\, l\, x = C + x\, l\, x - x - \tfrac{1}{2}\, l\, x + a\, x^{-1} + b\, x^{-3} + c\, x^{-5} + \&c.$$

The series

$$a,\ b,\ c \ldots k,\ l,$$

having for first differences

$$a',\ b',\ c' \ldots k'$$

we have

$$b = a + a'$$
$$c = b + b$$
$$d = c + c'$$
$$\&c.$$
$$l = k + k'$$

equations whose sum is

$$l = a + a' + b' + c' + \ldots k'.$$

If the numbers a′, b′, c′, &c. are known, we may consider them as being the first differences of another series a, b, c, &c. since it is easy to compose the latter by means of the first, and the first term a. By definition we know that any term whatever l′, taken in the given series a′, b′, c′, &c. is nothing else than Δ l, for $l' = m - 1$; integrating

$$l' = \Delta\, l$$

we have

$$\Sigma\, l' = l$$

or

$$\Sigma\, l' = a' + b' + c' \ldots + k',$$

supposing the initial a is comprised in the constant due to the integration. Consequently

The integral of any term whatever of a series, we obtain the sum of all the terms that precede it, and have

$$\Sigma\, y_x = y_0 + y_1 + y_2 + \cdots y_{x-1}.$$

In order to get the sum of a series, we must add y_x to the integral; or which is the same, in it must change x into x + 1, before we integrate. The arbitrary constant is determined by finding the value of the sum y_0 when

$$x = 1.$$

We know therefore how to find the summing term of every series whose general term is known in a rational and integer function of x.

Let

$$y_x = A\, x^m - B\, x^n + C$$

m and n being positive and integer, and we have

$$A\, \Sigma\, x^m - B\, \Sigma\, x^n + C\, \Sigma\, x^0$$

for the sum of the terms as far as y_x exclusively. This integral being once found by equation D, we shall change x into x + 1, and determine the constant agreeably.

For example, let

$$y_x = x(2x - 1);$$

changing x into x + 1, and integrating the result, we shall find

$$2\Sigma x^3 + 3\Sigma x + \Sigma x^0 = \frac{4x^3 + 3x^2 - x}{2.3}$$

$$= x.\frac{x+1}{2}.\frac{4x-1}{3};$$

there being no constant, because when x = 0, the sum = 0.

The series

$$1^m, 2^m, 3^m \ldots$$

of the mth powers of the natural numbers is found by taking Σx^m (equation D); but we must add afterwards the xth term which is x^m; that is to say, it is sufficient to change $-\frac{1}{2}x^m$, the second term of the equation (D), into $\frac{1}{2}x^m$; it then remains to determine the constant from the term we commence from.

For example, to find

$$S = 1 + 2^2 + 3^2 + 4^2 + \ldots x^2$$

we find Σx^2, changing the sign of the second term, and we have

$$S = \frac{x^3}{3} + \frac{x^2}{2} + \frac{x}{6} = x.\frac{x+1}{3}.\frac{2x+1}{2};$$

the constant is 0, because the sum is 0 when x = 0. But if we wish to find the sum

$$S' = (n+1)^2 + (n+2)^2 + \ldots x^2$$

S' = 0, whence x = n − 1, and the constant is

$$-n.\frac{n-1}{2}.\frac{2n-1}{3},$$

which of course must be added to the former; thus giving

$$S' = (n+1)^2 + (n+2)^2 + \ldots x^2$$

$$= x.\frac{x+1}{3}.\frac{2x+1}{2} - n.\frac{n-1}{2}.\frac{2n-1}{3}$$

$$= \frac{1}{6} \times \{x.(x+1).(2x+1) - n.(n-1)(2n-1)$$

$$= \frac{1}{6} \times \{2(x^3 - n^3) + 3(x^2 + n^2) + x - n\}.$$

This theory applies to the summation of *figurate numbers*, of the different orders:—

First order,	1 . 1 . 1 . 1 . 1 . 1 . 1 , &c.
Second order,	1 . 2 . 3 . 4 . 5 . 6 . 7 , &c.
Third order,	1 . 3 . 6 . 10 . 15 . 21 . 28 , &c.
Fourth order,	1 . 4 . 10 . 20 . 35 . 56 . 84 , &c.
Fifth order,	1 . 5 . 15 . 35 . 70 . 126 . 210, &c.

and so on.

The law which every term follows being the sum of the one immediately over it added to the preceding one. The general terms are

First, 1

Second, x

Third, $\frac{x.(x+1)}{2}$

Fourth, $\frac{x(x+1)(x+2)}{2.3}$

&c.

p^{th} $\frac{x.(x+1)(x+2)\ldots x+p-2}{1.2.3\ldots p-1}.$

To sum the Pyramidal numbers, we have

$$S = 1 + 4 + 10 + 20 + \&c.$$

Now the general or x^{th} term in this is

$$y_x = \frac{1}{6}.x(x+1)(x+2).$$

But we find for the $(x-1)^{th}$ term of numbers of the next order

$$\frac{1}{24}(x-1)x(x+1)(x+2);$$

finally changing x into $x+1$, we have for the required form

$$S = \frac{1}{24}x.(x+1)(x+2)(x+3).$$

Since $S = 1$, when $x = 1$, we have

$1 = 1 +$ constant, consequently

$\therefore$ constant $= 0$.

Hence it appears that the sum of x terms of the fourth order, is the xth term or general term of the fifth order, and *vice versâ*; and in like manner, it may be shown that the xth term of the $(n+1)^{th}$ order is the sum of x terms of the nth order.

Inverse figurate numbers are fractions which have 1 for the numerator, and a figurate series for the denominator. Hence the xth term of the pth order is

$$\frac{1.2.3\ldots(p-1)}{x(x+1)\ldots x+p-2}$$

and the integral of this is

$$C - \frac{1.2.3\ldots(p-1)}{(p-2)\,x\,(x+1)\ldots(x+p-3)}.$$

Changing x into x + 1, then determining the constant by making x = 0, which gives the sum = 0, we shall have

$$c = \frac{p-1}{p-2};$$

and the sum of the x first terms of this general series is

$$\frac{p-1}{p-2} - \frac{1.2.3\ldots(p-1)}{(p-2)(x+1)(x+2)\ldots(x+p-2)}.$$

In this formula make

$$p = 3, 4, 5\ldots$$

and we shall get

$$\frac{1}{1} + \frac{1}{3} + \frac{1}{6} + \frac{1}{10} + \ldots \frac{1.2}{x(x+1)} = \frac{2}{1} - \frac{2}{x+1}$$

$$\frac{1}{1} + \frac{1}{4} + \frac{1}{10} + \frac{1}{20} + \ldots \frac{1.2.3}{x(x+1)(x+2)} = \frac{3}{2} - \frac{3}{(x+1)(x+2)}$$

$$\frac{1}{1} + \frac{1}{5} + \frac{1}{10} + \frac{1}{35} + \ldots \frac{1.2.3.4}{x(x+1)(x+2)(x+3)} = \frac{4}{3} - \frac{2.4}{(x+1)\ldots(x+3)}$$

$$\frac{1}{1} + \frac{1}{6} + \frac{1}{21} + \frac{1}{56} + \ldots \frac{1.2.3.4.5}{x(x+1)\ldots(x+4)} = \frac{5}{4} - \frac{2.3.5}{(x+1)\ldots(x+4)}$$

and so on. To obtain the whole sum of these series continued to infinity, we must make

$$x = \infty$$

which gives for the sum required the general value

$$\frac{p-1}{p-2}$$

which in the above particular cases, becomes

$$\frac{2}{1}, \frac{3}{2}, \frac{4}{3}, \frac{5}{4}, \&c.$$

To sum the series

$$\sin. a + \sin.(a+h) + \sin.(a+2h) + \ldots \sin.(a + \overline{x-1\,h})$$

we have

$$\Sigma \sin.(a + xh) = C - \frac{\cos.\left(a + hx - \frac{h}{2}\right)}{2\sin.\frac{h}{2}}$$

changing x into x + 1, and determining C by the condition that x = − 1 makes the sum = zero, we find for the *summing-term*.

$$\frac{\cos.\left(a - \frac{h}{2}\right) - \cos.\left(a + hx + \frac{h}{2}\right)}{2\sin.\frac{h}{2}}.$$

or

$$\frac{\sin.\left(a+\frac{h}{2}x\right)\sin.\frac{h(x+1)}{2}}{\sin.\frac{h}{2}}$$

In a similar manner, if we wish to sum the series

$$\cos. a + \cos.(a+h) + \cos.(a+2h) + \ldots \cos.(a+\overline{x-1}\,h)$$

we easily find the *summing-term* to be

$$\frac{\sin.\left(a-\frac{h}{2}\right)-\sin.\left(a+h\,x+\frac{h}{2}\right)}{2\sin.\frac{h}{2}}$$

or

$$\frac{\cos.\left(a+\frac{h}{2}x\right)\sin.\frac{h(x+1)}{2}}{\sin.\frac{h}{2}}.$$

A COMMENTARY

ON

NEWTON'S PRINCIPIA.

SUPPLEMENT

TO

SECTION XI.

460 PROP. LVII, depends upon Cor. 4 to the Laws of Motion, which is

If any number of bodies mutually *attract each other, their center of gravity will either remain at rest or will move uniformly in a straight line.*

First let us prove this for two bodies.

Let them be referred to a fixed point by the rectangular coordinates

$$x, y;\ x', y',$$

and let their masses be

$$\mu, \mu'.$$

Also let their distance be ϱ, and $f(\varrho)$ denote the law according to which they attract each other.

Then

$$\mu f(\varrho),\ \mu' f(\varrho)$$

will be their respective actions, and resolving these parallel to the axes of abscissas and ordinates, we have (46)

$$\left.\begin{aligned} \frac{d^2 x}{d t^2} &= \mu' f(\varrho) \frac{x' - x}{\varrho} \\ \frac{d^2 y}{d t^2} &= \mu' f(\varrho) \frac{y' - y}{\varrho} \end{aligned}\right\} \quad \ldots\ldots\ldots\ldots \quad (1)$$

$$\left.\begin{aligned}\frac{d^2x'}{dt^2} &= -\mu f(\rho)\frac{x'-x}{\rho}\\ \frac{d^2y'}{dt^2} &= -\mu f(\rho)\frac{y'-y}{\rho}\end{aligned}\right\} \quad . \quad . \quad . \quad . \quad . \quad . \quad . \quad . \quad . \quad (2)$$

Hence multiplying equations (1) by μ and those marked (2) by μ' and adding, &c. we get

$$\frac{\mu\, d^2x + \mu' d^2x'}{dt^2} = 0,$$

and

$$\frac{\mu\, d^2y + \mu' d^2y'}{dt^2} = 0$$

and integrating

$$\mu \cdot \frac{dx}{dt} + \mu' \cdot \frac{dx'}{dt} = c$$

$$\mu \cdot \frac{dy}{dt} + \mu' \cdot \frac{dy'}{dt} = c'.$$

Now if the coordinates of the center of gravity be denoted by

$$\bar{x},\ \bar{y},$$

we have by Statics

$$\bar{x} = \frac{\mu x + \mu' x'}{\mu + \mu'}$$

$$\bar{y} = \frac{\mu y + \mu' y'}{\mu + \mu'}$$

$$\therefore \frac{d\bar{x}}{dt} = \frac{1}{\mu + \mu'} \cdot \left(\mu \cdot \frac{dx}{dt} + \mu' \frac{dx'}{dt}\right) = \frac{c}{\mu + \mu'}$$

and

$$\frac{d\bar{y}}{dt} = \frac{1}{\mu + \mu'}\left(\mu \frac{dy}{dt} + \mu' \frac{dy'}{dt}\right) = \frac{c'}{\mu + \mu'}.$$

But

$$\frac{d\bar{x}}{dt},\ \frac{d\bar{y}}{dt}$$

represent the velocity of the center of gravity resolved parallel to the axes of coordinates, and these resolved parts have been shown to be constant. Hence it easily appears by composition of motion, that the actual velocity of the center of gravity is uniform, and also that it moves in a straight line, viz. in that produced which is the diagonal of the rectangular parallelogram whose two sides are $d\bar{x}$, $d\bar{y}$.

If

$$c = 0,\ c' = 0$$

then the center of gravity remains quiescent.

461 The general proposition is similarly demonstrated, thus.

Let the bodies whose masses

$$\mu', \mu'', \mu''', \&c.$$

be referred to three rectangular axes, issuing from a fixed point by the coordinates

$$\begin{array}{lll} x', & y', & z' \\ x'', & y'', & z'' \\ x''', & y''', & z''' \\ \&c. \end{array}$$

Also let

$$\begin{array}{ll} \rho_{1,2} \text{ be the distance of } & \mu', \ \mu'' \\ \rho_{1,3} \text{ - - - - - - - - - } & \mu', \ \mu''' \\ \rho_{2,3} \text{ - - - - - - - - - } & \mu'', \ \mu''' \\ \&c. \ \&c. \end{array}$$

and suppose the law of attraction to be denoted by

$$f.(\rho_{1,2}), f(\rho_{1,3}), f(\rho_{2,3}), \&c.$$

Now resolving the attractions or forces

$$\begin{array}{l} \mu' \ f(\rho_{1,2}) \\ \mu'' \ f(\rho_{1,2}) \\ \mu''' f(\rho_{1,3}) \\ \&c. \end{array}$$

parallel to the axes, and collecting the parts we get

$$\frac{d^2 x'}{d t^2} = \mu'' f(\rho_{1,2}) \frac{x' - x''}{\rho_{1,2}} + \mu''' f(\rho_{1,3}) \frac{x' - x'''}{\rho_{1,3}} + \&c.$$

$$\frac{d^2 x''}{d t^2} = - \mu' f(\rho_{1,2}) \frac{x' - x''}{\rho_{1,2}} + \mu''' f(\rho_{2,3}) \frac{x'' - x'''}{\rho_{2,3}} + \&c.$$

$$\frac{d^2 x'''}{d t^2} = - \mu' f(\rho_{1,3}) \frac{x' - x'''}{\rho_{1,3}} - \mu'' f(\rho_{2,3}) \frac{x'' - x'''}{\rho_{2,3}} + \&c.$$

$$\&c. = \&c.$$

Hence multiplying the first of the above equations by μ', the second by μ'', and so on, and adding, we get

$$\frac{\mu' d^2 x' + \mu'' d^2 x'' + \mu''' d^2 x''' + \&c.}{d t^2} = 0.$$

Again, since it is a matter of perfect indifference whether we collect the forces parallel to the other axes or this; or since all the circumstances are similar with regard to these independent axes, the results arising from similar operations must be similar, and we therefore have also

$$\frac{\mu' d^2 y' + \mu'' d^2 y'' + \mu''' d^2 y''' + \&c.}{d t^2} = 0,$$

$$\frac{\mu' d^2 z' + \mu'' d^2 z'' + \mu''' d^2 z''' + \&c.}{d t^2} = 0.$$

Hence by integration

$$\mu' . \frac{d x'}{d t} + \mu'' \frac{d x''}{d t} + \mu''' \frac{d x'''}{d t} + \&c. = c$$

$$\mu' . \frac{d y'}{d t} + \mu'' \frac{d y''}{d t} + \mu''' \frac{d y'''}{d t} + \&c = c'$$

$$\mu' . \frac{d z'}{d t} + \mu'' \frac{d z''}{d t} + \mu''' \frac{d z'''}{d t} + \&c. = c''.$$

But $\bar{x}$, $\bar{y}$, $\bar{z}$ denoting the coordinates of the center of gravity, by statics we have

$$\bar{x} = \frac{\mu' x' + \mu'' x'' + \mu''' x''' + \&c.}{\mu' + \mu'' + \mu''' + \&c.}$$

$$\bar{y} = \frac{\mu' y' + \mu'' y'' + \mu''' y''' + \&c.}{\mu' + \mu'' + \mu''' + \&c.}$$

$$\bar{z} = \frac{\mu' z' + \mu'' z'' + \mu''' z''' + \&c.}{\mu' + \mu'' + \mu''' + \&c.}$$

and hence by taking the differentials, &c. we get

$$\frac{d \bar{x}}{d t} = \frac{c}{\mu' + \mu'' + \mu''' + \&c.}$$

$$\frac{d \bar{y}}{d t} = \frac{c'}{\mu' + \mu'' + \mu''' + \&c.}$$

$$\frac{d \bar{z}}{d t} = \frac{c''}{\mu' + \mu'' + \mu''' + \&c.}$$

that is, the velocity of the center of gravity resolved parallel to any three rectangular axes is constant. Hence by composition of motion the actual velocity of the center of gravity is constant and uniform, and it easily appears also that its path is a straight line, scil. the diagonal of the rectangular parallelopiped whose sides are $d\bar{x}$, $d\bar{y}$, $d\bar{z}$.

462. We will now give another demonstration of Prop. LXI. or that *Of two bodies the motion of each about the center of gravity, is the same as if that center was the center of force, and the law of force the same as that of their mutual attractions.*

Supposing the coordinates of the two bodies referred to the center of gravity to be

$$x_{,}, y_{,}; x_{,,}, y_{,,}$$

we have

$$\left.\begin{array}{l} x = \bar{x} + x_{,} \\ y = \bar{y} + y_{,} \end{array}\right\} \quad \left.\begin{array}{l} x' = \bar{x} + x_{,,} \\ y' = \bar{y} + y_{,,} \end{array}\right\}$$

Hence since

$$\frac{d \bar{x}}{d t}, \frac{d \bar{y}}{d t}$$

are constant as it has been shown, and therefore

$$\frac{d^2\bar{x}}{dt^2} = 0, \quad \frac{d^2\bar{y}}{dt^2} = 0$$

we have

$$\frac{d^2x}{dt^2} = \frac{d^2x_{,}}{dt^2}$$

$$\frac{d^2y}{dt^2} = \frac{d^2y_{,}}{dt^2}$$

and we therefore get (46)

$$\frac{d^2x_{,,}}{dt^2} = -\mu f(\varrho)\frac{x_{,,} - x_{,}}{\varrho}$$

$$\frac{d^2y_{,,}}{dt^2} = -\mu f(\varrho)\frac{y_{,,} - y_{,}}{\varrho}.$$

But by the property of the center of gravity

$$\varrho = \frac{\mu + \mu'}{\mu} . \delta'$$

δ being the distance of μ' from the center of gravity. We also have

$$\frac{x_{,,} - x_{,}}{\varrho} = \frac{x_{,,}}{\delta'}.$$

Hence by substitution the equations become

$$\frac{d^2x_{,,}}{dt^2} = -\mu f\left(\frac{\mu + \mu'}{\mu}\delta'\right)\frac{x_{,,}}{\delta'},$$

$$\frac{d^2y_{,,}}{dt^2} = -\mu f\left(\frac{\mu + \mu'}{\mu}\delta'\right)\frac{y_{,,}}{\delta'}.$$

Similarly we should find

$$\frac{d^2x_{,}}{dt^2} = \mu' f\left(\frac{\mu + \mu'}{\mu'}\delta\right)\frac{x_{,}}{\delta}$$

and

$$\frac{d^2y_{,}}{dt^2} = \mu' f\left(\frac{\mu + \mu'}{\mu'}\delta\right)\frac{y_{,}}{\delta}.$$

Hence if the force represented by

$$\mu f\left(\frac{\mu + \mu'}{\mu}\delta'\right)$$

were placed in the center of gravity, it would cause μ' to move about it as a fixed point; and if

$$\mu' f\left(\frac{\mu + \mu'}{\mu'}\delta\right)$$

were there residing, it would cause μ to centripetate in like manner.

Moreover if

$$f(\varrho) = \varrho^n$$

then these forces vary as

$$\delta'^n, \delta^n;$$

so that the law of force &c. &c.

ANOTHER PROOF OF PROP. LXII.

463. Let μ, μ' denote the two bodies. Then since μ has no motion round G (G being the center of gravity), it will descend in a straight line to G. In like manner μ' will fall to G in a straight line.

Also since the accelerating forces on μ, μ' are inversely as μ, μ' or directly as G μ, G μ', the velocities will follow the same law and corresponding portions of G μ, G μ' will be described in the same times; that is, the whole will be described in the same time. Moreover after they meet at G, the bodies will go on together with the same constant velocity with which G moved before they met.

Since here

$$f(\varrho) = \frac{1}{\varrho^2}$$

μ will move towards G as if a force

$$\mu'\left(\frac{\mu + \mu'}{\mu'}\delta\right)^{-2}$$

or

$$\frac{\mu'^3}{(\mu + \mu')^2} \times \frac{1}{\delta^2}.$$

Hence by the usual methods it will be found that if a be the distance at which μ begins to fall, the time to G is

$$\frac{(\mu + \mu')\, a^{\frac{3}{2}}}{\mu'^{\frac{3}{2}}} \cdot \frac{\pi}{2\sqrt{2}}$$

and if a' be the original distance of μ', the time is

$$\frac{(\mu + \mu')\, a'^{\frac{3}{2}}}{\mu^{\frac{3}{2}}} \cdot \frac{\pi}{2\sqrt{2}}.$$

But

$$a : a' :: \mu' : \mu$$

therefore these times are equal, which has just been otherwise shown.

ANOTHER PROOF OF PROP. LXIII.

464. We know from (461) that the center of gravity moves uniformly in a straight line; and that (Prop. LVII,) μ and μ' will describe about G similar figures, μ moving as though actuated by the force

$$\frac{\mu'^3}{(\mu+\mu')^2} \cdot \frac{1}{\delta^2}$$

and Q as if by

$$\frac{\mu^3}{(\mu+\mu')^2} \cdot \frac{1}{\delta'^2}.$$

Hence the curves described will be similar ellipses, with the center of force G in the focus. Also if we knew the original velocities of μ and μ' about G, the ellipse would easily be determined.

The velocities of μ and μ' at any time are composed of two velocities, viz. the progressive one of the center of gravity and that of each round G. *Hence having given the whole original velocities required to find the separate parts of them,*

is a problem which we will now resolve.

Let

$$V, V'$$

be the original velocities of μ, μ', and suppose their directions to make with the straight line $\mu\,\mu'$ the angles

$$\alpha, \alpha'.$$

Also let the velocity of the center of gravity be

$$\bar{v}$$

and the direction of its motion to make with $\mu\,\mu'$ the angle

$$\bar{\alpha}.$$

Moreover let

$$v, v'$$

be the velocities of μ, μ' around G and the common inclination of their directions to be

$$\theta.$$

Now V resolved parallel to $\mu\,\mu'$ is

$$V \cos. \alpha.$$

But since it is composed of $\bar{v}$ and of v it will also be

$$\bar{v} \cos. \bar{\alpha} + v \cos. \theta$$

$$\therefore V \cos. \alpha = \bar{v} \cos. \bar{\alpha} + v \cos. \theta.$$

In like manner we get

$$V \sin. \alpha = \bar{v} \sin. \bar{\alpha} + v \sin. \theta.$$

and also

$$V' \cos. \alpha' = \bar{v} \cos. \bar{\alpha} - v' \cos. \theta$$

$$V \sin. \alpha' = \bar{v} \sin. \bar{\alpha} - v' \sin. \theta.$$

Hence multiplying by μ, μ', adding and putting

$$\mu v = \mu' v'$$

we get

$$\left.\begin{aligned} &\mu V \cos. \alpha + \mu' V' \cos. \alpha' = (\mu + \mu') \bar{v} \cos. \bar{\alpha} \\ &\text{and} \\ &\mu V \sin. \alpha + \mu' V' \sin. \alpha' = (\mu + \mu') \bar{v} \sin. \bar{\alpha} \end{aligned}\right\}$$

Squaring these and adding them, we get

$$\mu^2 V^2 + \mu'^2 V'^2 + 2 \mu \mu' V V' \cos. (\alpha - \alpha') = (\mu + \mu')^2 \bar{v}^2$$

which gives

$$\bar{v} = \frac{\sqrt{\{\mu^2 V^2 + \mu'^2 V'^2 + 2 \mu \mu' V V' \cos. (\alpha - \alpha')\}}}{\mu + \mu'}.$$

By division we also have

$$\tan. \bar{\alpha} = \frac{\mu V \sin. \alpha + \mu' V' \sin. \alpha'}{\mu V \cos. \alpha + \mu' V' \cos. \alpha'}.$$

Again, from the first four equations by subtraction we also have

$$V \cos. \alpha - V' \cos. \alpha' = (v + v') \cos. \theta = v . \frac{\mu + \mu'}{\mu'} \cos. \theta$$

$$V \sin. \alpha - V' \sin. \alpha' = (v + v') \sin. \theta = v . \frac{\mu + \mu'}{\mu'} \sin. \theta$$

and adding the squares of these

$$V^2 + V'^2 - 2 V V' \cos. (\alpha - \alpha') = v^2 \left(\frac{\mu + \mu'}{\mu'}\right)^2$$

whence

$$v = \frac{\mu'}{\mu + \mu'} . \sqrt{\{V^2 + V'^2 - 2 V V' \cos. (\alpha - \alpha')\}}$$

$$v' = \frac{\mu}{\mu + \mu'} \sqrt{\{V^2 + V'^2 - 2 V V' \cos. (\alpha - \alpha')\}}$$

and by division

$$\tan. \theta = \frac{V \sin. \alpha - V' \sin. \alpha'}{V \cos. \alpha - V' \cos. \alpha'}.$$

Whence are known the velocity and direction of projection of μ about G and (by Sect. III. or Com.) the conic section can therefore be found; and combining the motion in this orbit with that of the center of gravity, which is given above, we have also that of μ.

465. Hence since the orbit of μ round μ' is similar to the orbit of μ round G, if A be the semi-axis of the ellipse which μ describes round

G, and a that of the ellipse which it describes relatively to μ' which is also in motion; we shall have

$$A : a :: \mu' : \mu + \mu'.$$

466. Hence also since an ellipse whose semi-axis is A, is described by the force

$$\frac{\mu'^3}{(\mu + \mu')^2} \times \frac{1}{\delta^2}$$

we shall have (309) the periodic time, viz.

$$T = \frac{2 A^{\frac{3}{2}} \pi}{\sqrt{\frac{\mu'^3}{(\mu + \mu')^2}}} = \frac{2 \pi A^{\frac{3}{2}} (\mu + \mu')}{\mu'^{\frac{3}{2}}}$$

$$= \frac{2 \pi a^{\frac{3}{2}}}{\sqrt{(\mu + \mu')}}.$$

467. Hence we easily get Prop. LIX.

For if μ were to revolve round μ' at rest, its semi-axis would be a, and periodic time

$$T' = \frac{2 \pi a^{\frac{3}{2}}}{\sqrt{\mu'}}$$

$$\therefore T : T' :: \sqrt{\mu'} : \sqrt{(\mu + \mu')}.$$

468. Prop. LX is also hence deducible. For if μ revolve round μ' at rest, in an ellipse whose semi-axis is a', we have

$$T'' = \frac{2 \pi a'^{\frac{3}{2}}}{\sqrt{\mu'}}$$

and equating this with T in order to give it the same time about μ' at rest as about μ' in motion, we have

$$\frac{2 \pi a'^{\frac{3}{2}}}{\sqrt{\mu'}} = \frac{2 \pi a^{\frac{3}{2}}}{\sqrt{(\mu + \mu')}};$$

$$\therefore a : a' :: (\mu + \mu')^{\frac{1}{3}} : \mu'^{\frac{1}{3}}.$$

ANOTHER PROOF OF PROP. LXIV.

469. *Required the motions of the bodies whose masses are*

$$\mu, \mu', \mu'', \mu''', \text{\&c.}$$

and which mutually attract each other with forces varying directly as the distance.

Let the distance of any two of them as μ, μ', be ϱ; then the force of μ' on μ is

$$\mu' \varrho$$

and the part resolved parallel to x is

$$\mu' \rho . \frac{x - x'}{\rho} = \mu' (x - x').$$

In like manner the force of μ'' on μ, resolved parallel to x, is

$$\mu'' (x - x'')$$

and so on for the rest of the bodies and for their respective forces resolved parallel to the other axes of coordinates.

Hence

$$\frac{d^2 x}{d t^2} = \mu' (x - x') + \mu'' (x - x'') + \&c.$$

$$\frac{d^2 x'}{d t^2} = \mu (x' - x) + \mu'' (x' - x'') + \&c.$$

$$\frac{d^2 x''}{d t^2} = \mu (x'' - x) + \mu' (x'' - x') + \&c.$$

$$\&c. = \&c.$$

which give

$$\frac{d^2 x}{d t^2} = (\mu + \mu' + \mu'' + \&c.)\, x - (\mu x + \mu' x' + \&c.)$$

$$\frac{d^2 x'}{d t^2} = (\mu + \mu' + \mu'' + \&c.)\, x' - (\mu x + \mu' x' + \&c.)$$

$$\frac{d^2 x''}{d t^2} = (\mu + \mu' + \mu'' + \&c.)\, x'' - (\mu x + \mu' x' + \&c.)$$

$$\&c. = \&c.$$

Or since

$$\mu x + \mu' x' + \&c. = (\mu + \mu' + \&c.)\, \bar{x}$$

making the coordinates of the center of gravity

$$\bar{x}, \bar{y}, \bar{z},$$

we have

$$\frac{d^2 x}{d t^2} = (\mu + \mu' + \&c.)\, (x - \bar{x})$$

$$\frac{d^2 x'}{d t^2} = (\mu + \mu' + \&c.)\, (x' - \bar{x})$$

$$\frac{d^2 x''}{d t^2} = (\mu + \mu' + \&c.)\, (x'' - \bar{x}).$$

$$\&c. = \&c.$$

In like manner, we easily get

$$\frac{d^2 y}{d t^2} = (\mu + \mu' + \&c.)\, (y - \bar{y})$$

$$\frac{d^2 y'}{d t^2} = (\mu + \mu' + \&c.)\, (y' - y)$$

$$\frac{d^2 y''}{d t^2} = (\mu + \mu' + \&c.)(y'' - \bar{y})$$

$$\&c. = \&c.$$

and also

$$\frac{d^2 z}{d t^2} = (\mu + \mu' + \&c.)(z - \bar{z})$$

$$\frac{d^2 z'}{d t^2} = (\mu + \mu' + \&c.)(z' - \bar{z})$$

$$\frac{d^2 z''}{d t^2} = (\mu + \mu' + \&c.)(z'' - \bar{z})$$

$$\&c. = \&c.$$

Again,

$$x - \bar{x},\ y - \bar{y},\ z - \bar{z}$$

$$x' - \bar{x},\ y' - \bar{y},\ z' - \bar{z}$$

$$\&c. \quad \&c. \quad \&c.$$

are the coordinates of μ, μ', μ'', &c. when measured from the center of gravity, and it has been shown already that

$$\frac{d^2 (x - \bar{x})}{d t^2} = \frac{d^2 x}{d t^2}$$

$$\frac{d^2 (y - \bar{y})}{d t^2} = \frac{d^2 y}{d t^2}.$$

$$\frac{d^2 (z - \bar{z})}{d t^2} = \frac{d^2 z}{d t^2}$$

and so on for the other bodies. Hence then it appears, that the motions of the bodies about the center of gravity, are the same as if there were but one force, scil.

$$(\mu + \mu' + \&c.) \times \text{distance}$$

and as if this force were placed in the center of gravity.

Hence the bodies will all describe ellipses about the center of gravity, as a center; and their periodic times will all be the same. But their magnitudes, excentricities, the positions of the planes of their orbits, and of the major axes, may be of all varieties.

Moreover the motion of any one body relative to any other, will be governed by the same laws as the motion of a body relative to a center of force, which force varies directly as the distance; for if we take the equations

$$\frac{d^2 x}{d t^2} = (\mu + \mu' + \&c.)(x - \bar{x})$$

$$\frac{d^2 x'}{d t^2} = (\mu + \mu' + \&c.)(x' - \bar{x})$$

and subtract them we get

$$\frac{d^2(x-x')}{dt^2} = (\mu + \mu' + \&c.)(x-x')$$

and similarly

$$\frac{d^2(y-y')}{dt^2} = (\mu + \mu' + \&c.)(y-y')$$

and

$$\frac{d^2(z-z')}{dt^2} = (\mu + \mu' + \&c.)(z-z').$$

Hence by composition and the general expression for force $\left(\frac{d^2 \varrho}{dt^2}\right)$ it readily appears that the motion of μ about μ', is such as was asserted.

470. Thus far relates merely to the motions of two bodies; and these can be accurately determined. But the operations of Nature are on a grander scale, and she presents us with Systems composed of Three, and even more bodies, mutually attracting each other. In these cases the equations of motion cannot be integrated by any methods hitherto discovered, and we must therefore have recourse to methods of approximation.

In this portion of our labours we shall endeavour to lay before the reader such an exposition of the Lunar, Planetary and Cometary Theorics, as may afford him a complete succedaneum to the discoveries of our author.

471. Since relative motions are such only as can be observed, we refer the motions of the Planets and Comets, to the center of the sun, and the motions of the Satellites to the center of their planets. Thus to compare theory with observations,

It is required to determine the relative motion of a system of bodies, about a body considered as the center of their motions.

Let M be this last body, μ, μ', μ'', &c. being the other bodies of which is required the relative motion about M. Also let

$$\zeta, \varpi, \gamma$$

be the rectangular coordinates of M;

$$\zeta + x, \varpi + y, \gamma + z;$$
$$\zeta + x' \varpi + y', \gamma + z';$$
&c.

those of μ, μ', &c. Then it is evident that

$$x, y, z;$$
$$x', y', z'$$
&c.

will be the coordinates of μ, μ', &c. referred to M.

Call ϱ, ϱ', &c.

the distances of μ, μ', &c. from M; then we have

$$\varrho = \sqrt{(x^2 + y^2 + z^2)}$$
$$\varrho' = \sqrt{(x'^2 + y'^2 + z'^2)}.$$

ϱ, ϱ', &c. being the diagonals of rectangular parallelopipeds, whose sides are

$$x, y, z$$
$$x', y', z'$$
$$\&c.$$

Now the actions of μ, μ', μ'', &c. upon M are

$$\frac{\mu}{\varrho^2}, \frac{\mu'}{\varrho'^2}, \frac{\mu''}{\varrho''^2}, \&c.$$

and these resolved parallel to the axis of x, are

$$\frac{\mu x}{\varrho^3}, \frac{\mu' x'}{\varrho'^3}, \frac{\mu'' x''}{\varrho''^3}, \&c.$$

Therefore to determine ζ, we have

$$\frac{d^2 \zeta}{dt^2} = \frac{\mu x}{\varrho^3} + \frac{\mu' x'}{\varrho'^3} + \frac{\mu'' x''}{\varrho''^3} + \&c.$$
$$= \Sigma . \frac{\mu x}{\varrho^3}$$

the symbol Σ denoting the sum of such expressions.

In like manner to determine Π, γ we have

$$\frac{d^2 \Pi}{dt^2} = \Sigma . \frac{\mu y}{\varrho^3};$$

$$\frac{d^2 \gamma}{dt^2} = \Sigma . \frac{\mu z}{\varrho^3}.$$

The action of M upon μ, resolved parallel to the axis of x, and in the contrary direction, is

$$-\frac{M x}{\varrho^3}.$$

Also the actions of μ', μ'', &c. upon μ resolved parallel to the axis of x are, in like manner,

$$\frac{\mu' (x' - x)}{\varrho^3_{0,1}}, \frac{\mu'' (x'' - x)}{\varrho^3_{0,2}}, \frac{\mu''' (x''' - x)}{\varrho^3_{0,3}}, \&c.$$

$\varrho_{n,m}$ generally denoting the distance between $\mu''' \cdots {}^{n}$ and $\mu''' \cdots {}^{m}$

But

$$\varrho_{0,1} = \sqrt{(x' - x)^2 + (y' - y)^2 + (z' - z)^2}$$
$$\varrho_{0,2} = \sqrt{(x'' - x)^2 + (y'' - y)^2 + (z'' - z)^2}$$
$$\&c. = \&c.$$

$$\varrho_{1,2} = \sqrt{(x''-x')^2 + (y''-y')^2 + (z''-z')^2}$$

and so on.

Hence if we assume

$$\lambda = \frac{\mu \cdot \mu'}{\varrho_{0,1}} + \frac{\mu\,\mu''}{\varrho_{0,2}} + \&c.$$
$$+ \frac{\mu'\,\mu''}{\varrho_{1,2}} + \frac{\mu'\,\mu'''}{\varrho_{1,3}}$$
$$+ \frac{\mu''\,\mu'''}{\varrho_{2,3}} + \&c.$$
$$\&c.$$

and taking the Partial Difference upon the supposition that x is the only variable, we have

$$\frac{1}{\mu} \cdot \left(\frac{d\,\lambda}{d\,x}\right) = \frac{\mu' \times (x'-x)}{\varrho^3_{0,1}} + \frac{\mu''(x''-x)}{\varrho^3_{0,2}} + \&c.$$

the parenthesis ( ) denoting the Partial Difference. Hence the sum of all the actions of μ', μ'', &c. on μ is

$$\frac{1}{\mu} \cdot \left(\frac{d\,\lambda}{d\,x}\right).$$

Hence then the whole action upon μ parallel to x is

$$\frac{d.^2(\zeta + x)}{d\,t^2} = \frac{1}{\mu} \cdot \left(\frac{d\,\lambda}{d\,x}\right) - \frac{M\,x}{\varrho^3};$$

But

$$\frac{d^2\,\zeta}{d\,t^2} = \Sigma \frac{\mu\,x}{\varrho^3}$$

$$\therefore \frac{d^2\,x}{d\,t^2} = \frac{1}{\mu}\left(\frac{d\,\lambda}{d\,x}\right) - \frac{M\,x}{\varrho^3} - \Sigma . \frac{\mu\,x}{\varrho^3} \quad \dots\dots \quad (1)$$

Similarly, we have

$$\frac{d^2\,y}{d\,t^2} = \frac{1}{\mu}\left(\frac{d\,\lambda}{d\,y}\right) - \frac{M\,y}{\varrho^3} - \Sigma . \frac{\mu\,y}{\varrho^3} \quad \dots\dots \quad (2)$$

$$\frac{d^2\,z}{d\,t^2} = \frac{1}{\mu}\left(\frac{d\,\lambda}{d\,z}\right) - \frac{M\,z}{\varrho^3} - \Sigma . \frac{\mu\,z}{\varrho^3} \quad \dots\dots \quad (3)$$

If we change successively in the equations (1), (2), (3) the quantities μ, x, y, z into

μ', x', y', z';
μ'', x'', y'', z'';
&c.

and reciprocally; we shall have all the equations of motion of the bodies μ', μ'', &c. round M.

If we multiply the equations involving ζ by $M + \Sigma . \mu$; that in x, by μ; that in x′, by μ', and so on; and add them together, we shall have

$$(M + \Sigma . \mu)\frac{d^2 \zeta}{d t^2} = \left(\frac{d \lambda}{d x}\right) + \left(\frac{d \lambda}{d x'}\right) + \left(\frac{d \lambda}{d x''}\right) + \&c. - \Sigma . \mu \frac{d^2 x}{d t^2}$$

But since

$$\left(\frac{d \lambda}{d x}\right) = \frac{\mu \mu' (x' - x)}{\varrho^3_{0,1}} + \&c.$$

$$\left(\frac{d \lambda}{d x'}\right) = - \frac{\mu \mu' (x' - x)}{\varrho^3_{0,1}} + \&c.$$

and so on in pairs, it will easily appear that

$$\left(\frac{d \lambda}{d x}\right) + \left(\frac{d \lambda}{d x'}\right) + \&c. = 0.$$

$$\therefore (M + \Sigma . \mu) \frac{d^2 \zeta}{d t^2} = - \Sigma . \mu \frac{d^2 x}{d t^2};$$

whence by integrating we get

$$(M + \Sigma . \mu) \frac{d \zeta}{d t} = c - \Sigma . \mu \frac{d x}{d t}.$$

$$\therefore d \zeta = \frac{c}{M + \Sigma . \mu} . d t - \frac{\Sigma . \mu d x}{M + \Sigma . \mu}$$

and again integrating

$$\zeta = a + b t - \frac{\Sigma . \mu x}{M + \Sigma . \mu},$$

a and b being arbitrary constants.

Similarly, it is found that

$$\Pi = a' + b' t - \frac{\Sigma . \mu y}{M + \Sigma . \mu}$$

$$\gamma = a'' + b'' t - \frac{\Sigma . \mu z}{M + \Sigma . \mu}$$

These three equations, therefore, give the *absolute* motion of M in space, when the *relative* motions around it of μ, μ', μ'', &c. are known.

Again, if we multiply the equations in x and y by

$$- \mu y + \mu . \frac{\Sigma . \mu y}{M + \Sigma . \mu},$$

and

$$\mu x - \mu . \frac{\Sigma . \mu x}{M + \Sigma . \mu};$$

in like manner the equations in x′ and y′ by

$$- \mu' y' + \mu' . \frac{\Sigma . \mu y}{M + \Sigma . \mu},$$

and

$$\mu' x' - \mu' . \frac{\Sigma . \mu x}{M + \Sigma . \mu};$$

and so on.

And if we add all these results together, observing that from the nature of λ, (which is easily shown)

$$\Sigma . x . \left(\frac{d \lambda}{d y}\right) = \Sigma . y \left(\frac{d \lambda}{d x}\right)$$

and that (as we already know)

$$\Sigma . \left(\frac{d \lambda}{d x}\right) = 0, \Sigma . \left(\frac{d \lambda}{d y}\right) = 0,$$

we have

$$\Sigma . \mu . \frac{x d^2 y - y d^2 x}{d t^2} = \frac{\Sigma . \mu x}{M + \Sigma \mu} . \Sigma . \mu . \frac{d^2 y}{d t^2}$$
$$- \frac{\Sigma . \mu y}{M + \Sigma \mu} . \Sigma . \mu . \frac{d^2 x}{d t^2}$$

and integrating, since

$$\int (x d^2 y - y d^2 x) = \int x d^2 y - \int y d^2 x$$
$$= x d y - \int d x d y - (y d x - \int d x d y)$$
$$= x d y - y d x,$$

we have

$$\Sigma . \mu . \frac{x d y - y d x}{d t} = \text{const.} + \frac{\Sigma . \mu x}{M + \Sigma . \mu} . \Sigma . \mu . \frac{d y}{d t}$$
$$- \frac{\Sigma . \mu y}{M + \Sigma . \mu} . \Sigma . \mu . \frac{d x}{d t}$$

Hence

$$c = M . \Sigma . \mu . \frac{x d y - y d x}{d t} + \Sigma . \mu \times \Sigma \mu . \frac{x d y - y d x}{d t} + \Sigma . \mu y \times \Sigma . \mu \frac{d x}{d t}$$
$$- \Sigma . \mu x \times \Sigma . \mu \frac{d y}{d t}$$

$$= M . \Sigma . \mu . \frac{x d y - y d x}{d t} + \Sigma . \mu \mu' . \left\{\frac{(x'-x)(d y'-d y)-(y'-y)(d x'-d x)}{d t}\right\} \quad . . (4)$$

c being an arbitrary constant.

In the same manner we arrive at these two integrals,

$$c' = M . \Sigma . \mu . \frac{x d z - z d x}{d t} + \Sigma . \mu \mu' \left\{\frac{(x'-x)(d z'-d z)-(z'-z)(d x'-d x)}{d t}\right\} \quad . . (5)$$

$$c'' = M . \Sigma . \mu . \frac{y d z - z d y}{d t} + \Sigma . \mu \mu' \left\{\frac{(y'-y)(d z'-d z)-(z'-z)(d y'-d y)}{d t}\right\} \quad . . (6)$$

c′ and c″ being two other arbitrary constants.

Again, if we multiply the equation in x by

$$2\,\mu\,d\,x - 2\,\mu\,.\,\frac{\Sigma\,.\,\mu\,d\,x}{M + \Sigma\,.\,\mu};$$

the equation in y by

$$2\,\mu\,d\,y - 2\,\mu\,.\,\frac{\Sigma\,.\,\mu\,d\,y}{M + \Sigma\,.\,\mu};$$

the equation in z by

$$2\,\mu\,d\,z - 2\,\mu\,.\,\frac{\Sigma\,.\,\mu\,d\,z}{M + \Sigma\,.\,\mu};$$

if in like manner we multiply the equations in x', y', z' by

$$2\,\mu'\,d\,x' - 2\,\mu'\,.\,\frac{\Sigma\,.\,\mu\,d\,x}{M + \Sigma\,.\,\mu}$$

$$2\,\mu'\,d\,y' - 2\,\mu'\,.\,\frac{\Sigma\,.\,\mu\,d\,y}{M + \Sigma\,.\,\mu}$$

$$2\,\mu'\,d\,z' - 2\,\mu'\,.\,\frac{\Sigma\,.\,\mu\,d\,z}{M + \Sigma\,.\,\mu};$$

respectively, and so on for the rest; and add the several results, observing that

$$\Sigma\,.\left(\frac{d\,\lambda}{d\,x}\right) = 0;$$

$$\Sigma\,.\left(\frac{d\,\lambda}{d\,y}\right) = 0;$$

$$\Sigma\,.\left(\frac{d\,\lambda}{d\,z}\right) = 0;$$

we get

$$2\,.\,\Sigma\,.\,\mu\,\frac{d\,x\,d^2\,x + d\,y\,d^2\,y + d\,z\,d^2\,z}{d\,t^2} = \frac{2\,\Sigma\,.\,\mu\,d\,x}{M + \Sigma\,\mu}\,.\,\Sigma\,.\,\frac{\mu\,d^2\,x}{d\,t^2}$$

$$+ \frac{2\,\Sigma\,.\,\mu\,d\,y}{M + \Sigma\,\mu}\,.\,\Sigma\,.\,\frac{\mu\,d^2\,y}{d\,t^2} + \frac{2\,\Sigma\,.\,\mu\,d\,z}{M + \Sigma\,\mu}\,.\,\Sigma\,.\,\frac{\mu\,d^2\,z}{d\,t^2} -$$

$$2\,M\,.\,\Sigma\,.\,\frac{\mu\,d\,\varrho}{\varrho^2} + 2\,d\,\lambda;$$

and integrating, we have

$$\Sigma\,.\,\mu\,\frac{d\,x^2 + d\,y^2 + d\,z^2}{d\,t^2} = \text{const.} + \frac{(\Sigma\,.\,\mu\,d\,x)^2 + (\Sigma\,.\,\mu\,d\,y)^2 + (\Sigma\,.\,\mu\,d\,z)^2}{(M + \Sigma\,\mu)\,d\,t^2}$$

$$+ 2\,M\,\Sigma\,\frac{\mu}{\varrho} + 2\,\lambda,$$

which gives

$$h = M.\ \Sigma.\mu\,\frac{d\,x^2 + d\,y^2 + d\,z^2}{d\,t^2} + \Sigma.\,\mu\,\mu'.\left\{\frac{(d\,x' - d\,x)^2 + (d\,y' - d\,y)^2 + (d\,z' - d\,z)^2}{d\,t^2}\right\}$$

$$- \left\{2\,M\,\Sigma.\,\frac{\mu}{\varrho} + 2\,\lambda\right\}(M + \Sigma\,\mu)\ .\ .\ .\ .\ .\ .\ .\ .\ .\ .\ . \quad (7)$$

h being an arbitrary constant.

These integrals being the only ones attainable by the present state of analysis, we are obliged to have recourse to Methods of Approximation, and for this object to take advantage of the facilities afforded us by the constitution of the system of the World. One of the principal of these is due to the fact, that the Solar System is composed of Partial Systems, formed by the Planets and their Satellites: which systems are such, that the distances of the Satellites from their Planet, are small in comparison with the distance of the Planet from the Sun: whence it results, that the action of the Sun being nearly the same upon the Planet as upon its Satellites, these latter move nearly the same as if they obeyed no other action than that of the Planet. Hence we have this remarkable property, namely,

472. *The motion of the Center of Gravity of a Planet and its Satellites, is very nearly the same as if all the bodies formed one in that Center.*

Let the mutual distances of the bodies μ, μ', μ'', &c. be very small compared with that of their center of gravity from the body M. Let also

$$x = \bar{x} + x_{,};\ y = \bar{y} + y_{,};\ z = \bar{z} + z_{,}.$$
$$x' = \bar{x} + x_{,}';\ y' = y + y_{,}';\ z' = \bar{z} + z_{,}';$$
&c.

$\bar{x}$, $\bar{y}$, $\bar{z}$ being the coordinates of the center of gravity of the system of bodies μ, μ', μ'', &c.; the origin of these and of the coordinates x, y, z; x', y', z', &c. being at the center of M. It is evident that $x_{,}$, $y_{,}$, $z_{,}$; $x_{,}'$, $y_{,}'$, $z_{,}'$, &c. are the coordinates of μ, μ', &c. relatively to their center of gravity; we will suppose these, compared with $\bar{x}$, $\bar{y}$, $\bar{z}$, as small quantities of the first order. This being done, we shall have, as we know by Mechanics, the force which sollicits the center of gravity of the system parallel to any straight line, by taking the sum of the forces which act upon the bodies parallel to the given straight line, multiplied respectively by their masses, and by dividing this sum by the sum of the masses. We also know (by Mech.) that the mutual action of the bodies upon one another, does not alter the motion of the center of gravity of the system; nor does their mutual attraction. It is sufficient, therefore, in estimating the forces which animate the center of gravity of a system, merely to regard the action of the body M which forms no part of the system.

The action of M upon μ, resolved parallel to the axis of x is

$$-\frac{M x}{\rho^3};$$

the whole force which sollicits the center of gravity parallel to this straight line is, therefore,

$$-\frac{M.\Sigma.\frac{\mu x}{\varrho^3}}{\Sigma \mu}.$$

Substituting for x and ϱ their values

$$\frac{x}{\varrho^3} = \frac{\bar{x} + x_{,}}{\{(\bar{x} + x_{,})^2 + (\bar{y} + y_{,})^2 + (\bar{z} + z_{,})^2\}^{\frac{3}{2}}}.$$

If we neglect small quantities of the second order, scil. the squares and products of

$$x_{,}, y_{,}, z_{,}; x_{,}', y_{,}', z_{,}'; \&c.$$

and put

$$\bar{\varrho} = \sqrt{(\bar{x}^2 + \bar{y}^2 + \bar{z}^2)}$$

the distance of the center of gravity from M, we have

$$\frac{x}{\varrho^3} = \frac{\bar{x}}{\bar{\varrho}^3} + \frac{x_{,}}{\bar{\varrho}^3} - \frac{3\,x\,(x\,x_{,} + \bar{y}\,y_{,} + z\,z_{,})}{\bar{\varrho}^3}$$

for omitting x^2, y^2 &c., we have

$$\frac{x}{\varrho^3} = (\bar{x} + x_{,}) \times \{(\bar{\varrho})^2 + 2\,(\bar{x}\,x_{,} + \bar{y}\,y_{,} + \bar{z}\,z_{,})\}^{-\frac{3}{2}} \text{ nearly}$$

$$= (\bar{x}+x_{,}) \times \{(\bar{\varrho})^{-3} - 3\,(\bar{\varrho})^{-5}\,(\bar{x}\,x_{,} + y\,y_{,} + \bar{z}\,z_{,}\} \text{ nearly}$$

$$= \frac{\bar{x} + x_{,}}{(\bar{\varrho})^3} - \frac{3\,\bar{x}}{(\bar{\varrho})^5}.(\bar{x}\,x_{,} + \bar{y}\,y_{,} + \bar{z}\,z_{,}) \text{ nearly.}$$

Again, marking successively the letters $x_{,}$, $y_{,}$, $z_{,}$, with one, two, three, &c. dashes or accents, we shall have the values of

$$\frac{x'}{\varrho'^3}, \frac{x''}{\varrho''^3}, \&c.$$

But from the nature of the center of gravity

$$\Sigma.\mu x = 0,\ \Sigma.\mu y = 0,\ \Sigma.\mu z = 0$$

we shall therefore have

$$-\frac{M.\Sigma.\frac{\mu x}{\varrho^3}}{\Sigma \mu} = -\frac{M\bar{x}}{\bar{\varrho}^3} \text{ nearly.}$$

Thus the center of gravity of the system is sollicited parallel to the axis of x, by the action of the body M, very nearly as if all the bodies of the system were collected into one at the center. The same result evidently takes place relatively to the axes of y and z; so that the forces, by which the center of gravity of the system is animated parallel to these axes, by the action of M, are respectively

$$-\frac{M\bar{y}}{(\bar{\varrho})^3} \text{ and } -\frac{M\bar{z}}{(\bar{\varrho})^3}.$$

When we consider the relative motion of the center of gravity of the system about M, the direction of the force which sollicits M must be changed. This force resulting from the action of μ, μ', &c. upon M, and resolved parallel to x, in the contrary direction from the origin, is

$$\Sigma.\frac{\mu x}{\varrho^3};$$

if we neglect small quantities of the second order, this function becomes, after what has been shown, equal to

$$\frac{\bar{x}\,\Sigma.\mu}{\bar{\varrho}^3}.$$

In like manner, the forces by which M is actuated arising from the system, parallel to the axes of y, and of z, in the contrary direction, are

$$\frac{\bar{y}\,\Sigma.\mu}{(\bar{\varrho})^3}, \text{ and } \frac{\bar{z}\,\Sigma.\mu}{(\bar{\varrho})^3}.$$

It is thus perceptible, that the action of the system upon the body M, is very nearly the same as if all the bodies were collected at their common center of gravity. Transferring to this center, and with a contrary sign, the three preceding forces; this point will be sollicited parallel to the axes of x, y and z, in its relative motion about M, by the three following forces, scil.

$$-(M+\Sigma\mu)\frac{\bar{x}}{(\bar{\varrho})^3},\ -(M+\Sigma\mu)\frac{\bar{y}}{(\bar{\varrho})^3},\ -(M+\Sigma\mu)\frac{\bar{z}}{(\bar{\varrho})^3}.$$

These forces are the same as if all the bodies μ, μ', μ'', &c. were collected at their common center of gravity; *which center, therefore, moves nearly (to small quantities of the second order) as if all the bodies were collected at that center.*

Hence it follows, that if there are many systems, whose centers of gravity are very distant from each other, relatively to the respective distances of the bodies of each system; these centers will be moved very nearly, as if the bodies of each system were there collected; for the action of the first system upon each body of the second system, is the same very nearly as if the bodies of the first system were collected at their common center of gravity; the action of the first system upon the center of gravity of the second, will be therefore, by what has preceded, the same as on this hypothesis; whence we may conclude generally *that the reciprocal action of different systems upon their respective centers of gravity, is the same as if all the bodies of each system were there collected; and also that these centers move as on that supposition.*

It is clear that this result subsists equally, whether the bodies of each system be free, or connected together in any way whatever; for their mutual action has no influence upon the motion of their common center of gravity.

The system of a planet acts, therefore, upon the other bodies of the Solar system, very nearly the same as if the Planet and its Satellites, were collected at their common center of gravity; and this center itself is attracted by the different bodies of the Solar system, as it would be on that hypothesis.

Having given the equations of motion of a system of bodies submitted to their mutual attraction, it remains to integrate them by successive approximations. In the solar system, the celestial bodies move *nearly* as if they obeyed only the principal force which actuates them, and the perturbing forces are inconsiderable; we may, therefore, in a first approximation consider only the mutual action of two bodies, scil. that of a planet or of a comet and of the sun, in the theory of planets and comets; and the mutual action of a satellite and of its planet, in the theory of satellites. We shall begin by giving a rigorous determination of the motion of two attracting bodies: this first approximation will conduct us to a second in which we shall include the first powers of small quantities or the perturbing forces; next we shall consider the squares and products of these forces; and continuing the process, we shall determine the motions of the heavenly bodies with all the accuracy that observations will admit of.

FIRST APPROXIMATION.

473. We know already that a body attracted towards a fixed point, by a force varying reciprocally as the square of the distance, describes a conic section; or in the relative motion of the body μ, round M, this latter body being considered as fixed, we must transfer in a direction contrary to that of μ, the action of μ upon M; so that in this relative motion, μ is sollicited towards M, by a force equal to the sum of the masses M, and μ divided by the square of their distance. All this has been ascertained already. But the importance of the subject in the Theory of the system of the world, will be a sufficient excuse for representing it under another form.

First transform the variables x, y, z into others more commodious for astronomical purposes. ϱ being the distance of the centers of μ and M, call (v) the angle which the projection of ϱ upon the plane of x, y makes with the axis of x; and (θ) the inclination of ϱ to the same plane; we shall have

$$\left.\begin{array}{l} x = \varrho \cos.\theta \cos. v; \\ y = \varrho \cos.\theta \sin. v; \\ z = \varrho \sin.\theta. \end{array}\right\} \quad \ldots\ldots\ldots \quad (1)$$

Next putting

$$Q = \frac{M+\mu}{\varrho} - \Sigma.\frac{\mu'(x x' + y y' + z z')}{\varrho'^3} + \frac{\lambda}{\mu}$$

we have

$$\left(\frac{dQ}{dx}\right) = \frac{1}{\mu}\left(\frac{d\lambda}{dx}\right) - \frac{M+\mu}{\varrho^3} - \Sigma.\frac{\mu' x'}{\varrho'^3}$$

$$= \frac{1}{\mu}\left(\frac{d\lambda}{dx}\right) - \frac{M}{\varrho^3} - \Sigma.\frac{\mu x}{\varrho^3}.$$

Similarly

$$\left(\frac{dQ}{dy}\right) = \frac{1}{\mu}\left(\frac{d\lambda}{dy}\right) - \frac{M}{\varrho^3} - \Sigma.\frac{\mu y}{\varrho^3}$$

$$\left(\frac{dQ}{dz}\right) = \frac{1}{\mu}\left(\frac{d\lambda}{dz}\right) - \frac{M}{\varrho^3} - \Sigma.\frac{\mu z}{\varrho^3}.$$

Hence equations (1), (2), (3) of number 471, become

$$\frac{d^2x}{dt^2} = \left(\frac{dQ}{dx}\right);\ \frac{d^2y}{dt^2} = \left(\frac{dQ}{dy}\right);\ \frac{d^2z}{dt^2} = \left(\frac{dQ}{dz}\right).$$

Now multiplying the first of these equations by cos. θ. cos. v; the second by cos. θ. sin. v; the third by sin. θ, we get, by adding them

$$\frac{d^2\varrho}{dt^2} - \frac{\varrho\, dv^2}{dt^2}.\cos.^2\theta - \frac{\varrho\, d\theta^2}{dt^2} = \left(\frac{dQ}{d\varrho}\right) \quad \ldots\ldots \quad (2)$$

In like manner, multiplying the first of the above equations by $-\varrho \cos.\theta \times$ sin. v; the second by ϱ cos. θ cos. v and adding them, &c. we have

$$\frac{d\left(\varrho^2 \frac{dv}{dt}\cos.^2\theta\right)}{dt} = \left(\frac{dQ}{dv}\right) \quad \ldots\ldots\ldots \quad (3).$$

And lastly multiplying the first by $-\varrho$ sin. θ. cos. v; the second by $-\varrho$ sin. θ. cos. v and adding them to the third multiplied by cos. θ. we have

$$\varrho^2.\frac{d^2\theta}{dt^2} + \varrho^2\frac{dv^2}{dt^2}.\sin.\theta\cos.\theta + \frac{2\varrho\, d\varrho\, d\theta}{dt^2} = \left(\frac{dQ}{d\theta}\right) \quad \ldots \quad (4)$$

To render the equations (2), (3), (4), still better adapted for use, let

$$u = \frac{1}{\varrho\cos.\theta}$$

and

$$s = \tan.\,\theta$$

u being unity divided by the projection of the radius ϱ upon the plane of x, y; and s the tangent of the latitude of μ from that same plane.

If we multiply equation (3) by $\varrho^2\, d\,v \cos.^2\theta$ and integrate, we get

$$\left(\frac{d\,v}{u^2\,d\,t}\right)^2 = h^2 + 2\int\left(\frac{d\,Q}{d\,v}\right)\cdot\frac{d\,v}{u^2};$$

h being the arbitrary constant.

Hence

$$d\,t = \frac{d\,v}{u^2\sqrt{\left(h^2 + 2\int\left(\frac{d\,Q}{d\,v}\right)\cdot\frac{d\,v}{u^2}\right)}} \quad \ldots\ldots \quad (5)$$

If we add equation (2) multiplied by $-\cos.\,\theta$ to equation (4) multiplied by $\frac{\sin.\,\theta}{\varrho}$, we shall have

$$-\frac{d^2\frac{1}{u}}{d\,t^2} + \frac{1}{u}\cdot\frac{d\,v^2}{d\,t^2} = u^2\left(\frac{d\,Q}{d\,u}\right) + u\,s\left(\frac{d\,Q}{d\,s}\right)$$

whence

$$d\,.\left(\frac{d\,u}{u^2\,d\,t}\right) + \frac{d\,v^2}{u\,d\,t} = u^2\,d\,t\left\{\left(\frac{d\,Q}{d\,u}\right) + \frac{s}{u}\left(\frac{d\,Q}{d\,s}\right)\right\}.$$

Substituting for d t, its foregoing value, and making d v constant, we shall have

$$0 = \frac{d^2\,u}{d\,v^2} + u + \frac{\left(\frac{d\,Q}{d\,v}\right)\frac{d\,u}{u^2\,d\,v} - \left(\frac{d\,Q}{d\,u}\right) - \frac{s}{u}\left(\frac{d\,Q}{d\,s}\right)}{h^2 + 2\int\left(\frac{d\,Q}{d\,v}\right)\frac{d\,v}{u^2}} \quad \ldots \quad (6)$$

In the same way making d v constant, equation (4) will become

$$0 = \frac{d^2\,s}{d\,v^2} + s + \frac{\frac{d\,s}{d\,v}\left(\frac{d\,Q}{d\,v}\right) - u\,s\left(\frac{d\,Q}{d\,u}\right) - (1+s^2)\left(\frac{d\,Q}{d\,s}\right)}{u^2\left\{h^2 + 2\int\left(\frac{d\,Q}{d\,v}\right)\cdot\frac{d\,v}{u^2}\right\}} \quad \ldots \quad (7)$$

Now making $M + \mu = m$, we have (in this case)

$$Q = \frac{m}{\varrho} \text{ or } = \frac{m\,u}{\sqrt{(1+s^2)}}$$

and the equations (5), (6), (7) will become

$$\left.\begin{aligned} d\,t &= \frac{d\,v}{h\,.\,u^2}; \\ 0 &= \frac{d^2\,u}{d\,v^2} + u - \frac{\mu}{h^2\,(1+s^2)^{\frac{3}{2}}}; \\ 0 &= \frac{d^2\,s}{d\,v^2} + s. \end{aligned}\right\} \quad \ldots\ldots\ldots \quad (8)$$

(These equations may be more simply deduced directly 124 and Woodhouse's Phys. Astron.)

The area described during the element of time d t, by the projection of the radius-vector is $\frac{1}{2}\frac{dv}{u^2}$; the first of equations (8) show that this area is proportional to that element, and also that in a finite time it is proportional to the time.

Moreover integrating the last of them (by 122) or by multiplying by 2 d s, we get

$$s = \gamma \sin. (v - \theta) \quad \ldots\ldots\ldots\ldots \quad (9)$$

γ and θ being two arbitrary constants.

Finally, the second equation gives by integration

$$u = \frac{\mu}{h^2(1+\gamma^2)}\{\sqrt{1+s^2} + e \cos. (v - \varpi)\} = \frac{\sqrt{1+s^2}}{\varrho}; \ldots \quad (10)$$

e and ϖ being two new arbitraries.

Substituting for s in this expression, its value in terms of v, and then this expression in the equation

$$d t = \frac{d v}{h u^2};$$

the integral of this equation will give t in terms of v; thus we shall have v, u and s in functions of the time.

This process may be considerably simplified, by observing that the value of s indicates the orbit to lie wholly in one plane, the tangent of whose inclination to a fixed plane is γ, the longitude of the node θ being reckoned from the origin of the angle v. In referring, therefore, to this plane the motion of μ; we shall have

$$s = 0 \text{ and } \gamma = 0,$$

which give

$$u = \frac{1}{\varrho} = \frac{\mu}{h^2}\{1 + e \cos. (v - \varpi)\}.$$

This equation is that of an ellipse in which the origin of ϱ is at the focus:

$$\frac{h^2}{\mu(1 - e^2)}$$

is the semi-axis-major which we shall designate by a; e is the ratio of the excentricity to the semi-axis-major; and lastly ϖ is the longitude of the perihelion. The equation

$$d t = \frac{d v}{h u^2}$$

hence becomes

$$dt = \frac{a^{\frac{3}{2}}(1-e^2)^{\frac{3}{2}}}{\sqrt{\mu}} \times \frac{dv}{\{1+e\cos.(v-\varpi)\}^2}.$$

Develope the second member of this equation, in a series of the angle $v - \varpi$ and of its multiples. For that purpose, we will commence by developing

$$\frac{1}{1+e\cos.(v-\varpi)}$$

in a similar series. If we make

$$\lambda = \frac{e}{1+\sqrt{(1-e^2)}};$$

we shall have

$$\frac{1}{1+e\cos.(v-\varpi)} = \frac{1}{\sqrt{1-e^2}}\left\{\frac{1}{1+\lambda c^{(v-\varpi)\sqrt{-1}}} - \frac{\lambda . c^{-(v-\varpi)\sqrt{-1}}}{1+\lambda c^{-(v-\varpi)\sqrt{-1}}}\right\};$$

c being the number whose hyperbolic is unity. Developing the second member of this equation, in a series; namely the first term relatively to powers of $c^{(v-\varpi)\sqrt{-1}}$, and the second term relatively to powers of $c^{-(v-\varpi)\sqrt{-1}}$ and then substituting, instead of imaginary exponentials, their expressions in terms of sine and cosine; we shall find

$$\frac{1}{1+e\cos.(v-\varpi)} = \frac{1}{\sqrt{1-e^2}} \times$$

$$\{1 - 2\lambda\cos.(v-\varpi) + 2\lambda^2\cos.2(v-\varpi) - 2\lambda^3\cos.3(v-\varpi) + \&c.\}$$

Calling φ the second member of this equation, and making $q = \frac{1}{e}$; we shall have generally

$$\frac{1}{\{1+e\cos.(v-\varpi)\}^{m+1}} = \frac{\pm e^{-m-1} d^m . \left(\frac{\varphi}{q}\right)}{1.2.3\ldots\ldots m . dq^m},$$

for putting

$$\frac{\varphi}{q} = \frac{1}{q+R}$$

R being $= \cos.(v-\varpi)$

$$\frac{d.\left(\frac{\varphi}{q}\right)}{dq} = -\frac{1}{(q+R)^2}$$

$$\frac{d^2.\left(\frac{\varphi}{q}\right)}{dq^2} = \frac{2}{(q+R)^3}$$

&c. = &c.

$$\frac{d^m\left(\frac{\rho}{q}\right)}{d\,q^m} = \pm\frac{2.3\ldots.m}{(q+R)^{m+1}}$$

$$\therefore \frac{\pm d^m\left(\frac{\rho}{q}\right)}{d\,q^m} \times \frac{e^{-m-1}}{2.3\ldots m} = \frac{q^{m+1}}{(q+R)^{m+1}} = \rho^{m+1}$$

$$= \frac{1}{\{1+e\cos.(v-\varpi)\}^{m+1}}.$$

Hence it is easy to conclude that if we make

$$\frac{1}{\{1+e\cos.(v-\varpi)\}^2} = (1-e^2)^{-\frac{3}{2}} \times$$

$$\{1+E^{(1)}.\cos.(v-\varpi)+E^{(2)}.\cos.2(v-\varpi)+\&c.\}$$

we shall have generally whatever be the number (i)

$$E^{(i)} = \pm\frac{2e^i\{1+i\sqrt{1-e^2}\}}{(1+\sqrt{1-e^2})^i};$$

the signs $\pm$ being used according as i is even or odd; supposing therefore that $u = a^{-\frac{3}{2}}\sqrt{m}$, we have

$$n\,d\,t = d\,v\{1+E^{(1)}\cos.(v-\varpi)+E^{(2)}\cos.2(v-\varpi)+\&c.\}$$

and integrating

$$n\,t+\epsilon = v+E^{(1)}\sin.(v-\varpi)+\tfrac{1}{2}E^{(2)}\sin.2(v-\varpi)+\&c.$$

ϵ being an arbitrary constant. This expression for $n\,t+\epsilon$ is very convergent when the orbits are of small excentricity, such as are those of the Planets and of the Satellites; and by the Reversion of Series we can find v in terms of t: we shall proceed to this presently.

474. When the Planet comes again to the same point of its orbit, v is augmented by the circumference 2π; naming therefore T the time of the whole revolution, we have (see also 159)

$$T = \frac{2\pi}{n} = \frac{2\pi a^{\frac{3}{2}}}{\sqrt{m}}.$$

This could be obtained immediately from the expression

$$T = \frac{\int \rho^2 d\,v}{h}$$

$$= \frac{2\ \text{area of Ellipse}}{h} = \frac{2\pi a b}{h}$$

But by 157

$$h^2 = m\,a\,(1-e^2)$$

$$\therefore T = \frac{2\pi a^{\frac{3}{2}}}{\sqrt{m}}.$$

If we neglect the masses of the planets relatively to that of the sun we have

$$\sqrt{\mu} = \sqrt{M}$$

which will be the same for all the planets; T is therefore proportional in that hypothesis to $a^{\frac{3}{2}}$, and consequently the squares of the Periods are as the cubes of the major axes of the orbits. We see also that the same law holds with regard to the motion of the satellites around their planet, provided their masses are also deemed inconsiderable compared with that of the planet.

475. The equations of motion of the two bodies M and μ may also be integrated in this manner.

Resuming the equations (1), (2), (3), of 471, and putting $M+\mu=m$, we have for these two bodies

$$\left.\begin{aligned} 0 &= \frac{d^2 x}{d t^2} + \frac{m x}{\varrho^3} \\ 0 &= \frac{d^2 y}{d t^2} + \frac{m y}{\varrho^3} \\ 0 &= \frac{d^2 z}{d t^2} + \frac{m z}{\varrho^3} \end{aligned}\right\} \quad (0)$$

The integrals of these equations will give in functions of the time t, the three coordinates x, y, z of the body μ referred to the center of M; we shall then have (471) the coordinates ζ, Π, γ of the body M, referred to a fixed point by means of the equations

$$\left.\begin{aligned} \zeta &= a + b\,t - \frac{\mu x}{m}; \\ \Pi &= a' + b'\,t - \frac{\mu y}{m}; \\ \gamma &= a'' + b''\,t - \frac{\mu z}{m}. \end{aligned}\right\}$$

Lastly, we shall have the coordinates of μ, referred to the same fixed point, by adding x to ζ, y to Π, and z to γ: We shall also have the relative motion of the bodies M and μ, and their absolute motion in space.

476. To integrate the equations (0) we shall observe that if amongst the (n) variables $x^{(1)}$, $x^{(2)}$ . . . . . $x^{(n)}$ and the variable t, whose difference is supposed constant, a number n of equations of the following form

$$0 = \frac{d^i x^{(s)}}{d t^i} + A\,\frac{d^{i-1} x^{(s)}}{d t^{i-1}} + B\,\frac{d^{i-2} x^{(s)}}{d t^{i-2}} \ldots\ldots H \,.\, x^{(s)}$$

in which we suppose s successively equal to 1, 2, 3 . . . . . n; A, B . . . . . H being functions of the variables $x^{(1)}$, $x^{(2)}$, &c. and of t symmetrical with regard to the variables $x^{(1)}$, $x^{(2)}$, &c. that is to say, such that they remain the same, when we change any one of these variables to any other and reciprocally; suppose

$$x^{(1)} = a^{(1)} x^{(n-i+1)} + b^{(1)} x^{(x-i+2)} + \dots\dots h^{(1)} x^{(n)},$$
$$x^{(2)} = a^{(2)} x^{(n-i+1)} + b^{(2)} x^{(n-i+2)} + \dots\dots h^{(2)} x^{n}.$$

- - - - - - - - - - - - - - - - - -

$$x^{(n-i)} = a^{(n-i)} x^{(n-i+1)} + b^{(n-i)} x^{(n-i+2)} \dots\dots + h^{(n-i)} x.^{(n)}$$

$a^{(1)}$, $b^{(1)}$, ... $h^{(1)}$; $a^{(2)}$, $b^{(2)}$, &c. being the arbitraries of which the number is i (n — i). It is clear that these values satisfy the proposed system of equations: Moreover these equations are thereby reduced to i equations involving the i variables $x^{(n-i+1)} \dots\dots x^{(n)}$. Their integrals will introduce i^2 new arbitraries, which together with the i (n — i) preceding ones will form i n arbitraries which ought to give the integration of the equations proposed.

477. To apply the above Theorem to equations (0); we have

$$z = a x + b y$$

a and b being two arbitrary constants, this equation being that of a plane passing through the origin of coordinates; also the orbit of μ is wholly in one plane.

The equations (0) give

$$\left.\begin{array}{l} 0 = d\left(\varrho^3 \cdot \dfrac{d^2 x}{d t^2}\right) + m\, d x \\ 0 = d\left(\varrho^3 \cdot \dfrac{d^2 y}{d t^2}\right) + m\, d y \\ 0 = d\left(\varrho^3 \cdot \dfrac{d^2 z}{d t^2}\right) + m\, d z \end{array}\right\}; \quad (0')$$

Also since

$$\varrho^2 = x^2 + y^2 + z^2$$

and

$$\therefore \varrho\, d \varrho = x\, d x + y\, d y + z\, d z$$

and differentiating twice more, we have

$$\varrho\, d^3 \varrho + 3\, d \varrho\, d^2 \varrho = x\, d^3 x + y\, d^3 y + z\, d^3 z$$
$$+ 3\,(d x\, d^2 x + d y\, d^2 y + d z\, d^2 z),$$

and consequently

$$d \cdot \left(\varrho^3 \frac{d^2 \varrho}{d t^2}\right) = \varrho^2 \left\{x \frac{d^3 x}{d t^2} + y \frac{d^3 y}{d t^2} + z \frac{d^3 z}{d t^2}\right\}$$
$$+ 3 \varrho^2 \left\{d x \frac{d^2 x}{d t^2} + d y \frac{d^2 y}{d t^2} + d z \frac{d^2 z}{d t^2}\right\}.$$

Substituting in the second member of this equation for $d^3 x$, $d^3 y$, $d^3 z$

their values given by equations (0′), and for $d^2 x$, $d^2 y$, $d^2 z$ their values given by equations (0); we shall find

$$0 = d\left(\rho^3 \frac{d^2 \rho}{d t^2} + m\, d\, \rho\right).$$

If we compare this equation with equations (0′), we shall have in virtue of the preceding Theorem, by considering $\frac{d x}{d t}$, $\frac{d y}{d t}$, $\frac{d z}{d t}$, $\frac{d \rho}{d t}$, as so many particular variables $x^{(1)}$, $x^{(2)}$, $x^{(3)}$, $x^{(4)}$, and ρ as a function of the time t;

$$d \rho = \lambda\, d x + \gamma\, d y;$$

λ and γ being constants; and integrating

$$= \frac{h^2}{m} + \lambda x + \gamma y,$$

$\frac{h^2}{m}$ being a constant. This equation combined with

$$z = a x + b y; \quad \rho^2 = x^2 + y^2 + z^2$$

gives an equation of the second degree in terms of x, y, or in terms of x, z, or of y, z; whence it follows that the three projections of the curve described by μ about M, are lines of the second order, and therefore that the curve itself (lying in one plane) is a line of the second order or a conic section. It is easy to perceive from the nature of conic sections that, the radius-vector ρ being expressed by a linear function of x, y, the origin of x, y ought to be in the focus. But the equation

$$\rho = \frac{h^2}{m} + \lambda x + \gamma y$$

gives by means of equations (0)

$$0 = \frac{d^2 \rho}{d t^2} + \mu \frac{\left(\rho - \frac{h^2}{m}\right)}{\rho^3}.$$

Multiplying this by $d \rho$ and integrating we get

$$\rho^2 \frac{d \rho^2}{d t^2} - 2 m \rho + \frac{m \rho^2}{a'} + h^2 = 0,$$

a' being an arbitrary constant. Hence

$$d t = \frac{\rho\, d \rho}{m \sqrt{\left(2 \rho - \frac{\rho^2}{a'} - \frac{h^2}{m}\right)}};$$

which will give ρ in terms of t; and since x, y, z are given above in terms of ρ, we shall have the coordinates of μ in functions of the times.

478. We can obtain these results by the following method, which has the advantage of giving the arbitrary constants in terms of the coordinates x, y, z and of their first differences; which will presently be of great use to us.

Let V = constant, be an integral of the first order of equations (0), V being a function of x, y, z, $\frac{dx}{dt}$, $\frac{dy}{dt}$, $\frac{dz}{dt}$. Call the three last quantities x′, y′, z′. Then V = constant will , by taking the differential,

$$0 = \left(\frac{dV}{dx}\right) \cdot \frac{dx}{dt} + \left(\frac{dV}{dy}\right) \cdot \frac{dy}{dt} + \left(\frac{dV}{dz}\right) \cdot \frac{dz}{dt}$$

$$+ \left(\frac{dV}{dx'}\right) \cdot \frac{dx'}{dt} + \left(\frac{dV}{dy'}\right) \cdot \frac{dy'}{dt} + \left(\frac{dV}{dz'}\right) \cdot \frac{dz'}{dt}$$

But equations (0) give

$$\frac{dx'}{dt} = -\frac{mx}{\varrho^3}, \quad \frac{dy'}{dt} = -\frac{my}{\varrho^3}, \quad \frac{dz'}{dt} = -\frac{mz}{\varrho^3};$$

we have therefore the equation of Partial Differences

$$0 = x' \left(\frac{dV}{dx}\right) + y' \left(\frac{dV}{dy}\right) + z' \left(\frac{dV}{dz}\right)$$

$$- \frac{m}{\varrho^3} \left\{ x \left(\frac{dV}{dx'}\right) + y \left(\frac{dV}{dy'}\right) + z \left(\frac{dV}{dz'}\right) \quad . \quad . \quad . \quad \text{(I)}\right.$$

It is evident that every function of x, y, z, x′, y′, z′ which, when substituted for V in this equation, satisfies it, becomes, by putting it equal to an arbitrary constant, an integral of the first order of the equations (0).

Suppose

$$V = U + U' + U'' + \&c.$$

U being a function of x, y, z; U′ a function of x, y, z, x′, y′, z′ but of the first order relatively to x′, y′, z′; U″ a function of x, y, z, x′, y′, z′ and of the second order relatively to x′, y′, z′, and so on. Substitute this value of V in the equation (I) and compare separately 1. the terms without x′, y′, z′; 2. those which contain their first powers; 3. those involving their squares and products, and so on; and we shall have

$$0 = x \left(\frac{dU'}{dx'}\right) + y \left(\frac{dU'}{dy'}\right) + z \left(\frac{dU'}{dz'}\right);$$

$$x' \left(\frac{dU}{dx}\right) + y' \left(\frac{dU}{dy}\right) + z' \left(\frac{dU}{dz}\right) = \frac{m}{\varrho^3} \left\{ x \left(\frac{dU''}{dx'}\right) + y \left(\frac{dU''}{dy'}\right) + z \left(\frac{dU''}{dz'}\right) \right.$$

$$x' \left(\frac{dU'}{dx}\right) + y' \left(\frac{dU'}{dx}\right) + z' \left(\frac{dU'}{dz}\right) = \frac{m}{\varrho^3} \left\{ x \left(\frac{dU'''}{dx'}\right) + y \left(\frac{dU'''}{dy'}\right) + z \left(\frac{dU'''}{dz'}\right) \right\}$$

$$x' \left(\frac{dU''}{dx}\right) + y' \left(\frac{dU''}{dx}\right) + z' \left(\frac{dU''}{dz}\right) = \frac{m}{\varrho^3} \left\{ x \left(\frac{dU''''}{dx'}\right) + y \left(\frac{dU''''}{dy'}\right) + z \left(\frac{dU''''}{dz'}\right) \right\}$$

&c.

which four equations call (I′).

The integral of the first of them is

$$U' = \text{funct.} \{x y' - y x', x z' - z x', y z' - z y', x, y, z\}$$

The value of U' is linear with regard to x', y', z'; suppose it of this form

$$U' = A(x y' - y x') + B(x z' - z x') + C(y z' - z y');$$

A, B, C being arbitrary constants. Make

$$U''', \&c. = 0;$$

then the third of the equations (I') will become

$$0 = x'\left(\frac{d U'}{d x}\right) + y'\left(\frac{d U'}{d y}\right) + z'\left(\frac{d U'}{d z}\right)$$

The preceding value of U' satisfies also this equation.

Again, the fourth of the equations (I') becomes

$$0 = x'\left(\frac{d U''}{d x}\right) + y'\left(\frac{d U''}{d y}\right) + z'\left(\frac{d U''}{d z}\right);$$

of which the integral is

$$U'' = \text{funct.}\ \{x y' - y x', x z' - z x', y z' - z y', x', y', z'\}.$$

This function ought to satisfy the second of equations (I'), and the first member of this equation multiplied by d t is evidently equal to d U. The second member ought therefore to be an exact differential of a function of x, y, z; and it is easy to perceive that we shall satisfy at once this condition, the nature of the function U'', and the supposition that this function ought to be of the second order, by making

$$U'' = (D y' - E x').(x y' - y x') + (D z' - F x')(x z' - z x') + (E z' - F y')(y z' - z y') + G(x'^2 + y'^2 + z'^2);$$

D, E, F, G being arbitrary constants; and then ϱ being $= \sqrt{x^2+y^2+z^2}$, we have

$$U = -\frac{m}{\varrho}(D x + E y + F z + 2 G);$$

Thus we have the values of

$$U, U', U'';$$

and the equation V = constant will become

$$\text{const.} = -\frac{m}{\varrho}\{D x + E y + F z + 2 G\} + (A + D y' - E x')(x y' - y x') + (B + D z' - F x')(x z' - z x') + (C + E z' - F y')(y z' - z y') + G(x'^2 + y'^2 + z'^2).$$

This equation satisfies equation (I) and consequently the equations (0) whatever may be the arbitrary constants A, B, C, D, E, F, G. Supposing all these = 0, 1. except A, 2. except B, 3. except C, &c. and putting

$$\frac{d x}{d t}, \frac{d y}{d t}, \frac{d z}{d t} \text{ for } x', y', z',$$

we shall have the integrals

$$
(P)\left\{\begin{array}{l}
c=\frac{x\,d\,y-y\,d\,x}{d\,t},\ c'=\frac{x\,d\,z-z\,d\,x}{d\,t},\ c''=\frac{y\,d\,z-z\,d\,y}{d\,t}\\
0=f+x\left\{\frac{m}{\varrho}-\frac{d\,y^2+d\,z^2}{d\,t^2}\right\}+\frac{y\,d\,y.d\,x}{d\,t^2}+\frac{z\,d\,z.d\,x}{d\,t^2}\\
0=f'+y\left\{\frac{m}{\varrho}-\frac{d\,x^2+d\,z^2}{d\,t^2}\right\}+\frac{x\,d\,x.d\,y}{d\,t^2}+\frac{z\,d\,z.d\,y}{d\,t^2}\\
0=f''+z\left\{\frac{m}{\varrho}-\frac{d\,x^2+d\,y^2}{d\,t^2}\right\}+\frac{x\,d\,x.d\,z}{d\,t^2}+\frac{y\,d\,y.d\,z}{d\,t^2}\\
0=\frac{m}{a}-\frac{2\,m}{\varrho}+\frac{d\,x^2+d\,y^2+d\,z^2}{d\,t^2}
\end{array}\right.
$$

c, c', c'', f, f', f'' and a being arbitrary constants.

The equations (0) can have but six distinct integrals of the first order, by means of which, if we eliminate $d\,x$, $d\,y$, $d\,z$, we shall have the three variables x, y, z in functions of the time t; we must therefore have at least one of the seven integrals (P) contained in the six others. We also perceive *à priori*, that two of these integrals ought to enter into the five others. In fact, since it is the element only of the time which enters these integrals, they cannot give the variables x, y, z in functions of the time, and therefore are insufficient to determine completely the motion of μ about M. Let us examine how it is that these integrals make but five distinct integrals.

If we multiply the fourth of the equations (P) by $\frac{z\,d\,y-y\,d\,z}{d\,t}$, and add the product to the fifth multiplied by $\frac{x\,d\,z-z\,d\,x}{d\,t}$, we shall have

$$
0=f.\frac{z\,d\,y-y\,d\,z}{d\,t}+f'.\frac{x\,d\,z-z\,d\,x}{d\,t}+z.\frac{x\,d\,y-y\,d\,x}{d\,t}\left\{\frac{m}{\varrho}-\frac{d\,x^2+d\,y^2}{d\,t^2}\right\}
+\frac{x\,d\,y-y\,d\,x}{d\,t}\left\{\frac{x\,d\,x.d\,z}{d\,t^2}+\frac{y\,d\,y.d\,z}{d\,t^2}\right\}.
$$

Substituting for $\frac{x\,d\,y-y\,d\,x}{d\,t}$, $\frac{x\,d\,z-z\,d\,x}{d\,t}$, $\frac{y\,d\,z-z\,d\,y}{d\,t}$, their values given by the three first of the equations (P), we shall have

$$
0=\frac{f'\,c'-f\,c''}{c}+z\left\{\frac{m}{\varrho}-\frac{d\,x^2+d\,y^2}{d\,t^2}\right\}+\frac{x\,d\,x.d\,z}{d\,t^2}+\frac{y\,d\,y.d\,z}{d\,t^2}.
$$

This equation enters into the sixth of the integrals P, by making $f''=\frac{f'\,c'-f\,c''}{c}$ or $0=f\,c''-f'\,c'+f''\,c$. Also the sixth of these integrals results from the five first, and the six arbitraries c, c', c'', f, f', f'' are connected by the preceding equation.

If we take the squares of f, f′, f″ given by the equations (P), then add them together, and make $f^2 + f'^2 + f''^2 = l^2$, we shall have

$$l^2 - m^2 = \left\{ \varrho^2 . \frac{dx^2 + dy^2 + dz^2}{dt^2} - \left(\frac{\varrho d\varrho}{dt}\right)^2 \right\} . \left\{ \frac{dx^2 + dy^2 + dz^2}{dt^2} - \frac{2m}{\varrho} \right\};$$

but if we square the values of c, c′, c″, given by the same equations, and make $c^2 + c'^2 + c''^2 = h^2$; we get

$$\varrho^2 . \frac{dx^2 + dy^2 + dz^2}{dt^2} - \left(\frac{\varrho d\varrho}{dt}\right)^2 = h^2;$$

the equation above thus becomes

$$0 = \frac{dx^2 + dy^2 + dz^2}{dt^2} - \frac{2m}{\varrho} + \frac{m^2 - l^2}{h^2}.$$

Comparing this equation with the last of equations (P), we shall have the equation of condition,

$$\frac{m^2 - l^2}{h^2} = \frac{m}{a}.$$

The last of equations (P) consequently enters the six first, which are themselves equivalent only to five distinct integrals, the seven arbitrary constants, c, c′, c″, f, f′, f″, and a being connected by the two preceding equations of condition. Whence it results that we shall have the most general expression of V, which will satisfy equation (I) by taking for this expression an arbitrary function of the values of c, c′, c″, f, and f′, given by the five first of the equations (P).

479. Although these integrals are insufficient for the determination of x, y, z in functions of the time; yet they determine the nature of the curve described by μ about M. In fact, if we multiply the first of the equations (P) by z, the second by — y, and the third by x, and add the results, we shall have

$$0 = cz - c'y + c''x,$$

the equation to a plane whose position depends upon the constants c, c′, c″.

If we multiply the fourth of the equations (P) by x, the fifth by y, and the sixth by z, we shall have

$$0 = fx + f'y + f''z + m\varrho - \varrho^2 . \frac{dx^2 + dy^2 + dz^2}{dt^2} + \frac{\varrho^2 d\varrho^2}{dt^2};$$

but by the preceding number

$$\varrho^2 . \frac{dx^2 + dy^2 + dz^2}{dt^2} - \frac{\varrho^2 d\varrho^2}{dt^2} = h^2;$$

$$\therefore 0 = m\varrho - h^2 + fx + f'y + f''z.$$

This equation combined with

$$0 = c''x - c'y + cz$$

and

$$\rho^2 = x^2 + y^2 + z^2$$

gives the equation to conic sections, the origin of ρ being at the focus. The planets and comets describe therefore round the sun very nearly conic sections, the sun being in one of the foci; and these stars so move that their radius-vectors describe areas proportional to the times. In fact, if d v denote the elemental angle included by ρ, $\rho + d\rho$, we have

$$dx^2 + dy^2 + dz^2 = \rho^2 dv^2 + d\rho^2$$

and the equation

$$\rho^2 . \frac{dx^2 + dy^2 + dz^2}{dt^2} - \frac{\rho^2 d\rho^2}{dt^2} = h^2$$

becomes

$$\rho^4 dv^2 = h^2 dt^2;$$

$$\therefore dv = \frac{h\,dt}{\rho^2}.$$

Hence we see that the elemental area $\frac{1}{2}\rho^2 dv$, described by ρ, is proportional to the element of time d t; and the area described in a finite time is therefore also proportional to that time. We see also that the angular motion of μ about M, is at every point of the orbit, as $\frac{1}{\rho^2}$; and *since without sensible error we may take very short times for those indefinitely small, we shall have, by means of the above equation, the horary motions of the planets and comets, in the different points of their orbits.*

The elements of the section described by μ, are the arbitrary constants of its motion; these are functions of the arbitraries c, c′, c″, f, f′, f″, and $\frac{m}{a}$. Let us determine these functions.

Let θ be the angle which the intersection of the planes of the orbit and of (x, y) makes with the axis of x, this intersection being called the *line of the nodes;* also let φ be the inclination of the planes. If x′, y′ be the coordinates of μ referred to the line of the nodes as the axis of abscissas, then we have

$$x' = x \cos.\theta + y \sin.\theta$$
$$y' = y \cos.\theta - x \sin.\theta.$$

Moreover

$$z = y' \tan.\varphi$$
$$\therefore z = y \cos.\theta \tan.\varphi - x \sin.\theta \tan.\varphi.$$

Comparing this equation with the following one

$$0 = c'' x - c' y + c z$$

we shall have

$$c' = c \cos. \theta . \tan. \varphi$$
$$c'' = c \sin. \theta \tan. \varphi$$

whence

$$\tan. \theta = \frac{c''}{c'}$$

and

$$\tan. \varphi = \frac{\sqrt{(c'^2 + c''^2)}}{c}.$$

Thus are determined the position of the nodes and the inclination of the orbit, in functions of the arbitrary constants c, c', c''.

At the perihelion, we have

$$\varrho \, d \varrho = 0, \text{ or } x \, d x + y \, d y + z \, d z = 0.$$

Let X, Y, Z be the coordinates of the planet at this point; the fourth and the fifth of the equations (P) will give

$$\frac{Y}{X} = \frac{f'}{f}.$$

But if I be called the longitude of the projection of the perihelion upon the plane of x, y this longitude being reckoned from the axis of x, we have

$$\frac{Y}{X} = \tan. I;$$

$$\therefore \tan. I = \frac{f'}{f},$$

which determines the position of the major axis of the conic section.

If from the equation

$$\varrho^2 . \frac{d x^2 + d y^2 + d z^2}{d t^2} - \frac{\varrho^2 d \varrho^2}{d t^2} = h^2$$

we eliminate $\frac{d x^2 + d y^2 + d z^2}{d t^2}$, by means of the last of the equations (P), we shall have

$$2 m \varrho - \frac{m \varrho^2}{a} - \frac{\varrho^2 d \varrho^2}{d t^2} = h^2$$

but d ϱ is 0 at the extremities of the axis major; we therefore have at these points

$$0 = \varrho^2 - 2 a \varrho + \frac{a h^2}{m}.$$

The sum of the two values of ϱ in this equation, is the axis major, and their difference is double the excentricity; thus a is the semi-axis major of the orbit, or the mean distance of μ from M; and

$$\sqrt{\left(1 - \frac{h^2}{m a}\right)}$$

is the ratio of the excentricity to the semi-axis major. Let

$$e = \sqrt{\left(1 - \frac{h^2}{m\,a}\right)}$$

and having by the above

$$\frac{m}{a} = \frac{m^2 - l^2}{h^2};$$

we shall get

$$m\,e = l.$$

Thus we know all the elements which determine the nature of the conic section and its position in space.

480. The three finite equations found above between x, y, z and ρ give x, y, z in functions of ρ; and to get these coordinates in functions of the time it is sufficient to obtain ρ in a similar function; which will require a new integration. For that purpose take the equation

$$2\,m\,\rho - \frac{m\,\rho^2}{a} - \frac{\rho^2\,d\,\rho^2}{d\,t^2} = h^2.$$

But we have above

$$h^2 = \frac{a}{m}\,(m^2 - l^2) = a\,m\,(1 - e^2);$$

$$\therefore d\,t = \frac{\rho\,d\,\rho}{\sqrt{m}\,\sqrt{\left\{2\,\rho - \frac{\rho^2}{a} - a\,(1 - e^2)\right\}}}.$$

whose integral (237) is

$$t + T = \frac{a^{\frac{3}{2}}}{\sqrt{m}}\,(u - e\,\sin.\,u) \quad . \; . \; . \; . \; . \; . \; . \; . \quad (S)$$

u being $= \cos.^{-1}\left(\frac{1}{e} - \frac{\rho}{a\,e}\right)$, and T an arbitrary constant.

This equation gives u and therefore ρ in terms of t; and since x, y, z are given in functions of ρ, we shall have the values of the coordinates for any instants whatever.

We have therefore completely integrated the equations (0) of 475, and thereby introduced the six arbitrary constants a, e, I, θ, φ, and T. The two first depend upon the nature of the orbit; the three next depend upon its position in space, and the last relates to the position of the body μ at any given epoch; or which amounts to the same, depends upon the instant of its passing the perihelion.

Referring the coordinates of the body μ, to such as are more commodious for astronomical uses, and for that, naming v the angle which the radius-vector makes with the major axis setting out from the perihelion, the equation to the ellipse is

$$\varrho = \frac{a\,(1 - e^2)}{1 + e\cos. v}.$$

The equation

$$\varrho = a\,(1 - e\cos. u)$$

indicates that u is 0 at the perihelion, so that this point is the origin of two angles u and v; and it is easy hence to conclude that the angle u is formed by the axis major, and by the radius drawn from its center to the point where the circumference described upon the axis major as a diameter, is met by the ordinate passing through the body μ at right angles to the axis major. Hence as in (237) we have

$$\tan. \frac{v}{2} = \sqrt{\frac{1+e}{1-e}}\,.\tan. \frac{u}{2}\,.$$

We therefore have (making T = 0, &c.)

$$\left.\begin{aligned} n\,t &= u - e\sin. u \\ \varrho &= a\,(1 - e\cos. u) \\ \text{and}\qquad \tan. \frac{v}{2} &= \sqrt{\frac{1+e}{1-e}}\,.\tan. \frac{u}{2} \end{aligned}\right\} \quad \dots\dots\dots \quad (f)$$

n t being the *Mean Anomaly*,

n the *Excentric Anomaly*,

v the *True Anomaly*.

The first of these equations gives u in terms of t, and the two others will give ϱ and v when u shall be determined. The equation between u and t is transcendental, and can only be resolved by approximation. Happily the circumstances attending the motions of the heavenly bodies present us with rapid approximations. In fact the orbits of the stars are either nearly circular or nearly parabolical, and in both cases, we can determine u in terms of t by series very convergent, which we now proceed to develope. For this purpose we shall give some general Theorems upon the reduction of functions into series, which will be found very useful hereafter.

481. Let u be any function whatever of α, which we propose to develope into a series proceeding by the powers of α. Representing this series by

$$u = u + \alpha\,.\,q_1 + \alpha^2.\,q_2 + \dots\dots \alpha^n.\,q_n + \alpha^{n+1}.\,q_{n+} + \&c.$$

u, q_1, q_2, &c. being quantities independent of α, it is evident that u is what u will become when we suppose $\alpha = 0$; and that whatever n may be

$$\left(\frac{d^n u}{d\alpha^2}\right) = 1.2\ldots\ldots n.q_n + 2.3\ldots\ldots(n+1).\alpha.q_{n+1} + \&c.$$

the difference $\left(\frac{d^n u}{dx^n}\right)$ being taken on the supposition that every thing in u varies with α. Hence if we suppose after the differentiations, that $\alpha = 0$, in the expression $\left(\frac{d^n u}{d\alpha^n}\right)$ we have

$$q_n = \left(\frac{d^n u}{da^n}\right) \times \frac{1}{1.2\ldots\ldots n}.$$

This is Maclaurin's Theorem (see 32) for one variable.

Again, if u be a function of two quantities α, α', let it be put

$$\begin{aligned} u = u + \alpha.q_{1,0} + \alpha^2.q_{2,0} &+ \&c. \\ + \alpha'.q_{0,1} + \alpha\alpha'.q_{1,1} &+ \&c. \\ + \alpha'^2.q_{0,2} &+ \&c. \end{aligned}$$

the general term being

$$\alpha^n \alpha'^{n'} q_{n,n'}.$$

Then if generally

$$\left(\frac{d^{n+n'} u}{d\alpha^n.d\alpha'^{n'}}\right)$$

denotes the $(n + n')^{th}$ difference of u, the operation being performed (n) times, on the supposition that α is the only variable, and then n′ times on that of α' being the only variable, we have

$$\begin{aligned} \left(\frac{du}{d\alpha}\right) = q_{1,0} + 2\alpha.q_{2,0} + 3\alpha^2.q_{3,0} + 4\alpha^3 q_{4,0} + 5\alpha^4.q_{5,0} + \&c. \\ + \alpha' q_{1,1} + 2\alpha.\alpha' q_{2,1} + 3\alpha^2\alpha' q_{3,1} + 4\alpha^3\alpha' q_{4,1} + \&c. \\ + \alpha'^2 q_{1,2} + 2\alpha\alpha'^2 q_{2,2} + 3\alpha^2\alpha'^2 q_{3,2} + \&c. \\ + \alpha'^3 q_{1,3} + 2\alpha\alpha'^3 q_{2,3} + \&c. \\ + \alpha'^4 q_{1,4} + \&c. \end{aligned}$$

$$\begin{aligned} \left(\frac{d^2\alpha}{d\alpha^2}\right) = 2q_{2,0} + 3.2\alpha q_{3,0} + 4.3\alpha^2 q_{4,0} + 5.4\alpha^3 q_{5,0} + \&c. \\ + 2\alpha' q_{2,1} + 3.2\alpha\alpha' q_{3,1} + 4.3\alpha^2\alpha' q_{4,1} + \&c. \\ + 2\alpha'^2 q_{2,2} + 3.2\alpha\alpha'^2 q_{3,2} + \&c. \\ + 2\alpha'^3 q_{2,3} + \&c. \end{aligned}$$

$$\begin{aligned} \left(\frac{d^3\alpha}{d\alpha^2.d\alpha'}\right) = 2q_{2,1} + 3.2\alpha q_{3,1} + \&c. \\ + 2\alpha' q_{2,2} + \&c. \end{aligned}$$

and continuing the process it will be found that

$$\left(\frac{d^{n+n'} u}{da^n.d\alpha'^n}\right) = 2.3\ldots\ldots n \times 2.3\ldots\ldots n' \times q_{n,n'} + \left\{\begin{matrix} Q \times \alpha \\ Q' \times \alpha' \end{matrix}\right\}$$

so that when α, $\acute{\alpha}$ both equal 0, we have

$$q_{n, n'} = \frac{\left(\frac{d^{n+n'} u}{d\alpha^n . d\alpha'^{n'}}\right)}{2.3\ldots . n \times 2.3\ldots . n'} \quad \ldots\ldots\ldots\ldots (1)$$

And generally, if u be a function of α, $\acute{\alpha}$, α'', &c. and in developing it into a series, if the coefficient of $\alpha^n . \acute{\alpha}^{n'} . \alpha''^{n''}$. &c. be denoted by $q_{n, n', n'', \&c.}$ we shall have, in making α, $\acute{\alpha}$, α'', &c. all equal 0,

$$q_{n, n', n'', \&c.} = \frac{\left(\frac{d^{n+n'+n''+\&c.} u}{d\alpha^n . d\acute{\alpha}^{n'} . d\alpha''^{n''}, \&c.}\right)}{2.3\ldots . n \times 2.3\ldots . n' \times 2.3\ldots . n'' \times \&c.} \ldots (2)$$

This is Maclaurin's Theorem made general.

482. Again let u be any function of $t + \alpha$, $t' + \acute{\alpha}$, $t'' + \alpha''$, &c. and put

$$u = \varphi (t + a, t' + \acute{\alpha}, t'' + \alpha'', \&c.)$$

then since t and α are similarly involved it is evident that

$$\left(\frac{d^{n+n'+n''+\&c.} . u}{d\alpha^n . d\acute{\alpha}^{n'} . d\alpha''^{n''} \&c.}\right) = \left(\frac{d^{n+n'+n''+\&c.} . u}{dt^n . dt'^{n'} . dt''^{n''} . \&c.}\right)$$

and making

$$\alpha, \acute{\alpha}, \alpha'', \&c. = 0,$$

or

$$u = \varphi (t, t', t'', \&c.)$$

by (2) of the preceding article we have

$$q_{n, n', n'', \&c.} = \frac{\left(\frac{d^{n+n'+n'', \&c.} . \varphi (t, t', t'', \&c.)}{dt^n . dt'^{n'} . dt''^{n''} \&c}\right)}{2.3\ldots . n \times 2.3\ldots . n' \times 2.3\ldots . n'' \times \&c.} \ldots (1)$$

which gives Taylor's Theorem in all its generality (see 32).

Hence when

$$u = \varphi . (t + \alpha)$$

$$q_n = \frac{d^n . \varphi (t)}{2.3\ldots . n . dt^n}$$

and we thence get

$$\varphi (t + \alpha) = \varphi (t) + \alpha \frac{d . \varphi (t)}{dt} + \frac{\alpha^2}{2} . \frac{d^2 \varphi (t)}{dt^2} + \&c. \ldots\ldots (i)$$

483. Generally, suppose that u is a function of α, $\acute{\alpha}$, α'', &c. and of t, t', t'', &c. Then, if by the nature of the function or by an equation of Partial Differences which represents it, we can obtain

$$\left(\frac{d^{n+n'+\&c.} . u}{d\alpha^n . d\acute{\alpha}^{n'} . \&c.}\right)$$

in a function of u, and of its Differences taken with regard to t, t', &c.

calling it F when for u we put *u* or make α, α', α'', &c. $= 0$; it is evident we have

$$q_{n, n', n'', \&c.} = \frac{F}{2.3\ldots n \times 2.3\ldots n' \times 2.3\ldots n'', \times \&c.}$$

and therefore the law of the series into which u is developed.

For instance, let u, instead of being given immediately in terms of α, and t, be a function of x, x itself being deducible from the equation of Partial Differences

$$\left(\frac{dx}{d\alpha}\right) = X\left(\frac{dx}{dt}\right)$$

in which X is any function whatever of x. That is

Given

$$u = \text{function } (x)$$
$$\left(\frac{dx}{d\alpha}\right) = X.\left(\frac{dx}{dt}\right)$$

to develope u *into a series ascending by the powers of* α.

First, since d

$$\left(\frac{u}{d\alpha}\right) = \left(\frac{du}{dx}\right).\left(\frac{dx}{d\alpha}\right) = X\left(\frac{du}{dx}\right).\left(\frac{dx}{dt}\right)$$

$$\therefore \left(\frac{du}{d\alpha}\right) = \left(\frac{d\int X\,du}{dt}\right) \quad \ldots\ldots\ldots\ldots \quad (k)$$

Hence

$$\left(\frac{d^2u}{d\alpha^2}\right) = \left(\frac{d^2\int X\,du}{d\alpha.dt}\right);$$

But by equation (k), changing u into $\int X\,du$

$$\left(\frac{d.\int X\,du}{d\alpha}\right) = \left(\frac{d.\int X^2\,du}{dt}\right);$$

$$\therefore \left(\frac{d^2u}{d\alpha^2}\right) = \left(\frac{d^2\int X^2\,du}{dt^2}\right).$$

Again

$$\left(\frac{d^3u}{d\alpha^3}\right) = \left(\frac{d^3\int X^2\,du}{d\alpha.dt^2}\right).$$

But by equation k, and changing u into $\int X^2\,du$

$$\left(\frac{d\int X^2\,du}{d\alpha}\right) = \left(\frac{d\int X^3\,du}{dt}\right)$$

$$\therefore \left(\frac{d^3u}{d\alpha^3}\right) = \left(\frac{d^3.\int X^3\,du}{dt^3}\right).$$

Thus proceeding we easily conclude generally that

$$\left(\frac{d^nu}{d\alpha^n}\right) = \left(\frac{d^n\int X^n\,du}{dt^n}\right) = \left(\frac{d^{n-1}.X^n\left(\frac{du}{dt}\right)}{dt^{n-1}}\right) \quad \ldots \quad (l)$$

Now, when $\alpha = 0$, let

$$x = \text{function of } t = T$$

and substitute this value of x in X and u; and let these then become X and u respectively. Then we shall have

$$\left(\frac{d^n u}{d \alpha^n}\right) = \frac{d^{n-1}. X^n. \frac{d u}{d t}}{d t^{n-1}},$$

and

$$\therefore q_n = \frac{d^{n-1}. X^n. \frac{d u}{d t}}{2.3 \ldots . n d t^{n-1}} \quad \ldots\ldots\ldots\ldots\ldots \quad (2)$$

which gives

$$u = u + \alpha X . \frac{d u}{d t} + \frac{a^2}{2} . \frac{d\left(X^2 . \frac{d u}{d t}\right)}{d t} + \frac{\alpha^3}{2.3} . \frac{d^2\left(X^3 . \frac{d u}{d t}\right)}{d t^2} + \&c. \ldots (p)$$

which is Lagrange's Theorem.

To determine the value of x in terms of t and α, we must integrate

$$\left(\frac{d x}{d \alpha}\right) = X . \left(\frac{d x}{d t}\right).$$

In order to accomplish this object, we have

$$d x = \left(\frac{d x}{d t}\right) d t + \left(\frac{d x}{d \alpha}\right) d \alpha,$$

and substituting

$$\left(\frac{d x}{d \alpha}\right) \text{ for } X . \left(\frac{d x}{d t}\right)$$

we shall have

$$d x = \left(\frac{d x}{d t}\right) \{d t + X d \alpha\}$$

$$= \frac{d x}{d t}\left\{d (t + \alpha X) - \alpha\left(\frac{d X}{d x}\right) d x\right\};$$

$$\therefore d x = \frac{\left(\frac{d x}{d t}\right) . d . (t + \alpha X)}{1 + \alpha\left(\frac{d X}{d x}\right) . \left(\frac{d x}{d t}\right)};$$

which by integration, gives

$$x = \varphi (t + \alpha X) \quad \ldots\ldots\ldots\ldots \quad (2)$$

φ denoting an arbitrary function.

Hence whenever we have an equation reducible to this form $x = \varphi (t + \alpha X)$, the value of u will be given by the formula (p), in a series of the powers of α.

By an extension of the process, the Theorem may be generalized to the case, when

$$u = \text{function } (x, x', x'', \&c.)$$

and

$$x = \varphi (t + \alpha X)$$
$$x' = \varphi' (t' + \alpha' X')$$
$$x'' = \varphi'' (t'' + \alpha'' X'')$$
$$\&c. = \&c.$$

484. *Given* (237)

$$u = n t + e \sin. u$$

required to develope u *or any function of it according to the powers of* e.

Comparing the above form with

$$x = \varphi (t + \alpha X)$$

x, t, α, X become respectively

u, n t, e, sin. u.

Hence the formula (p) 483. gives

$$\psi (u) = \psi (n t) + e \psi' (n t) \sin. n t + \frac{e^2}{2} \cdot \frac{d \{\psi' (n t) \sin.^2 n t\}}{n d t}$$
$$+ \frac{e^3}{2.3} \cdot \frac{d^2 \{\psi' (n t) \sin.^3 n t\}}{n^2 d t^2} + \&c. \quad . \quad . \quad . \quad . \quad . \quad . \quad . \quad . \quad (q)$$
$$\psi' (n t) \text{ being} = \frac{d . \psi (n t)}{n d t}.$$

To farther develope this formula we have generally (see Woodhouse's Trig.)

$$\sin.^i (n t) = \left(\frac{c^{n t \sqrt{-1}} - c^{-n t \sqrt{-1}}}{2 \sqrt{-1}}\right)^i ; \cos.^i (n t) = \left(\frac{c^{n t \sqrt{-1}} + c^{-n t \sqrt{-1}}}{2}\right)^i;$$

c being the hyperbolic base, and i any number whatever. Developing the second members of these equations, and then substituting

$$\cos. r n t + \sqrt{-1} \sin. r n t, \text{ and } \cos. r n t - \sqrt{-1} \sin. r n t$$

for $c^{r n t \sqrt{-1}}$, and $c^{-r n t \sqrt{-1}}$, r being any number whatever, we shall have the powers i of sin. n t, and of cos. n t expressed in sines and cosines of n t and its multiples; hence we find

$$P = \sin. n t + \frac{e}{2} \sin^2 n t + \frac{e^2}{2.3} \sin.^3 n t + \&c.$$
$$= \sin. n t - \frac{e}{2.2} . \{\cos. 2 n t - 1\}$$
$$- \frac{e^2}{2.3.2^2} . \{\sin. 3 n t - 3 \sin. n t\}$$
$$+ \frac{e^3}{2.3.4.2^3} . \left\{\cos. 4 n t - 4 \cos. 2 n t + \frac{1}{2} . \frac{4.3}{1.2}\right\}$$
$$+ \frac{e^4}{2.3.4.5.2^4} \left\{\sin. 5 n t - 5 \sin. 3 n t + \frac{5.4}{1.2} \sin. n t\right\}$$

$$-\frac{e^5}{2.3.4.5.6.2^5}.\left\{\cos.6nt-6\cos.4nt+\frac{6.5}{1.2}\cos.2nt-\frac{1}{2}.\frac{6.5.4}{1.2.3}\right\}$$

— &c.

Now multiply this function by ψ' (n t), and differentiate each of its terms relatively to t a number of times indicated by the power of e which multiplies it, d t being supposed constant; and divide these differentials by the corresponding power of n d t. Then if P′ be the sum of the quotients, the formula (q) will become

$$\psi\,(u) = \psi\,(n\,t) + e\,P'.$$

By this method it is easy to obtain the values of the angle u, and of the sine and cosine of its multiples. Supposing for example, that

$$\psi\,u = \sin.\,i\,u$$

we have

$$\psi'\,(n\,t) = i\cos.\,i\,n\,t.$$

Multiply therefore the preceding value of P, by i. cos. i n t, and develope the product into sines and cosines of n t and its multiples. The terms multiplied by the even powers of e, are sines, and those multiplied by the odd powers of e, are cosines. We change therefore any term of the form, $K\,e^{2r}\sin.\,s\,n\,t$, into $\pm K\,e^{2r}\,s^{2r}\sin.\,s\,n\,t$, + or — obtaining according as r is even or odd. In like manner, we change any term of the form, $K\,e^{2r+1}\cos.\,s\,n\,t$, into $\mp K\,e^{2r+1}.\,s^{2r+1}.\sin.\,s\,n\,t$, — or + obtaining according as r is even or odd. The sum of all these terms will be P′ and we shall have

$$\sin.\,i\,u = \sin.\,i\,n\,t + e\,P'.$$

But if we suppose

$$\psi\,(u) = u;$$

then

$$\psi\,(n\,t) = 1$$

and we find by the same process

$$u = n\,t + e\sin.\,n\,t + \frac{e^2}{2.\,2}.\,2\sin.\,2\,n\,t$$

$$+\frac{e^3}{2.\,3.\,2^2}.\{3^2\sin.\,3\,n\,t - 3\sin.\,n\,t\}$$

$$+\frac{e^4}{2.\,3.\,4.\,2^3}.\{4^3\sin.\,4\,n\,t - 4\,.\,2^3\sin.\,2\,n\,t\}$$

$$+\frac{e^5}{2.\,3.\,4.\,5.\,2^4}.\left\{5^4\sin.\,5\,n\,t - 5.\,3^4\sin.\,3\,n\,t + \frac{5.\,4}{1.\,2}\sin.\,n\,t\right\}$$

+ &c.

a formula which expresses the Excentric Anomaly in terms of the Mean Anomaly.

This series is very convergent for the Planets. Having thus determined u for any instant, we could thence obtain by means of (237), the corresponding values of ϱ and v. But these may be found directly as follows, also in convergent series.

485. *Required to express ϱ in terms of the Mean Anomaly.*

By (237) we have

$$\varrho = a(1 - e \cos. u).$$

Therefore if in formula (q) we put

$$\psi(u) = 1 - e \cos. u$$

we have

$$\psi'(n t) = e \sin. n t,$$

and consequently

$$1 - e \cos. u = 1 - e \cos. n t + e^2 \sin.^2 n t + \frac{e^3}{2} \cdot \frac{d. \sin.^3 n t}{n d t} + \&c.$$

Hence, by the above process, we shall find

$$\frac{\varrho}{a} = 1 + \frac{e^2}{2} - e \cos. n t - \frac{e^2}{2} \cos. 2 n t$$

$$- \frac{e^3}{2. 2^2} . \{3 \cos. 3 n t - 3 \cos. n t\}$$

$$- \frac{e^4}{2. 3. 2^3} . \{4^2 \cos. 4 n t - 4. 2^2. \cos. 2 n t\}$$

$$- \frac{e^5}{2. 3. 4. 2^4} . \left\{5^3 \cos. 5 n t - 5. 3^3 \cos. 3 n t + \frac{5. 4}{1. 2} . \cos. n t\right\}$$

$$- \frac{e^6}{2.3.4.5.2^5} \left\{6^4 \cos. 6 n t - 6. 4^4 \cos. 4 n t + \frac{6.5}{1.2} . 2^4 \cos. 2 n t\right\}$$

$$- \&c.$$

486. *To express the True Anomaly in terms of the Mean.*

First we have (237)

$$\frac{\sin. \frac{v}{2}}{\cos. \frac{v}{2}} = \sqrt{\frac{1+e}{1-e}} . \frac{\sin. \frac{u}{2}}{\cos. \frac{u}{2}}$$

∴ substituting the imaginary expressions

$$\frac{c^{v\sqrt{-1}} - 1}{c^{v\sqrt{-1}} + 1} = \sqrt{\frac{1+e}{1-e}} \times \frac{c^{u\sqrt{-1}} - 1}{c^{u\sqrt{-1}} + 1};$$

and making

$$\lambda = \frac{e}{1 + \sqrt{(1 - e^2)}}$$

we shall have

$$c^{v\sqrt{-1}} = c^{u\sqrt{-1}} \times \frac{1 - \lambda\, c^{-u\sqrt{-1}}}{1 - \lambda\, c^{u\sqrt{-1}}};$$

and therefore

$$v = u + \frac{\log.(1 - \lambda . c^{-u\sqrt{-1}}) - \log.(1 - \lambda\, c^{u\sqrt{-1}})}{\sqrt{-1}}$$

whence expanding the logarithms into series (see p. 28), and putting sines and cosines for their imaginary values, we have

$$v = u + 2\,\lambda \sin. u + \frac{2\,\lambda^2}{2} \sin. 2\,u + \frac{2\,\lambda^3}{3} \sin. 3\,u + \&c.$$

But by the foregoing process we have u, sin. u, sin. 2 u, &c. in series ordered by the powers of e, and developed into sines and cosines of n t and its multiples. There is nothing else then to be done, in order to express v in a similar series, but to expand λ into a like series.

The equation, (putting $u = 1 + \sqrt{1 - e^2}$)

$$u = 2 - \frac{e^2}{u}$$

will give by the formula (p) of No. (483)

$$\frac{1}{u^i} = \frac{1}{2^i} + \frac{i.\, e^2}{2^{i+2}} + \frac{i\,(i+3)}{2} \cdot \frac{e^4}{2^{i+4}} + \frac{i\,(i+3)\,(i+5)}{2.\,3} \cdot \frac{e^6}{2^{i+6}} \&c.$$

and since

$$u = 1 + \sqrt{1 - e^2}$$

we have

$$\lambda^i = \frac{e^i}{2^i}\left\{1 + i\left(\frac{e}{2}\right)^2 + \frac{i\,(i+3)}{1.\,2} \cdot \left(\frac{e}{2}\right)^4 + \&c.\right\}$$

These operations being performed we shall find

$$\begin{aligned} v = n\,t &+ \left\{2\,e - \frac{1}{4}\,e^3 + \frac{5}{96}\,e^5\right\} \sin. n\,t \\ &+ \left\{\frac{5}{4}\,e^2 - \frac{11}{24}\,e^4 + \frac{17}{192}\,e^6\right\} \sin. 2\,n\,t \\ &+ \left\{\frac{13}{12}\,e^3 - \frac{43}{64}\,e^5\right\} \sin. 3\,n\,t \\ &+ \left\{\frac{103}{96}\,e^4 - \frac{451}{480}\,e^6\right\} \sin. 4\,n\,t \\ &+ \frac{1097}{960}\,e^5 \sin. 5\,n\,t \\ &+ \frac{1223}{960}\,e^6 \sin. 6\,n\,t, \end{aligned}$$

the approximation being carried on to quantities of the order e^6 inclusively.

487. The angles v and n t are here reckoned from the Perihelion; but if we wish to compute from the Aphelion, we have only to make e negative. It would, therefore, be sufficient to augment the angle n t by π, in order to render negative the sines and cosines of the odd multiples of n t; then to make the results of these two methods identical; we have only in the expressions for ϱ and v, to multiply the sines and cosines of odd multiples of n t by odd powers of e; and the even multiples by the even powers. This is confirmed, in fact, by the process, *a posteriori.*

488. Suppose that instead of reckoning v from the perihelion, we fix its origin at any point whatever; then it is evident that this angle will be augmented by a constant, which we shall call ϖ, and which will express the Longitude of the Perihelion. If instead of fixing the origin of t at the instant of the passage over the perihelion, we make it begin at any point, the angle n t will be augmented by a constant which we will call $\varepsilon - \varpi$; and then the foregoing expressions for $\frac{\varrho}{a}$ and v, will become

$$\frac{\varrho}{a} = 1 + \frac{1}{2}e^2 - (e - \frac{3}{8}e^3)\cos.(nt + \varepsilon - \varpi) - (\frac{1}{2}e^2 - \frac{1}{3}e^4)\cos.2(nt + \varepsilon - \varpi) + \&c.$$

$$v = nt + \varepsilon + (2e - \frac{1}{4}e^3)\sin.(nt + \varepsilon - \varpi) + (\frac{5}{4}e^2 - \frac{11}{24}e^4)\sin.2(nt + \varepsilon - \varpi) + \&c.$$

where v is the true longitude of the planet and n t + ε its mean longitude, these being measured on the plane of the orbit.

Let, however, the motion of the planet be referred to a fixed plane a little inclined to that of the orbit, and φ be the mutual inclination of the two planes, and θ the longitude of the Ascending Node of the orbit, measured upon the fixed plane; also let β be this longitude measured upon the plane of the orbit, so that θ is the projection of β, and lastly let $v_{,}$ be the projection of v upon the fixed plane. Then we shall have

$$v_{,} - \theta, \quad v - \beta,$$

making the two sides of a right angled spherical triangle, v — β being opposite the right angle, and φ the angle included between them, and therefore by Napier's Rules

$$\tan.(v_{,} - \theta) = \cos.\varphi \tan.(v - \beta) \quad . \quad . \quad . \quad . \quad . \quad . \quad . \quad (1)$$

This equation gives $v_{,}$ in terms of v and reciprocally; but we can express cither of them in terms of the other by a series very convergent after this manner.

By what has preceded, we have the series

$$\frac{1}{2}v = \frac{1}{2}u + \lambda \sin.u + \frac{\lambda^2}{2}\sin.2u + \frac{\lambda^3}{3}\sin.3u + \&c.$$

from

$$\tan. \frac{1}{2} v = \sqrt{\frac{1+e}{1-e}} . \tan. \frac{1}{2} u,$$

by making

$$\lambda = \frac{\sqrt{\frac{1+e}{1-e}} - 1}{\sqrt{\frac{1+e}{1-e}} + 1}.$$

If we change $\frac{1}{2}$ v into $v_{\prime} - \theta$, and $\frac{1}{2}$ u into $v - \beta$, and $\sqrt{\frac{1+e}{1-e}}$ into cos. φ, we have

$$\lambda = \frac{\cos. \varphi - 1}{\cos. \varphi + 1} = - \tan.^2 \frac{\varphi}{2}; \quad . \; . \; . \; . \; . \; . \; . \quad (1)$$

The equation between $\frac{1}{2}$ v and $\frac{1}{2}$ u will change into the equation between $v_{\prime} - \theta$ and $v - \beta$, and the above series will give

$$v_{\prime} - \theta = v - \beta - \tan^2 \frac{1}{2} \varphi . \sin. 2 (v - \beta) + \frac{1}{2} \tan.^4 \frac{1}{2} \varphi . \sin. 4 (v - \beta)$$
$$- \frac{1}{3} \tan.^6 \frac{1}{2} \varphi \sin. 6 (v - \beta) + \&c. \quad . \; . \; . \; . \; . \; . \; . \quad (2)$$

If in the equation between $\frac{v}{2}$ and $\frac{u}{2}$, we change $\frac{1}{2}$ v into $v - \beta$ and $\frac{1}{2}$ u into $v_{\prime} - \theta$, and $\sqrt{\frac{1+e}{1-e}}$ into $\frac{1}{\cos. \varphi}$, we shall have

$$\lambda = \tan.^2 \frac{1}{2} \varphi, \quad . \; . \; . \; . \; . \; . \; . \; . \; . \; . \; . \quad (3)$$

and

$$v - \beta = v_{\prime} - \theta + \tan.^2 \frac{1}{2} \varphi . \sin. 2 (v_{\prime} - \theta)$$
$$+ \frac{1}{2} \tan.^4 \frac{1}{2} \varphi . \sin. 4 (v_{\prime} - \theta)$$
$$+ \frac{1}{3} \tan.^6 \frac{1}{2} \varphi . \sin. 6 (v_{\prime} - \theta) \quad . \; . \; . \; . \; . \quad (4)$$

Thus we see that the two preceding series reciprocally interchange, by changing the sign of $\tan.^2 \frac{1}{2} \varphi$, and by changing $v_{\prime} - \theta$, $v - \beta$ the one for the other. We shall have $v_{\prime} - \theta$ in terms of the sine and cosine of n t and its multiples, by observing that we have, by what precedes

$$v = n t + \epsilon + e Q,$$

Q being a function of the sine of the angle $n t + \epsilon - \varpi$, and its multiples; and that the formula (i) of number (482) gives, whatever is i,

$$\sin. i (v - \beta) = \sin. i (n t + \epsilon - \beta + e Q)$$

$$= \left\{1 - \frac{i^2 e^2 Q^2}{1.2} + \frac{i^4 e^4 Q^4}{1.2.3.4.} - \&c.\right\} \sin. i (n t + \epsilon - \beta)$$

$$+ \left\{i e Q - \frac{i^3 e^3 Q}{1.2.3} + \frac{i^5 e^5. Q^5}{2.3.4.5.} - \&c.\right\} \cos. i (n t + \epsilon - \beta).$$

Lastly, s being the tangent of the latitude of the planet above the fixed plane, we have

$$s = \tan. \varphi \sin. (v_{,} - \theta);$$

and if we call $\varrho_{,}$ the radius-vector projected upon the fixed plane, we shall have

$$\varrho_{,} = \varrho (1 + s^2)^{-\frac{1}{2}} = \varrho \left\{1 - \tfrac{1}{2} s^2 + \frac{3}{8} s^4 - \&c.\right\},$$

we shall therefore be able to determine $v_{,}$, s and $\varrho_{,}$ in converging series of the sines and cosines of the angle n t and of its multiples.

489. Let us now consider very excentric orbits or such as are those of the Comets.

For this purpose resume the equations of No. (237), scil.

$$\varrho = \frac{a (1 - e^2)}{1 + e \cos. v}$$

$$n t = u - e \sin. u$$

$$\tan. \tfrac{1}{2} v = \sqrt{\frac{1 + e}{1 - e}} . \tan. \tfrac{1}{2} u.$$

In this case e differs very little from unity; we shall therefore suppose

$$1 - e = \alpha$$

α being very small compared with unity.

Calling D the perihelion distance of the Comet, we shall have

$$D = u (1 - e) = \alpha a;$$

and the expression for ϱ will become

$$\varrho = \frac{(2 - \alpha) D}{2 \cos.^2 \frac{1}{2} v - \alpha \cos. v} = \frac{D}{\cos.^2 \frac{1}{2} v \left\{1 + \frac{\alpha}{2 - \alpha} \tan.^2 \frac{1}{2} v\right\}},$$

which gives, by reduction into a series

$$\varrho = \frac{D}{\cos.^2 \frac{1}{2} v} \left\{1 - \frac{\alpha}{2 - \alpha} \tan.^2 \frac{1}{2} v + \left(\frac{\alpha}{2 - \alpha}\right)^2 \tan.^4 \frac{1}{2} v - \&c.\right\}$$

To get the relation of v to the time t, we shall observe that the expression of the arc in terms of the tangent gives

$$u = 2 \tan. \frac{1}{2} u \left\{1 - \frac{1}{3} \tan.^2 \frac{1}{2} u + \frac{1}{5} \tan.^4 \frac{1}{2} u - \&c.\right\}$$

But

$$\tan. \frac{1}{2} u = \sqrt{\frac{\alpha}{2 - \alpha}} \tan. \frac{1}{2} u;$$

we therefore have

$$u = 2\sqrt{\frac{\alpha}{2-\alpha}}\tan.\frac{1}{2}v\left\{1-\frac{1}{3}\left(\frac{\alpha}{2-\alpha}\right)\tan.^2\frac{1}{2}v+\frac{1}{5}\left(\frac{\alpha}{2-\alpha}\right)^2\tan.^4\frac{1}{2}v-\&c.\right\}$$

Next we have

$$\sin. u = \frac{2\tan.\frac{1}{2}u}{1+\tan.^2\frac{1}{2}u} = 2\tan.\frac{1}{2}u\left\{1-\tan.^2\frac{u}{2}+\tan.^4\frac{u}{2}-\&c.\right\}$$

Whence we get

$$e\sin. u = 2(1-\alpha)\sqrt{\frac{\alpha}{2-\alpha}}\tan.\frac{1}{2}v.\left\{1-\frac{\alpha}{2-\alpha}\tan.^2\frac{1}{2}v\right.$$
$$\left.+\left(\frac{\alpha}{2-\alpha}\right)^2.\tan.^4\frac{1}{2}v-\&c.\right\}.$$

Substituting these values of u, and e sin. u in the equation $n t = u - e \sin. u$, we shall have the time t in a function of the anomaly v, by a series very convergent; but before we make this substitution, we shall observe that (237)

$$n = a^{-\frac{3}{2}}.\sqrt{m},$$

and since

$$D = \alpha a,$$

we have

$$\frac{1}{n} = \frac{D^{\frac{3}{2}}}{\alpha^{\frac{3}{2}}\sqrt{m}}.$$

Hence we find

$$t = \frac{2D^{\frac{3}{2}}}{\sqrt{(2-\alpha)m}}\tan.\frac{1}{2}v\left\{1+\frac{\frac{2}{3}-\alpha}{2-\alpha}\tan.^2\frac{1}{2}v-\frac{\frac{4}{5}-\alpha}{(2-\alpha)^2}\tan.^4\frac{1}{2}v+\&c.\right\}.$$

If the orbit is parabolic

$$\alpha = 0$$

and consequently

$$\varrho = \frac{D}{\cos.\frac{1}{2}v};$$

$$t = \frac{D^{\frac{3}{2}}\sqrt{2}}{\sqrt{m}}\left\{\tan.\frac{v}{2}+\frac{1}{3}\tan.^3\frac{1}{2}v\right\}$$

which expression may also be got at once from (237).

The time t, the distance D and sum m of the masses of the sun and comet, are heterogeneous quantities, to compare which, we must divide each by the units of their species. We shall suppose therefore that the mean distance of the sun from the Earth is the unit of distance, so that D is expressed in parts of that distance. We may next observe that if T represent the time of a sidereal revolution of the Earth, setting off from the perihelion; we shall have in the equation

$$n\, t = u - e \sin. u$$

$u = 0$ at the beginning of the revolution, and $u = 2\,\pi$ at the end of it.

Hence

$$n\, T = 2\,\pi.$$

But we have

$$n = a^{-\frac{3}{2}} \sqrt{m} = \sqrt{m},$$

$$\therefore \sqrt{m} = \frac{2\,\pi}{T}.$$

The value of m is not exactly the same for the Earth as for the Comet, for in the first case it expresses the sum of the masses of the sun and earth; whereas in the second it implies the sum of the masses of the sun and comet: but the masses of the Earth and Comet being much smaller than that of the sun, we may neglect them, and suppose that m is the same for all Planets and all Comets and that it expresses the mass of the sun merely. Substituting therefore for $\sqrt{m}$ its value $\frac{2\,\pi}{T}$ in the preceding expression for t; we shall have

$$t = \frac{D^{\frac{3}{2}} . T}{\pi \sqrt{2}} \left\{ \tan. \frac{1}{2} v + \frac{1}{3} \tan.^3 \frac{1}{2} v \right\}.$$

This equation contains none but quantities comparable with each other; it will give t very readily when v is known; but to obtain v by means of t, we must resolve a Cubic Equation, which contains only one real root. We may dispense with this resolution, by making a table of the values of v corresponding to those of t, in a parabola of which the perihelion distance is unity, or equal to the mean distance of the earth from the sun. This table will give the time corresponding to the anomaly v, in any parabola of which D is the perihelion distance, by multiplying by $D^{\frac{3}{2}}$, the time which corresponds to the same anomaly in the Table. We also get the anomaly v corresponding to the time t, by dividing t by $D^{\frac{3}{2}}$, and seeking in the table, the anomaly which corresponds to the quotient arising from this division.

490. Let us now investigate the anomaly, corresponding to the time t, in an ellipse of great excentricity.

If we neglect quantities of the order α^2, and put $1 - e$ for α, the above expression of t in terms of v in an ellipse, will give

$$t = \frac{D^{\frac{3}{2}} \sqrt{2}}{\sqrt{m}} \left\{ \begin{array}{l} \tan. \frac{1}{2} v + \frac{1}{3} \tan.^3 \frac{1}{2} v \\ + (1 - e) \tan.^2 \frac{1}{2} v \left\{ \frac{1}{4} - \frac{1}{4} \tan.^2 \frac{1}{2} v - \frac{1}{5} \tan.^4 \frac{1}{2} v \right\} \end{array} \right\}.$$

Then, find by the table of the motions of the comets, the anomaly corresponding to the time t, in a parabola of which D is the perihelion distance. Let U be this anomaly and U + x the true anomaly in an ellipse corresponding to the same time, x being a very small angle. Then if we substitute in the above equation U + x for v, and then transform the second member into a series of powers of x, we shall have, neglecting the square of x, and the product of x by 1 — e,

$$t = \frac{D^{\frac{3}{2}} \sqrt{2}}{\sqrt{m}} \cdot \left\{ \begin{array}{l} \tan. \frac{1}{2} U + \frac{1}{3} \tan.^3 \frac{1}{2} U + \frac{x}{2 \cos.^4 \frac{1}{2} U} \\ + \frac{1-e}{4} . \tan. \frac{1}{2} U \{1 - \tan.^2 \frac{1}{2} U - \frac{4}{5} \tan.^4 \frac{1}{2} U\} \end{array} \right\}.$$

But by supposition

$$t = \frac{D^{\frac{3}{2}} \sqrt{2}}{\sqrt{m}} \{\tan. \tfrac{1}{2} U + \tfrac{1}{3} \tan.^3 \tfrac{1}{2} U\}.$$

Therefore, substituting for x its sine and substituting for $\sin.^4 \frac{1}{2} U$ its value $(1 - \cos.^2 \frac{1}{2} U)^2$, &c.

$$\sin. x = \tfrac{1}{10} (1 - e) \tan. \tfrac{1}{2} U \{4 - 3 \cos.^2 \tfrac{1}{2} U - 6 \cos.^4 \tfrac{1}{2} U\}.$$

Hence, in forming a table of logarithms of the quantity

$$\tfrac{1}{10} \tan. \tfrac{1}{2} U \{4 - 3 \cos.^2 \tfrac{1}{2} U - 6 \cos.^4 \tfrac{1}{2} U\}$$

it will be sufficient to add the logarithm of 1 — e, in order to have that of sin. x; consequently we have the correction of the anomaly U, estimated from the parabola, to obtain the corresponding anomaly in a very excentric ellipse.

491. *To find the masses of such planets as have satellites.*

The equation

$$T = \frac{2 \pi a^{\frac{3}{2}}}{\sqrt{m}}$$

gives a very simple method of comparing the mass of a planet, having satellites, with that of the sun. In fact, M representing the mass of the sun, if μ the mass of the planet be neglected, we have

$$T = \frac{2 \pi a^{\frac{3}{2}}}{\sqrt{M}}$$

If we next consider a satellite of any planet μ', and call its mass p, and mean distance from the center of μ', h, and T its periodic time, we shall have

$$\mathit{T} = \frac{2 \pi h^{\frac{3}{2}}}{\sqrt{\mu' + p}}$$

$$\therefore \frac{\mu' + p}{M} = \frac{h^3}{a^3} \times \frac{T^2}{\mathit{T}^2}.$$

This equation gives the ratio of the sum of the masses of the planet μ' and its satellite to that of the sun. Neglecting therefore the mass of the satellite, as small compared with that of the planet, or supposing their ratio known, we have the ratio of the mass of the planet to that of the sun.

492. *To determine the Elements of Elliptical Motion.*

After having exposed the General Theory of Elliptical Motion and Method of Calculating by converging series, in the two cases of nature, that of orbits almost circular, and the case of orbits greatly excentric, it remains to determine the *Elements* of those orbits. In fact if we call V the velocity of μ in its relative motion about M, we have

$$V^2 = \frac{dx^2 + dy^2 + dz^2}{dt^2}$$

and the last of the equations (P) of No. 478, gives

$$V^2 = m\left\{\frac{2}{\varrho} - \frac{1}{a}\right\}.$$

To make m disappear from this expression, we shall designate by U the velocity which μ would have, if it described about M, a circle whose radius is equal to the unity of distance. In this hypothesis, we have

$$\varrho = a = 1,$$

and consequently

$$U^2 = m.$$

Hence

$$V^2 = U^2\left\{\frac{2}{\varrho} - \frac{1}{a}\right\}.$$

This equation will give the semi-axis major a of the orbit, by means of the primitive velocity of μ and of its primitive distance from M. But a is positive in the ellipse, and infinite in the parabola, and negative in the hyperbola. Thus *the orbit described by* μ *is an ellipse, a parabola, or hyperbola, according as* V *is* $<$ $=$ *or* $>$ *than* $U\sqrt{\frac{2}{\varrho}}$. It is remarkable that *the direction of primitive motion has no influence upon the species of conic section.*

To find the excentricity of the orbit, we shall observe that if ε represents the angle made by the direction of the relative motion of μ with the radius-vector, we have

$$\frac{d\varrho^2}{dt^2} = V^2 \cos.^2 \varepsilon.$$

Substituting for V^2 its value $m\left\{\frac{2}{\varrho} - \frac{1}{a}\right\}$, we have

$$\frac{d\varrho^2}{dt^2} = m\left(\frac{2}{\varrho} - \frac{1}{a}\right)\cos.^2 \varepsilon;$$

But by 480

$$2 m \rho - \frac{m \rho^2}{a} - \frac{\rho^2 d \rho^2}{d t^2} = m a (1 - e^2):$$

$$\therefore a (1 - e)^2 = \rho^2 \sin.^2 \epsilon \left(\frac{2}{\rho} - \frac{1}{a}\right);$$

whence we know the excentricity a e of the orbit.

To find v or the true anomaly, we have

$$\rho = \frac{a (1 - e^2)}{1 + e \cos. v}$$

$$\therefore \cos. v = \frac{a (1 - e^2) - \rho}{e \rho}.$$

This gives the position of the Perihelion. Equations (f) of No. 480 will then give u and by its means the instant of the Planet's passing its perihelion.

To get the position of the orbit, referred to a fixed plane passing through the center of M, supposed immoveable, let φ be the inclination of the orbit to this plane, and β the angle which the radius ρ makes with the Line of the Nodes. Let, Moreover, z be the primitive elevation of μ above the fixed plane, supposed known. Then we shall have, C A D being the fixed plane, A D the line of the nodes, A B = ρ, &c. &c.

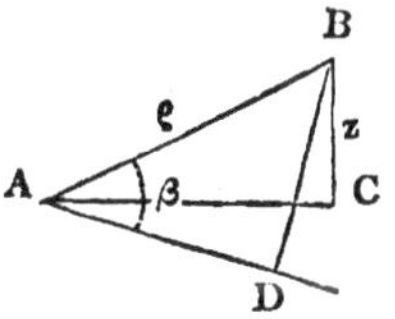

$$z = B D . \sin. \varphi = \rho \sin. \beta \sin. \varphi;$$

so that the inclination of the orbit will be known when we shall have determined β. For this purpose, let λ be the known angle which the primitive direction of the relative motion of μ makes with the fixed plane; then if we consider the triangle formed by this direction produced to meet the line of the nodes, by this last line and by the radius ρ, calling l the side of the triangle opposite to β, we have

$$l = \frac{\rho \sin. \beta}{\sin. (\beta + \epsilon)}.$$

Next we have

$$\frac{z}{l} = \sin. \lambda$$

consequently

$$\tan. \beta = \frac{z \sin. \epsilon}{\rho \sin. \lambda - z \cos. \epsilon}.$$

The elements of the Planetary Orbit being determined by these formulas, in terms of ρ and z, of the velocity of the planet, and of the direction of its motion, we can find the variation of these elements corresponding to the supposed variations in the velocity and its direction; and it will be easy, by methods about to be explained, from hence to obtain the differential variations of the Elements, due to the action of perturbing forces.

Taking the equation

$$V^2 = U^2\left\{\frac{2}{\varrho} - \frac{1}{a}\right\}.$$

In the circle $a = \varrho$ and $\therefore$

$$V = U\sqrt{\frac{1}{\varrho}}$$

so that the velocities of the planets in different circles are reciprocally as the squares of their radii (see Prop. IV of Princip.)

In the parabola, $a = \infty$,

$$\therefore V = U\sqrt{\frac{2}{\varrho}};$$

the velocities in the different points of the orbit, are therefore in this case reciprocally as the squares of the radius-vectors; and the velocity at each point, is to that which the body would have if it described a circle whose radius = the radius-vector ϱ, as $\sqrt{2} : 1$ (see 160)

An ellipse indefinitely diminished in breadth becomes a straight line, and in this case V expresses the velocity of μ, supposing it to descend in a straight line towards M. Let μ fall from rest, and its primitive distance be ϱ; also let its velocity at the distance ϱ' be V'; the above expression will give

$$0 = \frac{2}{\varrho} - \frac{1}{a};\ V'^2 = U^2\left\{\frac{2}{\varrho'} - \frac{1}{a}\right\};$$

whence

$$V' = U\sqrt{\frac{2(\varrho - \varrho')}{\varrho\varrho'}}.$$

Many other results, which have already been determined after another manner, may likewise be obtained from the above formula.

493. The equation

$$0 = \frac{dx^2 + dy^2 + dz^2}{dt^2} - m\left(\frac{2}{\varrho} - \frac{1}{a}\right)$$

is remarkable from its giving the velocity independently of the excentricity. It is also shown from a more general equation which subsists between the axis-major of the orbit, the chord of the elliptic arc, the sum of the extreme radius-vectors, and the time of describing this arc.

To obtain this equation, we have

$$\varrho = \frac{a(1 - e^2)}{1 + e\cos. v}$$

$$\rho = a\,(1 - e \cos.\, u)$$
$$t = a^{\frac{3}{2}}\,(u - e \sin.\, a);$$

in which suppose ρ, v, u, and t to correspond to the first extremity of the elliptic arc, and that ρ', v', u', t' belong to the other extremity; so that we also have

$$\rho' = \frac{a\,(1 - e^2)}{1 + e \cos.\, v'}$$
$$\rho' = a\,(1 - e \cos.\, u')$$
$$t' = a^{\frac{3}{2}}\,(u' - e \sin.\, u').$$

Let now

$$t - t' = T;\ \frac{u' - u}{2} = \beta;$$
$$\frac{u' + u}{2} = \beta';\ \rho' + \rho = R;$$

then, if we take the expression of t from that of t', and observe that

$$\sin.\, u' - \sin.\, u = 2 \sin.\, \beta \cos.\, \beta'$$

we shall have

$$T = 2\, a^{\frac{3}{2}}\, \{\beta - e \sin.\, \beta \cos.\, \beta'\}.$$

If we add them together taking notice that

$$\cos.\, u' + \cos.\, u = 2 \cos.\, \beta.\ \cos.\, \beta'$$

we shall get

$$R = 2\, a\,(1 - e \cos.\, \beta \cos.\, \beta').$$

Again, if c be the chord of the elliptic arc, we have

$$c^2 = \rho^2 + \rho'^2 - 2\,\rho\,\rho' \cos.\,(v - v')$$

but the two equations

$$\rho = \frac{a\,(1 - e^2)}{1 + e \cos.\, v};\ \rho = a\,(1 - e \cos.\, u)$$

give these

$$\cos.\, v = a\,\frac{\cos.\, u - e}{\rho};\ \sin.\, v = \frac{a\sqrt{1 - e^2}.\ \sin.\, u}{\rho};$$

and in like manner we have

$$\cos.\, v' = a\,.\,\frac{\cos.\, u' - e}{\rho'};\ \sin.\, v' = \frac{a\sqrt{1 - e^2}\ \sin.\, u'}{\rho'};$$

whence, we get

$$\rho\,\rho' \cos.\,(v - v') = a^2\,(e - \cos.\, u)\,(e - \cos.\, u') + a^2\,(1 - e^2) \sin.\, u \sin.\, u;$$

and consequently

$$c^2 = 2\,a^2\,(1 - e^2)\,\{1 - \sin.\, u \sin.\, u' - \cos.\, u \cos.\, u'\}$$
$$+\ a^2\, e^2\,(\cos.\, u - \cos.\, u')^2;$$

But

$$\sin. u \sin. u' + \cos. u \cos. u' = 2 \cos.^2 \beta - 1$$

$$\cos. u - \cos. u' = 2 \sin. \beta \sin. \beta'$$

$$\therefore c^2 = 4 a^2 \sin.^2 \beta (1 - e^2 \cos.^2 \beta').$$

We therefore have these three equations, scil.

$$R = 2 a \{1 - e \cos. \beta \cos. \beta'\};$$

$$T = 2 a^{\frac{3}{2}} \{\beta - e \sin. \beta \cos. \beta'\},$$

$$c^2 = 4 a^2 \sin.^2 \beta (1 - e^2 \cos.^2 \beta).$$

The first of them gives

$$e \cos. \beta' = \frac{2 a - R}{2 a \cos. \beta}$$

and substituting this value of e cos. β' in the two others, we shall have

$$T = 2 a^{\frac{3}{2}} \left\{\beta + \frac{R-2 a}{2 a} \tan. \beta\right\};$$

$$c^2 = 4 a^2 \tan.^2 \beta \left\{\cos.^2 \beta - \left(\frac{2 a - R}{2 a}\right)^2\right\}.$$

These two equations do not involve the excentricity e, and if in the first we substitute for β its value given by the second, we shall get T in a function c, R, and a. Thus we see that the time T depends only on the semi-axis major, the chord c and the sum R of the extreme radius-vectors.

If we make

$$z = \frac{2 a - R + c}{2 a}; \quad z' = \frac{2 a - R - c}{2 a};$$

the last of the preceding equations will give

$$\cos. 2 \beta = z z' + \sqrt{(1 - z^2) . (1 - z'^2)};$$

whence

$$2 \beta = \cos.^{-1} z' - \cos.^{-1} z$$

(for cos. (A — B) = cos. A cos. B + sin. A sin. B).

Consequently

$$\tan. \beta = \frac{\sin. (\cos.^{-1} z') - \sin. (\cos.^{-1} z)}{z + z'};$$

we have also

$$z + z' = \frac{2 a - R}{a}.$$

Hence the expression of T will become, observing that if T is the duration of the sidereal revolution, whose mean distance from the sun is taken for unity, we have

$$T = 2\,\pi,$$

$$T = \frac{a^{\frac{3}{2}}\,T}{2\,\pi}\,\{\cos.^{-1} z' - \cos.^{-1} z - \sin.\,(\cos.^{-1} z') + \sin.(\cos^{-1} z)\} \ldots (a)$$

Since the same cosines may belong to many arcs, this expression is ambiguous, and we must take care to distinguish the arcs which correspond to z, z'.

In the parabola, the semi-axis major is infinite, and we have

$$\cos.^{-1} z' - \sin.\,(\cos.^{-1} z') = \frac{1}{6}\left(\frac{R + c}{a}\right)^{\frac{3}{2}}.$$

And making c negative we shall have the value of

$$\cos.^{-1} z - \sin.\,(\cos.^{-1} z);$$

hence the formula (a) will give the time T employed to describe the arc subtending the chord c, scil.

$$T = \frac{T}{12\,\pi}\{(\varrho + \varrho' + c)^{\frac{3}{2}} \mp (\varrho + \varrho' - c)^{\frac{3}{2}}\};$$

the sign — being taken, when the two extremities of the parabolic arc are situated on the same side of the axis of the parabola.

Now T being = 365.25638 days, we have

$$\frac{T}{12\,\pi} = 9.688754 \text{ days.}$$

The formula (a) gives the time of a body's descent in a straight line towards the focus, beginning from a given distance; for this, it is sufficient to suppose the axis-minor of the ellipse indefinitely diminished. If we suppose, for example, that the body falls from rest at the distance 2 a from the focus and that it is required to find the time (T) of falling to the distance c, we shall have

$$R = 2\,a + \varrho, \quad \varrho = 2\,a - c.$$

whence

$$z' = -1, \quad z = \frac{c - a}{a}$$

and the formula gives

$$T = \frac{a^{\frac{3}{2}}\,T}{2\,\pi}\left\{\pi - \cos.^{-1}\frac{c - a}{a} + \sqrt{\frac{2\,a\,c - c}{a^2}}\right\}.$$

There is, however, an essential difference between elliptical motion towards the focus, and the motion in an ellipse whose breadth is indefinitely small. In the first case, the body having arrived at the focus, passes beyond it, and again returns to the same distance at which it departed; but in the second case, the body having arrived at the focus immediately returns to the point of departure. A tangential velocity at the aphelion, however small, suffices to produce this difference which has no influence upon the time of the body's descent to the center, nor upon the velocity resolved parallel to the axis-major. Hence the principles of the 7th Section of Newton give accurately the Times and Velocities, although they do not explain all the circumstances of motion. For it is clear that if there be absolutely no tangential velocity, the body having reached the center of force, will proceed beyond it to the same distance from which it commenced its motion, and then return to the center, pass through it, and proceed to its first point of departure, the whole being performed in just double the time as would be required to return by moving in the indefinitely small ellipse.

494. Observations not conducting us to the circumstances of the primitive motion of the heavenly bodies; by the formulas of No. 492 we cannot determine the elements of their orbits. It is necessary for this end to compare together their respective positions observed at different epochs, which is the more difficult from not observing them from the center of their motions. Relatively to the planets, we can obtain, by means of their oppositions and conjunctions, their Heliocentric Longitude. This consideration, together with that of the smallness of the excentricity and inclination of their orbits to the ecliptic, affords a very simple method of determining their elements. But in the present state of astronomy, the elements of these orbits need but very slight corrections; and as the variations of the distances of the planets from the earth are never so great as to elude observation, we can rectify, by a great number of observations, the elements of their orbits, and even the errors of which the observations themselves are susceptible. But with regard to the Comets, this is not feasible; we see them only near their perihelion: if the observations we make on their appearance prove insufficient for the determination of their elements, we have then no means of pursuing them, even by thought, through the immensity of space, and when after the lapse of ages, they again approach the sun, it is impossible for us to recognise them. It becomes therefore important to find a method of determining, by observations alone during the appearance of one Comet, the elements of its orbit. But this problem considered rigorously surpasses the powers of analysis, and we are obliged to have recourse to approximations, in order to obtain the first values of the elements, these being afterwards to be corrected to any degree of accuracy which the observations permit.

If we use observations made at remote intervals, the eliminations will lead to impracticable calculations; we must therefore be content to consider only near observations; and with this restriction, the problem is abundautly difficult.

It appears, that instead of directly making use of observations, it is better to get from them the data which conduct to exact and simple results. Those in the present instance, which best fulfil that condition, are the geocentric longitude and latitude of the Comet at a given instant, and their first and second differences divided by the corresponding powers of the element of time; for by means of these data, we can determine rigorously and with ease, the elements, without having recourse to a single integration, and by the sole consideration of the differential equations of the orbit. This way of viewing the problem, permits us moreover, to employ a great number of near observations, and to comprise also a considerable interval between the extreme observations, which will be found of great use in diminishing the influence of such errors, as are due to observations from the nebulosity by which Comets are enveloped. Let us first present the formulas necessary to obtain the first differences, of the longitude and latitude of any number of near observations; and then determine the elements of the orbit of a Comet by means of these differences; and lastly expose the method which appears the simplest, of correcting these elements by three observations made at remote intervals.

495. At a given epoch, let α be the geocentric longitude of a Comet, and θ its north geocentric latitude, the south latitudes being supposed negative. If we denote by s, the number of days elapsed from this epoch, the longitude and latitude of the Comet, after that interval, will, by using Taylor's Theorem (481), be expressed by these two series

$$\alpha + s \, . \left(\frac{d\,\alpha}{d\,s}\right) + \frac{s^2}{1.2}\left(\frac{d^2\,\alpha}{d\,s^2}\right) + \&c.$$

$$\theta + s \, . \left(\frac{d\,\theta}{d\,s}\right) + \frac{s^2}{1.2}\left(\frac{d^2\,\theta}{d\,s^2}\right) + \&c.$$

We must determine the values of

$$\alpha, \left(\frac{d\,\alpha}{d\,s}\right), \left(\frac{d^2\,\alpha}{d\,s^2}\right), \&c., \; \theta, \left(\frac{d\,\theta}{d\,s}\right), \&c.$$

by means of several observed geocentric longitudes and latitudes. To do this most simply, consider the infinite series which expresses the geocentric longitude. The coefficients of the powers of s, in this series, ought to be determined by the condition, that by it is represented each observed longitude; we shall thus have as many equations as observations; and if their number is n, we shall be able to find from them, in series, the n quantities α, $\left(\frac{d\,\alpha}{d\,s}\right)$, &c. But it ought to be observed that s being supposed very small, we may neglect all terms multiplied by s^n, s^{n+1}, &c. which will reduce the infinite series to its n first terms; which by n observations we shall be able to determine. These are only approximations, and their accuracy will depend upon the smallness of the terms which are omitted. They will be more exact in proportion as s is more diminutive, and as we employ a greater number of observations. The theory of interpolations is used therefore *To find a rational and integer function of* s *such, that in substituting therein for* s *the number of days which correspond to each observation, it shall become the observed longitude.*

Let β, β', β'', &c. be the observed longitudes of the comet, and by i, i′, i″, &c. the corresponding numbers of days from the given epoch, the numbers of the days prior to the given epoch being supposed negative. If we make

$$\frac{\beta' - \beta}{i' - i} = \delta\,\beta;\ \frac{\beta'' - \beta'}{i'' - i'} = \delta\,\beta';\ \frac{\beta''' - \beta''}{i''' - i''} = \delta\,\beta'';\ \&c.$$

$$\frac{\delta\,\beta' - \delta\,\beta}{i'' - i} = \delta^2\,\beta;\ \frac{\delta\,\beta'' - \delta\,\beta'}{i''' - i'}\ \delta^2\,\beta';\ \&c.$$

$$\frac{\delta^2\,\beta' - d^2\,\beta}{i''' - i} = \delta^3\,\beta;\ \&c.$$

$$\&c.;$$

the required functions will be

$$\beta + (s - i)\,.\,\delta\,\beta + (s - i)\,(s - i')\,.\,\delta^2\beta + (s - i)\,(s - i')\,(s - i'')\,\delta^3\,\beta,\ \&c.$$

for it is easy to perceive that if we make successively $s = i$, $s = i'$, $s = i''$, &c. it will change itself into β, β', β'', &c.

Again, if we compare the preceding function with this

$$\alpha + s\,.\left(\frac{d\,\alpha}{d\,s}\right) + \frac{s^2}{2}\,.\left(\frac{d^2\,\alpha}{d\,s^2}\right) + \&c.$$

we shall have by equating coefficients of homogeneous terms.

$$\alpha = \beta - i\,\delta\,\beta + i\,.\,i'\,.\,\delta^2\,\beta - i\,.\,i'\,.\,i''\,\delta^3\,\beta + \&c.$$

$$\left(\frac{d\,\alpha}{d\,s}\right) = \delta\,\beta - (i + i')\,\delta^2\,\beta + (i\,i' + i\,i'' + i'\,i'')\,\delta^3\,\beta - \&c.$$

$$\frac{1}{2}\left(\frac{d^2\,\alpha}{d\,s^2}\right) = \delta^2\,\beta - (i + i' + i'')\,\delta^3\,\beta + \&c.$$

The higher differences of α will be useless. The coefficients of these expressions are alternately positive and negative; the coefficient of $\delta^r\,\beta$ is, disregarding the sign, the product of r and r together of r quantities i, i′, ... $i^{(r-1)}$ in the value of α; it is the sum of the products of the same quantities, r — 1 together in the value of $\left(\frac{d\,\alpha}{d\,s}\right)$; lastly it is the sum of the products of these quantities r — 2, together in the value of $\frac{1}{2}\left(\frac{d^2\,\alpha}{d\,s^2}\right)$.

If γ, γ', γ'', &c. be the observed geocentric latitudes, we shall have the values of θ, $\left(\frac{d\,\theta}{d\,s}\right)$, $\left(\frac{d^2\,\theta}{d\,s^2}\right)$, &c. by changing in the preceding expressions for α $\left(\frac{d\,\alpha}{d\,s}\right)$, $\left(\frac{d^2\,\alpha}{d\,s^2}\right)$, &c. the quantities β, β', β'' into γ, γ', γ''.

These expressions are the more exact, the greater the number of observations and the smaller the intervals between them. We might, therefore, employ all the near observations made at a given epoch, provided they were accurate; but the errors of which they are always susceptible will conduct to imperfect results. So that, in order to lessen the influence of these errors, we must augment the interval between the extreme observations, employing in the investigation a greater number of them. In this way with five observations we may include an interval of thirty-five or forty degrees, which would give us very near approximations to the geocentric longitude and latitude, and to their first and second differences.

If the epoch selected were such, that there were an equal number of observations before and after it, so that each successive longitude may have a corresponding one which succeeds the epoch. This condition will give values still more correct of α, $\left(\frac{d\,\alpha}{d\,s}\right)$ and $\left(\frac{d^2\,\alpha}{d\,s^2}\right)$, and it easily appears that new observations taken at equal distances from either side of the epoch, would only add to these values, quantities which, with regard to their last terms, would be as $s^2\left(\frac{d^2\,\alpha}{d\,t^2}\right)$ to α. This symmetrical arrangement takes place, when all the observations being equidistant, we fix the epoch at the middle of the interval which they comprise. It is therefore advantageons to employ observations of this kind.

In general, it will be advantageous to fix the epoch near the middle of this interval; because the number of days included between the extreme observations being less considerable, the approximations will be more convergent. We can simplify the calculus still more by fixing the epoch at the instant of one of the observations; which gives immediately the values of α, and θ.

When we shall have determined as above the values of

$$\left(\frac{d\alpha}{ds}\right),\ \left(\frac{d^2\alpha}{ds^2}\right),\ \left(\frac{d\theta}{ds}\right),\ \text{and}\ \left(\frac{d^2\theta}{ds^2}\right)$$

we shall then obtain as follows the first and second differences of α, and θ divided by the corresponding powers of the elements of time. If we neglect the masses of the planets and comets, that of the sun being the unit of mass; if, moreover, we take the distance of the sun from the earth for the unit of distance; the mean motion of the earth round the sun will be the measure of the time t. Let therefore λ be the number of seconds which the earth describes in a day, by reason of its mean sidereal motion; the time t corresponding to the number of days will be λ s; we shall, therefore, have

$$\left(\frac{d\alpha}{dt}\right) = \frac{1}{\lambda}\left(\frac{d\alpha}{ds}\right)$$

$$\left(\frac{d^2\alpha}{dt^2}\right) = \frac{1}{\lambda^2}\left(\frac{d^2\alpha}{ds^2}\right).$$

Observations give hy the Logarithmic Tables,

$$\log.\lambda = 4.0394622$$

and also

$$\log.\lambda^2 = \log.\lambda + \log.\frac{\lambda}{R}$$

R being the radius of the circle reduced to seconds; whence

$$\log.\lambda^2 = 2.2750444;$$

$\therefore$ if we reduce to seconds, the values of $\left(\frac{d\alpha}{ds}\right)$, and of $\left(\frac{d^2\alpha}{ds^2}\right)$, we shall have the logarithms of $\left(\frac{d\alpha}{dt}\right)$, and of $\left(\frac{d^2\alpha}{dt^2}\right)$ by taking from the logarithms of these values the logarithms of 4.039422, and 2.2750444. In like manner we get the logarithms of $\left(\frac{d\theta}{dt}\right)$, $\left(\frac{d^2\theta}{dt^2}\right)$, after subtracting the same logarithms, from the logarithms of their values reduced to seconds.

On the accuracy of the values of

$$\alpha,\ \left(\frac{d\alpha}{dt}\right),\ \left(\frac{d^2\alpha}{dt^2}\right),\ \theta,\ \left(\frac{d\theta}{d}\right)\ \text{and}\ \left(\frac{d^2\theta}{dt^2}\right)$$

depends that of the following results; tand since their formation is very simple, we must select and multiply observations so as to obtain them with the greatest exactness possible. We shall determine presently, by means of these values, the elements of the orbit of a Comet, and to generalize these results, we shall

496. *Investigate the motion of a system of bodies sollicited by any forces whatever.*

Let x, y, z be the rectangular coordinates of the first body; x′, y′, z′ those of the second body, and so on. Also let the first body be sollicited parallel to the axes of x, y, z by the forces X, Y, Z, which we shall suppose tend to diminish these variables. In like manner suppose the second body sollicited parallel to the same axes by the forces X′, Y′, Z′, and so on. The motions of all the bodies will be given by differential equations of the second order

$$0 = \frac{d^2 x}{d t^2} + X; \quad 0 = \frac{d^2 y}{d t^2} + Y; \quad 0 = \frac{d^2 z}{d t^2} + Z;$$

$$0 = \frac{d^2 x'}{d t^2} + X'; \quad 0 = \frac{d^2 y'}{d t^2} + Y'; \quad 0 = \frac{d^2 z'}{d t^2} + Z'.$$

&c. = &c.

If the number of the bodies is n, that of the equations will be 3 n; and their finite integrals will contain 6 n arbitrary constants, which will be the elements of the orbits of the different bodies.

To determine these elements by observations, we shall transform the coordinates of each body into others whose origin is at the place of the observer. Supposing, therefore, a plane to pass through the eye of the observer, and of which the situation is always parallel to itself, whilst the observer moves along a given curve, call r, r′ r″, &c. the distances of the observer from the different bodies, projected upon the plane; $\alpha, \alpha', \alpha''$, &c. the apparent longitudes of the bodies, referred to the same plane, and $\theta, \theta', \theta''$, &c. their apparent latitudes. The variables x, y, z will be given in terms of r, α, θ, and of the coordinates of the observer. In like manner, x′, y′, z′ will be given in functions of r′, α', θ', and of the coordinates of the observer, and so on. Moreover, if we suppose that the forces X, Y, Z; X′, Y′, Z′, &c. are due to the reciprocal action of the bodies of the system, and independent of attractions; they will be given in functions of r, r′, r″, &c.; $\alpha, \alpha', \alpha''$, &c.; $\theta, \theta', \theta''$, &c. and of known quantities. The preceding differential equations will thus involve these new variables and their first and second differences. But observations make known, for a given instant, the values of

$$\alpha, \left(\frac{d\alpha}{dt}\right), \left(\frac{d^2\alpha}{dt^2}\right), \theta, \left(\frac{d\theta}{dt}\right), \left(\frac{d^2\theta}{dt^2}\right); \alpha', \left(\frac{d\alpha'}{dt}\right), \text{ \&c.}$$

There will hence of the unknown quantities only remain r, r′, r″, &c. and their first and second differences. These unknowns are in number 3 n, and since we have 3 n differential equations, we can determine them.

At the same time we shall have the advantage of presenting the first and second differences of r, r′, r″, &c. under a linear form.

The quantities α, θ, r, α', θ', r′, &c. and their first differences divided by d t, being known; we shall have, for any given instant, the values of x, y, z, x′, y′, z′, &c. and of their first differences divided by d t. If we substitute these values in the 3 n finite integrals of the preceding equations, and in the first differences of these integrals; we shall have 6 n equations, by means of which we shall be able to determine the 6 n arbitrary constants of the integrals, or the elements of the orbits of the different bodies.

497. To apply this method to the motion of the Comets,

We first observe that the principal force which actuates them is the attraction of the sun; compared with which all other forces may be neglected. If, however, the Comet should approach one of the greater planets so as to experience a sensible perturbation, the preceding method will still make known its velocity and distance from the earth; but this case happening but very seldom, in the following researches, we shall abstain from noticing any other than the action of the sun.

If the sun's mass be the unit, and its mean distance from the earth the unit of distance; if, moreover, we fix the origin of the coordinates x, y, z of a Comet, whose radius-vector is ρ; the equations (0) of No. 475 will become, neglecting the mass of the Comet,

$$\left.\begin{aligned} 0 &= \frac{d^2 x}{d t^2} + \frac{x}{\rho^3}; \\ 0 &= \frac{d^2 y}{d t^2} + \frac{y}{\rho^3}; \\ 0 &= \frac{d^2 z}{d t^2} + \frac{z}{\rho^3}. \end{aligned}\right\} \quad (k)$$

Let the plane of x, y be the plane of the ecliptic. Also let the axis of x be the line drawn from the center of the sun to the first point of aries, at a given epoch; the axis of y the line drawn from the center of the sun to the first point of cancer, at the same epoch; and finally the positive values of z be on the same side as the north pole of the ecliptic. Next call x′, y′ the coordinates of the earth and R its radius-vector. This being supposed, transfer the coordinates x, y, z to others relative to the observer; and to do this let α be the geocentric longitude, and r its distance from the center of the earth projected upon the ecliptic; then we shall have

$$x = x' + r \cos. \alpha; \; y = y + r \sin. \alpha; \; z = r \tan. \theta.$$

If we multiply the first of equations (k) by sin. α, and take from the result the second multiplied by cos. α, we shall have

$$0 = \sin. \alpha \frac{d^2 x}{d t^2} - \cos. \alpha . \frac{d^2 y}{d t^2} + \frac{x \sin. \alpha - y \cos. \alpha}{\varrho^3};$$

whence we derive, by substituting for x, y their values given above,

$$0 = \sin. \alpha . \frac{d^2 x'}{d t^2} - \cos. \alpha . \frac{d^2 y'}{d t^2} + \frac{x' \sin. \alpha - y \cos. \alpha}{\varrho^3}$$

$$- z . \left(\frac{d r}{d t}\right)\left(\frac{d \alpha}{d t}\right) - r \left(\frac{d^2 \alpha}{d t^2}\right).$$

The earth being retained in its orbit like a comet, by the attraction of the sun, we have

$$0 = \frac{d^2 x'}{d t^2} + \frac{x'}{R^3}, \; 0 = \frac{d^2 y'}{d t^2} + \frac{y'}{R^3};$$

which give

$$\sin. \alpha \frac{d^2 x'}{d t^2} - \cos. \alpha . \frac{d^2 y'}{d t^2} = \frac{y' \cos. \alpha - x' \sin. \alpha}{R^3}$$

We shall, therefore, have

$$0 = (y' \cos. \alpha - x' \sin. \alpha) \left\{ \frac{1}{R^2} - \frac{1}{\varrho^3} \right\} - z \left(\frac{d r}{dt}\right) . \left(\frac{d \alpha}{d t}\right) - r \left(\frac{d^2 \alpha}{d t^2}\right).$$

Let A be the longitude of the earth seen from the sun; we shall have

$$x' = R \cos. A; \; y' = R \sin. A;$$

therefore

$$y' \cos. \alpha - x' \sin. \alpha = R \sin. (A - \alpha);$$

and the preceding equation will give

$$\left(\frac{d r}{d t}\right) = \frac{R \sin. (A - \alpha)}{2 \left(\frac{d \alpha}{d t}\right)} . \left\{ \frac{1}{R^3} - \frac{1}{\varrho^3} \right\} - \frac{r . \left(\frac{d^2 \alpha}{d t^2}\right)}{2 \left(\frac{d \alpha}{d t}\right)} \quad . \; . \; . \; (1)$$

Now let us seek a second expression for $\left(\frac{d r}{d t}\right)$. For this purpose we will multiply the first of equations (k) by tan. θ . cos. α, the second by tan. θ sin. α, and take the third equation from the sum of these two products; we shall thence obtain

$$0 = \tan. \theta \left\{ \cos. \alpha \frac{d^2 x}{d t^2} + \sin. \alpha \frac{d^2 y}{d t^2} \right\}$$

$$+ \tan. \theta . \frac{x \cos. \alpha + y \sin. \alpha}{\varrho^3} - \frac{d^2 z}{d t^2} - \frac{z}{\varrho^3}.$$

This equation will become by substitution for x, y, z

$$0 = \tan. \theta \left\{ \left(\frac{d^2 x'}{d t^2} + \frac{x'}{\varrho^3}\right) \cos. \alpha + \left(\frac{d^2 y'}{d t^2} + \frac{y'}{\varrho^3}\right) \sin. \alpha \right\}$$

$$-\frac{2\left(\frac{d\theta}{dt}\right)\left(\frac{dr}{dt}\right)}{\cos.^2\theta}-r\left\{\frac{\left(\frac{d^2\theta}{dt^2}\right)}{\cos.^2\theta}+\frac{2\left(\frac{d\theta}{dt}\right)^2\sin.\theta}{\cos.^3\theta}+\left(\frac{d\alpha}{dt}\right)^2\tan.\theta\right\}$$

But

$$\left(\frac{d^2x'}{dt^2}+\frac{x'}{\varrho^3}\right)\cos.\alpha+\left(\frac{d^2y'}{dt^2}+\frac{y'}{\varrho^3}\right)\sin.\alpha=(x'\cos.\alpha+y'\sin.\alpha)\left(\frac{1}{\varrho^3}-\frac{1}{R^3}\right)$$

$$=R\cos.(A-\alpha)\left\{\frac{1}{\varrho^3}-\frac{1}{R^3}\right\};$$

Therefore,

$$\left(\frac{dr}{dt}\right)=-\tfrac{1}{2}r\left\{\frac{\left(\frac{d^2\theta}{dt^2}\right)}{\left(\frac{d\theta}{dt}\right)}+2\left(\frac{d\theta}{dt}\right)\tan.\theta+\frac{\left(\frac{d\alpha}{dt}\right)^2\sin.\theta\cos.\theta}{\left(\frac{d\theta}{dt}\right)}\right\}$$

$$+\frac{R\sin.\theta\cos.\theta\cos.(A-\alpha)}{2\left(\frac{d\theta}{dt}\right)}\cdot\left\{\frac{1}{\varrho^3}-\frac{1}{R^3}\right\}.\ \ .\ \ .\ \ .\ \ .\ \ (2)$$

If we take this value of $\left(\frac{d\varrho}{dt}\right)$ from the first and suppose

$$\mu'=\frac{\left(\frac{d\alpha}{dt}\right)\left(\frac{d^2\theta}{dt^2}\right)-\left(\frac{d\theta}{dt}\right)\left(\frac{d^2\alpha}{dt^2}\right)+2\left(\frac{d\alpha}{dt}\right)\left(\frac{d\theta}{dt}\right)^2\tan.\theta+\left(\frac{d\alpha}{dt}\right)^3\sin.\theta\cos.\theta}{\left(\frac{d\alpha}{dt}\right)\sin.\theta\cos.\theta\cos.(A-\alpha)+\left(\frac{d\theta}{dt}\right)\sin.(A-\alpha)}$$

we shall have

$$r=\frac{R}{\mu'}\cdot\left\{\frac{1}{\varrho^3}-\frac{1}{R^3}\right\}\ \ .\ \ .\ \ .\ \ .\ \ .\ \ .\ \ (3)$$

The projected distance r of the comet from the earth, being always positive, this equation shows that the distance ϱ of the comet from the sun, is less or greater than the distance R of the sun from the earth, according as μ' is positive or negative; the two distances are equal if $\mu'=0$.

By inspection alone of a celestial globe, we can determine the sign of μ'; and consequently whether the comet is nearer to or farther from the Earth. For that purpose imagine a great circle which passes through two Geocentric positions of the Comet infinitely near to one another. Let γ be the inclination of this circle to the ecliptic, and λ the longitude of its ascending node; we shall have

$$\tan.\gamma\sin.(\alpha-\lambda)=\tan.\theta;$$

whence

$$d\theta\sin.(\alpha-\lambda)=d\alpha\sin.\theta\cos.\theta\cos.(\alpha-\lambda).$$

Differentiating, we have, also

$$0 = \left(\frac{d\,\alpha}{d\,t}\right)\left(\frac{d^2\,\theta}{d\,t^2}\right) - \left(\frac{d\,\theta}{d\,t}\right)\left(\frac{d^2\,\alpha}{d\,t^2}\right) + 2\left(\frac{d\,\alpha}{d\,t}\right)\left(\frac{d\,\theta}{d\,t}\right)^2 \tan.\,\theta$$

$$+ \left(\frac{d\,\alpha}{d\,t}\right)^3 \sin.\,\theta\,\cos.\,\theta;$$

$d^2\,\theta_{,}$ being the value of $d^2\,\theta$, which would take place, if the apparent motion of the Comet continued in the great circle. The value of μ' thus becomes, by substituting for $d\,\theta$ its value

$$\frac{d\,\alpha\,\sin.\,\theta\,\cos.\,\theta\,\cos.\,(\alpha - \lambda)}{\sin.\,(\alpha - \lambda)},$$

$$\mu' = \frac{\left\{\left(\frac{d^2\,\theta}{d\,t^2}\right) - \left(\frac{d^2\,\theta_{,}}{d\,t^2}\right)\right\}\sin.\,(\alpha - \lambda)}{\sin.\,\theta\,\cos.\,\theta\,\sin.\,(A - \lambda)}.$$

The function $\frac{\sin.\,(\alpha - \lambda)}{\sin.\,\theta\,\cos.\,\theta}$ is constantly positive; the value of μ is therefore positive or negative, according as $\left(\frac{d^2\,\theta}{d\,t^2}\right) - \left(\frac{d^2\,\theta_{,}}{d\,t^2}\right)$ has the same or a different sign from that of $\sin.\,(A - \lambda)$. But $A - \lambda$ is equal to two right angles plus the distance of the sun from the ascending node of the great circle. Whence it is easy to conclude that μ' will be positive or negative, according as in a third geocentric position of the comet, indefinitely near to the two first, the comet departs from the great circle on the same or the opposite side on which is the sun. Conceive, therefore, that we make a great circle of the sphere pass through the two geocentric positions of the comet; then according as, in a third consecutive geocentric position, the comet departs from this great circle, on the same side as the sun or on the opposite one, it will be nearer to or farther from the sun than the Earth. If it continues to appear in this great circle, it will be equally distant from both; so that the different deflections of its apparent path points out to us the variations of its distance from the sun.

To eliminate ϱ from equation (3), and to reduce this equation so as to contain no other than the unknown r, we observe that $\varrho^2 = x^2 + y^2 + z^2$ in substituting for x, y, z, their values in terms of

r, α, and θ;

and we have

$$\varrho^2 = x'^2 + y'^2 + 2\,r\{x'\cos.\,\alpha + y'\sin.\,\alpha\} + \frac{r^2}{\cos.^2\,\theta'};$$

but we have

$$x' = R\cos.\,A,\ y' = R\sin.\,A;$$

$$\therefore \varrho^2 = \frac{r^2}{\cos.^2\,\theta} + 2\,R\,r\cos.\,(A - \alpha) + R^2;$$

But

$$x' = R \cos. A; \; y' = R \sin. A$$

$$\therefore \varrho^2 = \frac{r^2}{\cos.^2 \theta} + 2 R r \cos. (A - \alpha) + R^2.$$

If we square the two members of equation (3) put under this form

$$\varrho^3 \{\mu' R^2 r + 1\} = R^3$$

we shall get, by substituting for ϱ^2,

$$\left\{\frac{r^2}{\cos.^2 \theta} + 2 R r \cos. (A - \alpha) + R^2\right\}^3 . \{\mu' R^2 r + 1\}^2 = R^6 \ldots (4)$$

an equation in which the only unknown quantity is r, and which will rise to the seventh degree, because a term of the first member being equal to R^6, the whole equation is divisible by r. Having thence determined r, we shall have $\left(\frac{d r}{d t}\right)$ by means of equations (1) and (2). Substituting, for example, in equation (1), for $\frac{1}{\varrho^3} - \frac{1}{R^3}$ its value $\frac{\mu' r}{R}$, given by equation (3); we shall have

$$\left(\frac{d r}{d t}\right) = - \frac{r}{2\left(\frac{d \alpha}{d t}\right)} . \left\{\left(\frac{d^2 \alpha}{d t^2}\right) + \mu' \sin. (A - \alpha)\right\} .$$

The equation (4) is often susceptible of many real and positive roots; reducing it and dividing by r, its last term will be

$$2 R^5 \cos.^6 \theta \{\mu' R^3 + 3 \cos. (A - \alpha)\}.$$

Hence the equation in r being of the seventh degree or of an odd degree, it will have at least two real positive roots if $\mu' R^3 + 3. \cos. (A - \alpha)$ is positive; for it ought always, by the nature of the problem, to have one positive root, and it cannot then have an odd number of positive roots. Each real and positive value of r gives a different conic section, for the orbit of the comet; we shall, therefore, have as many curves which satisfy three near observations, as r has real and positive values; and to determine the true orbit of the comet, we must have recourse to a new observation.

498. The value of r, derived from equation (4) would be rigorously exact, if

$$\alpha, \left(\frac{d \alpha}{d t}\right), \left(\frac{d^2 \alpha}{d t^2}\right), \theta, \left(\frac{d \theta}{d t}\right), \left(\frac{d^2 \theta}{d t^2}\right)$$

were exactly known; but these quantities are only approximate. In fact, by the method above exposed, we can approximate more and more, merely by making use of a great number of observations, which presents the advantage of considering intervals sufficiently great, and of making the errors arising from observations compensate one another. But this method has the analytical inconvenience of employing more than three observations, in a problem where three are sufficient. This may be obviated, and the solution rendered as approximate as can be wished by three observations only, after the following manner.

Let α and θ, representing the geocentric longitude and latitude of the intermediate; if we substitute in the equations (k) of the preceding No. instead of x, y, z their values $x' + r \cos. \alpha$; $y' + r \sin. \alpha$; and $r \tan. \theta$; they will give $\left(\frac{d^2 r}{d t^2}\right)$, $\left(\frac{d^2 \alpha}{d t^2}\right)$ and $\left(\frac{d^2 \theta}{d t^2}\right)$ in functions of r, α, and θ, of their first differences a known quantities. If we differentiate these, we shall have $\left(\frac{d^3 r}{d t^3}\right)$, $\left(\frac{d^3 \alpha}{d t^3}\right)$ and $\left(\frac{d^3 \theta}{d t^3}\right)$ in terms of r, α, θ, and of their first and second differences. Hence by equation (2) of 497 we may eliminate the second difference of r by means of its value and its first difference. Continuing to differentiate successively the values of $\left(\frac{d^3 \alpha}{d t^3}\right)$, $\left(\frac{d^3 \theta}{d t^3}\right)$, and eliminating the differences of α, and of θ superior to second differences, and all the differences of r, we shall have the values of

$$\left(\frac{d^3 \alpha}{d t^3}\right), \quad \left(\frac{d^4 \alpha}{d t^4}\right), \text{ \&c.}$$

$$\left(\frac{d^3 \theta}{d t^3}\right), \quad \left(\frac{d^4 \theta}{d t^4}\right), \text{ \&c. in terms of}$$

$$r, \alpha, \left(\frac{d \alpha}{d t}\right), \quad \left(\frac{d^2 \alpha}{d t^2}\right), \theta, \left(\frac{d \theta}{d t}\right), \quad \left(\frac{d^2 \theta}{d t^2}\right);$$

this being supposed, let

$$\alpha_{,}, \alpha, \alpha',$$

be the three geocentric observed longitudes of the Comet; $\theta_{,}$, θ, θ' its three corresponding geocentric latitudes; let i be the number of days which separate the first from the second observation, and i' the interval between the second and third observation; lastly let λ be the arc which the earth describes in a day, by its mean sidereal motion; then by (481) we have

$$\alpha_{,} = \alpha - i . \lambda \left(\frac{d \alpha}{d t}\right) + \frac{i^2 . \lambda^2}{1.\ 2}\left(\frac{d^2 \alpha}{d t^2}\right) - \frac{i^3 \lambda^3}{1.2.3}\left(\frac{d^3 \alpha}{d t^3}\right) + \text{\&c.};$$

$$\alpha' = \alpha + i' . \lambda \left(\frac{d \alpha}{d t}\right) + \frac{i'^2 . \lambda^2}{1.\ 2}\left(\frac{d^2 \alpha}{d t^2}\right) + \frac{i'^3 . \lambda^3}{1.2.3}\left(\frac{d^3 \alpha}{d t^3}\right) + \text{\&c.},$$

$$\theta_{,} = \theta - i . \lambda \left(\frac{d \theta}{d t}\right) + \frac{i'^2 . \lambda^2}{1.\ 2}\left(\frac{d^2 \theta}{d t^2}\right) - \frac{i^3 \lambda^3}{1.2.3}\left(\frac{d^3 \alpha}{d t^3}\right) + \text{\&c.},$$

$$\theta' = \theta + i' . \lambda \left(\frac{d \theta}{d t}\right) + \frac{i'^2 . \lambda^2}{1.\ 2}\left(\frac{d^2 \theta}{d t^2}\right) + \frac{i'^3 . \lambda^3}{1.2.3}\left(\frac{d^3 \theta}{d t^3}\right) + \text{\&c.}$$

If we substitute in these series for

$$\left(\frac{d^3 \alpha}{d t^3}\right), \quad \left(\frac{d^4 \alpha}{d t^4}\right) \&c. \left(\frac{d^3 \theta}{d t^3}\right), \quad \left(\frac{d^4 \theta}{d t^4}\right), \&c.$$

their values obtained above, we shall have four equations between the five unknown quantities

$$r, \left(\frac{d \alpha}{d t}\right), \quad \left(\frac{d^2 \alpha}{d t^2}\right), \quad \left(\frac{d \theta}{d t}\right), \quad \left(\frac{d^2 \theta}{d t^2}\right).$$

These equations will be the more exact in proportion as we consider a greater number of terms in the series. We shall thus have

$$\left(\frac{d \alpha}{d t}\right), \quad \left(\frac{d^2 \alpha}{d t^2}\right), \quad \left(\frac{d \theta}{d t}\right), \quad \left(\frac{d^2 \theta}{d t^2}\right)$$

in terms of r and known quantities; and substituting in equation (4) of the preceding No. it will contain the unknown r only. As to the rest, this method, which shows how to approximate to r by employing three observations only, would require in practice, laborious calculations, and it is a more exact and simple process to consider a greater number of observations by the method of No. 495.

499. When the values of r and $\left(\frac{d r}{d t}\right)$ shall be determined, we shall have those of

$$x, y, z, \left(\frac{d x}{d t}\right), \quad \left(\frac{d y}{d t}\right) \text{ and } \left(\frac{d z}{d t}\right),$$

by means of the equations

$$x = R \cos. A + r \cos. \alpha$$
$$y = R \sin. A + r \sin. \alpha$$
$$z = r \tan. \theta$$

and of their differentials divided by d t, viz.

$$\left(\frac{d x}{d t}\right) = \left(\frac{d R}{d t}\right) \cos. A - R \left(\frac{d A}{d t}\right) \sin. A + \left(\frac{d r}{d t}\right) \cos. \alpha - r \left(\frac{d \alpha}{d t}\right) \sin. \alpha;$$

$$\left(\frac{d y}{d t}\right) = \left(\frac{d R}{d t}\right) \sin. A + R \left(\frac{d A}{d t}\right) \cos. A + \left(\frac{d r}{d t}\right) \sin. \alpha + r \left(\frac{d \alpha}{d t}\right) \cos. \alpha;$$

$$\left(\frac{d z}{d t}\right) = \left(\frac{d r}{d t}\right) \tan. \theta + \frac{r \left(\frac{d \theta}{d t}\right)}{\cos.^2 \theta}.$$

The values of $\left(\frac{d A}{d t}\right)$ and of $\left(\frac{d R}{d t}\right)$ are given by the Theory of the motion of the Earth:

To facilitate the investigation let E be the excentricity of the earth's orbit, and H the longitude of its perihelion; then by the nature of elliptical motion we have

$$\left(\frac{d\,A}{d\,t}\right) = \frac{\sqrt{(1 - E^2)}}{R^2}; \quad - R = \frac{1 - E^2}{1 + E \cos.(A - H)}.$$

These two equations give

$$\left(\frac{d\,R}{d\,t}\right) = \frac{E \sin.(A - H)}{\sqrt{(1 - E^2)}}.$$

Let R′ be the radius-vector of the earth corresponding to the longitude A of this planet augmented by a right angle; we shall have

$$R' = \frac{1 - E^2}{1 - E \sin.(A - H)};$$

whence is derived

$$E \sin.(A - H) = \frac{R' - 1 + E^2}{R'};$$

$$\therefore \left(\frac{d\,R}{d\,t}\right) = \frac{R' + E^2 - 1}{R' - \sqrt{(1 - E^2)}}.$$

If we neglect the square of the excentricity of the earth's orbit, which is very small, we shall have

$$\left(\frac{d\,A}{d\,t}\right) = \frac{1}{R^2}; \quad \left(\frac{d\,R}{d\,t}\right) = R' - 1;$$

the preceding values of $\left(\frac{d\,x}{d\,t}\right)$ and $\left(\frac{d\,y}{d\,t}\right)$ will hence become

$$\left(\frac{d\,x}{d\,t}\right) = (R' - 1)\cos.A - \frac{\sin.A}{R} + \left(\frac{d\,r}{d\,t}\right)\cos.\alpha - r\left(\frac{d\,\alpha}{d\,t}\right)\sin.\alpha;$$

$$\left(\frac{d\,y}{d\,t}\right) = (R' - 1)\sin.A + \frac{\cos.A}{R} + \left(\frac{d\,r}{d\,t}\right)\sin.\alpha + r\left(\frac{d\,\alpha}{d\,t}\right)\cos.\alpha;$$

R, R′, and A being given immediately by the tables of the sun, the estimate of the six quantities x, y, z, $\left(\frac{d\,x}{d\,t}\right)$, $\left(\frac{d\,y}{d\,t}\right)$, $\left(\frac{d\,z}{d\,t}\right)$ will be easy when r and $\left(\frac{d\,r}{d\,t}\right)$ shall be known. Hence we derive the elements of the orbit of the comet after this mode.

The indefinitely small sector, which the projection of the radius-vector and the comet upon the plane of the ecliptic describes during the element of time d t, is $\frac{x\,d\,y - y\,d\,x}{2}$; and it is evident that this sector is positive or negative, according as the motion of the comet is direct or retrograde. Thus in forming the quantity $x\left(\frac{d\,y}{d\,t}\right) - y\left(\frac{d\,x}{d\,t}\right)$, it will indicate by its sign, the direction of the motion of the comet.

To determine the position of the orbit, call φ its inclination to the ecliptic, and I the longitude of the node, which would be ascending if the motion of the comet were direct or progressive. We shall have

$$z = y \cos. I \tan. \varphi - x \sin. I \tan. \varphi$$

These two equations give

$$\tan. I = \frac{y\left(\frac{dz}{dt}\right) - z\left(\frac{dy}{dt}\right)}{x\left(\frac{dz}{dt}\right) - z\left(\frac{dx}{dt}\right)};$$

$$\tan. \varphi = \frac{y\left(\frac{dz}{dt}\right) - z\left(\frac{dy}{dt}\right)}{\sin. I\left\{x\left(\frac{dy}{dt}\right) - y\left(\frac{dx}{dt}\right)\right\}}.$$

Wherein since φ ought always to be positive and less than a right angle, the sign of sin. I is known. But the tangent of I and the sign of its sine being determined, the angle I is found completely. This angle is the longitude of the ascending node of the orbit, if the motion is progressive; but to this we must add two right angles, in order to get the longitude of the node when the motion is retrograde. It would be more simple to consider only progressive motions, by making vary φ, the inclination of the orbits, from zero to two right angles; for it is evident that then the retrograde motions correspond to an inclination greater than a right angle.

In this case, tan. φ has the same sign as $x\left(\frac{dy}{dt}\right) - y\left(\frac{dx}{dt}\right)$, which will determine sin. I, and consequently the angle I, which always expresses the longitude of the ascending node.

If a, a e be the semi-axis major and the excentricity of the orbit, we have (by 492) in making m = 1,

$$\frac{1}{a} = \frac{2}{\varrho} - \left(\frac{dx}{dt}\right)^2 - \left(\frac{dy}{dt}\right)^2 - \left(\frac{dz}{dt}\right)^2;$$

$$a(1 - e^2) = 2\varrho - \frac{\varrho^2}{a} - \left\{x\left(\frac{dx}{dt}\right) + y\left(\frac{dy}{dt}\right) + z\left(\frac{dz}{dt}\right)\right\}^2.$$

The first of these equations gives the semi-axis major, and the second the excentricity. The sign of the function $x\left(\frac{dx}{dt}\right) + y\left(\frac{dy}{dt}\right) + z\left(\frac{dz}{dt}\right)$ shows whether the comet has already passed its perihelion; for it approaches if this function is negative; and in the contrary case, the comet recedes from that point.

Let T be the interval of time comprised between the epoch and passage of the comet over the perihelion; the two first of equations (f) (480) will give, observing that m being supposed unity we have $n = a^{-\frac{3}{2}}$,

$$\varrho = a\,(1 - e \cos. u)$$
$$T = a^{\frac{3}{2}}\,(u - e \cos. u).$$

The first of these equations gives the angle u, and the second T. This time added to or subtracted from the epoch, according as the comet approaches or leaves its perihelion, will give the instant of its passage over this point. The values of x, y, determine the angle which the projection of the radius-vector ϱ makes with the axis of x; and since we know the angle I, formed by this axis and by the line of the nodes, we shall have the angle which this last line forms with the projection of ϱ; whence we derive by means of the inclination φ of the orbit, the angle formed by the line of the nodes and the radius ϱ. But the angle u being known, we shall have by means of the third of the equations (f), the angle v which this radius forms with the line of the apsides; we shall therefore have the angle comprised between the two lines of the apsides and of the nodes, and consequently, the position of the perihelion. All the elements of the orbit will thus be determined.

500. These elements are given, by the preceding investigations, in terms of r, $\left(\frac{d r}{d t}\right)$ and known quantities; and since $\left(\frac{d r}{d t}\right)$ is given in terms of r by No. 497, the elements of the orbit will be functions of r and known quantities. If one of them were given, we should have a new equation, by means of which we might determine r; this equation would have a common divisor with equation (4) of No. 497; and seeking this divisor by the ordinary methods, we shall obtain an equation of the first degree in terms of r; we should have, moreover, an equation of condition between the data of the observations, and this equation would be that which ought to subsist, in order that the given element may belong to the orbit of the comet.

Let us apply this consideration to the case of nature. First suppose that the orbits of the comets are ellipses of great excentricity, and are nearly parabolas, in the parts of their orbits in which these stars are visible. We may therefore without sensible error suppose $a = \infty$, and consequently $\frac{1}{a} = 0$; the expression for $\frac{1}{a}$ of the preceding No. will therefore give

$$0 = \frac{2}{\varrho} - \frac{d x^2 + d y^2 + d z^2}{d t^2}.$$

If we then substitute for $\left(\frac{d x}{d t}\right)$, $\left(\frac{d y}{d t}\right)$ and $\left(\frac{d z}{d t}\right)$ their values found in the same No., we shall have after all the reductions and neglecting the square of R′ — 1,

$$0 = \left(\frac{d r}{d t}\right)^2 + r^2 \left(\frac{d \alpha}{d t}\right)^2 + \left\{\frac{d r}{d t} \tan. \theta + \frac{r\left(\frac{d \theta}{d t}\right)}{\cos.^2 \theta}\right\}^2$$

$$+ 2\left(\frac{d r}{d t}\right) \cdot \left\{(R'-1) \cos. (A-\alpha) - \frac{\sin. (A-\alpha)}{R}\right\} \cdot \cdot (5)$$

$$+ 2 r \left(\frac{d \alpha}{d t}\right) \left\{(R'-1) \sin. (A-\alpha) + \frac{\cos. (A-\alpha)}{R}\right\}$$

$$+ \frac{1}{R^2} - \frac{2}{\varrho}.$$

Substituting in this equation for $\left(\frac{d r}{d t}\right)$ its value

$$-\frac{r}{2\left(\frac{d \alpha}{d t}\right)} \cdot \left\{\left(\frac{d^2 \alpha}{d t^2}\right) + \mu' \sin. (A - \alpha)\right\},$$

found in No. 497, and then making

$$4\left(\frac{d \alpha}{d t}\right)^2 . B = 4\left(\frac{d \alpha}{d t}\right)^4 + \left\{\left(\frac{d^2 \alpha}{d t^2}\right) + \mu' \sin. (A - \alpha)\right\}^2$$

$$+ \left\{\tan. \theta . \left(\frac{d^2 \alpha}{d t^2}\right) + \mu' \tan. \theta \sin. (A - \alpha) - \frac{2\left(\frac{d \alpha}{d t}\right) . \left(\frac{d \theta}{d t}\right)}{\cos.^2 \theta}\right\}^2,$$

and

$$C = \frac{\left\{\left(\frac{d^2 \alpha}{d t^2}\right) + \mu' \sin. (A-\alpha)\right\}}{\left(\frac{d \alpha}{d t}\right)} . \left\{\frac{\sin. (A-\alpha)}{R} - (R' - 1) \cos. (A - \alpha)\right\}$$

$$+ 2\left(\frac{d \alpha}{d t}\right) \left\{(R' - 1) \sin. (A - \alpha) + \frac{\cos. (A - \alpha)}{R}\right\},$$

we shall have

$$0 = B r^2 + C r + \frac{1}{R^2} - \frac{2}{\varrho}$$

and consequently

$$\varrho^2 \left\{B r^2 + C r + \frac{1}{R^2}\right\}^2 = 4.$$

This equation rising only to the sixth degree, is in that respect, more simple than equation (4) of No. (497); but it belongs to the parabola alone, whereas the equation (4) equally regards every species of conic section.

501. We perceive by the foregoing investigation, that the determination of the parabolic orbits of the comets, leads to more equations than unknown quantities; and that, therefore, in combining these equations in different ways, we can form as many different methods of calculating the orbits. Let us examine those which appear to give the most exact results, or which seem least susceptible of the errors of observations.

It is principally upon the values of the second differences $\left(\frac{d^2 \alpha}{d t^2}\right)$ and $\left(\frac{d^2 \theta}{d t^2}\right)$, that these errors have a sensible influence. In fact, to determine them, we must take the finite differences of the geocentric longitudes and latitudes of the comet, observed during a short interval of time. But these differences being less than the first differences, the errors of observations are a greater aliquot part of them; besides this, the formulas of No. 495 which determine, by the comparison of observations, the values of α, θ, $\left(\frac{d \alpha}{d t}\right)$, $\left(\frac{d \theta}{d t}\right)$, $\left(\frac{d^2 \alpha}{d t^2}\right)$ and $\left(\frac{d^2 \theta}{d t^2}\right)$ give with greater precision the four first of these quantities than the two last. It is, therefore, desirable to rest as little as possible upon the second differences of α and θ; and since we cannot reject both of them together, the method which employs the greater, ought to give the more accurate results. This being granted let us resume the equations found in Nos. 497, &c.

$$\rho^2 = \frac{r^2}{\cos.^2 \theta} + 2 R r \cos. (A - \alpha) + R^2;$$

$$\left(\frac{d r}{d t}\right) = \frac{R \sin. (A - \alpha)}{2 \left(\frac{d \alpha}{d t}\right)} \cdot \left\{\frac{1}{R^3} - \frac{1}{\rho^3}\right\} - \frac{r \cdot \left(\frac{d^2 \alpha}{d t^2}\right)}{2 \left(\frac{d \alpha}{d t}\right)}, \quad (L)$$

$$\left(\frac{d r}{d t}\right) = -\frac{1}{2} r \left\{ \frac{\left(\frac{d^2 \theta}{d t^2}\right)}{\left(\frac{d \theta}{d t}\right)} + 2 \left(\frac{d \theta}{d t}\right) \tan. \theta + \frac{\left(\frac{d \alpha}{d t}\right)^2 \sin. \theta \cos. \theta}{\left(\frac{d \theta}{d t}\right)} \right\}$$

$$+ \frac{R \sin. \theta \cos. \theta \cos. (A - \alpha)}{2 \left(\frac{d \theta}{d t}\right)} \cdot \left\{\frac{1}{\rho^3} - \frac{1}{R^3}\right\}$$

$$0 = \left(\frac{d r}{d t}\right)^2 + r^2 \left(\frac{d \alpha}{d t}\right)^2 + \left\{\left(\frac{d r}{d t}\right) \tan. \theta + \frac{r \left(\frac{d \theta}{d t}\right)}{\cos.^2 \theta}\right\}^2$$

$$+ 2\left(\frac{dr}{dt}\right).\left\{(R'-1)\cos.(A-\alpha) - \frac{\sin.(A-\alpha)}{R}\right\}$$

$$+ 2r\left(\frac{d\alpha}{dt}\right)\left\{(R'-1)\sin.(A-\alpha) + \frac{\cos.(A-\alpha)}{R}\right\}$$

$$+ \frac{1}{R^2} - \frac{2}{\varrho}.$$

If we wish to reject $\left(\frac{d^2\theta}{dt^2}\right)$, we consider only the first, second and fourth of those equations. Eliminating $\left(\frac{dr}{dt}\right)$ from the last by means of the second, we shall form an equation which cleared of fractions, will contain a term multiplied by $\varrho^6 r^2$, and other terms affected with even and odd powers of r and ϱ. If we put into one side of the equation all the terms affected with even powers of ϱ, and into the other all those which involve its odd powers, and square both sides, in order to have none but even powers of ϱ, the term multiplied by $\varrho^6 r^2$ will produce one multiplied by $\varrho^{12} r^4$. Substituting, therefore, instead of ϱ^2, its value given by the first of equations (L), we shall have a final equation of the sixteenth degree in r. But instead of forming this equation in order afterwards to resolve it, it will be more simple to satisfy by trial the three preceding ones.

If we wish to reject $\left(\frac{d^2a}{dt^2}\right)$, we must consider the first, third and fourth of equations (L). These three equations conduct us also to a final equation of the sixteenth degree in r; and we can easily satisfy by trial.

The two preceding methods appear to be the most exact, which we can employ in the determination of the parabolic orbits of the comets. It is at the same time necessary to have recourse to them, if the motion of the comet in longitude or latitude is insensible, or too small for the errors of observations sensibly to alter its second difference. In this case, we must reject that of the equations (L), which contains this difference. But although in these methods, we employ only three equations, yet the fourth is useful to determine amongst all the real and positive values of r, which satisfy the system of three equations, that which ought to be selected.

502. The elements of the orbit of a comet, determined by the above process, would be exact, if the values of α, θ and their first and second differences, were rigorous; for we have regarded, after a very simple manner, the excentricity of the terrestrial orbit, by means of the radius-vector R' of the earth, corresponding to its true anomaly + a right angle; we are therefore permitted only to neglect the square of this excen-

tricity, as too small a fraction to produce by its omission a sensible influence upon the results. But θ, α and their differences, are always susceptible of any degree of inaccuracy, both because of the errors of observations, and because these differences are only obtained approximately. It is therefore necessary to correct the elements, by means of three distant observations, which can be done in many ways; for if we know nearly, two quantities relative to the motion of a comet, such that the radius-vector corresponding to two observations, or the position of the node, and the inclination of the orbit; calculating the observations, first with these quantities and afterwards with others differing but little from them, the law of the differences between the results, will easily show the necessary corrections. But amongst the combinations taken two and two, of the quantities relative to the motion of comets, there is one which ought to produce greatest simplicity, and which for that reason should be selected. It is of importance, in fact, in a problem so intricate, and complicated, to spare the calculator all superfluous operations. The two elements which appear to present this advantage, are the perihelion distance, and the instant when the comet passes this point. They are not only easy to be derived from the values of r and $\left(\frac{d\,r}{d\,t}\right)$; but it is very easy to correct them by observations, without being obliged for every variation which they undergo, to determine the other corresponding elements of the orbit.

Resuming the equation found in No. 492

$$a\,(1 - e^2) = 2\,\varrho - \frac{\varrho^2}{a} - \frac{\varrho^2 . \, d\,\varrho^2}{d\,t^2};$$

a $(1 - e^2)$ is the semi-parameter of the conic section of which a is the semi axis-major, and a e the excentricity. In the parabola, where a is infinite, and e equal to unity, a $(1 - e^2)$ is double the perihelion distance: let D be this distance: the preceding equation becomes relatively to this curve

$$D = \varrho - \frac{1}{2}\left(\frac{\varrho\,d\,\varrho}{d\,t}\right)^2.$$

$\frac{\varrho\,d\,\varrho}{d\,t}$ is equal to $\frac{\frac{1}{2}\,d\,\varrho^2}{d\,t^2}$; in substituting for ϱ^2 its value $\frac{r^2}{\cos.^2\theta} + 2\,R\,r \times \cos.\,(A - \alpha) + R^2$, and for $\left(\frac{d\,R}{d\,t}\right)$ and $\left(\frac{d\,A}{d\,t}\right)$, their values found in No. 499, we shall have

$$\frac{\varrho\,d\,\varrho}{d\,t} = \frac{r}{\cos.^2\theta}\left\{\left(\frac{d\,r}{d\,t}\right) + r\left(\frac{d\,\theta}{d\,t}\right)\tan.\,\theta\right\} + R\left(\frac{d\,r}{d\,t}\right)\cos.\,(A - \alpha)$$

$$+ r\left\{(R' - 1)\cos.(A - \alpha) - \frac{\sin.(A - \alpha)}{R}\right\}$$

$$+ r\,R\left(\frac{d\,\alpha}{d\,t}\right)\sin.(A - \alpha) + R\,(R' - 1).$$

Let P represent this quantity; if it is negative, the radius-vector decreases, and consequently, the comet tends towards its perihelion. But it goes off into the distance, if P is negative. We have then

$$D = \varrho - \frac{1}{2}P^2;$$

the angular distance v of the comet from its perihelion, will be determined from the polar equation to the parabola,

$$\text{cor.}^2\,\frac{1}{2}\,v = \frac{D}{\varrho};$$

and finally we shall have the time employed to describe the angle v, by the table of the motion of the comets. This time added to or subtracted from that of the epoch, according as P is negative or positive, will give the instant when the comet passes its perihelion.

503. Recapitulating these different results, we shall have the following method to determine the parabolic orbits of the comets.

General method of determining the orbits of the comets.

This method will be divided into two parts; in the first, we shall give the means of obtaining approximately, the perihelion distance of the comet and the instant of its passage over the perihelion; in the second, we shall determine all the elements of the orbit on the supposition that the former are known.

Approximate determination of the Perihelion distance of the comet, and the instant of its passage over the perihelion.

We shall select three, four, five, &c. observations of the comet equally distant from one another as nearly as possible; with four observations we shall be able to consider an interval of 30°; with five, an interval of 36°, or 40° and so on for the rest; but to diminish the influence of their errors, the interval comprised between the observations must be greater, in proportion as their number is greater. This being supposed,

Let β, β', β'', &c. be the successive geocentric longitudes of the comet, γ, γ', γ'' the corresponding latitudes, these latitudes being supposed positive or negative according as they are north or south. We shall divide the difference $\beta' - \beta$, by the number of days between the first and second observation; we shall divide in like manner the difference $\beta'' - \beta'$ by the number of days between the second and third observation; and so on. Let $\delta\beta$, $\delta\beta'$, $\delta\beta''$, &c. be these quotients.

We next divide the difference $\delta\beta' - \delta\beta$ by the number of days between the first observation and the third; we divide, in like manner, the difference $\delta\beta'' - \delta\beta'$ by the number of days between the second and fourth observations; similarly we divide the difference $\delta\beta''' - \delta\beta''$ by the number of days between the third and fifth observation, and so on. Let $\delta^2\beta$, $\delta^2\beta'$, $\delta^2\beta''$, &c. denote these quotients.

Again, we divide the difference $\delta^2\beta' - \delta^2\beta$ by the number of days which separate the first observation from the fourth; we divide in like manner $\delta^2\beta'' - \delta^2\beta'$ by the number of days between the second observation and the fifth, and so on. Make $\delta^3\beta$, $\delta^3\beta'$, &c. these quotients. Thus proceeding, we shall arrive at $\delta^{n-1}\beta$, n being the number of observations employed.

This being done, we proceed to take as near as may be a mean epoch between the instants of the two extreme observations, and calling i, i′, i″, &c. the number of days, distant from each observation, i, i′, i″, &c. ought to be supposed negative for the observations made prior to this epoch; the longitude of the comet, after a small number z of days reckoned from the Epoch will be expressed by the following formula:

$$\left\{\begin{array}{l} \beta - i\,\delta\beta + i\,i'\,\delta^2\beta - i\,i'\,i''\,\delta^3\beta + \&c. \\ +z\{\delta\beta - (i+i')\delta^2\beta + (i\,i' + i\,i'' + i'\,i'')\delta^3\beta - (i\,i'\,i'' + i\,i'\,i''' + i\,i''\,i''' + .. \quad (p) \\ i'\,i''\,i''')\,\delta^4\beta + \&c.\} \\ +z^2\{\delta^2\beta - (i+i'+i'')\,\delta^3\beta + (i\,i' + i\,i'' + i\,i''' + i'\,i''' + i'' + i''')\,\delta^4\beta - \&c.\} \end{array}\right.$$

The coefficients of $-\delta\beta$, $+\delta^2\beta$, $-\delta^3\beta$, &c. in the part independent of z are 1st the numbers i and i′, secondly the sum of the products two and two of the three numbers i, i′, i″; thirdly the sum of the products three and three, of the four numbers i, i′, i″, i‴, &c.

The coefficients of $-\delta^3\beta$, $+\delta^4\beta$, $-\delta^5\beta$, &c. in the part multiplied by z^2, are first, the sum of the three numbers i, i′, i″; secondly of the products two and two of the four numbers i, i′, i″, i‴; thirdly the sum of the products three and three of the five numbers i, i′, i″, i‴, i⁗, &c.

Instead of forming these products, it is as simple to develope the function $\beta + (z - i)\,\delta\beta + (z - i)(z - i')\,\delta^2\beta + (z - i)(z - i')(z - i'') \times \delta^3\beta +$ &c. rejecting the powers of z superior to the square. This gives the preceding formula.

If we operate in a similar manner upon the observed geocentric latitudes of the comet; its geocentric latitude, after the number z of days from the epoch, will be expressed by the formula (p) in changing β into γ. Call (q) the equation (p) thus altered. This being done,

α will be the part independent of z in the formula (p); and θ that in the formula (q).

Reducing into seconds the coefficient of z in the formula (p), and taking from the tabular logarithm of this number of seconds, the logarithm 4,0394622, we shall have the logarithm of a number which we shall denote by a.

Reducing into seconds the coefficients of z^2 in the same formula, and taking from the logarithm of this number of seconds, the logarithm 1.9740144, we shall have the logarithm of a number, which we shall denote by b.

Reducing in like manner into seconds the coefficients of z and z^2 in the formula (q) and taking away respectively from the logarithms of these numbers of seconds, the logarithms, 4,0394622 and 1,9740144, we shall have the logarithms of two numbers, which we shall name h and l.

Upon the accuracy of the values of a, b, h, l, depends that of the method; and since their formation is very simple, we must select and multiply observations, so as to obtain them with all the exactness which the observations will admit of. It is perceptible that these values are only the quantities $\left(\frac{d\alpha}{dt}\right)$, $\left(\frac{d^2\alpha}{dt^2}\right)$, $\left(\frac{d\theta}{dt}\right)$, $\left(\frac{d^2\theta}{dt^2}\right)$, which we have expressed more simply by the above letters.

If the number of observations is odd, we can fix the Epoch at the instant of the mean observation; which will dispense with calculating the parts independent of z in the two preceding formulas; for it is evident, that then these parts are respectively equal to the longitude and latitude of the mean observation.

Having thus determined the values of α, a, b, θ, h, and l, we shall determine the longitude of the sun, at the instant we have selected for the epoch, R the corresponding distance of the Earth from the sun, and R′ the distance which answers to E augmented by a right angle. We shall have the following equations

$$\varrho^2 = \frac{x^2}{\cos.^2\theta} - 2\,R\,x\cos.(E-\alpha) + R^2 \;.\;.\;.\;.\;.\;.\;. \quad (1)$$

$$y = R\,\frac{\sin.(E-\alpha)}{2\,a}\left\{\frac{1}{\varrho^3} - \frac{1}{R^3}\right\} - \frac{b\,x}{2\,a} \;.\;.\;.\;.\;.\;.\;. \quad (2)$$

$$\left.\begin{aligned} y = -x\left\{h\tan.\theta + \frac{1}{2\,h} + \frac{a^2\sin.\theta.\cos.\theta}{2\,h}\right\} \\ + \frac{R\sin.\theta\cos.\theta}{2\,h}\cos.(E-\alpha)\left\{\frac{1}{R^3} - \frac{1}{\varrho^3}\right\} \end{aligned}\right\} \;.\;.\;.\;. \quad (3).$$

$$0 = y^2 + a^2x^2 + \left(y\tan.\theta + \frac{h\,x}{\cos.^2\theta}\right)^2 + 2\,y\left\{\frac{\sin.(E-\alpha)}{R}\right.$$

$$- (R' - 1) \cos. (E - \alpha)\Big\} - 2\, a\, x \Big\{(R' - 1) \sin. (E - \alpha) + \frac{\cos. (E - \alpha)}{R}\Big\} + \frac{1}{R^2} - \frac{2}{\varrho} \quad \ldots \ldots \quad (4)$$

To derive from these equations the values of the unknown quantities x, y, ϱ, we must consider, signs being neglected, whether b is greater or less than l. In the first case we shall make use of equation (1), (2), and (4). We shall form a first hypothesis for x, supposing it for instance equal to unity; and we then derive by means of equations (1), (2), the values of ϱ and of y. Next we substitute these values in the equation (4); and if the result is 0, this will be a proof that the value of x has been rightly chosen. But if it be negative we must augment the value of x, and diminish it if the contrary. We shall thus obtain, by means of a small number of trials the values of x, y and ϱ. But since these unknown quantities may be susceptible of many real and positive values, we must seek that which satisfies exactly or nearly so the equation (3).

In the second case, that is to say, if l be greater than b, we shall use the equations (1), (3), (4), and then equation (2) will give the verification.

Having thus the values of x, y, ϱ, we shall have the quantity

$$P = \frac{x}{\cos.^2 \theta} . \{y + h\, x \tan. \theta\} - R\, y \cos. (E - \alpha)$$
$$+ x \left\{\frac{\sin. (E - \alpha)}{R} - (R' - 1) \cos. (E - \alpha)\right\} - R\, a\, x \sin (E - \alpha)$$
$$+ R . (R' - 1).$$

The Perihelion distance D of the comet will be

$$D = \varrho - \frac{1}{2} P^2;$$

the cosine of its anomaly v will be given by the equation

$$\cos^2 \frac{1}{2} v = \frac{D}{\varrho};$$

and hence we obtain, by the table of the motion of the comets, the time employed to describe the angle v. To obtain the instant when the comet passes the perihelion, we must add this time to, or subtract it from the epoch according as P is negative or positive. For in the first case the comet approaches, and in the second recedes from, the perihelion.

Having thus nearly obtained the perihelion distance of the comet, and the instant of its passage over the perihelion; we are enabled to correct them by the following method, which has the advantage of being independent of the approximate values of the other elements of the orbit.

An exact Determination of the elements of the orbit, when we know approximate values of the perihelion distance of the comet, and of the instant of its passage over the perihelion.

We shall first select three distant observations of the comet; then taking the perihelion distance of the comet, and the instant of its crossing the perihelion, determined as above, we shall calculate the three anomalies of the comet and the corresponding radius-vectors corresponding to the instants of the three observations. Let v, v′, v″ be these anomalies, those which precede the passage over the perihelion being supposed negative. Also let ϱ, ϱ' ϱ'' be the corresponding radius-vectors of the comet; then v′ — v, v″ — v will be the angles comprised by ϱ and ϱ' and by ϱ, ϱ''. Let U be the first of these angles, U′ the second. Again, call α, α' α'' the three observed geocentric longitudes of the comet, referred to a fixed equinox; θ, θ', θ'' its three geocentric latitudes, the south latitudes being negative. Let β, β', β'' be the three corresponding heliocentric longitudes and ϖ, ϖ', ϖ'', its three heliocentric latitudes. Lastly call E, E′, E″ the three corresponding longitudes of the sun, and R, R′, R″ its three distances to the center of the earth.

Conceive that the letter S indicates the center of the sun, T that of the earth, and C that of the comet, C′ that of its projection upon the plane of the ecliptic. The angle S T C′ is the difference of the geocentric longitudes of the sun and of the comet. Adding the logarithm of the cosine of this angle, to the logarithm of the cosine of the geocentric latitude of the comet, we shall have the logarithm of the cosine of the angle S T C. We know, therefore, in the triangle S T C, the side S T or R, the side S C or ϱ, and the angle S T C, to find the angle C S T. Next we shall have the heliocentric latitude ϖ of the comet, by means of the equation

$$\sin. \varpi = \frac{\sin. \theta \sin. \text{C S T}}{\sin. \text{C T S}}.$$

The angle T S C′ is the side of a spherical right angled triangle, of which the hypothenuse is the angle T S C, and of which one of the sides is the angle ϖ. Whence we shall easily derive the angle T S C′, and consequently the heliocentric longitude β of the comet.

We shall have after the same manner ϖ', β'; ϖ'', β''; and the values of β, β', β'' will show whether the motion of the comet be direct or retrograde.

If we imagine the two arcs of latitude ϖ, ϖ', to meet at the pole of the ecliptic, they would make there an angle equal to $\beta' - \beta$; and in the spherical triangle formed by this angle, and by the sides $\frac{\pi}{2} - \varpi$, $\frac{\pi}{2} - \varpi'$ π being the semi-circumference, the side opposite to the angle $\beta' - \beta$ will be the angle at the sun comprised between the radius-vectors ϱ, and ϱ'. We shall easily determine this by spherical Trigonometry, or by the formula

$$\sin.^2 \frac{1}{2} V = \cos.^2 \frac{1}{2} (\varpi + \varpi') - \cos^2 \frac{1}{2} (\beta' - \beta) \cos. \varpi \cos. \varpi',$$

in which V represents this angle; so that if we call A the angle of which the sine squared is

$$\cos^2 \frac{1}{2} (\beta' - \beta) \cos. \varpi . \cos. \varpi',$$

and which we shall easily find by the tables, we shall have

$$\sin.^2 \frac{1}{2} V = \cos. \left(\frac{1}{2} \varpi + \frac{1}{2} \varpi' + A\right) \cos. \left(\frac{1}{2} \varpi + \frac{1}{2} \varpi' - A\right).$$

If in like manner we call V′ the angle formed by the two radius-vectors ϱ, ϱ'', we have

$$\sin.^2 \frac{1}{2} V' = \cos. \left(\frac{1}{2} \varpi + \frac{1}{2} \varpi' + A'\right) \cos. \left(\frac{1}{2} \varpi + \frac{1}{2} \varpi' - A'\right)$$

A′ being what A becomes, when ϖ', β' are changed into ϖ'', β''.

If, however, the perihelion distance and the instant of the comet's crossing the perihelion, were exactly determined, and if the observations were rigorously exact, we should have

$$V = U, \ V' = U';$$

But since that is hardly ever the case, we shall suppose

$$m = U - V; \ m' = U' - V'.$$

We shall here observe that the revolution of the triangle S T C, gives for the angle C S T two different values: for the most part the nature of the motion of the comets, will show that which we ought to use, and the more plainly if the two values are very different; for then the one will place the comet more distant from the earth, than the other; and it will be easy to judge, by the apparent motion of the comet at the instant of observation, which ought to be preferred. But if there remains any uncertainty, we can always remove it, by selecting the value which renders V and V′ least different from U and U′.

We next make a second hypothesis in which, retaining the same passage over the perihelion as before, we shall suppose the perihelion distance to vary by a small quantity; for instance, by the fiftieth part of its value, and we shall investigate on this hypothesis, the values of U—V, U′ — V′. Let then

$$n = U - V;\ n' = U' - V'.$$

Lastly, we shall frame a third hypothesis, in which, retaining the same perihelion distance as in the first, we shall suppose the instant of the passage over the perihelion to vary by a half-day, or a day more or less. In this new hypothesis we must find the values of

$$U - V \text{ and of } U' - V';$$

which suppose to be

$$p = U - V,\ p' = U' - V'.$$

Again, if we suppose u the number by which we ought to multiply the supposed variation in the perihelion distance in order to make it the true one, and t the number by which we ought to multiply the supposed variation of the instant when the comet passes over the perihelion in order to make it the true instant, we shall have the two following equations:

$$(m - n)\,u + (m - p)\,t = m;$$
$$(m' - n')\,u + (m' - p')\,t = m';$$

whence we derive u and t and consequently the perihelion distance corrected, and the true instant of the comet's passing its perihelion.

The preceding corrections suppose the elements determined by the first approximation, to be sufficiently near the truth for their errors to be regarded as infinitely small. But if the second approximation should not even suffice, we can have recourse to a third, by operating upon the elements already corrected as we did upon the first; provided care be taken to make them undergo smaller variations. It will also be sufficient to calculate by these corrected elements the values of U — V, and of U′ — V′. Calling them M, M′, we shall substitute them for m, m′ in the second members of the two preceding equations. We shall thus have two new equations which will give the values of u and t, relative to the corrections of these new elements.

Thus having obtained the true perihelion distance and the true instant of the comet's passing its perihelion, we obtain the other elements of the orbit in this manner.

Let j be the longitude of the node which would be ascending if the motion of the comet were direct, and φ the inclination of the orbit. We shall have by comparison of the first and last observation,

$$\tan. j = \frac{\tan. \varpi \sin. \beta' - \tan. \varpi' \sin. \beta}{\tan. \varpi \cos. \beta'' - \tan. \varpi'' \cos. \beta};$$

$$\tan. \varphi = \frac{\tan. \varpi''}{\sin. (\beta'' - j)}.$$

Since we can compare thus two and two together, the three observations, it will be more correct to select those which give to the above fractions, the greatest numerators and the greatest denominators.

Since tan. j may equally belong to j and π + j, j being the smallest of the positive angles containing its value, in order to find that which we ought to fix upon, we shall observe that φ is positive and less than a right angle; and that sin. $(\beta'' - j)$ ought to have the same sign as tan. ϖ''. This condition will determine the angle j, and this will be the position of the ascending node, if the motion of the comet is direct; but if retrograde we must add two right angles to the angle j to get the position of the node.

The hypothenuse of the spherical triangle whose sides are $\beta'' - j$ and ϖ'', is the distance of the comet from its ascending node in the third observation; and the difference between v'' and this hypothenuse is the interval between the node and the perihelion computed along the orbit.

If we wish to give to the theory of a comet all the precision which observations will admit of, we must establish it upon an *aggregate* of the best observations; which may be thus done. Mark with one, two, &c. dashes or strokes the letters m, n, p relative to the second observation, the third, &c. all being compared with the first observation. Hence we shall form the equations

$$\begin{aligned}(m - n)\,u + (m - p)\,t &= m\\ (m' - n')\,u + (m' - p')\,t &= m'\\ (m'' - n'')\,u + (m'' - p'')\,t &= m''\\ \&c. &= \&c.\end{aligned}$$

Again, combining these equations so as to make it easier to determine u and t, we shall have the corrections of the perihelion distance and of the instant of the comet's passing its perihelion, founded upon the *aggregate* of these observations. We shall have the values of

$$\beta, \beta', \beta'', \&c.\ \varpi, \varpi', \varpi'', \&c.,$$

and obtain

$$\tan. j = \frac{\tan. \varpi\,(\sin. \beta' + \sin. \beta'' + \&c.) - \sin. \beta\,(\tan. \varpi' + \tan. \varpi'' + \&c.)}{\tan. \varpi\,(\cos. \beta' + \cos. \beta'' + \&c.) - \cos. \beta\,(\tan. \varpi' + \tan. \varpi'' + \&c.)}$$

$$\tan. \varphi = \frac{\tan. \varpi' + \tan. \varpi'' + \&c.}{\sin. (\beta' - j) + \sin. (\beta'' - j) + \&c.}.$$

504. There is a case, very rare indeed, in which the orbit of a comet can be determined rigorously and simply; it is that where the comet has been observed in its two nodes. The straight line which joins these

two observed positions, passes through the center of the sun and coincides with the line of the nodes. The length of this straight line is determined by the time elapsed between the two observations. Calling T this time reduced into decimals of a day, and denoting by c the straight line in question, we shall have (No. 493)

$$c = \frac{1}{2}\sqrt[3]{\frac{T^2}{(9.688724)^2}}.$$

Let β be the heliocentric longitude of the comet, at the moment of the first observation; ϱ its radius-vector; r its distance from the earth; and α its geocentric longitude. Let, moreover, R be the radius of the terrestrial orbit, at the same instant, and E the corresponding longitude of the sun. Then we shall have

$$\varrho \sin. \beta = r \sin. \alpha - R \sin. E;$$
$$\varrho \cos. \beta = r \cos. \alpha - R \cos. E.$$

Now $\pi + \beta$ will be the heliocentric longitude of the comet at the instant of the second observation; and if we distinguish the quantities ϱ, α, r, R, and E relative to this instant by a dash, we shall have

$$\varrho' \sin. \beta = R' \sin. E' - r' \sin. \alpha';$$
$$\varrho' \cos. \beta = R' \cos. E' - r' \cos. \alpha'.$$

These four equations give

$$\tan. \beta = \frac{r \sin \alpha - R \sin. E}{r \cos. \alpha - R \cos. E} = \frac{r' \sin. \alpha' - R' \sin. E'}{r' \cos. \alpha' - R' \cos. E'};$$

whence we obtain

$$r' = \frac{R R' \sin. (E - E') - R r \sin. (\alpha - E')}{r \sin. (\alpha' - \alpha) - R \sin. (\alpha' - E)}.$$

We have also

$$(\varrho + \varrho') \sin. \beta = r \sin. \alpha - r' \sin. \alpha' - R \sin. E + R' \sin. E'$$
$$(\varrho + \varrho') \cos. \beta = r \cos. \alpha - r' \cos. \alpha' - R \cos. E + R' \cos. E'.$$

Squaring these two equations, and adding them together, and substituting c for $\varrho + \varrho'$, we shall have

$$\begin{aligned} c^2 = {} & R^2 - 2 R R' \cos. (E' - E) + R'^2 \\ & + 2 r \{R' \cos. (\alpha - E') - R \cos. (\alpha - E)\} \\ & + 2 r' \{R \cos. (\alpha' - E) - R' \cos. (\alpha' - E')\} \\ & + r^2 - 2 r r' \cos. (\alpha' - \alpha) + r'^2. \end{aligned}$$

If we substitute in this equation for r' its preceding value in terms of r, we shall have an equation in r of the fourth degree, which can be resolved by the usual methods. But it will be more simple to find values of r, r' by trial such as will satisfy the equation. A few trials will suffice for that purpose.

By means of these quantities we shall have β, ϱ and ϱ'. If v be the angle which the radius ϱ makes with the perihelion distance called D; π — v will be the angle formed by this same distance, and by the radius ϱ'. We shall thus have by the equation to the parabola

$$\varrho = \frac{D}{\cos.^2 \frac{1}{2} v}; \quad \varrho' = \frac{D}{\sin.^2 \frac{1}{2} v};$$

which give

$$\tan.^2 \frac{1}{2} v = \frac{\varrho}{\varrho'}; \quad D = \frac{\varrho \varrho'}{\varrho + \varrho'}.$$

We shall therefore have the anomaly v of the comet, at the instant of the first observation, and its perihelion distance D, whence it is easy to find the position of the perihelion, at the instant of the passage of the comet over that point. Thus, of the five elements of the orbit of the comet, four are known, namely, the perihelion distance, the position of the perihelion, the instant of the comet's passing the perihelion, and the position of the node. It remains to learn the inclination of the orbit; but for that purpose it will be necessary to have recourse to a third observation, which will also serve to select from amongst the real and positive roots of the equation in r, that which we ought to make use of.

505. The supposition of the parabolic motion of comets is not rigorous; it is, at the same time, not at all probable, since compared with the cases that give the parabolic motion, there is an infinity of those which give the elliptic or hyperbolic motions. Besides, a comet moving in either a parabolic or hyperbolic orbit, will only once be visible; thus we may with reason suppose these bodies, if ever they existed, long since to have disappeared; so that we shall now observe those only which, moving in orbits returning into themselves, shall, after greater or less incursions into the regions of space, again approach their center the sun. By the following method, we shall be able to determine, within a few years, the period of their revolutions, when we have given a great number of very exact observations, made before and after the passage over the perihelion.

Let us suppose we have four or a greater number of good observations, which embrace all the visible part of the orbit, and that we have determined, by the preceding method, the parabola, which nearly satisfies these observations. Let v, v', v'', v''', &c. be the corresponding anomalies; ϱ, ϱ', ϱ'', ϱ''', &c. the radius-vectors. Let also

$$v' - v = U, \quad v'' - v = U', \quad v''' - v = U'', \ \&c.$$

Then we shall estimate, by the preceding method with the parabola already found, the values of U, U′, U″, &c., V, V′, V″, &c. Make

$$m = U - V, \quad m' = U' - V', \quad m'' = U'' - V'', \text{ \&c.}$$

Next, let the perihelion distance in this parabola vary by a very small quantity, and on this hypothesis suppose

$$n = U - V; \quad n' = U' - V'; \quad n'' = U'' - V'', \text{ \&c.}$$

We will form a third hypothesis, in which the perihelion distance remaining the same as in the first, we shall make the instant of the comet's passing its perihelion vary by a very small quantity; in this case let

$$p = U - V; \quad p' = U' - V'; \quad p'' = U'' - V''; \text{ \&c.}$$

Lastly, we shall calculate the angle v and radius ϱ, with the perihelion distance, and instant over the perihelion on the first hypothesis, supposing the orbit an ellipse, and the difference 1 — e between its excentricity and unity a very small quantity, for instance $\frac{1}{50}$. To get the angle v, in this hypothesis, it will suffice (489) to add to the anomaly v, calculated in the parabola of the first hypothesis, a small angle whose sine is

$$\frac{1}{10}(1 - e) \tan. \frac{1}{2} v \left\{4 - 3 \cos.^2 \frac{1}{2} v - 6 \cos.^4 \frac{1}{2} v\right\}.$$

Substituting afterwards in the equation

$$\varrho = \frac{D}{\cos.^2 \frac{1}{2} v}\left\{1 - \frac{1 - e}{2} \tan.^2 \frac{1}{2} v\right\};$$

for v, this anomaly, as calculated in the ellipse, we shall have the corresponding radius-vector ϱ. After the same manner, we shall obtain v′, ϱ', v″, ϱ'', &c. Whence we shall derive the values of U, U′, U″, &c. and (by 503) of V, V′, V″, &c.

In this case let

$$q = U - V; \quad q' = U' - V'; \quad q'' = U'' - V'', \text{ \&c.}$$

Finally, call u the number by which we ought to multiply the supposed variation in the perihelion distance, to make it the true one; t the number by which we ought to multiply the supposed variation in the instant over the perihelion, to make it the true instant; and s that by which we should multiply the supposed value of 1 — e, in order to get the true one; and we shall obtain these equations:

$$(m - n)\,u + (m - p)\,t + (m - q)\,s = m;$$
$$(m' - n')\,u + (m' - p')\,t + (m' - q')\,s = m;$$
$$(m'' - n'')\,u + (m'' - p'')\,t + (m'' - q'')\,s = m'';$$
$$(m''' - n''')\,u + (m''' - p''')\,t + (m''' - q''')\,s = m''';$$
&c.

We shall determine, by means of these equations, the values of u, t, s; whence will be derived the true perihelion distance, the true instant over the perihelion, and the true value of 1 — e. Let D be the perihelion distance, and a the semi-axis major of the orbit; then we shall have $a = \frac{D}{1 - e}$; the time of a sidereal revolution of the comet, will be expressed by a number of sidereal years equal to $a^{\frac{3}{2}}$ or to $\left(\frac{D}{1-e}\right)^{\frac{3}{2}}$, the mean distance of the sun from the earth being unity. We shall then have (by 503) the inclination of the orbit and the position of the node.

Whatever accuracy we may attribute to the observations, they will always leave us in uncertainty as to the periodic times of the comets. To determine this, the most exact method is that of comparing the observations of a comet in two consecutive revolutions. But this is practicable, only when the lapse of time shall bring the comet back towards its perihelion.

Thus much for the motions of the planets and comets as caused by the action of the principal body of the system. We now come to

506. *General methods of determining by successive approximations, the motions of the heavenly bodies.*

In the preceding researches we have merely dwelt upon the elliptic motion of the heavenly bodies, but in what follows we shall estimate them as deranged by perturbing forces. The action of these forces requires only to be added to the differential equations of elliptic motion, whose integrals in finite terms we have already given, certain small terms. We must determine, however, by successive approximations, the integrals of these same equations when thus augmented. For this purpose here is a general method, let the number and degree of the equations be what they may.

Suppose that we have between the *n* variables y, y′, y″, &c. and the time t whose element d t is constant, the *n* differential equations

$$0 = \frac{d^i y}{d t^i} + P + \alpha . Q;$$

$$0 = \frac{d^i y'}{d t^i} + P' + \alpha . Q',$$

&c. = &c.

P, Q, P′, Q′, &c. being functions of t, y, y′, &c. and of the differences to the order i — 1 inclusively, and α being a very small constant coefficient, which, in the theory of celestial motions, is of the order of the perturbing forces. Then let us suppose we have the finite integrals of those equations when Q, Q′, &c. are nothing. Differentiating each $i - 1$ times successively, we shall form with their differentials $i\,n$ equations by means of which we shall determine by elimination, the arbitrary constants c, c′, c″, &c. in functions of t, y, y′, y″, &c. and of their differences to the order $i - 1$. Designating therefore by V, V′, V″, &c. these functions we shall have

$$c = V; \quad c' = V'; \quad c'' = V''; \text{ \&c.}$$

These equations are the $i\,n$ integrals of the $(i - 1)^{th}$ order, which the equations ought to have, and which, by the elimination of the differences of the variables, give their finite integrals.

But if we differentiate the preceding integrals of the order $i - 1$, we shall have

$$0 = d\,V; \quad 0 = d\,V'; \quad 0 = d\,V''; \text{ \&c.}$$

and it is clear that these last equations being differentials of the order i without arbitrary constants, they can only be the sums of the equations

$$0 = \frac{d^i y}{d t^i} + P$$

$$0 = \frac{d^i y'}{d t^i} + P'$$

$$0 = \text{\&c.}$$

each multiplied by proper factors, in order to make these sums exact differences. Calling, therefore, F d t, F′ d t′, &c. the factors which ought respectively to multiply them in order to make $0 = d\,V$; also in like manner making H d t, H′ d t′, &c. the factors which would make $0 = d\,V'$, and so on for the rest, we shall have

$$d V = F d t \left\{\frac{d^i y}{d t^i} + P\right\} + F' d t \left\{\frac{d^i y'}{d t^i} + P'\right\} + \text{\&c.}$$

$$d V' = H d t \left\{\frac{d^i y}{d t^i} + P\right\} + H' d t \left\{\frac{d^i y'}{d t^i} + P'\right\} + \text{\&c.}$$

&c.

F, F′, &c. H, H′, &c. are functions of t, y, y′, y″, &c. and of their differences to the order $i - 1$. It is easy to determine them when V, V′, &c. are known. For F is evidently the coefficient of $\frac{d^i y}{d t^i}$ in the differential of V; F′ is the coefficient of $\frac{d^i y'}{d t^i}$ in the same differential, and so on. In like manner, H, H′, &c. are the coefficients of $\frac{d^i y}{d t^i}$, $\frac{d^i y'}{d t^i}$, &c. in the differential of V′. Thus, since we may suppose V, V′, &c. known, by differentiating with regard to $\frac{d^{i-1}y}{dt^{i-1}}$, $\frac{d^{i-1}y'}{dt^{i-1}}$, &c. we shall have the factors by which we ought to multiply the differential equations

$$0 = \frac{d^i y}{dt^i} + P, \quad 0 = \frac{d^i y'}{dt^i} + P', \&c.$$

in order to make them exact differences.

Now resume the differential equations

$$0 = \frac{d^i y}{dt^i} + P + \alpha . Q; \quad 0 = \frac{d^i y'}{dt^i} + P' + \alpha . Q', \&c.$$

If we multiply the first by F d t, the second by F′ d t, and so on, we shall have by adding the results

$$0 = dV + \alpha\, dt \{F Q + F' Q' + \&c.\},$$

In the same manner, we shall have

$$0 = dV' + \alpha\, dt \{H Q + H' Q' + \&c.\}$$

&c.

whence by integration

$$c - \alpha \int dt \{F Q + F' Q' + \&c.\} = V;$$
$$c' - \alpha \int dt \{H Q + H' Q' + \&c.\} = V';$$

&c.

We shall thus have $i\,n$ differential equations, which will be of the same form as in the case when Q, Q′, &c. are nothing, with this only difference, that the arbitrary constants c, c′, c″, &c. must be changed into

$$c - \alpha \int dt \{F Q + F' Q' + \&c.\}, \; c - \alpha \int dt \{H Q + H' Q' + \&c.\} \&c.$$

But if, in the supposition of Q, Q′, &c. being equal to zero, we eliminate from the $i\,n$ integrals of the order i — 1, the differences of the variables y, y′, &c. we shall have n finite integrals of the proposed equations. We shall therefore have these same integrals when Q, Q′, &c. are not zero, by changing in the first integrals, c, c′, &c. into

$$c - \alpha \int dt \{F Q + \&c.\}, \; c' - \alpha \int dt \{H Q + \&c.\} \&c.$$

507. If the differentials

$$dt \{F Q + F' Q' + \&c.\}, \; dt \{H Q + H' Q' + \&c.\} \&c.$$

are exact, we shall have, by the preceding method, finite integrals of the proposed differentials. But this is not so, except in some particular cases, of which the most extensive and interesting is that in which they are linear. Thus let P, P′, &c. be linear functions of y, y′, &c. and of their differences up to the order i — 1, without any term independent of these variables, and let us first consider the case in which Q, Q′, &c. are nothing. The differential equations being linear, their successive integrals are likewise linear, so that $c = V$, $c' = V'$, &c. being the $i\,n$ integrals of the order $i - 1$, of the linear differential equations

$$0 = \frac{d^i y}{d t^i} + P; \quad 0 = \frac{d^i y'}{d t^i} + P'; \text{ \&c.}$$

V, V′, &c. may be supposed linear functions of y y′, &c. and of their differences to the order $i - 1$. To make this evident, suppose that in the expressions for y, y′, &c. the arbitrary constant c is equal to a determinate quantity plus an indeterminate δc; the arbitrary constant c′ equal to a determinate quantity plus an indeterminate $\delta c'$ &c.; then reducing these expressions according to the powers and products of δc, $\delta c'$, &c. we shall have by the formulas of No. 487

$$y = Y + \delta c \left(\frac{d Y}{d c}\right) + \delta c' \left(\frac{d Y}{d c'}\right) + \text{\&c.}$$
$$+ \frac{\delta c^2}{1.2} \cdot \left(\frac{d^2 Y}{d c^2}\right) + \text{\&c.}$$
$$y' = Y' + \delta c \left(\frac{d Y'}{d c}\right) + \delta c' \left(\frac{d Y'}{d c'}\right) + \text{\&c.}$$
$$+ \frac{\delta c^2}{1.2} \left(\frac{d^2 Y'}{d c^2}\right) + \text{\&c.}$$

&c.

$Y, Y', \left(\frac{d Y}{d c}\right)$, &c. being functions of t without arbitrary constants. Substituting those values, in the proposed differential equations, it is evident that δc, $\delta c'$, &c. being indeterminate, the coefficients of the first powers of such of them ought to be nothing in the several equations. But these equations being linear, we shall evidently have the terms affected with the first powers of δc, $\delta c'$, &c. by substituting for y, y′, &c. these quantities respectively

$$\left(\frac{d Y}{d c}\right) \delta c + \left(\frac{d Y}{d c'}\right) \cdot \delta c' + \text{\&c.}$$
$$\left(\frac{d Y'}{d c}\right) \delta c + \left(\frac{d Y'}{d c'}\right) \delta c' + \text{\&c.}$$

These expressions of y, y′, &c. satisfy therefore separately the proposed equations; and since they contain the $i\,n$ arbitraries δc, $\delta c'$, &c. they are complete integrals. Thus we perceive, that the arbitraries are under a linear form in the expressions of y, y′, &c. and consequently also in their differentials. Whence it is easy to conclude that the variables y, y′, &c. and their differences, may be supposed to be linear in the successive integrals of the proposed differential equations.

Hence it follows, that F, F′, &c. being the coefficients of $\frac{d^i y}{d t^i}$, $\frac{d^i y'}{d t^i}$,

&c. in the differential of V; H, H′, &c. being the coefficients of the same differences in the differential of V′, &c. these quantities are functions of variable t only. Therefore, if we suppose Q, Q′, &c. functions of t alone, the differentials

$$d\,t\,\{F\,Q + F'\,Q' + \&c.\}\,;\quad d\,t\,\{H\,Q + H'\,Q' + \&c.\}\,;\ \&c.$$

will be exact.

Hence there results a simple means of obtaining the integrals of any number whatever *n* of linear differential equations of the order i, and which contain any terms α Q, α Q′, &c. functions of one variable t, having known the integrals of the same equations in the case where Q, Q′, &c. are supposed nothing. For then if we differentiate their *n* finite integrals i — 1 times successively, we shall have *i n* equations which will give, by elimination, the values of the *i n* arbitrary constants c, c′, &c. in functions of t, y, y′, &c. and of their differences to the i — 1th order. We shall thus form the *i n* equations c = V, c′ = V′, &c. This being done, F, F′, &c. will be the coefficients of $\frac{d^{i-1}\,y}{d\,t^{i-1}}$, $\frac{d^{i-1}\,y'}{d\,t^{i-1}}$, &c. in V; H, H′, &c. will be the coefficients of the same differences in V′, and so on. We shall, therefore, have the finite integrals of the linear differential equations

$$0 = \frac{d^i\,y}{d\,t^i} + P + \alpha\,Q;\quad 0 = \frac{d^i\,y'}{d\,t^i} + P' + \alpha\,Q'\,;\ \&c.$$

by changing, in the finite integrals of these equations deprived of their last terms α Q, α Q′, &c: the arbitrary constants c, c′, &c. into

$$c - \alpha\int d\,t\,\{F\,Q + F'\,Q' + \&c\,\},\quad c' - \alpha\int d\,t\,\{H\,Q + H'\,Q' + \&c.\}\ \&c.$$

Let us take, for example, the linear equation

$$0 = \frac{d^2\,y}{d\,t^2} + P^2\,y + \alpha\,Q.$$

The finite integral of the equation

$$0 = \frac{d^2\,y}{d\,t^2} + a^2\,y$$

is (found by multiplying by cos. a t, and then *by parts* getting $\int \cos.\,a\,t\,.\,\frac{d^2\,y}{d\,t} = \cos.\,a\,t\,\frac{d\,y}{d\,t} + a\int \sin.\,a\,t\,\frac{d\,y}{d\,t}\,.\,d\,t = \cos.\,a\,t\,.\,\frac{d\,y}{d\,t} +$ $a \sin.\,a\,t\,.\,y - a^2\int \cos.\,a\,t\,.\,y\ \therefore\ c = a\cos.\,a\,t\,.\,\frac{d\,y}{d\,t} + a\sin.\,a\,t\,.\,y$, &c.)

$$y = \frac{c}{a}\sin.\,a\,t + \frac{c'}{a}\cos.\,a\,t,$$

c, c′ being arbitrary constants.

This integral gives by differentiation

$$\frac{d y}{d t} = c \cos. a t - c' \sin. a t.$$

If we combine this with the integral itself, we shall form two integrals of the first order

$$c = a y \sin. a t + \frac{d y}{d t} \cos. a t;$$

$$c' = a y \cos. a t - \frac{d y}{d t} \sin. a t;$$

and therefore shall have in this case

$$F = \cos. a t; \quad H = - \sin. a t,$$

and the complete integral of the proposed equation will therefore be

$$y = \frac{c}{a} \sin. a t + \frac{c'}{a} \cos. a t - \frac{\alpha \sin. a t}{a} \int Q d t \cos. a t$$

$$+ \frac{\alpha \cos. a t}{a} \int Q d t \sin. a t.$$

Hence it is easy to conclude that if Q is composed of terms of the form $K . \frac{\sin.}{\cos.} (m t + \epsilon)$ each of these terms will produce in the value of y the corresponding term

$$\frac{\alpha K}{m^2 - a^2} . \frac{\sin.}{\cos.} (m t + \epsilon).$$

If m be equal to a, the term $K \frac{\sin.}{\cos.} (m t + \epsilon)$ will produce in y, 1st. the term $- \frac{\alpha K}{4 a^2} . \frac{\sin.}{\cos.} (a t + \epsilon)$ which being comprised by the two terms $\frac{c}{a} \sin. a t + \frac{c'}{a} \cos. a t$, may be neglected; 2dly. the term $\pm \frac{\alpha K t}{2 a} . \frac{\cos.}{\sin.} (a t + \epsilon)$, + or — being used according as the term of Q is a sine or cosine. We thus perceive how the arc t produces itself in the values of y, y′, &c. without sines and cosines, by successive integrations, although the differentials do not contain it in that form. It is evident this will take place whenever the functions F Q, F′, Q′, &c. H Q, H′ Q′, &c. shall contain constant terms.

508. If the differences

$$d t \{F Q + \&c.\}, \quad d t \{H Q + \&c.\}$$

are not exact, the preceding analysis will not give their rigorous integrals. But it affords a simple process for obtaining them more and more nearly by approximation when α is very small, and when we have the values of

y, y', &c. on the supposition of α being zero. Differentiating these values, $i - 1$ times successively, we shall form the differential equations of the order $i - 1$, viz.

$$c = V; \quad c' = V', \&c.$$

The coefficients of $\frac{d^i y}{d t^i}$, $\frac{d^i y'}{d t^i}$, &c. in the differentials of V, V', &c. being the values of F, F', &c. H, H', &c. we shall substitute them in the differential functions

$$d t (F Q + F' Q' + \&c.); \quad d t (H Q + H' Q' + \&c); \&c.$$

Then, we shall substitute in these functions, for y, y', &c. their first approximate values, which will make these differences functions of t and of the arbitrary constants c, c', &c.

Let $T\,d t$, $T'\,d t$, &c. be these functions. If we change in the first approximate values of y, y', &c. the arbitrary constants c, c', &c. respectively into $c - \alpha \int T\,d t$, $c' - \alpha \int T'\,d t$, &c. we shall have the second approximate values of those variables.

Again substitute these second values in the differential functions

$$d t . (F Q + \&c.); \quad d t (H Q + \&c.) \&c.$$

But it is evident that these functions are then what $T\,d t$, $T'\,d t$, &c. become when we change the arbitrary constants c, c', &c. into $c - \alpha \int T\,d t$, $c' - \alpha \int T'\,d t$, &c. Let therefore $T_{,}$, $T'_{,}$, &c. denote what T, T', &c. become by these changes. We shall get the third approximate values of y, y', &c. by changing in the first c, c', &c. respectively into $c - \alpha \int T_{,}\,d t$, $c' - \int T'_{,}\,d t$, &c.

Calling $T_{,,}$, $T'_{,,}$, in like manner, what T, T', &c. become when we change c, c', &c. into $c - \alpha \int T_{,}\,d t$, $c' - \alpha \int T'_{,}\,d t$, &c. we shall have the fourth approximate values of y, y', &c. by changing in the first approximate values of these variables into $c - \alpha \int T_{,,}\,d t$, $c' - \alpha \int T'_{,,}\,d t$, &c. and so on.

We shall see presently that the determination of the celestial motions, depends almost always upon differential equations of the form

$$0 = \frac{d^2 y}{d t^2} + a^2 y + \alpha Q,$$

Q being a rational and integer function of y, of the sine and cosine of angles increasing proportionally with the time represented by t. The following is the easiest way of integrating this equation.

First suppose α nothing, and we shall have by the preceding No. a first value of y.

Next substitute this value in Q, which will thus become a rational and entire function of sines and cosines of angles proportional to the time. Then integrating the differential equation, we shall have a second value of y approximate up to quantities of the order α inclusively.

Again substitute this value in Q, and, integrating the differential equation, we shall have a third approximation of y, and so on.

This way of integrating by approximation the differential equations of the celestial motions, although the most simple of all, possesses the disadvantage of giving in the expressions of the variables y, y', &c. the arcs of a circle (symbols *sine* and *cosine*) in the very case where these arcs do not enter the rigorous values of these variables. We perceive, in fact, that if these values contain sines or cosines of angles of the order α t, these sines or cosines ought to present themselves in the form of series, in the approximate values found by the preceding method; for these last values are ordered according to the powers of α. This developement into series of the sine and cosine of angles of the order α t, ceases to be exact when, by lapse of time, the arc α t becomes considerable. The approximate values of y, y', &c. cannot extend to the case of an unlimited interval of time. It being important to obtain values which include both past and future ages, the reversion of arcs of a circle contained by the approximate values, into functions which produce them by their developement into series, is a delicate and interesting problem of analysis. Here follows a general and very simple method of solution.

509. Let us consider the differential equation of the order i,

$$0 = \frac{d^i y}{d t^i} + P + \alpha Q$$

α being very small, and P and Q algebraic functions of y, $\frac{dy}{dt}, \dots \frac{d^{i-1} y}{d t^{i-1}}$, and of sines and cosines of angles increasing proportionally with the time. Suppose we have the complete integral of this differential, in the case of $\alpha = 0$, and that the value of y given by this integral, does not contain the arc t, without the symbols *sine* and *cosine*. Also suppose that in integrating this equation by the preceding method of approximation, when α is not nothing, we have

$$y = X + t Y + t^2 Z + t^3 S + \&c.$$

X, Y, Z, &c. being periodic functions of t, which contain the i arbitraries c, c', c'', &c. and the powers of t in this expression of y, going on to infinity by the successive approximations. It is evident the coefficients of these powers will decrease with the greater rapidity, the less is α. In the theory of the motions of the heavenly bodies, α expresses the order of perturbing forces, relative to the principal forces which animate them.

If we substitute the preceding value of y in the function $\frac{d^i y}{d t^i} + P + \alpha Q$, it will take the form $k + k' t + k'' t^2 + \&c.$, k, k', k'', &c. being periodic functions of t; but by the supposition, the value of y satisfies the differential equation

$$0 = \frac{d^i y}{d t^i} + P + \alpha Q;$$

we ought therefore to have identically

$$0 = k + k' t + k'' t^2 + \&c.$$

If k, k', k'', &c. be not zero this equation will give by the reversion of series, the arc t in functions of sines and cosines of angles proportional to the time t. Supposing therefore α to be infinitely small, we shall have t equal to a finite function of sines and cosines of similar angles, which is impossible. Hence the functions k, k', &c. are identically nothing.

Again, if the arc t is only raised to the first power under the symbols sine and cosine, since that takes place in the theory of celestial motions, the arc will not be produced by the successive differences of y. Substituting, therefore, the preceding value of y, in the function $\frac{d^i y}{d t^i} + P + \alpha . Q$, the function of $k + k' t + \&c.$ to which it transforms, will not contain the arc t out of the symbols *sine* and *cosine*, inasmuch as it is already contained in y. Thus changing in the expression of y, the arc t, without the periodic symbols, into $t - \theta$, θ being any constant whatever, the function $k + k' t + \&c.$ will become $k + k' (t - \theta) + \&c.$ and since this last function is identically nothing by reason of the identical equations $k = 0$ $k' = 0$, it results that the expression

$$y = X + (t - \theta) Y + (t - \theta)^2 Z + \&c.$$

also satisfies the differential equation

$$0 = \frac{d^i y}{d t^i} + P + \alpha Q.$$

Although this second value of y seems to contain $i + 1$ arbitrary constants, namely, the i arbitraries c, c', c'', &c. and θ, yet it can only have i distinct ones. It is therefore necessary that by a proper change in the constants c, c', &c. the arbitrary θ be made to disappear, and thus the second value of y will coincide with the first. This consideration will furnish us with the means of making disappear the arc of a circle out of the periodic symbols.

Give the following form to the second expression for y:

$$y = X + (t - \theta) . R.$$

Then supposing θ to disappear from y, we have

$$\left(\frac{d y}{d \theta}\right) = 0$$

and consequently

$$R = \left(\frac{d X}{d \theta}\right) + (t - \theta)\left(\frac{d R}{d \theta}\right).$$

Differentiating successively this equation we shall have

$$2\left(\frac{d R}{d \theta}\right) = \left(\frac{d^2 X}{d \theta^2}\right) + (t - \theta)\left(\frac{d^2 R}{d \theta^2}\right)$$

$$3\left(\frac{d^2 R}{d \theta^2}\right) = \left(\frac{d^3 X}{d \theta^3}\right) + (t - \theta)\left(\frac{d^3 R}{d \theta^3}\right),$$

whence it is easy to obtain, by eliminating R and its differentials, from the preceding expression of y,

$$y = X + (t - \theta)\left(\frac{d X}{d \theta}\right) + \frac{(t-\theta)^2}{1.2}.\left(\frac{d^2 X}{d \theta^2}\right) + \frac{(t-\theta)^3}{2.3}.\left(\frac{d^3 X}{d t^3}\right) + \&c.$$

X is a function of t, and of the constants, c, c′, c″, &c. and since these constants are functions of θ, X is a function of t and of θ, which we can represent by φ (t, θ). The expression of y is by Taylor's Theorem the developement of the function φ (t, $\theta + t - \theta$), according to the powers of $t - \theta$. We have therefore $y = \varphi$ (t, t). Whence we shall have y by changing in X, θ into t. The problem thus reduces itself to determine X in a function of t and θ, and consequently to determine c, c′, c″, &c. in functions of θ.

To solve this problem, let us resume the equation

$$y = X + (t - \theta).Y + (t - \theta)^2.Z + \&c.$$

Since the constant θ is supposed to disappear from this expression of y, we shall have the identical equation

$$0 = \left(\frac{d X}{d \theta}\right) - Y + (t - \theta)\left\{\left(\frac{dY}{d \theta}\right) - 2Z\right\} + (t - \theta)^2\left\{\left(\frac{dZ}{d \theta}\right) - 3S\right\} + \&c. \ldots (a)$$

Applying to this equation the reasoning which we employed upon

$$0 = k + k' t + k'' t^2 + \&c.$$

we perceive that the coefficients of the successive powers of $t - \theta$ ought to be each zero. The functions X, Y, Z, &c. do not contain θ, inasmuch as it is contained in c, c′, &c. so that to form the partial differences $\left(\frac{d X}{d \theta}\right)$, $\left(\frac{d Y}{d \theta}\right)$, $\left(\frac{d Z}{d \theta}\right)$, &c. it is sufficient to make c, c′, &c. vary in these functions, which gives

$$\left(\frac{d X}{d \theta}\right) = \left(\frac{d X}{d c}\right)\frac{d c}{d \theta} + \left(\frac{d X}{d c'}\right)\frac{d c'}{d \theta} + \left(\frac{d X}{d c''}\right)\frac{d c''}{d \theta} + \&c.$$

$$\left(\frac{d\,Y}{d\,\theta}\right) = \left(\frac{d\,Y}{d\,c}\right)\frac{d\,c}{d\,\theta} + \left(\frac{d\,Y}{d\,c'}\right)\frac{d\,c'}{d\,\theta} + \left(\frac{d\,Y}{d\,c''}\right)\frac{d\,c''}{d\,\theta} + \&c.$$
$$\&c. = \&c.$$

Again, it may happen that some of the arbitrary constants c, c′, c″, &c. multiply the arc *t* in the periodic functions X, Y, Z, &c. The differentiation of these functions relatively to θ, or, which is the same thing, relatively to these arbitrary constants, will develope this arc, and bring it from without the symbols of the periodic functions. The differences $\left(\frac{d\,X}{d\,\theta}\right)$, $\left(\frac{d\,Y}{d\,\theta}\right)$, $\left(\frac{d\,Z}{d\,\theta}\right)$, &c. will be then of this form:

$$\left(\frac{d\,X}{d\,\theta}\right) = X' + t\,X'';$$
$$\left(\frac{d\,Y}{d\,\theta}\right) = Y' + t\,Y'';$$
$$\left(\frac{d\,Z}{d\,\theta}\right) = Z' + t\,Z'';$$
$$\&c.$$

X′, X″, Y′, Y″, Z′, Z″, &c. being periodic functions of t, and containing moreover the arbitrary constants c, c′, c″, &c. and their first differences divided by d θ, differences which enter into these functions only under a linear form; we shall have therefore

$$\left(\frac{d\,X}{d\,\theta}\right) = X' + \theta\,X'' + (t - \theta)\,X''$$
$$\left(\frac{d\,Y}{d\,\theta}\right) = Y' + \theta\,Y'' + (t - \theta)\,Y''$$
$$\left(\frac{d\,Z}{d\,\theta}\right) = Z' + \theta\,Z'' + (t - \theta)\,Z''$$
$$\&c.$$

Substituting these values in the equation (a) we shall have

$$\begin{aligned} 0 = {} & X' + \theta\,X'' - Y \\ & + (t - \theta)\,\{Y' + \theta\,Y'' + X'' - 2\,Z\} \\ & + (t - \theta)^2\,\{Z' + \theta\,Z'' + Y'' - 3\,S\} + \&c.; \end{aligned}$$

whence we derive, in equalling separately to zero, the coefficients of the powers of t — θ,

$$0 = X' + \theta\,X'' - Y$$
$$0 = Y' + \theta\,Y'' + X'' - 2\,Z$$
$$0 = Z' + \theta\,Z'' + Y'' - 3\,S;$$
$$\&c.$$

If we differentiate the first of these equations, $i - 1$ times successively relatively to t, we shall thence derive as many equations between the quantities c, c', c'', &c. and their first differences divided by $d\theta$. Then integrating these new equations relatively to θ, we shall obtain the constants in terms of θ.

Inspection alone of the first of the above equations will almost always suffice to get the differential equations in c, c', c'', &c. by comparing separately the coefficients of the sines and cosines which it contains. For it is evident that the values of c, c', &c. being independent of t, the differential equations which determine them, ought, in like manner, to be independent of it. The simplicity which this consideration gives to the process, is one of its principal advantages. For the most part these equations will not be integrable except by successive approximations, which will introduce the arc θ out of the periodic symbols, in the values of c, c', &c. at the same time that this arc does not enter the rigorous integrals. But we can make it disappear by the following method.

It may happen that the first of the preceding equations, and its $i - 1$ differentials in t, do not give a number i of distinct equations between the quantities c, c', c'', &c. and their differences. In this case we must have recourse to the second and following equations.

When we shall have thus determined c, c', c'', &c. in functions of θ, we shall substitute them in X, and changing afterwards θ into t, we shall obtain the value of y, without arcs of a circle or free from periodic symbols, when that is possible.

510. Let us now consider any number n of differential equations.

$$0 = \frac{d^i y}{d t^i} + P + \alpha Q;$$

$$0 = \frac{d^i y'}{d t^i} + P' + \alpha Q';$$

&c.

P, Q, P', Q' being functions of y, y', &c. of their differentials to the order $i - 1$, and of the sines and cosines of angles increasing proportionally with the variable t, whose difference is constant. Suppose the approximate integrals of these equations to be

$$y = X + t Y + t^2 Z + t^3 S + \&c.$$
$$y' = X_{,} + t Y_{,} + t^2 Z_{,} + t^3 S_{,} + \&c.$$

X, Y, Z, &c. $X_{,}, Y_{,}, Z_{,}$, &c. being periodic functions of t and containing $i\,n$ arbitrary constants c, c', c'', &c. We shall have as in the preceding No.

$$0 = X' + \theta X'' - Y;$$
$$0 = Y' + \theta Y'' + X'' - 2 Z;$$
$$0 = Z' + \theta Z'' + Y'' - 3 S;$$
&c.

The value of y' will give, in like manner, equations of this form

$$0 = X_{\prime}' + \theta X_{\prime}'' - Y_{\prime};$$
$$0 = Y_{\prime}' + \theta Y_{\prime}'' + X_{\prime}'' - 2 Z_{\prime};$$
&c.

The values of y'', y''', &c. will furnish similar equations. We shall determine by these different equations, selecting the most simple and approximable, the values of c, c', c'', &c. in functions of θ. Substituting these values in X, X', &c. and then changing θ into t, we shall have the values of y, y', &c. independent of arcs free from periodic symbols when that is possible.

511. Let us resume the method already exposed in No. 506. It thence results that, if instead of supposing the parameters c, c', c'', &c. constant, we make them vary so that we have

$$d c = - \alpha \, d t \{F Q + F' Q' + \&c\};$$
$$d c' = - \alpha \, d t \{H Q + H' Q' + \&c.\};$$

we shall always have the $i \, n$ integrals of the order i — 1,

$$c = V; \; c' = V'; \; c'' = V''; \; \&c.$$

as in the case of $\alpha = 0$. Whence it follows that not only the finite integrals, but also all the equations in which these enter the differences inferior to the order i, will preserve the same form, in the case of $\alpha = 0$, and in that where it is any quantity whatever; for these equations may result from the comparison alone of the preceding integrals of the order i — 1. We can, therefore, in the two cases equally differentiate i — 1 times successively the finite integrals, without causing c, c', &c. to vary; and since we are at liberty to make all vary together, there will thence result the equations of condition between the parameters c, c', &c. and their differences.

In the two cases where $\alpha = 0$, and α = any quantity whatever, the values of y, y', &c. and of their differences to the order i — 1 inclusively, are the same functions of t and of the parameters c, c', &c. Let Y be any function of the variables y, y', y'', &c. and of their differentials inferior to the order i — 1, and call T the function of t, which it becomes, when we substitute for these variables and their differences their values in t. We can differentiate the equation Y = T, regarding the parameters c, c', &c. constant; we can only, however, take the partial difference of Y relatively to one only or to many of the variables y, y', &c. provided we suppose what varies with these, to vary also in T. In all these differentiations, the parameters c, c', c'', &c. may always be treated as constants; since by substituting for y, y', &c. and their differences, their values in t, we shall have equations identically zero in the two cases of α nothing and of α any quantity whatever.

When the differential equations are of the order i — 1, it is no longer allowed, in differentiating them, to treat the parameters c, c', &c. as constants. To differentiate these equations, consider the equation $\varphi = 0$, φ being a differential function of the order i — 1, and which contains the parameters c, c', c'', &c. Let d φ be the difference of this function taken in regarding c, c', &c. constant, as also the differences $d^{i-1} y$, $d^{i-1} y'$, &c. Let S be the coefficient of $\frac{d^i y}{d t^{i-1}}$ in the entire difference of φ. Let S' be the coefficient of $\frac{d^i y'}{d t^{i-1}}$ in this same difference, and so on. The equation $\varphi = 0$ when differentiated will give

$$0 = \delta \varphi + \left(\frac{d \varphi}{d c}\right) d c + \left(\frac{d \varphi}{d c'}\right) d c' + \&c.$$

$$+ S \frac{d^i y}{d t^{i-1}} + S' \frac{d^i y'}{d t^{i-1}} + \&c.;$$

Substituting for $\frac{d^i y}{d t^{i-1}}$ its value $- d t \{P + \alpha Q\}$; for $\frac{d^i y'}{d t^{i-1}}$ its value $- d t \{P' + \alpha Q'\}$ &c. we shall have

$$0 = \delta \varphi + \left(\frac{d \varphi}{d c}\right) d c + \left(\frac{d \varphi}{d c'}\right) d c' + \&c.$$

$$- d t \{S P + S' P' + \&c.\} - \alpha d t \{S Q + S' Q' + \&c.\} \quad . \quad (t)$$

In the supposition of $\alpha = 0$, the parameters c, c', c'', &c. are constant. We have thus

$$0 = \delta \varphi - d t \{S P + S' P' + \&c.\}$$

If we substitute in this equation for c, c', c'', &c. their values V, V', V'', &c. we shall have differential equations of the order i — 1, without arbitraries, which is impossible, at least if this equation is to be identically nothing. The function

$$\delta \varphi - d t \{S P + S' P' + \&c.\},$$

becoming therefore identically nothing by reason of equations c = V, c' = V', &c. and since these equations hold still, when the parameters c, c', c'', &c. are variable, it is evident, that in this case, the preceding function is still identically nothing. The equation (t) therefore will become

$$0 = \left(\frac{d\,\varphi}{d\,c}\right) dc + \left(\frac{d\,\varphi}{d\,c'}\right) d\,c' + \&c.$$
$$- \alpha\, d\,t\, \{S\,Q + S'\,Q' + \&c.\} \quad \ldots \quad (x)$$

Thus we perceive that to differentiate the equation $\varphi = 0$, it suffices to vary the parameters c, c', &c. in φ and the differences $d^{i-1}\,y$, $d^{i-1}\,y'$, &c. and to substitute after the differentiations, for $-\alpha\,Q$, $\alpha\,Q'$, &c. the quantities $\frac{d^i\,y}{d\,t^i}$, $\frac{d^i\,y'}{d\,t^i}$, &c.

Let $\psi = 0$, be a finite equation between y, y', &c. and the variable t. If we designate by $\delta\,\psi$, $\delta^2\,\psi$, &c. the successive differences of ψ, taken in regarding c, c', &c. as constant, we shall have, by what precedes, in that case where c, c', &c. are variable, these equations:

$$\psi = 0;\ \delta\,\psi = 0;\ \delta^2\,\psi = 0 \ldots\ldots \delta^{i-1}\,\psi = 0;$$

changing therefore successively in the equation (x) the function φ into ψ, $\delta\,\psi$, $\delta^2\,\psi$, &c. we shall have

$$0 = \left(\frac{d\,\psi}{d\,c}\right) d\,c + \left(\frac{d\,\psi}{d\,c'}\right) . d\,c' + \&c.$$

$$0 = \left(\frac{d\,.\,\delta\,\psi}{d\,c}\right) d\,c + \left(\frac{d\,.\,\delta\,\psi}{d\,c'}\right) d\,c' + \&c.$$

- - - - - - - - - - - - -

$$0 = \left(\frac{d\,.\,\delta^{i-1}\,\psi}{d\,c}\right) . d\,c + \left(\frac{d\,.\,\delta^{i-1}\,\psi}{d\,c'}\right) d\,c' + \&c.$$
$$- \alpha\,d\,t \left\{ Q \left(\frac{d\,\psi}{d\,y}\right) + Q' \left(\frac{d\,\psi}{d\,y'}\right) + \&c. \right\}$$

Thus the equations $\psi = 0$, $\psi' = 0$, &c. being supposed to be the *n* finite integrals of the differential equations

$$0 = \frac{d^i\,y}{d\,t^i} + P$$

$$0 = \frac{d^i\,y'}{d\,t^i} + P'$$

&c.

we shall have *i n* equations, by means of which we shall be able to determine the parameters c, c', c'', &c. without which it would be necessary for that purpose to form the equations c = V, c' = V', &c. But when the integrals are under this last form, the determination will be more simple.

512. This method of making the parameters vary, is one of great utility

in analysis and in its applications. To exhibit a new use of it, let us take the differential equation

$$0 = \frac{d^i y}{d t^i} + P,$$

P being a function of t, y, of their differences to the order i — 1, and of the quantities q, q′, &c. which are functions of t. Suppose we have the finite integral of this differential equation of the supposition of q, q′, &c. being constant, and represent by $\varphi = 0$, this integral, which shall contain i arbitraries c, c′, &c. Designate by $\delta \varphi$, $\delta^2 \varphi$, $\delta^3 \varphi$, &c. the successive differences of φ taken in regarding q, q′, &c. constant, as also the parameters c, c′, c″, &c. If we suppose all these quantities to vary, the differences of φ will be d

$$\delta\varphi + \left(\frac{\varphi}{d c}\right) d c + \left(\frac{d \varphi}{d c'}\right) d c' + \&c. + \left(\frac{d \varphi}{d q}\right) d q + \left(\frac{d \varphi}{d q'}\right) d q' + \&c.$$

making therefore

$$0 = \left(\frac{d \varphi}{d c}\right) d c + \left(\frac{d \varphi}{d c'}\right) d c' + \&c. + \left(\frac{d \varphi}{d q}\right) d q + \left(\frac{d \varphi}{d q'}\right) d q' + \&c.$$

$\delta \varphi$ will be still the first difference of φ in the case of c, c′, &c. q, q′, &c. being variable. If we make, in like manner,

$$0 = \left(\frac{d \delta \varphi}{d c}\right) d c + \left(\frac{d \delta \varphi}{d c'}\right) d c' + \&c. + \left(\frac{d \delta \varphi}{d q}\right) d q + \left(\frac{d \delta \varphi}{d q'}\right) d q' + \&c.$$

- - - - - - - - - - - - - - - - - -

$$0 = \left(\frac{d \delta^{i-1} \varphi}{d c}\right) dc + \left(\frac{d \delta^{i-1} \varphi}{d c'} dc'\right) + \&c. + \left(\frac{\delta^{i-1} \varphi}{dd q}\right) dq + \left(\frac{d \delta^{i-1} \varphi}{d q'}\right) dq' + \&c.$$

$\delta^2 \varphi$, $\delta^3 \varphi$, ... $\delta^i \varphi$ will likewise be the second, third, &c. differences of φ when c, c′, &c. q, q′, &c. are supposed variable.

Again in the case of c, c′, &c. q, q′, &c. being constant, the differential equation

$$0 = \frac{d^i y}{d t^i} + P,$$

is the result of the elimination of the parameters c, c′, &c. by means of the equations $\varphi = 0$, $\delta \varphi = 0$, $\delta^2 \varphi = 0$, ... $d^i \varphi = 0$. Thus, these last equations still holding good when q, q′, &c. are supposed variable, the equation $\varphi = 0$ will also satisfy, in this case, the proposed differential equation, provided the parameters c, c′, &c. are determined by means of the i preceding differential equations; and since their integration gives i arbitrary constants, the function φ will contain these arbitraries, and the equation $\varphi = 0$ will be the complete integral of the proposed equation.

This method, the variation of parameters, may be employed with advantage when the quantities q, q', &c. vary very slowly. Because this consideration renders the integration by approximation of the differential equations which determine the variables c, c', c'', &c. in general much easier.

513. *Second Approximation of Celestial Motions.*

Let us apply the preceding method to the perturbations of celestial motions, in order thence to obtain the most simple expressions of their periodical and secular inequalities. For that purpose let us resume the differential equations (1), (2), (3) of No. 471, which determine the relative motion of μ about M. If we make

$$R = \frac{\mu' (x x' + y y' + z z')}{(x'^2 + y'^2 + z'^2)^{\frac{3}{2}}} + \frac{\mu'' (x x'' + y y'' + z z'')}{(x''^2 + y''^2 + z''^2)^{\frac{3}{2}}} + \&c. - \frac{\lambda}{\mu}$$

λ being by the No. cited equal to

$$\frac{\mu \mu'}{\{(x'-x)^2+(y'-y)^2+(z'-z)^2\}^{\frac{1}{2}}} + \frac{\mu \mu''}{\{(x''-x)^2+(y''-y)^2+(z''-z)^2\}^{\frac{1}{2}}} + \frac{\mu' \mu''}{\{(x''-x')^2+(y''-y')^2+(z''-z')^2\}^{\frac{1}{2}}} + \&c.$$

If, moreover, we suppose $M + \mu = m$ and

$$\varrho = \sqrt{x^2 + y^2 + z^2}$$
$$\varrho' = \sqrt{x'^2 + y'^2 + z'^2}$$

we shall have

$$\left.\begin{aligned} 0 &= \frac{d^2 x}{d t^2} + \frac{m x}{\varrho^3} + \left(\frac{d R}{d x}\right) \\ 0 &= \frac{d^2 y}{d t^2} + \frac{m y}{\varrho^3} + \left(\frac{d R}{d y}\right) \\ 0 &= \frac{d^2 z}{d t^2} + \frac{m z}{\varrho^3} + \left(\frac{d R}{d z}\right) \end{aligned}\right\}; \quad (P)$$

The sum of these three equations multiplied respectively by d x, d y, d z gives by integration

$$0 = \frac{d x^2 + d y^2 + d z^2}{d t^2} - \frac{2 m}{\varrho} + \frac{m}{a} + 2 \int d R \quad . \quad . \quad . \quad (Q)$$

the differential d R being only relative to the coordinates x, y, z of the body μ, and a being an arbitrary constant, which, when R = 0, becomes by No. 499, the semi-axis major of the ellipse described by μ about M.

The equations (P) multiplied respectively by x, y, z and added to the integral (Q) will give

$$0=\frac{1}{2}\frac{d^2\varrho^2}{dt^2}-\frac{m}{\varrho}+\frac{m}{a}+2\int dR+x\left(\frac{dR}{dx}\right)+y\left(\frac{dR}{dy}\right)+z\left(\frac{dR}{dz}\right);\quad (R)$$

We may conceive, however, the perturbing masses μ', μ'', &c. multiplied by a coefficient α; and then the value of ϱ will be a function of the time t and of α. If we develope this function according to the powers of α, and afterwards make $\alpha = 1$, it will be ordered according to the powers and products of the perturbing masses. Designate by the characteristic δ when placed before a quantity, this differential of it taken relatively to α, and divided by d α. When we shall have determined $\delta\varrho$ in a series ordered according to the powers of α, we shall have the radius ϱ by multiplying this series by d α, then integrating it relatively to α, and adding to the integral a function of t independent of α, a function which is evidently the value of ϱ in the case where the perturbing forces are nothing, and where the body μ describes a conic section. The determination of ϱ reduces itself, therefore, to forming and integrating the differential equation which determines $\delta\varrho$.

For that purpose, resume the differential equation (R) and make for the greater simplicity

$$x\left(\frac{dR}{dx}\right)+y\left(\frac{dR}{dy}\right)+z\left(\frac{dR}{dz}\right)=\varrho R';$$

differentiating this relatively to α, we shall have

$$0=\frac{d^2\varrho\delta\varrho}{dt^2}+\frac{m\varrho\delta\varrho}{\varrho^3}+2\int\delta dR+\delta.\varrho R' \quad .\ .\ .\ .\ .\ (S)$$

Call d v the indefinitely small arc intercepted between the two radius-vectors ϱ and $\varrho + d\varrho$; the element of the curve described by μ around M will be $\sqrt{d\varrho^2+\varrho^2 dv^2}$. We shall thus have

$$dx^2+dy^2+dz^2=d\varrho^2+\varrho^2dv^2,$$

and the equation (Q) will become

$$0=\frac{\varrho^2dv^2+d\varrho^2}{dt^2}-\frac{2m}{\varrho}+\frac{m}{a}+2\int dR.$$

Eliminating $\frac{m}{a}$ from this equation by means of equation (R) we shall have

$$\frac{\varrho^2dv^2}{dt^2}=\frac{\varrho d^2\varrho}{dt^2}+\frac{m}{\varrho}+\varrho R'$$

whence we derive, by differentiating relatively to α,

$$\frac{2\varrho^2dv.d\delta v}{dt^2}=\frac{\varrho d^2\delta\varrho-\delta\varrho d^2\varrho}{dt^2}-\frac{3m\varrho\delta\varrho}{\varrho^3}+\varrho\delta R'-R'\delta\varrho.$$

If we substitute in this equation for $\frac{m \varrho \delta \varrho}{\varrho^3}$ its value derived from equation (S), we shall have

$$d \delta v = \frac{d (d \varrho \delta \varrho + 2 \varrho d \delta \varrho) + d t^2 \{3 \int \delta d R + 2 \varrho \delta R' + R' \delta \varrho\}}{\varrho^2 d v} \quad (T)$$

By means of the equations (S), (T), we can get as exactly as we wish the values of $\delta \varrho$ and of δv. But we must observe that $d v$ being the angle intercepted between the radii ϱ and $\varrho + d \varrho$, the integral v of these angles is not wholly in one plane. To obtain the value of the angle described round M, by the projection of the radius-vector ϱ upon a fixed plane, denote by $v_{,}$, this last angle, and name s the tangent of the latitude of μ above this plane; then $\varrho (1 + s^2)^{-\frac{1}{2}}$ will be the expression of the projected radius-vector, and the square of the element of the curve described by μ will be

$$\frac{\varrho^2 d v_{,}^2}{1 + s^2} + d \varrho^2 + \frac{\varrho^2 d s^2}{(1 + s^2)^2};$$

But the square of this element is also $\varrho^2 d v^2 + d \varrho^2$; therefore we have, by equating these two expressions

$$d v_{,} = \frac{d v \sqrt{(1 + s^2)^2 - \frac{d s^2}{d v^2}}}{\sqrt{1 + s^2}}.$$

We shall thus determine $d v_{,}$ by means of $d v$, when s is known.

If we take for the fixed plane, that of the orbit of μ at a given epoch, s and $\frac{d s}{d v}$ will evidently be of the order of perturbing forces. Neglecting therefore the squares and the products of these forces, we shall have $v = v_{,}$. In the Theory of the planets and of the comets, we may neglect these squares and products with the exception of some terms of that order, which particular circumstances render of sensible magnitude, and which it will be easy to determine by means of the equations (S) and (T). These last equations take a very simple form, when we take into account the first power only of the disturbing forces. In fact, we may then consider $\delta \varrho$ and δv as the parts of ϱ and v due to these forces; δR, $\delta . \varrho R'$ are what R and $\varrho R'$ become, when we substitute for the coordinates of the bodies their values relative to the elliptic motion: We may designate them by these last quantities when subjected to that condition. The equation (S) thus becomes,

$$0 = \frac{d^2 . \varrho \delta \varrho}{d t^2} + \frac{m \varrho \delta \varrho}{\varrho^3} + 2 \int d R + \varrho R'.$$

The fixed plane of x, y being supposed that of the orbit of μ at a given epoch, z will be of the order of perturbing forces: and since we may neglect the square of these forces, we can also neglect the quantity $z\left(\frac{dR}{dz}\right)$. Moreover, the radius ϱ differs only from its projection by quantities of the order z^2. The angle which this radius makes with the axis of x, differs only from its projection by quantities of the same order. This angle may therefore be supposed equal to v and to quantities nearly of the same order

$$x = \varrho \cos. v; \; y = \varrho \sin. v;$$

whence we get

$$x\left(\frac{dR}{dx}\right) + y\left(\frac{dR}{dy}\right) = \varrho\left(\frac{dR}{d\varrho}\right);$$

and consequently $\varrho . R' = \varrho\left(\frac{dR}{d\varrho}\right)$. It is easy to perceive by differentiation, that if we neglect the square of the perturbing force, the preceding differential equation will become, by means of the two first equations (P)

$$\varrho \delta \varrho = \frac{x\int y\,dt\left\{2\int dR + \varrho\left(\frac{dR}{d\varrho}\right)\right\} - y\int x\,dt\left\{2\int dR + \varrho\left(\frac{dR}{d\varrho}\right)\right\}}{\left(\frac{x\,dy - y\,dx}{dt}\right)}.$$

In the second member of this equation the coordinates may belong to elliptic motion; this gives $\frac{x\,dy - y\,dx}{dt}$ constant and equal to $\sqrt{m\,a(1-e^2)}$, a e being the excentricity of the orbit of μ. If we substitute in the expression of $\varrho\,\delta\,\varrho$ for x and y, their values ϱ cos. v and ϱ sin. v, and for $\frac{x\,dy - y\,dx}{dt}$, the quantity $\sqrt{\mu\,a\,(1-e^2)}$; finally, if we observe that by No. (480)

$$m = n^2 a^3,$$

we shall have

$$\delta \varrho = \frac{\left\{\begin{array}{l} a \cos. v\int n\,dt . \varrho \sin. v\left\{2\int dR + \varrho\left(\frac{dR}{d\varrho}\right)\right\} \\ -a \sin. v\int n\,dt . \varrho \cos. v\left\{2\int dR + \varrho\left(\frac{dR}{d\varrho}\right)\right\}\end{array}\right\}}{m\sqrt{1-e^2}}. \quad (X)$$

The equation (T) gives by integration and neglecting the square of perturbing forces,

$$dv = \frac{\frac{2\varrho d.\delta\varrho + d\varrho.\delta\varrho}{a^2 n\,dt} + \frac{3a}{m}\iint n\,dt . dR + \frac{2a}{m}\int n\,dt . \varrho\left(\frac{dR}{d\varrho}\right)}{\sqrt{1-e^2}} \quad (Y)$$

This expression, when the perturbations of the radius-vector are known, will easily give those of the motion of μ in longitude.

It remains for us to determine the perturbations of the motion in latitude. For that purpose let us resume the third of the equations (P): integrating this in the same manner as we have integrated the equation (S), and making $z = \varrho\, \delta s$, we shall have

$$\delta s = \frac{a \cos. v \int n\, d t . \varrho \sin. v \left(\frac{d R}{d z}\right) - a \sin. v \int n\, d t . \varrho \cos. v \left(\frac{d R}{d z}\right)}{m \sqrt{1 - e^2}}; \quad (Z)$$

δs is the latitude of μ above the plane of its primitive orbit: if we wish to refer the motion of μ to a plane somewhat inclined to this orbit, by calling s its latitude, when it is supposed not to quit the plane of the orbit, $s + \delta s$ will be very nearly the latitude of μ above the proposed plane.

514. The formulas (X), (Y), (Z) have the advantage of presenting the perturbations under a finite form. This is very useful in the Cometary Theory, in which these perturbations can only be determined by quadratures. But the excentricity and inclination of the respective orbits of the planets being small, permits a developement of their perturbations into converging series of the sines and cosines of angles increasing proportionally to the time, and thence to make tables of them to serve for any times whatever. Then, instead of the preceding expressions of $\delta \varrho$, δs, it is more commodious to make use of differential equations which determine these variables. Ordering these equations according to the powers and products of the excentricities and inclinations of the orbits, we may always reduce the determination of the values of $\delta \varrho$, and of δs to the integration of equations of the form

$$0 = \frac{d^2 y}{d t^2} + n^2 y + Q$$

equations whose integrals we have already given in No. 509. But we can immediately reduce the preceding differential equations to this simple form, by the following method.

Let us resume the equation (R) of the preceding No., and abridge it by making

$$Q = 2 \int d R + \varrho \left(\frac{d R}{d \varrho}\right).$$

It thus becomes

$$0 = \frac{1}{2} \frac{d^2 . \varrho^2}{d t^2} - \frac{m}{\varrho} + \frac{m}{a} + Q \quad . \quad . \quad . \quad . \quad . \quad (R')$$

In the case of elliptic motion, where Q = 0, ρ^2 is by No. (488) a function of e cos. $(n t + \varepsilon - \varpi)$, a e being the excentricity of the orbit, and $n t + \varepsilon - \varpi$ the mean anomaly of the planet μ. Let e cos. $(n t + \varepsilon - \varpi) = u$, and suppose $\rho^2 = \varphi(u)$; we shall have

$$0 = \frac{d^2 u}{d t^2} + n^2 u.$$

In the case of disturbed motion, we can still suppose $\rho^2 = \varphi(u)$, but u will no longer be equal to e cos. $(n t + \varepsilon - \varpi)$. It will be given by the preceding differential equation augmented by a term depending upon the perturbing forces. To determine this term, we shall observe that if we make $u = \psi(\rho^2)$ we shall have

$$\frac{d^2 u}{d t^2} + n^2 u = \frac{d^2 . \rho^2}{d t^2} \psi'(\rho^2) + \frac{4 \rho^2 d \rho^2}{d t^2} \psi''(\rho^2) + n^2 . \psi(\rho^2),$$

$\psi'(\rho^2)$ being the differential of $\psi(\rho^2)$ divided by $d . \rho^2$ and $\psi''(\rho^2)$ the differential of $\psi'(\rho^2)$ divided by $d . \rho^2$. The equation (R') gives $\frac{d^2 . \rho^2}{d t^2}$ equal to a function of ρ plus a function depending upon the perturbing force. If we multiply this equation by $2 \rho d \rho$, and then integrate it, we shall have $\frac{\rho^2 d \rho^2}{d t^2}$ equal to a function of ρ plus a function depending upon the perturbing force. Substituting these values of $\frac{d^2 . \rho^2}{d t^2}$ and of $\frac{\rho^2 d \rho^2}{d t^2}$ in the preceding expression of $\frac{d^2 u}{d t^2} + n^2 u$, the function of ρ, which is independent of the perturbing force will disappear of itself, because it is identically nothing when that force is nothing. We shall therefore have the value of $\frac{d^2 u}{d t^2} + n^2 u$ by substituting for $\frac{d^2 . \rho^2}{d t^2}$, and $\frac{\rho^2 d \rho^2}{d t^2}$, the parts of their expressions which depend upon the perturbing force. But regarding these parts only, the equation (R') and its integral give

$$\frac{d^2 . \rho^2}{d t^2} = -2 Q;$$

$$\frac{4 \rho^2 d \rho^2}{d t^2} = -8 \int Q \rho d \rho.$$

Wherefore

$$\frac{d^2 u}{d t^2} + n^2 u = -2 Q \psi'(\rho^2) - 8 \psi''(\rho^2) \int Q . \rho d \rho.$$

Again, from the equation $u = \varphi(\rho^2)$, we derive $d u = 2 \rho d \rho \psi'(\rho^2)$; this $\rho^2 = \varphi(u)$ gives $2 \rho d \rho = d u . \varphi'(u)$ and consequently

$$\psi'(\varrho^2) = \frac{1}{\varphi'(u)}.$$

Differentiating this last equation and substituting $\varphi'(u)$ for $\frac{2\varrho\, d\varrho}{d u}$, we shall have

$$\psi''(\varrho^2) = -\frac{\varphi''(u)}{\varphi'(u)^3},$$

$\varphi''(u)$ being equal to $\frac{d.\varphi'(u)}{d u}$, in the same way as $\varphi'(u)$ is equal to $\frac{d.\varphi(u)}{d u}$. This being done; if we make

$$u = e\cos.(n t + \varepsilon - \varpi) + \delta u,$$

the differential equation in *u* will become

$$0 = \frac{d^2.\delta u}{d t^2} + n^2.\delta u - \frac{4.\varphi''(u)}{\varphi'(u)^3}\int Q\, d u.\varphi'(u) + \frac{2 Q}{\varphi'(u)};$$

and if we neglect the square of the perturbing force, u may be supposed equal to e cos. $(n t + \varepsilon - \varpi)$, in the terms depending upon Q.

The value of $\frac{\varrho}{a}$ found in No. (485) gives, including quantities of the order e^3

$$\varrho = a\left\{1 + e^2 - u\left(1 - \frac{3}{2}e^2\right) - u^2 - \frac{3}{2}u^3\right\}$$

whence we derive

$$\varrho^2 = a^2\left\{1 + 2e^2 - 2u\left(1 - \frac{1}{2}e^2\right) - u^2 - u^3\right\} = \varphi(u).$$

If we substitute this value of $\varphi(u)$ in the differential equation in δu, and restore to Q its value $2\int d R + \varrho\left(\frac{d R}{d\varrho}\right)$, and e cos. $(n t + \varepsilon - \varpi)$ for u, we shall have including quantities of the order e^3,

$$0 = \frac{d^2.\delta u}{d t^2} + n^2\delta u$$

$$-\frac{1}{a^2}\left\{1 + \frac{1}{4}e^2 - e\cos.(n t + \varepsilon - \varpi) - \frac{1}{4}e^2\cos.(2 n t + 2\varepsilon - 2\varpi)\right\}$$

$$\times\left\{2\int d R + \varrho\left(\frac{d R}{d\varrho}\right)\right\}$$

$$-\frac{2e}{a^2}\int n\, d t\left[\sin.(n t + \varepsilon - \varpi)\{1 + e\cos.(n t + \varepsilon - \varpi)\}\left\{2\int d R + \varrho\left(\frac{d R}{d\varrho}\right)\right\}\right] \quad (X')$$

When we shall have determined δu by means of this differential equation, we shall have $\delta \varrho$ by differentiating the expression of ϱ, relative to the characteristic δ, which gives

$$\delta \varrho = -a \delta u \left\{ 1 + \frac{3}{4} e^2 + 2 e \cos. (n t + \varepsilon - \varpi) + \frac{9}{4} e^2 \cos. (2 n t + 2 \varepsilon - 2 \varpi) \right\}.$$

This value of $\delta \varrho$ will give that of δv by means of formula (Y) of the preceding number.

It remains for us to determine δs; but if we compare the formulas (X) and (Z) of the preceding No. we perceive that $\delta \varrho$ changes itself into δs by substituting $\left(\frac{d R}{d z}\right)$ for $2 \int d R + \varrho \left(\frac{d R}{d \varrho}\right)$ in its expression. Whence it follows that to get δs, it suffices to make this change in the differential equation in δu, and then to substitute the value of δu given by this equation, and which we shall designate by $\delta u'$, in the expression of $\delta \varrho$. Thus we get

$$0 = \frac{d^2 \delta u'}{d t^2} + n^2 \delta u'$$

$$- \frac{1}{a^2} \left\{ 1 + \frac{1}{4} e^2 - e \cos. (n t + \varepsilon - \varpi) - \frac{1}{4} e^2 \cos. (2 n t + 2 \varepsilon - 2 \varpi) \right\} \left(\frac{d R}{d z}\right)$$

$$- \frac{2 e}{a^2} \int n d t \left\{ \sin. (n t + \varepsilon - \varpi) \{1 + e \cos. (n t + \varepsilon - \varpi)\}. \left(\frac{d R}{d z}\right) \right\}; \quad (Z')$$

$$\delta s = -a \delta u' \left\{ 1 + \frac{3}{4} e^2 + 2 e \cos. (n t + \varepsilon - \varpi) + \frac{9}{4} e^2 \cos. (2 n t + 2 e - 2 \varpi) \right\}$$

The system of equations (X'), (Y), (Z') will give, in a very simple manner, the perturbed motion of μ in taking into account only the first power of the perturbing force. The consideration of terms due to this power being in the Theory of Planets very nearly sufficient to determine their motions, we proceed to derive from them formulas for that purpose.

515. It is first necessary to develope the function R into a series. If we disregard all other actions than that of μ upon μ', we shall have by (513)

$$R = \frac{\mu' (x x' + y y' + z z')}{(x'^2 + y'^2 + z'^2)^{\frac{3}{2}}} - \frac{\mu'}{\{(x' - x)^2 + (y' - y)^2 + (z' - z)^2\}^{\frac{1}{2}}}.$$

This function is wholly independent of the position of the plane of x, y; for the radical $\sqrt{(x' - x)^2 + (y' - y)^2 + (z' - z)^2}$, expressing the distance of μ, μ', is independent of the position; the function $x^2 + y^2 + z^2 + x'^2 + y'^2 + z'^2 - 2 x x' - 2 y y' - 2 z z'$ is in like manner independent of it. But the squares $x^2 + y^2 + z^2$ and $x'^2 + y'^2 + z'^2$ of the radius-vectors, do not depend upon the position; and therefore the quantity $x x' + y y' + z z'$ does not depend upon it, and consequently

R is independent of the position of the plane of x, y. Suppose in this function

$$x = \varrho \cos. v; \quad y = \varrho \sin. v;$$
$$x' = \varrho' \cos. v'; \quad y' = \varrho' \sin. v';$$

we shall then have

$$R = \frac{\mu'\{\varrho\,\varrho' \cos. (v'-v) + z\,z'\}}{(\varrho'^2 + z'^2)^{\frac{3}{2}}} - \frac{\mu'}{\{\varrho^2 - 2\,\varrho\,\varrho' \cos. (v'-v) + \varrho'^2 + (z'-z)^2\}^{\frac{1}{2}}}.$$

The orbits of the planets being almost circular and but little inclined to one another, we may select the plane of x, y, so that z and z' may be very small. In this case ϱ and ϱ' are very little different from the semi-axis-majors a, a' of the elliptic orbits, we will therefore suppose

$$\varrho = a\,(1 + u_{,}); \quad \varrho' = a'\,(1 + u'_{,});$$

$u_{,}$ and $u'_{,}$ being small quantities. The angles v, v' differing but little from the mean longitudes $n\,t + \varepsilon$, $n'\,t + \varepsilon'$, we shall suppose

$$v = n\,t + \varepsilon + v_{,}; \quad v' = n'\,t + \varepsilon' + v'_{,};$$

v' and $v'_{,}$ being inconsiderable. Thus, reducing R into a series ordered according to the powers and products of $u_{,}$, $v_{,}$, z, $u'_{,}$, $v'_{,}$, and z', this series will be very convergent. Let

$$\frac{a}{a'^2} \cos. (n'\,t - n\,t + \varepsilon' - \varepsilon) - \{a^2 - 2\,a\,a' \cos. (n'\,t - n\,t + \varepsilon' - \varepsilon) + a'^2\}^{-\frac{1}{2}}$$

$$= \frac{1}{2}\,A^{(0)} + A^{(1)} \cos. (n'\,t - n\,t + \varepsilon' - \varepsilon) + A^{(2)} \cos. 2\,(n'\,t - n\,t + \varepsilon' - \varepsilon)$$

$$+ A^{(3)} \cos. 3\,(n'\,t - n\,t + \varepsilon' - \varepsilon) + \&c.;$$

We may give to this series the form $\frac{1}{2}\,\Sigma\,A^{(i)} \cos. i\,(n'\,t - n\,t + \varepsilon' - \varepsilon)$, the characteristic Σ of finite integrals, being relative to the number i, and extending itself to all whole numbers from $i = -\infty$ to $i = \infty$; the value $i = 0$, being comprised in this infinite number of values. But then we must observe that $A^{(-i)} = A^{(i)}$. This form has the advantage of serving to express after a very simple manner, not only the preceding series, but also the product of this series, by the sine or the cosine of any angle $f\,t + \varpi$; for it is perceptible that this product is equal to

$$\frac{1}{2}\,\Sigma\,A^{(i)} \begin{matrix}\sin.\\ \cos.\end{matrix} \{i\,(n'\,t - n\,t + \varepsilon' - \varepsilon) + f\,t + \varpi\}.$$

This property will furnish us with very commodious expressions for the perturbations of the planets. Let in like manner

$$\{a^2 - 2\,a\,a' \cos. (n'\,t - n\,t + \varepsilon' - \varepsilon) + a'^2\}^{-\frac{3}{2}}$$

$$= \frac{1}{2}\,\Sigma\,B^{i} \cos. i\,(n\,t - n\,t + \varepsilon - \varepsilon);$$

$B^{(-i)}$ being equal to $B^{(i)}$. This being done, we shall have by (483)

$$R = \frac{\mu'}{2} . \Sigma A^{(i)} \cos. i (n' t - n t + \epsilon' - \epsilon)$$
$$+ \frac{\mu'}{2} u_{,} \Sigma a \left(\frac{d A^{(i)}}{d a}\right) \cos. i (n' t - n t + \epsilon' - \epsilon)$$
$$+ \frac{\mu'}{2} u_{,}' \Sigma a' \left(\frac{d A^{(i)}}{d a'}\right) \cos. i (n' t - n t + \epsilon' - \epsilon)$$
$$- \frac{\mu'}{2} (v_{,}' - v_{,}) \Sigma . i A^{(i)} \sin. i (n' t - n t + \epsilon' - \epsilon)$$
$$+ \frac{\mu'}{4} . u_{,}^2 . \Sigma . a^2 \left(\frac{d^2 A^{(i)}}{d a^2}\right) \cos. i (n' t - n t + \epsilon' - \epsilon)$$
$$+ \frac{\mu'}{2} u_{,} u_{,}' \Sigma a a' \left(\frac{d^2 A^{(i)}}{d a d a'}\right) \cos. i (n' t - n t + \epsilon' - \epsilon)$$
$$+ \frac{\mu'}{4} u_{,}'^2 . \Sigma a'^2 \left(\frac{d^2 A^{(i)}}{d a'^2}\right) \cos. i (n' t - n t + \epsilon' - \epsilon)$$
$$- \frac{\mu'}{2} (v_{,}' - v_{,}) u_{,} \Sigma . i a \left(\frac{d A^{(i)}}{d a}\right) \sin. i (n' t - n t + \epsilon' - \epsilon)$$
$$- \frac{\mu'}{2} (v_{,}' - v_{,}) u_{,}' \Sigma . i a' \left(\frac{d A^{(i)}}{d a'}\right) \sin. i (n' t - n t + \epsilon' - \epsilon)$$
$$- \frac{\mu'}{4} (v_{,}' - v_{,})^2 . \Sigma . i^2 A^{(i)} \cos. i (n' t - n t + \epsilon' - \epsilon)$$
$$+ \frac{\mu' z z'}{a'^3} - \frac{3 \mu' a z'^2}{2 a'^4} \cos. (n' t - n t + \epsilon' - \epsilon)$$
$$+ \frac{\mu' (z' - z)^2}{4} \Sigma B^{(i)} \cos. i (n' t - n t + \epsilon' - \epsilon)$$
$$+ \&c.$$

If we substitute in this expression of R, instead of $u_{,}$, $u_{,}'$, $v_{,}$, $v_{,}'$, z and z', their values relative to elliptic motion, values which are functions of sines and cosines of the angles $n t + \epsilon$, $n' t + \epsilon'$ and of their multiples, R will be expressed by an infinite series of cosines of the form $\mu' k \cos. (i n' t - i n t + A)$, i and i' being whole numbers.

It is evident that the action of μ'', μ''', &c. upon μ will produce in R terms analogous to those which result from the action of μ', and we shall obtain them by changing in the preceding expression of R, all that relates to μ', in the same quantities relative to μ'', μ''', &c.

Let us consider any term $\mu' k \cos. (i' n' t - i n t + A)$ of the expression of R. If the orbits were circular, and in one plane we should have $i' = i$. Therefore i' cannot surpass i or be exceeded by it, except by means of the sines or cosines of the expression for $u_{,}$, $v_{,}$, z, $u_{,}'$, $v_{,}'$, z' which combined with the sines and cosines of the angle $n' t - n t + \epsilon' - \epsilon$

and of its multiples, produce the sines and cosines of angles in which i' is different from i.

If we regard the excentricities and inclinations of the orbits as very small quantities of the first order, it will result from the theorems of (481) that in the expressions of $u_{,}$, $v_{,}$, z or ϱ s, s being the tangent of the latitude of μ, the coefficient of the sine or of the cosine of an angle such as f. $(n t + \varepsilon)$, is expressed by a series whose first term is of the order f; second term of the order f + 2; third term of the order f + 4 and so on. The same takes place with regard to the coefficient of the sine or of the cosine of the angle $f'(n' t + \varepsilon')$ in the expressions of $u_{,}'$, $v_{,}'$, z'. Hence it follows that i, and i' being supposed positive and i' greater than i, the coefficient k in the term m' k cos. $(i' n' t - i n t + A)$ is of the order $i' - i$, and that in the series which expresses it, the first term is of the order $i' - i$ the second of the order $i' - i + 2$ and so on; so that the series is very convergent. If i be greater than i', the terms of the series will be successively of the orders $i - i'$, $i - i' + 2$, &c.

Call ϖ the longitude of the perihelion of the orbit of μ and θ that of its node, in like manner call ϖ' the longitude of the perihelion of μ', and θ' that of its node, these longitudes being reckoned upon a plane inclined to that of the orbits. It results from the Theorems of (481), that in the expressions of $u_{,}$, $v_{,}$, and z, the angle $n t + \varepsilon$ is always accompanied by $-\varpi$ or by $-\theta$; and that in the expressions of $u_{,}'$, $v_{,}'$, and z', the angle $n' t + \varepsilon'$ is always accompanied by $-\varpi'$, or by $-\theta'$; whence it follows that the term μ' k cos. $(i' n' t - i n t + A)$ is of the form

$$\mu' k \cos. (i n' t - i n t + i' \varepsilon' - i \varepsilon - g \varpi - g' \varpi' - g'' \theta - g''' \theta'),$$

g, g', g'', g''' being whole positive or negative numbers, and such that we have

$$0 = i' - i - g - g' - g'' - g'''.$$

It results also from this that the value of R, and its different terms are independent of the position of the straight line from which the longitudes are measured. Moreover in the Theorems of (No. 481) the coefficient of the sine and cosine of the angle ϖ, has always for a factor the excentricity e of the orbit of μ; the coefficient of the sine and of the cosine of the angle 2ϖ, has for a factor the square e^2 of this excentricity, and so on. In like manner, the coefficient of the sine and cosine of the angle θ, has for its factor tan. $\frac{1}{2}\varphi$, φ being the inclination of the orbit of μ upon the fixed plane. The coefficient of the sine, and of the cosine of the angle 2θ, has for its factor $\tan.^2 \frac{1}{2}\varphi$, and so on. Whence it results that the coefficient k has for its factor, $e^{g}. e'^{g'}. \tan.^{g''} (\frac{1}{2}\varphi) \tan.^{g'''} (\frac{1}{2}\varphi')$; the numbers g, g', g'', g''' being taken positively in the exponents of this factor. If all these numbers are positive, this factor will be of the order i' — i, by virtue of the equation

$$0 = i' - i - g - g' - g'' - g''';$$

but if one of them such as g, is negative and equal to — g, this factor will be of the order i' — i + 2 g. Preserving, therefore, amongst the terms of R, only those which depending upon the angle i' n' t — i n t are of the order i' — i, and rejecting all those which depending upon the same angle, are of the order i' — i + 2, i' — i + 4, &c.; the expression of R will be composed of terms of the form

$$H\, e^{g}.\, e'^{g'} \tan.^{g''}\left(\frac{1}{2}\varphi\right) \tan.^{g'''}.\left(\frac{1}{2}\varphi'\right) \cos.\, (i'\, n'\, t - i\, n\, t + i'\, \varepsilon' - i\, \varepsilon - g.\, \varpi - g'.\, \varpi' - g''.\, \theta - g'''.\, \theta'),$$

H being a coefficient independent of the excentricities, and inclinations of the orbits, and the numbers g, g', g'', g''' being all positive, and such that their sum is equal to i' — i.

If we substitute in R, a $(1 + u_{,})$, instead of ϱ, we shall have

$$\varrho\left(\frac{d\, R}{d\, \varrho}\right) = a\left(\frac{d\, R}{d\, a}\right).$$

If in this same function, we substitute instead of u', v' and z, their values given by the theorems of (481), we shall have

$$\left(\frac{d\, R}{d\, v}\right) = \left(\frac{d\, R}{d\, \varepsilon}\right),$$

provided that we suppose $\varepsilon - \varpi$, and $\varepsilon - \theta$ constant in the differential of R, taken relatively to ε; for then $u_{,}$, $v_{,}$ and z are constant in this differential, and since we have $v = n\, t + \varepsilon + v_{,}$, it is evident that the preceding equation still holds. We shall, therefore, easily obtain the values of $\varrho\left(\frac{d\, R}{d\, \varrho}\right)$, and of $\left(\frac{d\, R}{d\, v}\right)$, which enter into the differential equations of the preceding numbers, when we shall have the value of R developed into a series of angles increasing proportionally to the time t. The differential *d* R it will be in like manner easy to determine, observing to vary in R the angle n t, and to suppose n' t constant; for *d* R is the difference of R, taken in supposing constant, the coordinates of μ', which are functions of n' t.

516. The difficulty of the developement of R into a series, may be reduced to that of forming the quantities $A^{(i)}$, $B^{(i)}$, and their differences taken relatively to a and to a'. For that purpose consider generally the function

$$(a^2 - 2\, a\, a' \cos.\, \theta + a'^2)^{-s}$$

and develope it according to the cosine of the angle θ and its multiples.

If we make $\frac{a}{a'} = \alpha$, it will become

$$a'^{-2s} . \{1 - 2\alpha \cos. \theta + \alpha^2\}^{-s}.$$

Let

$$(1 - 2\alpha \cos. \theta + \alpha^2)^{-s} = \tfrac{1}{2} b_s^{(0)} + b_s^{(1)} \cos. \theta + b_s^{(2)} \cos. 2\theta + b_s^{(3)} \cos. 3\theta + \&c.$$

$b_s^{(0)}, b_s^{(1)}, b_s^{(2)}$, &c. being functions of α and of s. If we take the logarithmic differences of the two members of this equation, relative to the variable θ, we shall have

$$\frac{-2s\alpha \sin. \theta}{1 - 2\alpha \cos. \theta + \alpha^2} = \frac{-b_s^{(1)} \sin. \theta - 2b_s^{(2)} \sin. 2\theta - \&c.}{\tfrac{1}{2} b_s^{(0)} + b_s^{(1)} \cos. \theta + b_s^{(2)} \cos. 2\theta + \&c.}.$$

Multiplying this equation crosswise, and comparing similar cosines, we find generally

$$b_s^{(i)} = \frac{(i-1)(1+\alpha^2) b_s^{(i-1)} - (i+s-2) a b_s^{(i-2)}}{(i-s).\alpha} \quad . . . \quad (a)$$

We shall thus have $b_s^{(2)}, b_s^{(3)}$, &c. when $b_s^{(0)}$ and $b_s^{(1)}$ are known.

If we change s into s + 1, in the preceding expression of $(1 - 2\alpha \cos. \theta + \alpha^2)^{-s}$, we shall have

$$(1 - 2\alpha \cos. \theta + \alpha^2)^{-s-1} = \tfrac{1}{2} b_{s+1}^{(0)} + b_{s+1}^{(1)} \cos. \theta + b_{s+1}^{(2)} \cos. 2\theta + b_{s+1}^{(3)} \cos. 3\theta + \&c.$$

Multiplying the two members of this equation, by $1 - 2\alpha \cos. \theta + \alpha^2$, and substituting for $(1 - 2\alpha \cos. \theta + \alpha^2)^{-s}$ its value in series, we shall have

$$\tfrac{1}{2} b_s^{(0)} + b_s^{(1)} \cos. \theta + b_s^{(2)} \cos. 2\theta + \&c.$$

$$= (1 - 2\alpha \cos. \theta + \alpha^2)\{ b_{s+1}^{(0)} + b_{s+1}^{(1)} \cos. \theta + b_{s+1}^{(2)} \cos. 2\theta + \&c.\}$$

whence by comparing homogeneous terms, we derive

$$b_s^{(i)} = (1 + \alpha^2) b_{s+1}^{(i)} - \alpha b_{s+1}^{(i-1)} - \alpha b_{s+1}^{(i+1)}.$$

The formula (a) gives

$$b_{s+1}^{(i+1)} = \frac{i(1+\alpha^2) b_{s+1}^{(i)} - (i+s)\alpha b_{s+1}^{(i-1)}}{(i-s).\alpha};$$

The preceding expression of $b_s^{(i)}$ will thus become

$$b_s^{(i)} = \frac{2s.\alpha b_{s+1}^{(i-1)} - s(1+\alpha^2) b_{s+1}^{(i)}}{i-s}.$$

Changing i into i + 1 in this equation we shall have

$$b_s^{(i+1)} = \frac{2\, s\, \alpha\, b_{s+1}^{(i)} - s\,(1 + \alpha^2)\, b_{s+1}^{(i+1)}}{i - s + 1}$$

and if we substitute for $b_{s+1}^{(i+1)}$ its preceding value, we shall have

$$b_s^{(i+1)} = \frac{s\,(i + s)\,\alpha\,(1 + \alpha^2)\, b_{s+1}^{(i-1)} + s\{2\,(i - s)\,\alpha^2 - i\,(1 + \alpha^2)^2\}\, b_{s+1}^{(i)}}{(i - s)\,(i - s + 1)\,\alpha}$$

These two expressions of $b_s^{(i)}$ and $b_s^{(i+1)}$ give

$$b_{s+1}^{(i)} = \frac{\frac{(i + s)}{s} \cdot (1 + \alpha^2)\, b_s^{(i)} - 2 \cdot \frac{i - s + 1}{s}\, \alpha\, b_s^{(i+1)}}{(1 - \alpha^2)^2}; \quad \text{(b)}$$

substituting for $b_s^{(i+1)}$ its value derived from equation (a), we shall have

$$b_{s+1}^{(i)} = \frac{\frac{(s - i)}{s}\,(1 + \alpha^2)\, b_s^{(i)} + \frac{2\,(i + s - 1)}{s} \cdot \alpha \cdot b_s^{(i-1)}}{(1 - \alpha^2)^2}; \quad \text{(c)}$$

an expression which may be derived from the preceding by changing i into — i, and observing that $b^{(i)} = b^{(-i)}$. We shall therefore have by means of this formula, the values of $b_{s+1}^{(0)}$, $b_{s+1}^{(1)}$, $b_{s+1}^{(2)}$, &c. when those of $b_s^{(0)}$, $b_s^{(1)}$, $b_s^{(2)}$, &c. are known.

Let λ, for brevity, denote the function $1 - 2\,\alpha \cos. \theta + \alpha^2$. If we differentiate relatively to α, the equation

$$\lambda^{-s} = \tfrac{1}{2}\, b_s^{(0)} + b_s^{(1)} \cos. \theta + b_s^{(2)} \cos. 2\,\theta + \&c.$$

we shall have

$$-2\, s\,(\alpha - \cos. \theta)\, \lambda^{-s-1} = \tfrac{1}{2} \cdot \frac{d\, b_s^{(0)}}{d\,\alpha} + \frac{d\, b_s^{(1)}}{d\,\alpha} \cos. \theta + \frac{d\, b_s^{(2)}}{d\,\alpha} \cos. 2\,\theta + \&c.$$

But we have

$$-\alpha + \cos. \theta = \frac{1 - \alpha^2 - \lambda}{2\,\alpha};$$

We shall, therefore, have

$$\frac{s\,(1 - \alpha^2)}{\alpha}\, \lambda^{-s-1} - \frac{s\,\lambda^{-s}}{\alpha} = \tfrac{1}{2}\, \frac{d\, b_s^{(0)}}{d\,\alpha} + \frac{d\, b_s^{(1)}}{d\,\alpha} \cos. \theta + \&c.$$

whence generally we get

$$\frac{d\, b_s^{(i)}}{d\,\alpha} = \frac{s\,(1 - \alpha^2)}{\alpha}\, b_{s+1}^{(i)} - \frac{s\, b_s^{(i)}}{\alpha}.$$

Substituting for $b_{s+1}^{(i)}$ its value given by the formula (b), we shall have

$$\frac{d\, b_s^{(i)}}{d\,\alpha} = \frac{i + (i + 2\,s)\,\alpha^2}{\alpha\,(1 - \alpha^2)} \cdot b_s^{(i)} - \frac{2\,(i - s + 1)}{1 - \alpha^2} \cdot b_s^{i+1}.$$

If we differentiate this equation, we shall have

$$\frac{d^2 b_s^{(i)}}{d\alpha^2} = \frac{i + (i + 2s)\alpha^2}{\alpha(1 - \alpha^2)} \cdot \frac{d b_s^{(i)}}{d\alpha} + \left\{\frac{2(i + s)(1 + \alpha^2)}{(1 - \alpha^2)^2} - \frac{i}{\alpha^2}\right\} b_s^{(i)}$$

$$- \frac{2(i - s + 1)}{1 - \alpha^2} \cdot \frac{d b_s^{(i+1)}}{d\alpha} - \frac{4(i - s + 1)\alpha}{(1 - \alpha^2)^2} b_s^{(i+1)}.$$

Again differentiating, we shall get

$$\frac{d^3 b_s^{(i)}}{d\alpha^3} = \frac{i + (i + 2s)\alpha^2}{\alpha(1 - \alpha^2)} \cdot \frac{d^2 b_s^{(i)}}{d\alpha^2} + 2\left\{\frac{(i + s)(1 + \alpha^2)}{(1 - \alpha^2)^2} - \frac{i}{\alpha^2}\right\} \frac{d b_s^{(i)}}{d\alpha}$$

$$+ \left\{\frac{4(i + s)\alpha(3 + \alpha^2)}{(1 - \alpha^2)^3} + \frac{2i}{\alpha^3}\right\} b_s^{(i)} - \frac{2(i - s + 1)}{1 - \alpha^2} \cdot \frac{d^2 b_s^{(i+1)}}{d\alpha^2}$$

$$- \frac{8(i - s + 1)\alpha}{(1 - \alpha^2)^2} \cdot \frac{d b_s^{(i+1)}}{d\alpha} - \frac{4(i - s + 1)(1 + 3\alpha^2)}{(1 - \alpha^2)^3} b_s^{(i+1)}.$$

Thus we perceive that in order to determine the values of b_s and of its successive differences, it is sufficient to know those of $b_s^{(0)}$ and of $b_s^{(1)}$. We shall determine these two as follows:

If we call c the hyperbolic base, we can put the expression of λ^{-s} under this form

$$\lambda^{-s} = (1 - \alpha c^{\theta\sqrt{-1}})^{-s}(1 - \alpha c^{-\theta\sqrt{-1}})^{-s}.$$

Developing the second member of this equation relatively to the powers of $c^{\theta\sqrt{-1}}$, and $c^{-\theta\sqrt{-1}}$, it is evident the two exponentials $c^{i\theta\sqrt{-1}}$, $c^{-i\theta\sqrt{-1}}$ will have the same coefficient which we denote by k. The sum of the two terms $k \cdot c^{i\theta\sqrt{-1}}$ and $k\, c^{-i\theta\sqrt{-1}}$ is 2 k cos. i θ. This will be the value of $b_s^{(i)}$ cos. i θ. We have, therefore, $b_s^{(i)} = 2\,k$. Again the expression of λ^{-s} is equal to the product of the two series

$$1 + s\alpha c^{\theta\sqrt{-1}} + \frac{s(s + 1)}{1.2}\alpha^2 c^{2\theta\sqrt{-1}} + \&c.$$

$$1 + s\alpha c^{-\theta\sqrt{-1}} + \frac{s(s + 1)}{1.2}\alpha^2 c^{-2\theta\sqrt{-1}} + \&c.;$$

multiplying therefore these two together, we shall have when i = 0

$$k = 1 + s^2\alpha^2 + \left(\frac{s(s + 1)}{1.2}\right)^2 \alpha^4 + \&c.;$$

and in the case of i = 1,

$$k = \alpha\left\{s + s \cdot \frac{s(s + 1)}{1.2} \cdot \alpha^2 + \frac{s(s + 1)}{1.2} \cdot \frac{s(s + 1)(s + 2)}{1.2.3}\alpha^4 + \&c.\right\}$$

wherefore

$$b_s^{(0)} = 2\left\{1 + s^2\alpha^2 + \left(\frac{s(s + 1)}{1.2}\right)^2 \alpha^4 + \left(\frac{s(s + 1)(s + 2)}{1.2.3}\right)^2 \alpha^6 + \&c.\right\}$$

$$b_s^{(1)} = 2\alpha\left\{s + s.\frac{s(s+1)}{1.2}\alpha^2 + \frac{s(s+1)}{1.2}.\frac{s(s+1)(s+2)}{1.2.3}\alpha^4 + \&c.\right\}.$$

That these series may be convergent, we must have α less than unity, which can always be made so, unless $a = a'$; α being $= \frac{a}{a'}$, we have only to take the greater for the denominator.

In the theory of the motion of the bodies μ, μ', μ'', &c. we have occasion to know the values of $b_s^{(0)}$ and of $b_s^{(1)}$ when $s = \frac{1}{2}$ and $s = \frac{3}{2}$. In these two cases, these values have but little convergency unless α is a small fraction.

The series converge with greater rapidity when $s = -\frac{1}{2}$, and we have

$$\frac{1}{2}b_{-\frac{1}{2}}^{(0)} = 1 + \left(\frac{1}{2}\right)^2\alpha^2 + \left(\frac{1.1}{2.4}\right)^2\alpha^4 + \left(\frac{1.1.3}{2.4.6}\right)^2\alpha^6 + \left(\frac{1.1.3.5}{2.4.6.8}\right)^2\alpha^8 + \&c.$$

$$b_{-\frac{1}{2}}^{(1)} = -\alpha\left\{1 - \frac{1.1}{2.4}\alpha^2 - \frac{1}{4}.\frac{1.1.3}{2.4.6}\alpha^4 - \frac{1.3}{4.6}.\frac{1.1.3.5}{2.4.6.8}\alpha^6 - \frac{1.3.5}{4.6.8}.\frac{1.1.3\,..7}{2.3\ldots10}\right\}\alpha^8 + \&c.$$

In the Theory of the planets and satellites, it will be sufficient to take the sum of eleven or a dozen first terms, in neglecting the following terms or more exactly in summing them as a geometric progression whose common ratio is $1 - \alpha^2$. When we shall have thus determined $b_{-\frac{1}{2}}^{(0)}$ and $b_{-\frac{1}{2}}^{(1)}$, we shall have $b_{\frac{1}{2}}^{(0)}$ in making $i = 0$, and $s = -\frac{1}{2}$ in the formula (b), and we shall find

$$b_{\frac{1}{2}}^{(0)} = \frac{(1+\alpha^2)\,b_{-\frac{1}{2}}^{(0)} + 6\alpha\, b_{-\frac{1}{2}}^{(1)}}{(1-\alpha^2)^2}$$

If in the formula (c) we suppose $i = 1$ and $s = -\frac{1}{2}$ we shall have

$$b_{\frac{1}{2}}^{(1)} = \frac{2\alpha\, b_{-\frac{1}{2}}^{(0)} + 3(1+\alpha^2)\,b_{-\frac{1}{2}}^{(1)}}{(1-\alpha^2)^2}.$$

By means of these values of $b_{\frac{1}{2}}^{(0)}$ and of $b_{\frac{1}{2}}^{(1)}$ we shall have by the preceding forms the values of $b_{\frac{1}{2}}^{(i)}$ and of its partial differences whatever may be the number i; and thence we derive the values of $b_{\frac{3}{2}}^{(i)}$ and of its differences. The values of $b_{\frac{3}{2}}^{(0)}$ and of $b_{\frac{3}{2}}^{(1)}$ may be determined very simply, by the following formulæ

$$b^{(0)}_{\frac{3}{2}} = \frac{b^{(0)}_{-\frac{1}{2}}}{(1 - \alpha^2)^2}; \quad b^{(1)}_{\frac{3}{2}} = -3 . \frac{b^{(1)}_{-\frac{1}{2}}}{(1 - \alpha^2)^2}.$$

Again to get the quantities $A^{(0)}$, $A^{(1)}$, &c. and their differences, we must observe that by the preceding No., the series

$$\tfrac{1}{2} A^{(0)} + A^{(1)} \cos. \theta + A^{(2)} \cos. 2\,\theta + \&c.$$

results from the developement of the function

$$\frac{a \cos. \theta}{a'^2} - (a^2 - 2\, a\, a' \cos. \theta + a'^2)^{-\frac{1}{2}},$$

into a series of cosines of the angle θ and of its multiples. Making $\frac{a}{a'} = \alpha$, this same function becomes

$$-\frac{1}{2 a'} b^{(0)}_{\frac{1}{2}} + \left(\frac{a}{a'^2} - \frac{1}{a'} b^{(1)}_{\frac{1}{2}}\right) \cos. \theta - \frac{1}{a'} b^{(2)}_{\frac{1}{2}} \cos. 2\,\theta - \&c.$$

which gives generally

$$A^{(i)} = -\frac{1}{a'} . b^{(i)}_{\frac{1}{2}};$$

when i is zero, or greater than 1, abstraction being made of the sign.

In the case of $i = 1$, we have

$$A^{(i)} = \frac{a}{a'^2} - \frac{1}{a'} b^{(1)}_{\frac{1}{2}}.$$

We have next

$$\left(\frac{d A^{(i)}}{d a}\right) = -\frac{1}{a'} . \frac{d b^{(i)}_{\frac{1}{2}}}{d \alpha} \left(\frac{d \alpha}{d a}\right);$$

But we have $\frac{d \alpha}{d a} = \frac{1}{a'}$; therefore

$$\left(\frac{d A^{(i)}}{d a}\right) = -\frac{1}{a'^2} . \frac{d b^{(i)}_{\frac{1}{2}}}{d \alpha};$$

and in the case of $i = 1$, we have

$$\left(\frac{d A^{(1)}}{d \alpha}\right) = \frac{1}{a'^2} \left\{ 1 - \frac{d b^{(1)}_{\frac{1}{2}}}{d \alpha} \right\}.$$

Finally, we have, in the same case of $i = 1$

$$\left(\frac{d^2 A^{(1)}}{d a^2}\right) = -\frac{1}{a'^3} . \frac{d^2 b^{(i)}_{\frac{1}{2}}}{d \alpha^2};$$

$$\left(\frac{d^3 A^{(i)}}{d a^3}\right) = -\frac{1}{a'^4} \cdot \frac{d^3 b^{(i)}_{\frac{1}{2}}}{d \alpha^3};$$

&c.

To get the differences of $A^{(i)}$ relative to a′, we shall observe that $A^{(i)}$ being a homogeneous function in a and a′, of the dimension — 1, we have by the nature of such functions,

$$a\left(\frac{d A^{(i)}}{d a}\right) + a'\left(\frac{d A^{(i)}}{d a'}\right) = -A^{(i)};$$

whence we get

$$a'\left(\frac{d A^{(i)}}{d a'}\right) = -A^{(i)} - a\left(\frac{d A^{(i)}}{d a}\right);$$

$$a'\left(\frac{d^2 A^{(i)}}{d a\, d a'}\right) = -2\left(\frac{d A^{(i)}}{d a}\right) - a\left(\frac{d^2 A^{(i)}}{d a^2}\right);$$

$$a'^2\left(\frac{d^2 A^{(i)}}{d a'^2}\right) = 2 A^{(i)} + 4 a\left(\frac{d A^{(i)}}{d a}\right) + a^2\left(\frac{d^2 A^{(i)}}{d a^2}\right);$$

$$a'^2\left(\frac{d^3 A^{(i)}}{d a'^2 d a}\right) = 6\left(\frac{d A^{(i)}}{d a}\right) + 6 a\left(\frac{d^2 A^{(i)}}{d a^2}\right) + a^2\left(\frac{d^3 A^{(i)}}{d a^3}\right);$$

$$a'^3\left(\frac{d^3 A^{(i)}}{d a'^3}\right) = -6A^{(i)} - 18 a\left(\frac{dA^{(i)}}{d a}\right) - 9a^2\left(\frac{d^2A^{(i)}}{d a^2}\right) - a^3\left(\frac{d^3A^{(i)}}{d a^3}\right);$$

&c.

We shall get $B^{(i)}$ and its differences, by observing that by the No. preceding, the series

$$\tfrac{1}{2} B^{(0)} + B^{(1)} \cos.\theta + B^{(2)} \cos. 2\theta + \&c.$$

is the developement of the function

$$a'^{-3} (1 - 2\alpha \cos.\theta + \alpha^2)^{-\frac{3}{2}}$$

according to the cosine of the angle θ and its multiples. But this function thus developed is equal to

$$a'^{-3}\left\{\tfrac{1}{2} b^{(0)}_{\frac{3}{2}} + b^{(1)}_{\frac{3}{2}} \cos.\theta + b^{(2)}_{\frac{3}{2}} \cos. 2\theta + \&c.\right\};$$

therefore we have generally

$$B^{(i)} = \frac{1}{a'^3} b^{(i)}_{\frac{3}{2}};$$

Whence we derive

$$\left(\frac{d B^{(i)}}{d a}\right) = \frac{1}{a'^4} \cdot \frac{d b^{(i)}_{\frac{3}{2}}}{d a}; \quad \left(\frac{d^2 B^{(i)}}{d a^2}\right) = \frac{1}{a'^5} \cdot \frac{d^2 b^{(i)}_{\frac{3}{2}}}{d a^2}; \ \&c.$$

Moreover, $B^{(i)}$ being a homogeneous function of a and of a′, of the dimension — 3 we have

$$a\left(\frac{d B^{(i)}}{d a}\right) + a'\left(\frac{d B^{(i)}}{d a'}\right) = -3 B^{(i)};$$

whence it is easy to get the partial differences of $B^{(i)}$ taken relatively to a' by means of those in a.

In the theory of the Perturbations of μ', by the action of μ, the values of $A^{(i)}$ and of $B^{(i)}$, are the same as above with the exception of $A^{(1)}$ which in this theory becomes $\frac{a'}{a^2} - \frac{1}{a'} b^{(1)}_{\frac{1}{2}}$. Thus the estimate of the values of $A^{(i)}$, $B^{(i)}$, and their differences will serve also for the theories of the two bodies μ and μ'.

517. After this digression upon the developement of R into series, let us resume the differential equations (X'), (Y), (Z') of Nos. 513, 514; and find by means of them, the values of $\delta \varrho$, δv, and δs true to quantities of the order of the excentricities and inclinations of orbits.

If in the elliptic orbits, we suppose

$$\varrho = a(1 + u_{,}); \quad \varrho' = a'(1 + u_{,}');$$
$$v = n t + \varepsilon + v_{,}; \quad v' = n' t - \varepsilon' + v_{,}';$$

we shall have by No. (488)

$$u_{,} = - e \cos.(n t + \varepsilon - \varpi); \quad u_{,}' = - e' \cos.(n' t + \varepsilon' - \varpi');$$
$$v_{,} = 2 e \sin.(n t + \varepsilon - \varpi); \quad v_{,}' = 2 e' \sin.(n' t + \varepsilon' - \varpi');$$

$n t + \varepsilon$, $n' t + \varepsilon'$ being the mean longitudes of μ, μ'; a, a' being the semi-axis-majors of their orbits; e, e' the ratios of the excentricity to the semi-axis-major; and lastly ϖ, ϖ' being the longitudes of their perihelions. All these longitudes may be referred indifferently to the planes of the orbits, or to a plane which is but very little inclined to the orbits; since we neglect quantities of the order of the squares and products of the excentricities and inclinations. Substituting the preceding values in the expression of R in No. 515, we shall have

$$R = \frac{\mu'}{2} \Sigma A^{(i)} \cos. i(n' t - n t + \varepsilon' - \varepsilon)$$
$$- \frac{\mu'}{2} . \Sigma . \left\{ a \left(\frac{d A^{(i)}}{d a}\right) + 2 i A^{(i)} \right\} \times$$
$$e \cos. \{i(n' t - n t + \varepsilon' - \varepsilon) + n t + \varepsilon - \varpi\}$$
$$- \frac{\mu'}{2} \Sigma \left\{ a' \left(\frac{d A^{(i-1)}}{d a'}\right) - 2(i - 1) A^{(i-1)} \right\} \times$$
$$e' \cos. \{i(n' t - n t + \varepsilon' - \varepsilon) + n t + \varepsilon - \varpi'\};$$

the symbol Σ of finite integrals, extending to all the whole positive and negative values of i, not omitting the value i = 0.

Hence we obtain

$$2 \int d R + \varrho \left(\frac{d R}{d \varrho}\right) = 2 \mu' g + \frac{\mu'}{2} a \left(\frac{d A^{(0)}}{d a}\right)$$

$$+\frac{\mu'}{2}\Sigma\left\{a\left(\frac{d\,A^{(i)}}{d\,a}\right)+\frac{2\,n}{n-n'}\,A^{(i)}\right\}\cos.\,i\,(n'\,t-n\,t+\varepsilon'-\varepsilon)$$

$$-\frac{\mu'}{2}\left\{a^2\left(\frac{d^2\,A^{(0)}}{d\,a^2}\right)+3\,a\left(\frac{d\,A^{(0)}}{d\,a}\right)\right\}e\cos.\,(n\,t+\varepsilon-\varpi)$$

$$-\frac{\mu'}{2}\left\{a\,a'\left(\frac{d^2A^{(1)}}{d\,a\,d\,a'}\right)+2a\left(\frac{dA^{(1)}}{d\,a}\right)+2a'\left(\frac{dA^{(1)}}{d\,a'}\right)+4A^{(1)}\right\}.e'\cos.(nt+\varepsilon-\varpi')$$

$$-\frac{\mu}{2}\Sigma.\left\{\begin{matrix}a^2\left(\frac{d^2\,A^{(i)}}{d\,a^2}\right)+(2\,i+1)\,a\left(\frac{d\,A^{(i)}}{d\,a}\right)\\ +\frac{2(i-1)\,n}{i(n-n')-n}\left\{a\left(\frac{A^{(i)}}{d\,a}\right)+2iA^{(i)}\right\}\end{matrix}\right\}e\cos.\{i(n't-nt+\varepsilon'-\varepsilon)+nt+\varepsilon-\varpi\}$$

$$-\frac{\mu'}{2}\Sigma\left\{\begin{matrix}a\,a'\left(\frac{d^2\,A^{(i-1)}}{d\,a\,d\,a'}\right)-2\,(i-1)\,a\left(\frac{d\,A^{(i-1)}}{d\,a}\right)\\ +\frac{2\,(i-1)\,n}{i(n-n')-n}\left\{a,\left(\frac{dA^{(i-1)}}{d\,a'}\right)-2(i-1)A^{(i-1)}\right\}\end{matrix}\right\}e'\cos.\{i\,(n'\,t-n\,t+\varepsilon'-\varepsilon)$$

$$+\,n\,t+\varepsilon-\varpi'\};$$

the integral sign Σ extending, as in what follows, to all integer positive and negative values of i, the value $i=0$ being alone excepted, because we have brought from without this symbol, the terms in which $i=0$: $\mu'\,g$ is a constant added to the integral $\int d\,R$. Making therefore

$$C=\tfrac{1}{2}\,a^3\left(\frac{d^2\,A^{(0)}}{d\,a^2}\right)+3\,a^2\left(\frac{d\,A^{(0)}}{d\,a'}\right)+6\,a\,g;$$

$$D=\tfrac{1}{2}\,a^2\,a'\left(\frac{d^2\,A^{(1)}}{d\,a\,d\,a'}\right)+a^2\left(\frac{d\,A^{(1)}}{d\,a}\right)+a\,a'\left(\frac{d\,A^{(1)}}{d\,a'}\right)+2\,a\,A^{(1)}$$

$$C^{(i)}=\tfrac{1}{2}\,a^3\left(\frac{d^2\,A^{(i)}}{d\,a^2}\right)+\frac{2\,i+1}{2}\,a^2\left(\frac{d\,A^{(i)}}{d\,a}\right)$$

$$+\frac{i\,(n-n')-3\,n}{2\{i\,(n-n')-n\}}.\left\{a^2\left(\frac{d\,A^{(i)}}{d\,a}\right)+\frac{2\,n}{n-n'}\,a\,A^{(i)}\right\}$$

$$+\frac{(i-1)\,n}{i\,(n-n')-n}\left\{a^2\left(\frac{d\,A^{(i)}}{d\,a}\right)+2\,i\,a\,A^{(i)}\right\};$$

$$D^{(i)}=\tfrac{1}{2}\,a^2\,a'\left(\frac{d^2\,A^{(i-1)}}{d\,a\,d\,a'}\right)-(i-1)\,a^2\left(\frac{d\,A^{(i-1)}}{d\,a}\right)$$

$$+\frac{(i-1)\,n}{i\,(n-n')-n}\left\{a\,a'\left(\frac{d\,A^{(i-1)}}{d\,a'}\right)-2\,(i-1)\,a\,A^{(i-1)}\right\};$$

taking then for unity the sum of the masses $M+\mu$, and observing that (237) $\frac{M+\mu}{a^3}=n^2$, the equation (X′) will become

$$0=\frac{d^2.\,\delta\,u}{d\,t^2}+n^2\,\delta\,u-2\,n^2\,\mu'\,a\,g-\frac{n^2\,\mu'}{2}\,a^2\left(\frac{d\,A^{(0)}}{d\,a}\right)$$

$$-\frac{n^2\,\mu'}{2}\,\Sigma\left\{a^2\left(\frac{d\,A^{(i)}}{d\,a}\right)+\frac{2\,n}{n-n'}\,a\,A^{(i)}\right\}\cos.\,i\,(n'\,t-n\,t+\varepsilon'-\varepsilon)$$

$$+ n^2 \mu' C e \cos. (n t + \varepsilon - \varpi)$$
$$+ n^2 \mu' D e' \cos. (n t + \varepsilon - \varpi')$$
$$+ n^2 \mu' \Sigma C^{(i)} e \cos. \{i (n' t - n t + \varepsilon' - \varepsilon) + n t + \varepsilon - \varpi\}$$
$$+ n^2 \mu' \Sigma D^{(i)} e' \cos. \{i (n' t - n t + \varepsilon' - \varepsilon) + n t + \varepsilon - \varpi'\};$$

and integrating

$$\delta u = 2 \mu a g + \frac{\mu'}{2} a^2 \left(\frac{d A^{(0)}}{d a}\right)$$
$$- \frac{\mu'}{2} n^2 \Sigma . \frac{\left\{a^2 \left(\frac{d A^{(i)}}{d a}\right) + \frac{2 n}{n - n'} a A^{(i)}\right\}}{i^2 (n - n')^2 - n^2} \cos. i (n' t - n t + \varepsilon' - \varepsilon)$$
$$+ \mu' f_{,} e \cos. (n t + \varepsilon - \varpi) + \mu' f_{,}' e' \sin. (n t + \varepsilon - \varpi')$$
$$- \frac{\mu'}{2} C . n t . e \sin. (n t + \varepsilon - \varpi) - \frac{\mu'}{2} D . n t . e' \sin. (n t + \varepsilon - \varpi')$$
$$+ \mu' \Sigma . \frac{C^{(i)} n^2}{\{i (n - n') - n\}^2 - n^2} e \cos. \{i (n' t - n t + \varepsilon' - \varepsilon) + n t + \varepsilon - \varpi\}$$
$$+ \mu' \Sigma \frac{D^{(i)} n^2}{\{i (n - n') - n\}^2 - n^2} . e' \cos. \{i (n' t - n t + \varepsilon' - \varepsilon) + n t + \varepsilon - \varpi'\},$$

$f_{,}$ and $f_{,}'$ being two arbitraries. The expression of $\delta \varrho$ in terms δu, found in No. 514 will give

$$\frac{\delta \varrho}{a} = - 2 \mu' . a g - \frac{\mu'}{2} a^2 \left(\frac{d A^{(0)}}{d a}\right)$$
$$+ \frac{\mu'}{2} n^2 . \Sigma \left\{\frac{a^2 \left(\frac{d A^{(i)}}{d a}\right) + \frac{2 n}{n - n'} a A^{(i)}}{i^2 (n - n')^2 - n^2}\right\} \cos. i (n' t - n t + \varepsilon' - \varepsilon)$$
$$- \mu' f e \cos. (n t + \varepsilon - \varpi) - \mu' f' e' \cos. (n t + \varepsilon - \varpi')$$
$$+ \tfrac{1}{2} \mu' C n t e \sin. (n t + \varepsilon - \varpi) + \tfrac{1}{2} \mu' D n t e' \sin. (n t + \varepsilon - \varpi')$$
$$+ \mu' n^2 \Sigma . \left\{ \left\{ \frac{a^2 \left(\frac{d A^{(i)}}{d a}\right) + \frac{2 n}{n - n'} a A^{(i)}}{i^2 (n - n')^2 - n^2} - \frac{C^{(i)}}{\{i (n - n') - n\}^2 - n^2} \right\} \times e \cos. \{i (n' t - n t + \varepsilon' - \varepsilon) + n t + \varepsilon - \varpi\} \right\}$$
$$- \mu' . n^2 \Sigma . \frac{D^{(i)}}{\{i (n - n') - n\}^2 - n^2} e' \cos. \{i (n' t - n t + \varepsilon' - \varepsilon) + n t + \varepsilon - \varpi'\},$$

f and f' being arbitrary constants independent of $f_{,}$, $f_{,}'$.

This value of $\delta \varrho$, substituted in the formula (Y) of No. 513 will give δv or the perturbations of the planet in longitude. But we must observe that $n t$ expressing the mean motion of μ, the term proportional to the time, ought to disappear from the expression of δv. This condition determines the constant (g) and we find

$$g = - \frac{1}{3} a \left(\frac{d A^{(0)}}{d a}\right).$$

We might have dispensed with introducing into the value of $\delta \rho$ the arbitraries f, $f_{\prime}$, for they may be considered as comprised in the elements e and ϖ of elliptic motion. But then the expression of δv would include terms depending upon the mean anomaly, and which would not have been comprised in those which the elliptic motion gives: that is, it is more commodious to make these terms in the expression of the longitude disappear in order to introduce them into the expression of the radius-vector; we shall thus determine f, and $f_{\prime}$ so as to fulfil this condition. Then if we substitute for $a\left(\dfrac{d\,A^{(i-1)}}{d\,a'}\right)$ its value $-A^{(i-1)} - a\left(\dfrac{d\,A^{(i-1)}}{d\,a}\right)$, we shall have

$$C = a^2\left(\frac{d\,A^{(0)}}{d\,a}\right) + \tfrac{1}{2}a^3\left(\frac{d^2\,A^{(0)}}{d\,a^2}\right);$$

$$D = a\,A^{(1)} - a^2\left(\frac{d\,A^{(1)}}{d\,a}\right) - \tfrac{1}{2}a^3\left(\frac{d^2\,A^{(1)}}{d\,a^2}\right);$$

$$D^{(i)} = \frac{(i-1)(2i-1)n}{n-i(n-n')}\cdot a\,A^{(i-1)} + \frac{i^2(n-n')-n}{n-i(n-n')}a^2\left(\frac{d\,A^{(i-1)}}{d\,a}\right)$$
$$-\tfrac{1}{2}a^3\left(\frac{d^2\,A^{(i-1)}}{d\,a^2}\right);$$

$$f = \tfrac{2}{3}a^2\left(\frac{d\,A^{(0)}}{d\,a}\right) + \tfrac{1}{4}a^3\left(\frac{d^2\,A^{(0)}}{d\,a^2}\right);$$

$$f' = \tfrac{1}{4}\left\{a\,A^{(1)} - a^2\left(\frac{d\,A^{(1)}}{d\,a}\right) + a^3\left(\frac{d^2\,A^{(1)}}{d\,a^2}\right)\right\};$$

Moreover let

$$E^{(i)} = -\frac{3\,n}{n-n'}a\,A^{(i)} + \frac{i^2(n-n')\cdot\{n+i(n-n')\} - 3n^2}{i^2(n-n')^2 - n^2}$$
$$\times\left\{a^2\left(\frac{d\,A^{(i)}}{d\,a}\right) + \frac{2\,n}{n-n'}a\,A^{(i)}\right\} + \tfrac{1}{2}a^3\left(\frac{d^2\,A^{(i)}}{d\,a^2}\right);$$

$$F^{(i)} = \frac{(i-1)n}{n-n'}a\,A^{(i)} + \frac{\dfrac{i\,n}{2}\{n+i(n-n')\} - 3n^2}{i^2(n-n')^2 - n^2}$$
$$\times\left\{a^2\left(\frac{d\,A^{(i)}}{d\,a}\right) + \frac{2\,n}{n-n'}a\,A^{(i)}\right\} - \frac{2\,n^2\,E^{(i)}}{n^2 - \{n-i(n-n')\}^2};$$

$$G^{(i)} = \frac{(i-1)(2i-1)n\,a\,A^{(i-1)} + (i-1)n\,a^2\left(\dfrac{d\,A^{(i-1)}}{d\,a}\right)}{2\{n-i(n-n')\}}$$
$$- \frac{2\,n^2\,D^{(i)}}{n^2 - \{n-i(n-n')\}^2};$$

and we shall have

$$\frac{\delta \varrho}{a} = \frac{\mu'}{6} a^2 \left(\frac{d A^{(0)}}{d a}\right) + \frac{\mu' n^2}{2} \Sigma . \frac{a^2 \left(\frac{d A^{(i)}}{d a}\right) + \frac{2 n}{n - n'} a A^{(i)}}{i^2 (n - n')^2 - n^2} \times$$

$\cos. i (n' t - n t + \varepsilon' - \varepsilon)$

$- \mu' f e \cos. (n t + \varepsilon + \varpi) - \mu' f' e' \cos. (n t + \varepsilon - \varpi')$

$+ \frac{1}{2} \mu' C . n t e \sin. (n t + \varepsilon - \varpi) + \frac{1}{2} \mu' D n t e' \sin. (n t + \varepsilon - \varpi')$

$$+ n^2 \mu' \Sigma . \left\{ \begin{array}{l} \frac{E^{(i)}}{n^2 - \{n - i (n - n')\}^2} e \cos. \{i (n' t - n t + \varepsilon' - \varepsilon) + n t + \varepsilon - \varpi\} \\ \frac{D^{(i)}}{n^2 - \{n - i (n - n')\}^2} e' \cos. \{i (n' t - n t + \varepsilon' - \varepsilon) + n t + \varepsilon - \varpi'\} \end{array} \right\};$$

$$\delta v = \frac{\mu'}{2} \Sigma \left\{ \frac{n^2}{i (n - n')^2} a A^{(i)} + \frac{2 n^3 \left\{ a^2 \left(\frac{d A^{(i)}}{d a}\right) + \frac{2 n}{n - n'} a A^{(i)} \right\}}{i (n - n') . \{i^2 . (n - n')^2 - n^2\}} \right\} \sin. i$$

$(n' t - n t + \varepsilon' - \varepsilon)$

$+ \mu' . C . n t . e \cos. (n t + \varepsilon - \varpi) + \mu' D . n t . e' \cos. (n t + \varepsilon - \varpi')$

$$+ n \mu' \Sigma \left\{ \begin{array}{l} \frac{F^{(i)}}{n - i (n - n')} e \sin. \{i (n' t - n t + \varepsilon' - \varepsilon) + n t + \varepsilon - \varpi\} \\ \frac{G^{(i)}}{n - i (n - n')} e' \sin. \{i (n' t - n t + \varepsilon' - \varepsilon) + n t + \varepsilon - \varpi'\} \end{array} \right\};$$

the integral sign Σ extending in these expressions to all the whole positive and negative values of i, with the value $i = 0$ alone excepted.

Here we may observe, that even in the case where the series represented by

$$\Sigma . A^{(i)} \cos. i (n' t - n t + \varepsilon' - \varepsilon)$$

is but little convergent, these expressions of $\frac{\delta \varrho}{a}$ and of δv, become convergent by the divisors which they acquire. This remark is the more important, because, did this not take place, it would have been impossible to express analytically the mutual perturbations of the planets, of which the ratios of their distances from the sun are nearly unity.

These expressions may take the following form, which will be useful to us hereafter. Let

$$h = e \sin. \varpi; \; h' = e' \sin. \varpi';$$
$$l = e \cos. \varpi; \; l' = e' \cos. \varpi';$$

then we shall have

$$\frac{\delta \varrho}{a} = \frac{\mu'}{6} . a^2 \left(\frac{d A^{(0)}}{d a}\right) + \frac{\mu' n^2}{2} \Sigma \left\{ \frac{a^2 \left(\frac{d A^{(i)}}{d a}\right) + \frac{2 n}{n - n'} a A^{(i)}}{i^2 (n - n')^2 - n^2} \right\} \cos. i (n' t - n t + \varepsilon' - \varepsilon)$$

$- \mu' (h f + h' f') \cos. (n t + \varepsilon) - \mu' (l f + l' f') \sin. (n t + \varepsilon)$

$$+\frac{\mu'}{2}\{l\,C+l'\,D\}\,n\,t\sin.(n\,t+\varepsilon)-\frac{\mu'}{2}\{h\,C+h'\,D\}n\,t\cos.(n\,t+\varepsilon)$$

$$+n^2\mu'\,\Sigma.\left\{\begin{array}{l}\dfrac{h\,E^{(i)}+h'\,D^{(i)}}{n^2-\{n-i(n-n')\}^2}\sin.\{i(n'\,t-n\,t+\varepsilon'-\varepsilon)+n\,t+\varepsilon\}\\ +\dfrac{l\,E^{(i)}+l'\,D^{(i)}}{n^2-\{n-i(n-n')\}^2}\cos.\{i(n'\,t-n\,t+\varepsilon'-\varepsilon)+n\,t+\varepsilon\}\end{array}\right\}.$$

$$\delta\,v=\frac{\mu'}{2}\Sigma.\left\{\frac{n^2}{i(n-n')^2}a\,A^{(i)}+2\,n^3\,\frac{a^2\left(\dfrac{d\,A^{(i)}}{d\,a}\right)+\dfrac{2\,n}{n-n'}a\,A^{(i)}}{i(n-n')\{i^2.(n-n')^2-n^2\}}\right\}\times$$
$$\sin.i(n'\,t-n\,t+\varepsilon'-\varepsilon)$$
$$+\mu'\{h\,C+h'\,D\}.n\,t.\sin.(n\,t+\varepsilon)+\mu'\{l.C+l'.D\}\,n\,t.\cos.(n\,t+\varepsilon)$$

$$+n\,\mu'.\Sigma\left\{\begin{array}{l}\dfrac{l\,F^{(i)}+l'\,G^{(i)}}{n-i(n-n')}\sin.\{i(n'\,t-n\,t+\varepsilon'-\varepsilon)+n\,t+\varepsilon\}\\ -\dfrac{h\,F^{(i)}+h'\,G^{(i)}}{n-i(n-n')}\cos.\{i(n'\,t-n\,t+\varepsilon'-\varepsilon)+n\,t+\varepsilon\}\end{array}\right\};$$

Connecting these expressions of $\delta\,\varrho$ and $\delta\,v$ with the values of ϱ and v relative to elliptic motion, we shall have the entire values of the radius-vector of μ, and of its motion in longitude.

518. Now let us consider the motion of μ in latitude. For that purpose let us resume the formula (Z') of No. 514. If we neglect the product of the inclinations by the excentricities of the orbits it will become

$$0=\frac{d^2\,\delta\,u'}{d\,t^2}+n^2\,\delta\,u'-\frac{1}{a^2}\left(\frac{d\,R}{d\,z}\right);$$

the expression of R of No. 515 gives, in taking for the fixed plane that of the primitive orbit of μ,

$$\left(\frac{d\,R}{d\,z}\right)=\frac{\mu'\,z'}{a^3}-\frac{\mu'\,z'}{2}\Sigma.B^{(i)}\cos.i(n'\,t-n\,t+\varepsilon'-\varepsilon);$$

the value of i belonging to all whole positive and negative numbers including also $i=0$. Let γ be the tangent of the inclination of the orbit of μ', to the primitive orbit of μ, and Π the longitude of the ascending node of the first of these orbits upon the second; we shall have very nearly

$$z'=a'\,\gamma\sin.(n'\,t+\varepsilon'-\Pi);$$

which gives

$$\left(\frac{d\,R}{d\,z}\right)=\frac{\mu'}{a^2}.\gamma.\sin.(n'\,t+\varepsilon'-\Pi)-\frac{\mu'}{2}.a'\,B^{(i)}\gamma\sin.(n\,t+\varepsilon-\Pi)$$
$$\frac{\mu'}{2}a'\,\Sigma\,B^{(i-1)}\gamma\sin.\{i(n'\,t-n\,t+\varepsilon'-\varepsilon)+n\,t+\varepsilon-\Pi\}$$

the value here, as in what follows, extending to all whole positive and negative numbers, $i=0$ being alone excepted. The differential equation in δ a′ will become, therefore, when the value of $\left(\frac{d R}{d z}\right)$ is multiplied by $n^2 a^3$, which is equal to unity,

$$0 = \frac{d^2 \delta u'}{d t^2} + n^2 \delta u' - \mu' n^2 . \frac{a}{a'^2} \gamma \sin. (n' t + \varepsilon' - \Pi)$$

$$+ \frac{\mu' n^2}{2} a a' B^{(i)} \gamma \sin. (n t + \varepsilon - \Pi)$$

$$+ \frac{\mu' n^2}{2} a a' \Sigma B^{(i-1)} \gamma \sin. \{i (n' t - n t + \varepsilon' - \varepsilon) + n t + \varepsilon - \Pi\} ;$$

whence by integrating and observing that by 514

$$\delta s = - a \delta u',$$

$$\delta s = - \frac{\mu' n^2}{n^2 - n'^2} . \frac{a^2}{a'^2} \gamma \sin. (n' t + \varepsilon' - \Pi)$$

$$- \frac{\mu' . a^2 a'}{4} B^{(i)} . n t . \gamma \cos. (n t + \varepsilon - \Pi)$$

$$+ \frac{\mu' n^2 . a^2 a'}{2} \Sigma . \frac{B^{(i-1)}}{n^2 - \{n - i(n - n')\}^2} \gamma \sin. \{i(n't - nt + \varepsilon' - \varepsilon) + nt + \varepsilon - \Pi\}.$$

To find the latitude of μ above a fixed plane a little inclined to that of its primitive orbit, by naming φ the inclination of this orbit to the fixed plane, and θ the longitude of its ascending node upon the same plane; it will suffice to add to δ s the quantity tan. φ sin. (v — θ), or tan. φ sin. (n t + ε — θ), neglecting the excentricity of the orbit. Call $\varphi_{\prime}'$ and θ' what φ and θ become relatively to μ'. If μ were in motion upon the primitive orbits of μ', the tangent of its latitude would be tan. φ' sin. (n t + ε — θ'); this tangent would be tan. φ sin. (n t + ε — θ), if μ continued to move in its own primitive orbit. The difference of these two tangents is very nearly the tangent of the latitude of μ, above the plane of its primitive orbit, supposing it moved upon the primitive orbit of μ'; we have therefore

$$\tan. \varphi' \sin. (n t + \varepsilon - \theta') - \tan. \varphi \sin. (n t + \varepsilon - \theta) = \gamma \sin. (n t + \varepsilon - \Pi).$$

Let

$$\tan. \varphi \sin. \theta = p ; \quad \tan. \varphi \sin. \theta' = p' ;$$
$$\tan. \varphi \cos. \theta = q ; \quad \tan. \varphi' \cos. \theta' = q' ;$$

we shall have

$$\gamma \sin. \Pi = p' - p ; \quad \gamma \cos. \Pi = q' - q .$$

and consequently if we denote by s the latitude of μ above the fixed plane, we shall very nearly have

$$s = q \sin. (n t + \varepsilon) - p \cos. (n t + \varepsilon)$$

$$- \frac{\mu' a^2 a'}{4} (p' - p) B^{(i)} n t \sin. (n t + \varepsilon)$$

$$-\frac{\mu' a^2 a'}{4}(q'-q) B^{(i)} n t \cos.(n t + \epsilon)$$

$$-\frac{\mu' n^2}{n^2 - n'^2}\cdot\frac{a^2}{a'^2}\{(q'-q)\sin.(n' t+\epsilon') - (p'-p)\cos.(n' t+\epsilon')\}$$

$$+\frac{\mu' n^2 . a^2 a'}{z}\Sigma.\left\{\begin{array}{l}\frac{(q'-q) B^{(i-1)}}{n^2 - \{n - i(n-n')\}^2}\sin.\{i(n't - nt + \epsilon' - \epsilon) + nt + \epsilon\} \\ \frac{-(p'-p) B^{(i-1)}}{n^2 - \{n - i(n-n')\}^2}.\cos.\{i(n't - nt + \epsilon' - \epsilon) + nt + \epsilon\}\end{array}\right\}.$$

519. Now let us recapitulate. Call (ϱ) and (v) the parts of the radius-vector and longitude v upon the orbit, which depend upon the elliptic motion, we shall have

$$\varrho = (\varrho) + \delta\varrho;\ v = (v) + \delta v.$$

The preceding value of s, will be the latitude of μ above the fixed plane. But it will be more exact to employ, instead of its two first terms, which are independent of μ', the value of the latitude, which takes place in the case where μ quits not the plane of its primitive orbit. These expressions contain all the theory of the planets, when we neglect the squares and the products of the excentricities and inclinations of the orbits, which is in most cases allowable. They moreover possess the advantage of being under a very simple form, and which shows the law of their different terms.

Sometimes we shall have occasion to recur to terms depending on the squares and products of the excentricities and inclinations, and even to the superior powers and products. We can find these terms by the preceding analysis, the consideration which renders them necessary will always facilitate their determination. The approximations in which we must notice them, would introduce new terms which would depend upon new arguments. They would reproduce again the arguments, which the preceding approximations afford, but with coefficients still smaller and smaller, following that law which it is easy to perceive from the developement of R into a series, which was given in No. 515; *an argument which, in the successive approximations, is found for the first time among the quantities of any order whatever* r, *and is reproduced only by quantities of the orders* r+2, r+4, &c.

Hence it follows that the coefficients of the terms of the form $t.\frac{\sin.}{\cos.}.(n t + \epsilon)$, which enter into the expressions of ϱ, v, and s, are approximated up to quantities of the third order, that is to say, that the approximation in which we should have regard to the squares and products of the excentricities and inclinations of the orbits would add nothing to their values; they have therefore all the exactness that can be desired. This it is the more essential to observe, because the secular variations of the orbits depend upon these same coefficients.

The several terms of the perturbations of ϱ, v, s are comprised in the form

$$k \cdot \begin{matrix}\sin.\\ \cos.\end{matrix} \{i\,(n'\,t - n\,t + \varepsilon' - \varepsilon) + r\,n\,t + r\,\varepsilon\},$$

r being a whole positive number or zero, and k being a function of the excentricities and inclinations of the orbits of the order r, or of a superior order. Hence we may judge of what order is a term depending upon a given angle.

It is evident that the motion of the bodies μ'', μ''', &c. make it necessary to add to the preceding values of ϱ, v, and s, terms analogous to those which result from the action of μ'; and that neglecting the square of the perturbing force, the sums of all these terms will give the whole values of ϱ, v and s. This follows from the nature of the formulas (X'), (Y), (Z'), which are linear relatively to quantities depending on the disturbing force.

Lastly, we shall have the perturbations of μ', produced by the action of μ by changing in the preceding formulas, a, n, h, l, ε, ϖ, p, q, and μ' into a', n', h', l', ε', ϖ', p', q', and μ and reciprocally.

THE SECULAR INEQUALITIES OF THE CELESTIAL MOTIONS.

520. The perturbing forces of elliptical motion introduce into the expressions of ϱ, $\frac{d\,v}{d\,t}$, and s of the preceding Nos. the time t free from the symbols *sine* and *cosine*, or under the form of arcs of a circle, which by increasing indefinitely, must at length render the expressions defective. It is therefore essential to make these arcs disappear, and to obtain the functions which produce them by their developement into series. We have already given, for this purpose, a general method, from which it results that these arcs arise from the variations of elliptic motion, which are then functions of the time. These variations taking place very slowly have been denominated *Secular Inequalities.* Their theory is one of the most interesting subjects of the system of the world. We now proceed to expound it to the extent which its importance demands.

By what has preceded we have

$$\varrho = a \left\{ \begin{array}{l} 1 - h \sin.(n t + \varepsilon) - l \cos.(n t + \varepsilon) - \&c. \\ + \frac{\mu'}{2} \{l . C + l' . D\} . n t . \sin.(n t + \varepsilon) \\ - \frac{\mu'}{2} \{h . C + h' . D\} . n t . \cos.(n t + \varepsilon) + \mu' S. \end{array} \right\}$$

$$\frac{d v}{d t} = n + 2 n h \sin.(n t + \varepsilon) + 2 n l \cos.(n t + \varepsilon) + \&c.$$
$$- \mu' \{l C + l' D\} u^2 t \sin.(n t + \varepsilon)$$
$$+ \mu' \{h C + h' . D\} n^2 . t \cos.(n t + \varepsilon) + \mu' T;$$
$$s = q \sin.(n t + \varepsilon) - p \cos.(n t + \varepsilon) + \&c.$$
$$- \frac{\mu'}{4} a^2 a' (p' - p) B^{(1)} . n t . \sin.(n t + \varepsilon)$$
$$- \frac{\mu'}{4} a^2 a' (q' - q) B^{(1)} . n t . \cos.(n t + \varepsilon) + \mu' \chi;$$

S, T, χ being periodic functions of the time t. Consider first the expression of $\frac{d v}{d t}$, and compare it with the expression of y in 510. The arbitrary n multiplying the arc t, under the periodic symbols, in the expression of $\frac{d v}{d t}$; we ought then to make use of the following equations found in No. 510,

$$0 = X' + \theta . X'' - Y;$$
$$0 = Y' + \theta . Y'' + X'' - 2 Z;$$

Let us see what these X, X', X'', Y, &c. become. By comparing the expression of $\frac{d v}{d t}$ with that of y cited above, we find

$$X = n + 2 n h \sin.(n t + \varepsilon) + 2 n l \cos.(n t + \varepsilon) + \mu' T$$
$$Y = \mu' n^2 \{h C + h' D\} \cos.(n t + \varepsilon) - \mu' n^2 \{l C + l' D\} \sin.(n t + \varepsilon).$$

If we neglect the product of the partial differences of the constants by the perturbing masses, which is allowed, since these differences are of the order of the masses, we shall have by No. 510,

$$X' = \left(\frac{d n}{d \theta}\right) \{1 + 2 h \sin.(n t + \varepsilon) + 2 l \cos.(n t + \varepsilon)\}$$
$$+ 2 n \left(\frac{d \varepsilon}{d \theta}\right) \{h \cos.(n t + \varepsilon) - l \sin.(n t + \varepsilon)\}$$
$$+ 2 n \left(\frac{d h}{d \theta}\right) \sin.(n t + \varepsilon) + 2 n \left(\frac{d l}{d \theta}\right) \cos.(n t + \varepsilon);$$

$$X'' = 2 n \left(\frac{d n}{d \theta}\right) \{h \cos.(n t + \varepsilon) - l \sin.(n t + \varepsilon)\}$$

The equation $0 = X' + \theta X'' - Y$ will thus become

$$0 = \left(\frac{d\,n}{d\,\theta}\right)\{1 + 2\,h\,\sin.(n\,t + \epsilon) + 2\,l\,\cos.(n\,t + \epsilon)\}$$

$$+ 2\,n\left(\frac{d\,h}{d\,\theta}\right)\sin.(n\,t + \epsilon) + 2\,n\left(\frac{d\,l}{d\,\theta}\right)\cos.(n\,t + \epsilon)$$

$$+ 2\,n\left\{\theta\left(\frac{d\,n}{d\,\theta}\right) + \left(\frac{d\,\epsilon}{d\,\theta}\right)\right\}.\{h\,\cos.(n\,t + \epsilon) - l\,\sin.(n\,t + \epsilon)\}$$

$$- \mu'\,n^2\{h\,C + h'\,D\}\cos.(n\,t + \epsilon) + \mu'\,n^2\{l\,C + l'\,D\}\sin.(n\,t + \epsilon).$$

Equating separately to zero, the coefficients of like sines and cosines, we shall have

$$0 = \left(\frac{d\,n}{d\,\theta}\right)$$

$$0 = \left(\frac{d\,h}{d\,\theta}\right) - l\left(\frac{d\,\epsilon}{d\,\theta}\right) + \frac{\mu'\,n}{2}\{l.\,C + l'\,D\},$$

$$0 = \left(\frac{d\,l}{d\,\theta}\right) + h\left(\frac{d\,\epsilon}{d\,\theta}\right) - \frac{\mu'\,n}{2}\{h\,C + h'\,D\}.$$

If we integrated these equations, and if in their integrals we change θ into t, we shall have by No. 510, the values of the arbitraries in functions of t, and we shall be able to efface the circular arcs from the expressions of $\frac{d\,v}{d\,t}$ and of ϱ. But instead of this change, we can immediately change θ into t in these differential equations. The first of the equations shows us that n is constant, and since the arbitrary a of the expression for ϱ depends upon it, by reason of $n^2 = \frac{1}{a^3}$, a is likewise constant. The two other equations do not suffice to determine h, l, ϵ. We shall have a new equation in observing that the expression of $\frac{d\,v}{d\,t}$, gives, in integrating, $\int n\,d\,t$ for the value of the mean longitude of μ. But we have supposed this longitude equal to $n\,t + \epsilon$; we therefore have $n\,t + \epsilon = \int n\,d\,t$, which gives

$$t.\frac{d\,n}{d\,t} + \frac{d\,\epsilon}{d\,t} = 0;$$

and as we have $\frac{d\,n}{d\,t} = 0$, we have in like manner $\frac{d\,\epsilon}{d\,t} = 0$. Thus the two arbitraries n and ϵ are constants; the arbitraries h, l, will consequently be determined by means of the differential equations,

$$\frac{d\,h}{d\,t} = -\frac{\mu'\,n}{2}\{l\,C + l'\,D\}; \quad (1)$$

$$\frac{d\,l}{d\,t} = \frac{\mu'\,n}{2}\{h\,C + h'\,D\}; \quad (2)$$

The consideration of the expression of $\frac{d\,v}{d\,t}$ having enabled us to determine the values of n, a, h, l, and ε, we perceive *a priori*, that the differential equations between the same quantities, which result from the expression of ϱ, ought to coincide with those preceding. This may easily be shown *a posteriori*, by applying to this expression the method of **510**.

Now let us consider the expression of s. Comparing it with that of y cited above, we shall have

$$X = q \sin. (n\,t + \varepsilon) - p \cos. (n\,t + \varepsilon) + \mu' . \chi$$

$$Y = \frac{\mu' n}{4} . a^2 a' B^{(1)} (p - p') \sin. (n\,t + \varepsilon)$$

$$+ \frac{\mu' n}{4} . a^2 a' B^{(1)} (q - q') \cos. (n\,t + \varepsilon),$$

n and ε, by what precedes, being constants; we shall have by No. **510**,

$$X' = \left(\frac{d\,q}{d\,\theta}\right) \sin. (n\,t + \varepsilon) - \left(\frac{d\,p}{d\,\theta}\right) \cos. (n\,t + \varepsilon)$$

$$X'' = 0.$$

The equation $0 = X' + \theta X'' - Y$ hence becomes

$$0 = \left(\frac{d\,q}{d\,\theta}\right) \sin. (n\,t + \varepsilon) - \frac{d\,p}{d\,\theta} \cos. (n\,t + \varepsilon)$$

$$- \frac{\mu' n}{4} a^2 a' B^{(1)} (p - p') \sin. (n\,t + \varepsilon)$$

$$- \frac{\mu' n}{4} a^2 a' B^{(1)} (q - q') \cos. (n\,t + \varepsilon);$$

whence we derive, by comparing the coefficients of the like sines and cosines, and changing θ into t, in order to obtain directly p and q in functions of t,

$$\frac{d\,p}{d\,t} = - \frac{\mu' n}{4} . a^2 a' B^{(1)} . (q - q'); \quad (3)$$

$$\frac{d\,q}{d\,t} = \frac{\mu' n}{4} . a^2 a' B^{(1)} (p - p'); \quad (4)$$

When we shall have determined p and q by these equations, we shall substitute them in the preceding expression of s, effacing the terms which contain circular arcs, and we shall have

$$s = q \sin. (n\,t + \varepsilon) - p \cos. (n\,t + \varepsilon) + \mu' \chi.$$

521. The equation $\frac{d\,n}{d\,t} = 0$, found above, is one of great importance in the theory of the system of the world, inasmuch as it shows that the mean motions of the celestial bodies and the major-axes of their orbits are unalterable. But this equation is approximate to quantities of the order

$\mu' h$ inclusively. If quantities of the order $\mu' h^2$, and following orders, produce in $\frac{dv}{dt}$, a term of the form $2kt$, k being a function of the elements of the orbits of μ and μ'; there will thence result in the expression of v, the term kt^2, which by altering the longitude of μ, proportionally to the time, must at length become extremely sensible. We shall then no longer have

$$\frac{dn}{dt} = 0;$$

but instead of this equation we shall have by the preceding No.

$$\frac{dn}{dt} = 2k;$$

It is therefore very important to know whether there are terms of the form $k.t^2$ in the expression of v. We now demonstrate, that *if we retain only the first power of the perturbing masses, however far may proceed the approximation, relatively to the powers of the excentricities and inclinations of the orbits, the expression* v *will not contain such terms.*

For this object we will resume the formula (X) of No. 513,

$$\delta\varrho = \frac{a\cos.v\int ndt.\varrho\sin.v\left\{2\int dR+\varrho\left(\frac{dR}{d\varrho}\right)\right\} - a\sin.v\int ndt.\varrho\cos.v\left\{2\int dR+\varrho\left(\frac{dR}{d\varrho}\right)\right\}}{m\sqrt{1-e^2}}$$

Let us consider that part of $\delta\varrho$ which contains the terms multiplied by t^2, or for the greater generality, the terms which being multiplied by the sine or cosine of an angle $\alpha t + \beta$, in which α is very small, have at the same time a^2 for a divisor. It is clear that in supposing $\alpha = 0$, there will result a term multiplied by t^2, so that the second case shall include the first. The terms which have the divisor α^2, can evidently only result from a double integration; they can only therefore be produced by that part of $\delta\varrho$ which contains the double integral sign $\int$ Examine first the term

$$\frac{2a\cos.v\int ndt(\varrho\sin.v\int dR)}{m\sqrt{(1-e^2)}}.$$

If we fix the origin of the angle v at the perihelion, we have

$$\varrho = \frac{a(1-e^2)}{1+e\cos.v},$$

and consequently

$$\cos.v = \frac{a(1-e^2)-\varrho}{e\varrho};$$

whence we derive by differentiating,

$$\varrho^2 dv.\sin.v = \frac{a(1-e^2)}{e}.d\varrho;$$

but we have,

$$\varrho^2 d v = d t \sqrt{m a (1 - e^2)} = a^2 . n d t \sqrt{1 - e^2};$$

we shall, therefore, have

$$\frac{a n d t \varrho \sin. v}{\sqrt{1 - e^2}} = \frac{\varrho d \varrho}{e}.$$

The term

$$\frac{2 a \cos. v \int n d t . \{\varrho \sin. v \int d R\}}{m \sqrt{1 - e^2}}$$

will therefore become

$$\frac{2 \cos. v}{m e} \int (\varrho d \varrho \int d R), \text{ or } \frac{\cos. v}{m e} \{\varrho^2 \int d R - \int \varrho^2 . d R\}.$$

It is evident, this last function, no longer containing double integrals, there cannot result from it any term having the divisor α^2.

Now let us consider the term

$$-\frac{2 a \sin. v \int n d t \{\varrho \cos. v \int d R\}}{m \sqrt{1 - e^2}}$$

of the expression of $\delta \varrho$. Substituting for cos. v, its preceding value m ϱ, this term becomes

$$\frac{2 a \sin. v \int n d t . \{\varrho - a (1 - e^2)\} . \int d R}{m e \sqrt{1 - e^2}}.$$

We have

$$\varrho = a \{1 + \tfrac{1}{2} e^2 + e \chi'\},$$

χ' being an infinite series of cosines of the angle n t + ι, and of its multiples; we shall therefore have

$$\frac{\int n d t}{e} . \{\varrho - a (1 - e^2)\} \int d R = a \int n d t \{\tfrac{3}{2} e + \chi'\} \int d R.$$

Call χ'', the integral $\int \chi'$ n d t; we shall have

$$a \int n d t . \{\tfrac{3}{2} e + \chi'\} \int d R = \tfrac{3}{2} a e \int n d t \int d R + a \chi'' \int d R - a \int \chi'' . d R.$$

These two last terms not containing a double integral sign, there cannot thence result any term having α^2 for a divisor; reckoning only terms of this kind, we shall have

$$-\frac{2 a \sin. v \int n d t \{\varrho \cos. v \int d R\}}{m \sqrt{1 - e^2}} = \frac{3 a^2 e \sin. v \int n d t \int d R}{m \sqrt{1 - e^2}}$$

$$= \frac{d \varrho}{n d t} . \frac{3 a}{m} \int n d t \int d R;$$

and the radius ϱ will become

$$(\varrho) + \left(\frac{d \varrho}{n d t}\right) . \frac{3 a}{m} \int n d t . \int d R;$$

(ϱ) and $\left(\frac{d\varrho}{n\,d\,t}\right)$ being the expressions of ϱ and of $\frac{d\varrho}{n\,d\,t}$, relative to the elliptic motion. Thus, to estimate in the expression of the radius-vector, that part of the perturbations, which is divided by α^2, it is sufficient to augment by the quantity $\frac{3a}{m} \cdot \times \int n\,d\,t \,.\, \int d\,R$, the mean longitude $n\,t + \epsilon$, of this expression relative to the elliptic motion.

Let us see how we ought to estimate this part of the perturbations in the expression of the longitude v. The formula (Y) of No. 516 gives by substituting $\frac{3a}{m} \cdot \frac{d\varrho}{n\,d\,t} \cdot \int n\,d\,t \int d\,R$ for $\delta\,\varrho$ and retaining only the terms divided by α^2,

$$\delta\,v = \frac{\left\{\frac{2\,\varrho\,d^2\varrho + d\,\varrho^2}{a^2\,n^2\,d\,t^2} + 1\right\}}{\sqrt{1 - e^2}} \cdot \frac{3a}{m} \int n\,d\,t \int d\,R;$$

But we have by what precedes

$$d\,\varrho = \frac{a\,e\,.\,n\,d\,t\,.\,\sin.\,v}{\sqrt{1 - e^2}};\ \varrho^2\,d\,v = a^2\,n\,d\,t\,\sqrt{1 - e^2};$$

whence it is easy to obtain, by substituting for cos. v its preceding value in ϱ,

$$\frac{\frac{2\,\varrho\,d^2\,\varrho + d\,\varrho^2}{a^2\,n^2\,d\,t^2} + 1}{\sqrt{1 - e^2}} = \frac{d\,v}{n\,d\,t};$$

in estimating therefore only that part of the perturbations, which has the divisor α^2, the longitude v will become

$$(v) + \left(\frac{d\,v}{n\,d\,t}\right) \cdot \frac{3a}{m} \int n\,d\,t \int d\,R;$$

(v) and $\left(\frac{d\,v}{n\,d\,t}\right)$ being the parts of v and $\frac{d\,v}{n\,d\,t}$, relative to the elliptic motion. Thus, in order to estimate that part of the perturbations in the expression of the longitude of μ, we ought to follow the same rule which we have given with regard to the same in the expression of the radius-vector, that is to say, we must augment in the elliptic expression of the true longitude, the mean longitude $n\,t + \epsilon$ by the quantity $\frac{3a}{m} \int n\,d\,t \int d\,R$.

The constant part of the expression of $\left(\frac{d\,v}{n\,d\,t}\right)$ developed into a series of cosines of the angle $n\,t + \epsilon$ and of its multiples, being reduced (see 488) to unity, there thence results, in the expression of the longitude, the term $\frac{3a}{m}\int n\,d\,t\int d\,R$. If $d\,R$ contain a constant term $k\,\mu'.\,n\,d\,t$, this term will produce in the expression of the longitude v, the following one, $\frac{3}{2}.\frac{a\,\mu'}{m}k\,n^2\,t^2$. To ascertain the existence of such terms in this expression, we must therefore find whether $d\,R$ contains a constant term.

When the orbits are but little excentric and little inclined to one another, we have seen, No. 518, that R can always be developed into an infinite series of sines and cosines of angles increasing proportionally to the time. We can represent them generally by the term

$$k\,\mu'.\cos.\{i'\,n'\,t+i\,n\,t+A\},$$

i and i' being whole positive or negative numbers or zero. The differential of this term, taken solely relatively to the mean motion of μ, is

$$-i\,k.\,\mu'.\,n\,d\,t.\sin.\{i'\,n'\,t+i\,n\,t+A\};$$

this cannot be constant unless we have $0=i'\,n'+i\,n$, which supposes the mean motions of the bodies μ and μ' to be parts of one another; and since that does not take place in the solar system, we ought thence to conclude that the value of $d\,R$ does not contain constant terms, and that in considering only the first power of the perturbing masses, the mean motions of the heavenly bodies, are uniform, or which comes to the same thing, $\frac{d\,n}{d\,t}=0$. The value of a being connected to n by means of the equation $n^2=\frac{m}{a^3}$, it thence results that if we neglect the periodical quantities, the major-axes of the orbits are constant.

If the mean motions of the bodies μ and μ', without being exactly commensurable, approach, however, very nearly to that condition, there will exist in the theory of their motions, inequalities of a long period, and which, by reason of the smallness of the divisor α^2, will become very sensible. We shall see hereafter this is the case with regard to Jupiter and Saturn. The preceding analysis will give, in a very simple manner, that part of the perturbations which depend upon this divisor. It hence results that in this case it is sufficient to vary the mean longitude $n\,t+\epsilon$ or $\int n\,d\,t$ by the quantity $\frac{3a}{m}\int n\,d\,t\int d\,R$; or, which is the same, to augment n in the integral $\int n\,d\,t$ by the quantity $\frac{3\,a\,n}{m}\int d\,R$; but considering the orbit of μ as a variable ellipse, we have $n^2 = \frac{M}{r^3}$; the preceding variation of n introduces, therefore, in the semi-axis-major a of the orbit, the variation $-\frac{2\,a^2 \int d\,R}{m}$.

If we carry the approximation of the value $\frac{d\,v}{d\,t}$, to quantities of the order of the squares of the perturbing masses, we shall find terms proportional to the time; but considering attentively the differential equations of the motion of the bodies μ, μ', &c. we shall easily perceive that these terms are at the same time of the order of the squares and products of the excentricities and inclinations of the orbits. Since, however, every thing which affects the mean motion, may at length become very sensible, we shall now notice these terms, and perceive that they produce the secular equations observed in the motion of the moon.

522. Let us resume the equations (1) and (2) of No. 520, and suppose

$$(0, 1) = -\frac{\mu' . n . C}{2}; \quad \boxed{0, 1} = \frac{\mu' . n . D}{2};$$

they will become

$$\frac{d\,h}{d\,t} = (0, 1)\,l - \boxed{0, 1}\,l';$$

$$\frac{d\,l}{d\,t} = -(0, 1)\,h + \boxed{0, 1}\,h'.$$

The expression of (0, 1) and of $\boxed{0, 1}$ may be very simply determined in this way. Substituting, instead of C and D, their values determined in No. 517, we shall have

$$(0, 1) = -\frac{\mu' n}{2}\left\{a^2\left(\frac{d\,A^{(0)}}{d\,a}\right) + \tfrac{1}{2}\,a^3\left(\frac{d^2\,A^{(0)}}{d\,a^2}\right)\right\};$$

$$\boxed{0, 1} = \frac{\mu' n}{2}\left\{a\,A^{(1)} - a^2\left(\frac{d\,A^{(1)}}{d\,a}\right) - \tfrac{1}{2}\,a^3\left(\frac{d^2\,A^{(1)}}{d\,a^2}\right)\right\}.$$

We have by No. 516,

$$a^2\left(\frac{d\,A^{(0)}}{d\,a}\right) + \tfrac{1}{2}\,a^3\left(\frac{d^2\,A^{(0)}}{d\,a^2}\right) = -\alpha^2\,\frac{d\,b^{(0)}_{\frac{1}{2}}}{d\,\alpha} - \tfrac{1}{2}\,\alpha^3 . \frac{d^2\,b^{(0)}_{\frac{1}{2}}}{d\,\alpha^2};$$

and we shall easily obtain, by the same No. $\frac{d\,b^{(0)}_{\frac{1}{2}}}{d\,\alpha}$ and $\frac{d^2\,b^{(0)}_{\frac{1}{2}}}{d\,\alpha^2}$ in functions of $b^{(0)}_{\frac{1}{2}}$ and $b^{(1)}_{\frac{1}{2}}$; and these quantities are given in linear functions of $b^{(0)}_{-\frac{1}{2}}$

and of $b^{(1)}_{-\frac{1}{2}}$; this being done, we shall find

$$a^2\left(\frac{d\,A^{(0)}}{d\,a}\right) + \tfrac{1}{2}\,a^3\left(\frac{d^2\,A^{(0)}}{d\,a^2}\right) = \frac{3\,\alpha^2\,b^{(1)}_{-\frac{1}{2}}}{2\,(1-\alpha^2)^2};$$

wherefore

$$(0,1) = -\frac{3\,\mu'.\,n\,.\,\alpha^2.\,b^{(1)}_{-\frac{1}{2}}}{4\,(1-\alpha^2)^2}.$$

Let

$$(a^2 - 2\,a\,a'\cos.\theta + a'^2)^{\frac{1}{2}} = (a,a') + (a,a')'\cos.\theta + (a,a')''\cos.2\theta + \&c.$$

we shall have by No. 516.

$$(a,a') = \tfrac{1}{2}\,a'.\,b^{(0)}_{-\frac{1}{2}};\ (a,a')' = a'.\,b^{(1)}_{-\frac{1}{2}},\ \&c.$$

We shall, therefore, have

$$(0,1) = -\frac{3\,\mu'.\,n\,a^2\,a'.\,(a,a')'}{4\,(a'^2-a^2)^2}.$$

Next we have, by 516,

$$a\,A^{(1)} - a^2\left(\frac{d\,A^{(1)}}{l\,a}\right) - \tfrac{1}{2}\,a^3\left(\frac{d^2\,A^{(1)}}{d\,a^2}\right) = -\alpha\left\{b^{(1)}_{\frac{1}{2}} - \alpha.\frac{d\,b^{(1)}_{\frac{1}{2}}}{d\,\alpha} - \tfrac{1}{2}\,\alpha^2.\frac{d^2\,b^{(1)}_{\frac{1}{2}}}{d\,\alpha^2}\right\}.$$

Substituting for $b^{(1)}_{\frac{1}{2}}$ and its differences, their values in $b^{(0)}_{-\frac{1}{2}}$ and $b^{(1)}_{-\frac{1}{2}}$, we shall find the preceding function equal to

$$-\frac{3\,\alpha\left\{(1+\alpha^2)\,b^{(1)}_{-\frac{1}{2}} + \tfrac{1}{2}\,\alpha\,b^{(0)}_{-\frac{1}{2}}\right\}}{(1-\alpha^2)^2}.$$

therefore

$$\boxed{0,1} = -\frac{3\,\alpha\,.\,\mu'\,n\left\{(1+\alpha^2)\,b^{(1)}_{-\frac{1}{2}} + \tfrac{1}{2}\,\alpha\,.\,b^{(0)}_{-\frac{1}{2}}\right\}}{2\,(1-\alpha^2)^2}$$

or

$$\boxed{0,1} = -\frac{3\,\mu'.\,a\,n\{(a^2+a'^2)\,(a,a')' + a\,a'\,(a,a')\}}{2\,(a'^2-a^2)^2}.$$

We shall, therefore, thus obtain very simple expressions of (0, 1) and of $\boxed{0,1}$, and it is easy to perceive from the values in the series of $b^{(0)}_{-\frac{1}{2}}$ and of $b^{(1)}_{-\frac{1}{2}}$, given in the No. 516, that these expressions are positive, if n is positive, and negative if n is negative.

Call (0, 2) and $\boxed{0,2}$, what (0, 1) and $\boxed{0,1}$ become, when we change a' and μ' into a$''$ and μ''. In like manner let (0, 3), and (0, 3) be what the same quantities become, when we change a$'$ and μ' into a$'''$ and μ'''; and so on. Moreover let h$''$, l$''$; h$'''$, l$'''$, &c. denote the values of h and l relative to the bodies μ'', μ''', &c. Then, in virtue of the united actions of the different bodies μ', μ'', μ''', &c. upon μ, we shall have

$$\frac{d\,h}{d\,t} = \{(0,1) + (0,2) + (0,3) + \&c.\}\,l - \boxed{0,1}\,.\,l' - \boxed{0,2}\,.\,l'' - \&c.;$$

$$\frac{d\,l}{d\,t} = -\{(0,1) + (0,2) + (0,3) + \&c.\}\,h + \boxed{0,1}\,.\,h' + \boxed{0,2}\,.\,h'' + \&c.$$

It is evident that $\frac{d\,h'}{d\,t}$, $\frac{d\,l'}{d\,t}$; $\frac{d\,h''}{d\,t}$, $\frac{d\,l''}{d\,t}$; &c. will be determined by expressions similar to those of $\frac{d\,h}{d\,t}$ and of $\frac{d\,l}{d\,t}$; and they are easily obtained by changing successively what is relative to μ into that which relates to μ', μ'', &c. and reciprocally. Let therefore

$$(1,0),\ \boxed{1,0};\ (1,2),\ \boxed{1,2};\ \&c.$$

be what

$$(0,1),\ \boxed{0,1};\ (0,2),\ \boxed{0,2};\ \&c.$$

become, when we change that which is relative to μ, into what is relative to μ' and reciprocally. Let moreover

$$(2,0),\ \boxed{2,0};\ (2,1),\ \boxed{2,1};\ \&c.$$

be what

$$(0,2),\ \boxed{0,2};\ (0,1),\ \boxed{0,1};\ \&c.$$

become, when we change what is relative to μ into what is relative to μ'' and reciprocally; and so on. The preceding differential equations referred successively to the bodies μ, μ', μ'', &c. will give for determining h, l, h$'$, l$'$, h$''$, l$''$, &c. the following system of equations,

$$\left.\begin{aligned}
\frac{d\,h}{d\,t} &= \{(0,1) + (0,2) + \&c.\}\,l - \boxed{0,1}\,.\,l' - \boxed{0,2}\,l'' - \&c.\\
\frac{d\,l}{d\,t} &= -\{(0,1) + (0,2) + \&c.\}\,h + \boxed{0,1}\,h' + \boxed{0,2}\,h'' + \&c.\\
\frac{d\,h'}{d\,t} &= \{(1,0) + (1,2) + \&c.\}\,l' - \boxed{1,0}\,.\,l - \boxed{1,2}\,l'' - \&c.\\
\frac{d\,l'}{d\,t} &= -\{(1,0) + (1,2) + \&c.\}\,h' + \boxed{1,0}\,.\,h + \boxed{1,2}\,.\,h'' + \&c.\\
\frac{d\,h''}{d\,t} &= \{(2,0) + (2,1) + \&c.\}\,l'' - \boxed{2,0}\,.\,l - \boxed{2,1}\,.\,l' - \&c.\\
\frac{d\,l''}{d\,t} &= -\{(2,0) + (2,1) + \&c.\}\,.\,h'' + \boxed{2,0}\,h + \boxed{2,1}\,h' + \&c.
\end{aligned}\right\}\ (A)$$

&c.

The quantities (0, 1) and (1, 0), $\boxed{0, 1}$ and $\boxed{1, 0}$ have remarkable relations, which facilitate the operations, and will be useful hereafter. By what precedes we have

$$(0, 1) = -\frac{3\,\mu'.\,n\,a^2.\,a'\,(a, a')'}{4\,(a'^2 - a^2)^2}.$$

If in this expression of (0, 1) we change μ' into μ, n into n′, a into a′ and reciprocally, we shall have the expression of (1, 0), which will consequently be

$$(1, 0) = -\frac{3\,\mu\,.\,n'\,a'^2.\,a\,(a,' a)'}{4\,(a'^2 - a^2)^2};$$

but we have $(a, a')' = (a', a)'$, since both these quantities result from th developement of the function $(a^2 - 2\,a\,a'\cos\theta + a'^2)^{\frac{1}{2}}$ into a series ordered according to the cosine of θ and of its multiples. We shall, therefore, have

$$(0, 1).\,\mu.\,n'\,a' = (1, 0).\,\mu'.\,n\,a.$$

But, neglecting the masses μ, μ', &c. in comparison with M, we have

$$n^2 = \frac{M}{a^3};\; n'^2 = \frac{M}{a'^3};\; \&c.$$

Therefore

$$(0, 1)\,\mu\,\sqrt{a} = (1, 0)\,\mu'\,\sqrt{a'};$$

an equation from which we easily derive (1, 0) when (0, 1) is determined.

In the same manner we shall find,

$$\boxed{0, 1}\,\mu\,\sqrt{a} = \boxed{1, 0}\,\mu'\,\sqrt{a'}.$$

These two equations will also subsist in the case where n and n′ have different signs; that is to say, if the two bodies μ, μ' circulated in different directions; but then we must give the sign of n to the radical $\sqrt{a}$, and the sign of n′ to the radical $\sqrt{a'}$.

From the two preceding equations evidently result these

$$(0, 2)\,\mu\,\sqrt{a} = (2, 0)\,\mu''\,\sqrt{a''};\quad \boxed{0, 2}\,\mu\,\sqrt{a} = \boxed{2, 0}.\,\mu''\,\sqrt{a''};\; \&c.$$

$$(1, 2)\,\mu'\,\sqrt{a'} = (2, 1)\,\mu''\,\sqrt{a''};\quad \boxed{1, 2}\,\mu'\,\sqrt{a'} = \boxed{2, 1}.\,\mu''\,\sqrt{a''};\; \&c.$$

523. To integrate the equations (A) of the preceding No., we shall make

$$h = N.\sin.(g\,t + \beta);\; l = N.\cos.(g\,t + \beta);$$
$$h' = N'.\sin.(g\,t + \beta);\; l' = N'\cos.(g\,t + \beta);$$
&c.

Then substituting these values in the equations (A), we shall have

$$\left.\begin{aligned}
N\,g &= \{(0, 1) + (0, 2) + \&c.\}N - \boxed{0, 1}.\,N' - \boxed{0, 2}\,N'' - \&c.\\
N'\,g &= \{(1, 0) + (1, 2) + \&c.\}N' - \boxed{1, 0}.\,N - \boxed{1, 2}\,N'' - \&c.\\
N''g &= \{(2, 0) + (2, 1) + \&c.\}N'' - \boxed{2, 0}.\,N - \boxed{2, 1}\,N' - \&c.
\end{aligned}\right\};\quad (B)$$

If we suppose the number of the bodies μ, μ', μ'', &c. equal to i; these equations will be in number i, and eliminating from them the constants N, N', &c., we shall have a final equation in g, of the degree i, which we easily obtain as follows:

Let φ be the function

$$\begin{aligned}
& N^2 . \mu \sqrt{a}\{g - (0,1) - (0,2) - \&c.\} \\
+ & N'^2 \mu' \sqrt{a'}\{g - (1,0) - (1,2) - \&c.\} \\
+ & \&c. \\
+ & 2 N \mu \sqrt{a}\{\boxed{0,1} N' + \boxed{0,2} N'' + \&c.\} \\
+ & 2 N' \mu' \sqrt{a'}\{\boxed{1,2} N'' + \boxed{1,3} N''' + \&c.\} \\
+ & 2 N'' \mu'' \sqrt{a''}\{\boxed{2,3} N''' + \&c.\} \\
+ & \&c.
\end{aligned}$$

The equations (B) are reducible from the relations given in the preceding No. to these

$$\left(\frac{d\varphi}{dN}\right) = 0; \left(\frac{d\varphi}{dN'}\right) = 0; \left(\frac{d\varphi}{dN''}\right) = 0; \&c.$$

Considering therefore, N, N', N'', &c. as so many variables, φ will be a *maximum*. Moreover, φ being a homogeneous function of these variables, of the second dimension; we have

$$N\left(\frac{d\varphi}{dN}\right) + N'.\left(\frac{d\varphi}{dN'}\right) + \&c. = 2\varphi;$$

we have, therefore, $\varphi = 0$, in virtue of the preceding equations.

Thus we can determine the *maximum* of the function φ. We shall first differentiate this function relatively to N, and then substitute in φ, for N, its value derived from the equation $\left(\frac{d\varphi}{dN}\right) = 0$, a value which will be a linear function of the quantities N', N'', &c. In this manner we shall have a rational function whole and homogeneous of the second dimension in terms of N', N'', &c.: let $\varphi^{(1)}$ be this function. We shall differentiate $\varphi^{(1)}$ relatively to N', and we shall substitute in $\varphi^{(1)}$ for N' its value derived from the equation $\left(\frac{d\varphi^{(0)}}{dN'}\right) = 0$: we shall have a homogeneous function of the second dimension in N'', N''', &c.: let $\varphi^{(2)}$ be this function. Continuing thus, we shall arrive at a function $\varphi^{(i-1)}$ of the second dimension, in $N^{(i-1)}$ and which will consequently be of the form $(N^{(i-1)})^2 . k$, k being a function of g and constants. If we equal to zero, the differential of $\varphi^{(i-1)}$ taken relatively to $N^{(i-1)}$, we shall have $k = 0$; which will give an equation in g of the degree i, and whose different roots will give as many different systems for the indeterminates N, N', N'', &c.: the indeterminate $N^{(i-1)}$ will be the arbitrary of each system; and we shall immediately obtain, the relation of the other indeterminates N, N', &c. of the same system, to this one, by means of the preceding equations taken in an inverse order, viz.,

$$\left(\frac{d\,\varphi^{(i-1)}}{d\,N^{(i-2)}}\right) = 0;\ \left(\frac{d\,\varphi^{(i-3)}}{d\,N^{(i-3)}}\right) = 0;\ \&c.$$

Let g, g_1, g_2, &c. be the i roots of the equation in g: let N, N', N'', &c. be the system of indeterminates, relative to the root g: let $N_{,}$, $N_{,}'$, $N_{,}''$, &c. be the system of indeterminates relative to the root g_1, and so on: by the known theory of linear differential equations, we shall have

$$h = N \sin.(g\,t + \beta) + N_1 \sin.(g_1\,t + \beta_1) + N_2\,(g_2\,t + \beta_2) + \&c.;$$
$$h' = N' \sin.(g\,t + \beta) + N_1' \sin.(g_1\,t + \beta_1) + N_2'\,(g_2\,t + \beta_2) + \&c.;$$
$$h'' = N'' \sin.(g\,t + \beta) + N_1'' \sin.(g_1\,t + \beta_1) + N_2''(g_2\,t + \beta_2) + \&c.;$$
$$\&c.$$

β, β_1, β_2, &c. being arbitrary constants. Changing in these values of h, h', h'', &c. the sines into cosines; we shall have the values of l, l', l'', &c. These different values contain twice as many arbitraries as there are roots g, g_1, g_2, &c.; for each system of indeterminates contains an arbitrary, and moreover, it has i arbitraries β, β_1, β_2, &c.; these values are consequently the complete integrals of the equations (A) of the preceding No.

It is necessary, however, to determine only the constants N, N_1, &c.; N,' N_1', &c.; β, β_1, &c. Observations will not give immediately the constants, but they make known at a given epoch, the excentricities e, e', &c. of the orbits, and the longitudes ϖ, ϖ', &c. of their perihelions, and cousequently, the values of h, h', &c., l, l', &c.: we shall thus derive the values of the preceding constants. For that purpose, we shall observe that if we multiply the first, third, fifth, &c. of the differential equations (A) of the preceding No., respectively by $N.\mu.\sqrt{a}$, $N'.\mu'.\sqrt{a'}$, &c.; we shall have in virtue of equations (B), and the relations found in the preceding No. between (0, 1) and (1, 0), (0, 2), and (2, 0), &c.

$$N.\frac{d\,h}{d\,t}\mu\sqrt{a} + N'.\frac{d\,h'}{d\,t}\mu'\sqrt{a'} + N''.\frac{d\,h''}{d\,t}\mu''\sqrt{a''} + \&c.$$
$$= g\{N.l.\mu.\sqrt{a} + N'.l'.\mu'.\sqrt{a'} + N''.l''.\mu''.\sqrt{a''} + \&c.\}$$

If we substitute in this equation for h, h', &c. l, l', &c. their preceding values; we shall have by comparing the coefficients of the same cosines

$$0 = N.N_1.\mu\sqrt{a} + N'.N_1'.\mu'\sqrt{a'} + N''.N_1''.\mu''.\sqrt{a''} + \&c.;$$
$$0 = N.N_2.\mu\sqrt{a} + N'.N_2'.\mu'\sqrt{a'} + N''.N_2''.\mu''.\sqrt{a''} + \&c.$$

Again, if we multiply the preceding values of h, h', &c. respectively by

$$N . \mu . \sqrt{a},\ N' . \mu' . \sqrt{a'},\ \&c.$$

we shall have, in virtue of these last equations,

$$N . \mu\, h . \sqrt{a} + N' \mu' . h' . \sqrt{a'} + N'' . \mu''\, h'' . \sqrt{a''} + \&c.$$
$$= \{N^2 . \mu . \sqrt{a} + N'^2 . \mu' . \sqrt{a'} + N''^2 . \mu'' . \sqrt{a''} + \&c.\} \sin (g\, t + \beta)$$

In like manner, we have

$$N . \mu\, l . \sqrt{a} + N' . \mu'\, l' . \sqrt{a'} + N'' . \mu''\, l'' . \sqrt{a''} + \&c.$$
$$= \{N^2 . \mu . \sqrt{a} + N'^2 . \mu' . \sqrt{a'} + N''^2 . \mu'' . \sqrt{a''} + \&c.\} \cos. (g\, t + \beta).$$

By fixing the origin of the time t at the epoch for which the values of h, l, h', l', &c. are supposed known; the two preceding equations give

$$\tan. \beta = \frac{N . h\, \mu \sqrt{a} + N' . h'\, \mu' . \sqrt{a'} + N'' . h''\, \mu'' . \sqrt{a''} + \&c.}{N . l\, \mu . \sqrt{a} + N' . l'\, \mu' . \sqrt{a'} + N'' . l''\, \mu'' . \sqrt{a''} + \&c.}.$$

This expression of tan. β contains no indeterminate; for although the constants N, N', N'', &c. depend upon the indeterminate $N^{(i-1)}$, yet, as their relations to this indeterminate are known by what precedes, it will disappear from the expression of tan. β. Having thus determined β, we shall have $N^{(i-1)}$, by means of one of the two equations which give tan. β; and we thence obtain the system of indeterminates, N, N', N'', &c. relative to the root g. Changing, in the preceding expressions, this root into g_1, g_2, g_3, &c. we shall have the values of the arbitraries relative to each of these roots.

If we substitute these values in the expressions of h, l, h', l', &c.; we hence derive the values of the excentricities e, e', &c. of the orbits, and the longitudes of their perihelions, by means of the equations

$$e^2 = h^2 + l^2;\ e'^2 = h'^2 + l'^2;\ \&c.$$

$$\tan. \varpi = \frac{h}{l};\ \tan. \varpi' = \frac{h'}{l'};\ \&c.$$

we shall thus have

$$e^2 = N^2 + N_1^2 + N_2^2 + \&c. + 2\, N\, N_1 \cos. \{(g_1 - g)\, t + \beta_1 - \beta\}$$
$$+ 2\, N\, N_2 \cos. \{(g_2 - g)\, t + \beta_2 - \beta)\} + 2\, N_1\, N_2 \cos. \{(g_2 - g_1)\, t + \beta_2 - \beta_1\} + \&c.$$

This quantity is always less than $(N + N_1 + N_2 + \&c.)^2$, when the roots g, g_1, &c. are all real and unequal, by taking positively the quantities N, N_1, &c. In like manner, we shall have

$$\tan. \varpi = \frac{N \sin. (g\, t + \beta) + N_1 \sin. (g_1\, t + \beta_1) + N_2 \sin. (g_2\, t + \beta_2) + \&c.}{N \cos. (g\, t + \beta) + N_1 \cos. (g_1\, t + \beta_1) + N_2 \cos. (g_2\, t + \beta_2) + \&c.}$$

whence it is easy to get,

$$\tan. (\varpi - g\, t - \beta) = \frac{N_1 \sin. \{(g_1 - g)\, t + \beta_1 - \beta\} + N_2 \sin. \{(g_2 - g)\, t + \beta_2 - \beta\} + \&c.}{N + N_1 \cos. \{(g_1 - g)\, t + \beta_1 - \beta\} + N_2 \cos. \{(g_2 - g)\, t + \beta_2 - \beta\} + \&c.}$$

Whilst the sum $N_1 + N_2 +$ &c. of the coefficients of the cosines of the denominator, all taken positively, is less than N, tan. $(\varpi - g\,t - \beta)$ can never become infinite; the angle $\varpi - g\,t - \beta$ can never reach the quarter of the circumference; so that in this case the true mean motion of the perihelion is equal to g t.

524. From what has been shown it follows, that the excentricities of the orbits and the positions of their axis-majors, are subject to considerable variations, which at length change the nature of the orbits, and whose periods depending on the roots g, g_1, g_2, &c., embrace with regard to the planets, a great number of ages. We may thus consider the excentricities as variably elliptic, and the motions of the perihelions as not uniform. These variations are very sensible in the satellites of Jupiter, and we shall see hereafter, that they explain the singular inequalities, observed in the motion of the third satellite.

But it is of importance to examine whether the variations of the excentricities have limits, and whether the orbits are constantly almost circular. We know that if the roots of the equation in g are all real and unequal, the excentricity e of the orbit of μ is always less than the sum $N + N_1 + N_2 +$ &c. of the coefficients of the sines of the expression of h taken positively; and since the coefficients are supposed very small, the value of e will always be inconsiderable. By taking notice, therefore, of the secular variations only, we see that the orbits of the bodies μ, μ', μ'', &c. will only flatten more or less in departing a little from the circular form; but the positions of their axis-majors will undergo considerable variations. These axes will be constantly of the same length, and the mean motions which depend upon them will always be uniform, as we have seen in No. 521. The preceding results, founded upon the smallness of the excentricity of the orbits, will subsist without ceasing, and will extend to all ages past and future; so that we may affirm that at any time, the orbits of the planets and satellites have never been nor ever will be very excentric, at least whilst we only consider their mutual actions. But it would not be the same if any of the roots g, g_1, g_2, &c. were equal or imaginary: the sines and cosines of the expressions of h, l, h', l', &c. corresponding to these roots, would then change into circular arcs or exponentials; and since these quantities increase indefinitely with the time, the orbits would at length become very excentric; the stability of the planetary system would then be destroyed, and the results found above would cease to take place. It is therefore highly important to show that g, g_1, g_2, &c. are all real and unequal. This we will now demonstrate in a very simple manner, for the case of nature, in which the bodies μ, μ', μ'', &c. of the system, all circulate in the same direction.

Let us resume the equations (A) of No. 528. If we multiply the first by $\mu \,.\, \sqrt{a} \,.\, h$; the second, by $\mu \,.\, \sqrt{a} \,.\, l$; the third by $\mu' \,.\, \sqrt{a'} \,.\, h'$; the fourth by $\mu' \,.\, \sqrt{a'} \,.\, l'$, &c. and afterwards add the results together; the coefficients of h l, h′ l′, h″ l″, &c. will be nothing in this sum, the coefficients of $h' l - h l'$ will be $\boxed{0, 1} \,.\, \mu \,.\, \sqrt{a} - \boxed{1, 0} \,.\, \mu' \,.\, \sqrt{a'}$, and this will be nothing in virtue of the equation $\boxed{0, 1} \,.\, \mu \,.\, \sqrt{a} = \boxed{1, 0} \,.\, \mu' \,.\, \sqrt{a'}$ found in No. 522. The coefficients of $h'' l - h l''$, $h'' l' - h' l''$, &c. will be nothing for the same reason; the sum of the equations (A) thus prepared will therefore be reduced to

$$\frac{h\,d\,h + l\,d\,l}{d\,t} \,.\, \mu \,.\, \sqrt{a} + \frac{h'\,d\,h' + l'\,d\,l'}{d\,t} \,.\, \mu' \,.\, \sqrt{a'} + \&c. = 0;$$

and consequently to

$$0 = e\,d\,e \,.\, \mu \,.\, \sqrt{a} + e'\,d\,e' \,.\, \mu' \,.\, \sqrt{a'} + \&c.$$

Integrating this equation and observing that (No. 521) the semi-axis-majors are constant, we shall have

$$e^2 \,.\, \mu \,\sqrt{a} + e'^2 \,.\, \mu' \,.\, \sqrt{a'} + e''^2 \,.\, \mu'' \,.\, \sqrt{a''} + \&c. = \text{constant}; \quad (a)$$

The bodies μ, μ', μ'', &c. however being supposed to circulate in the same direction, the radicals $\sqrt{a}$, $\sqrt{a'}$, $\sqrt{a''}$, &c. ought to be taken positively in the preceding equation, as we have seen in No. 522; all the terms of the first member of this equation are therefore positive, and consequently, each of them is less than the constant of the second member. But by supposing at any epoch the excentricities to be very small, this constant will be very small; each of the terms of the equation will, therefore, remain always very small and cannot increase indefinitely; the orbits will always be very nearly circular.

The case which we have thus examined, is that of the planets and satellites of the solar system; since all these bodies circulate in the same direction, and at the present epoch their orbits have little excentricity. That no doubt may exist as to a result so important, we shall observe that if the equation which determines g, contained imaginary roots, some of the sines and cosines of the expressions of h, l, h′, l′, &c. would transform into exponentials; thus the expression of h would contain a finite number of terms of the form $P \,.\, c^{ft}$, c being the number of which the hyperbolic logarithm is unity, and P being a real quantity, because h or e sin. ϖ is a real quantity. Let

$$Q \,.\, c^{ft},\; P' \,.\, c^{ft},\; Q' \,.\, c^{ft},\; P'' \,.\, c^{ft},\; \&c.$$

be the corresponding terms of l, h′, l′, h″, &c.; Q, P′, Q′, P″, &c. being also real quantities: the expression of e^2 will contain the term (P^2+Q^2) c^{2ft}; the expression of e'^2 will contain the term $(P'^2+Q'^2)\,c^{2ft}$, and so on; the first member of the equation (u) will therefore contain the term

$$\{(P^2+Q^2)\,\mu.\sqrt{a}+(P'^2+Q'^2)\,\mu'\sqrt{a'}+(P''^2+Q''^2)\,\mu''\sqrt{a''}+\&c.\}\,c^{2ft}.$$

If, therefore, we suppose c^{ft} to be the greatest of the exponentials which contain h, l, h', l', &c. that is to say, that in which f is the most considerable, c^{2ft} will be the greatest of the exponentials which contain the first member of the preceding equation: the preceding term cannot therefore be destroyed by any other term of this first member; so that for this member to be reduced to a constant, the coefficient of c^{2ft} must be nothing, which gives

$$0=(P^2+Q^2)\,\mu\sqrt{a}+(P'^2+Q'^2)\,\mu'\sqrt{a'}+(P''^2+Q''^2)\,\mu''\sqrt{a''}+\&c.$$

When $\sqrt{a}$, $\sqrt{a'}$, $\sqrt{a''}$, &c. have the same sign, or which is tantamount, when the bodies μ, μ', μ'', &c. circulate in the same direction, this equation is impossible, provided we do not suppose $P=0$, $Q=0$, $P'=0$, &c.; whence it follows that the quantities h, l, h' l', &c. do not contain exponentials, and that the equation in g does not contain imaginary roots.

If this equation had equal roots, the expressions of h, l, h', l', &c. would contain as we know, circular arcs and in the expression of h, we should have a finite number of terms of the form $P\,t^r$. Let $Q\,t^r$, $P'\,t^r$, $Q'\,t^r$, &c. be the corresponding terms of l, h', l', &c. P, Q, P', Q', &c. being real quantities; the first member of the equation (u) will contain the term

$$\{(P^2+Q^2)\,\mu\sqrt{a}+(P'^2+Q'^2)\,\mu'\sqrt{a'}+(P''^2+Q''^2)\,\mu''\sqrt{a''}+\&c.\}.\,t^{2r}.$$

If t^r is the highest power of t, contained by the values of h, l, h' l', &c.; t^{2r} will be the highest power of t contained in the first member of the equation (u); thus, that this member may be reduced to a constant, we must have

$$0=(P^2+Q^2)\,\mu\sqrt{a}+(P'^2+Q'^2)\,\mu'\sqrt{a'}+\&c.$$

which gives

$$P=0,\ Q=0,\ P'=0,\ Q'=0,\ \&c.$$

The expressions of h, l, h', l', &c. contain therefore, neither exponentials nor circular arcs, and consequently all the roots of the equation in g are real and unequal.

The system of the orbits of μ, μ', μ'', &c. is therefore perfectly stable relatively to their excentricities; these orbits merely oscillate about a mean state of ellipticity, which they depart from but little by preserving the same major-axis: their excentricities are always subject to this condition, viz. *that the sum of their squares multiplied respectively by the masses of the bodies and by the square roots of the major-axes is always the same.*

525. When we shall have determined, by what precedes, the values of e and of ϖ; we shall substitute in all the terms of the expressions of ρ, and $\frac{dv}{dt}$, given in the preceding Nos., effacing the terms which contain the time t without the symbols *sine* and *cosine*. The elliptic part of these expressions will be the same as in the case of an orbit not disturbed, with this only difference, that the excentricity and the position of the perihelion are variable; but the periods of these variations being very long, by reason of the smallness of the masses μ, μ', μ'', &c. relatively to M; we may suppose these variations proportional to the time, during a great interval, which, for the planets, may extend to many ages before and after the given epoch.

It is useful, for astronomical purposes, to obtain under this form, the secular variations of the excentricities and perihelions of the orbits: we may easily get them from the preceding formulæ. In fact, the equation $e^2 = h^2 + l^2$, gives $e\,de = h\,dh + l\,dl$; but in considering only the action of μ', we have by No. 522,

$$\frac{dh}{dt} = (0, 1)\, l - \boxed{0, 1}\, l';$$

$$\frac{dl}{dt} = -(0, 1)\, h + \boxed{0, 1}.\, h';$$

wherefore

$$\frac{e\,de}{dt} = \boxed{0, 1}.\, \{h'\, l - h\, l'\};$$

but we have $h'\, l - h\, l' = e\, e' \sin. (\varpi' - \varpi)$; we, therefore, have

$$\frac{de}{dt} = \boxed{0, 1}.\, e' \sin. (\varpi' - \varpi);$$

thus, with regard to the reciprocal action of the different bodies μ', μ'', &c. we shall have

$$\frac{de}{dt} = \boxed{0, 1}.\, e' \sin. (\varpi' - \varpi) + \boxed{0, 2}.\, e'' \sin. (\varpi'' - \varpi) + \&c.$$

$$\frac{de'}{dt} = \boxed{1, 0}\, e \sin. (\varpi - \varpi') + \boxed{1, 2}\, e'' \sin. (\varpi'' - \varpi') + \&c.$$

$$\frac{de''}{dt} = \boxed{2, 0}\, e \sin. (\varpi - \varpi'') + \boxed{2, 1}\, e' \sin. (\varpi' - \varpi'') + \&c.$$

&c.

The equation $\tan. \varpi = \frac{h}{l}$, gives by differentiating

$$e^2 d\varpi = l\, dh - h\, dl.$$

With respect only to the action of μ', by substituting for d h and d l their values, we shall have

$$\frac{e^2 d\varpi}{dt} = (0, 1)(h^2 + l^2) - \boxed{0, 1}.\{h h' + l l'\};$$

which gives

$$\frac{d\varpi}{dt} = (0, 1) - \boxed{0, 1}.\frac{e'}{e}\cos.(\varpi' - \varpi);$$

we shall, therefore, have, through the reciprocal actions of the bodies μ, μ', μ'', &c.

$$\frac{d\varpi}{dt} = (0,1) + (0,2) + \&c. - \boxed{0, 1}.\frac{e'}{e}\cos.(\varpi' - \varpi) - \boxed{0, 2}.\frac{e''}{e}\cos.(\varpi'' - \varpi) - \&c.$$

$$\frac{d\varpi'}{dt} = (1,0) + (1,2) + \&c. - \boxed{1, 0}.\frac{e}{e'}\cos.(\varpi - \varpi') - \boxed{1, 2}\frac{e''}{e'}\cos.(\varpi'' - \varpi') - \&c.$$

$$\frac{d\varpi''}{dt} = (2,0) + (2,1) + \&c. - \boxed{2, 0}\frac{e}{e''}\cos.(\varpi - \varpi'') - \boxed{2, 1}.\frac{e'}{e''}\cos.(\varpi' - \varpi'') - \&c.$$

&c.

If we multiply these values of $\frac{de}{dt}, \frac{de'}{dt}$, &c. $\frac{d\varpi}{dt}, \frac{d\varpi'}{dt}$, &c. by the time t; we shall have the differential expressions of the secular variations of the excentricities and of the perihelions, and these expressions which are only rigorous whilst t is indefinitely small, will however serve for a long interval relatively to the planets. Their comparison with precise and distant observations, affords the most exact mode of determining the masses of the planets which have no satellites. For any time t we have the excentricity e, equal to

$$e + t.\frac{de}{dt} + \frac{t^2}{1.2}.\frac{d^2e}{dt^2} + \&c.$$

$e, \frac{de}{dt}, \frac{d^2e}{dt^2}$, &c. being relative to the origin of the time t or to the given epoch. The preceding value of $\frac{de}{dt}$ will give, by differentiating it, and observing that a, a', &c. are constant, the values of $\frac{d^2e}{dt^2}, \frac{d^3e}{dt^3}$, &c.; we can, therefore, thus continue as far as we wish, the preceding series, and by the same process, the series also relative to ϖ: but relatively to the planets, it will be sufficient, in comparing the most ancient observations which have come down to us, to take into account the square of the time, in the expressions of the series of e, e', &c. ϖ, ϖ', &c.

526. We will now consider the equations relative to the position of the orbits. For this purpose let us resume the equations (3) and (4) of No. 520,

$$\frac{dp}{dt} = -\frac{\mu' n}{4} . a^2 a' B^{(1)} . (q - q');$$

$$\frac{dq}{dt} = \frac{\mu' n}{4} . a^2 a' B^{(1)} . (p - p').$$

By No. 516, we have

$$a^2 a' . B^{(1)} = \alpha^2 . b^{(1)}_{\frac{3}{2}};$$

and by the same No.,

$$b^{(1)}_{\frac{3}{2}} = -\frac{3 b^{(1)}_{-\frac{1}{2}}}{(1 - \alpha^2)^2};$$

We shall therefore have

$$\frac{\mu' n}{4} a^2 a' B^{(1)} = -\frac{3 \mu' . n . \alpha^2 b^{(1)}_{-\frac{1}{2}}}{4 (1 - \alpha^2)^2}.$$

The second member of this equation is what we have denoted by (0, 1) in 522; we shall hence have

$$\frac{dp}{dt} = (0, 1) (q' - q);$$

$$\frac{dq}{dt} = (0, 1) (p - p');$$

Hence, it is easy to conclude that the values of q, p, q', p', &c. will be determined by the following system of differential equations:

$$\left.\begin{aligned}
\frac{dq}{dt} &= \{(0, 1) + (0, 2) + \&c.\} . p - (0, 1) p' - (0, 2) p'' - \&c. \\
\frac{dp}{dt} &= -\{(0, 1) + (0, 2) + \&c.\} . q + (0, 1) q' + (0, 2) q'' + \&c. \\
\frac{dq'}{dt} &= \{(1, 0) + (1, 2) + \&c.\} . p' - (1, 0) p - (1, 2) p'' - \&c. \\
\frac{dp'}{dt} &= -\{(1, 0) + (1, 2) + \&c.\} . q' + (1, 0) q + (1, 2) q'' + \&c. \\
\frac{dq''}{dt} &= \{(2, 0) + (2, 1) + \&c.\} . p'' - (2, 0) p - (2, 1) p' - \&c. \\
\frac{dp''}{dt} &= -\{(2, 0) + (2, 1) + \&c.\} . q'' + (2, 0) q + (2, 1) q' + \&c. \\
&\&c.
\end{aligned}\right\} ; \quad (C)$$

This system of equations is similar to that of the equations (A) of No. 522: it would entirely coincide with it, if in the equations of (A) we were to change h, l, h′, l′, &c. into q, p, q′, p′, &c. and if we were to suppose

$$\boxed{0,1} = (0, 1);$$

$$\boxed{1, 0} = (1, 0);$$

&c.

Hence, the process which we have used in No. 523 to integrate the equations (A) applies also to the equations (C). We shall therefore suppose

$$q = N \cos. (g t + \beta) + N_1 \cos. (g_1 t + \beta_1) + N_2 \cos. (g_2 t + \beta_2) + \&c.$$
$$p = N \sin. (g t + \beta) + N_1 \sin. (g_1 t + \beta_1) + N_2 \sin. (g_2 t + \beta_2) + \&c.$$
$$q' = N' \cos. (g t + \beta) + N_1' \cos. (g_1 t + \beta_1) + N_2' \cos. (g_2 t + \beta_2) + \&c.$$
$$p' = N' \sin. (g t + \beta) + N_1' \sin. (g_1 t + \beta_1) + N_2' \sin. (g_2 t + \beta_2) + \&c.$$
&c.

and by No. 523, we shall have an equation in g of the degree i, and whose different roots will be g, g_1, g_2, &c. It is easy to perceive that one of these roots is nothing; for it is clear we satisfy the equations (C) by supposing p, p′, p″, &c. equal and constant, as also q, q′, q″, &c. This requires one of the roots of the equation in g to be zero, and we can thence depress the equation to the degree i — 1. The arbitraries N, N_1, N', &c. β, β_1, &c. will be determined by the method exposed in No. 523. Finally, we shall find by the process employed in No. 524.

$$\text{const.} = (p^2 + q^2) \mu \sqrt{a} + (p'^2 + q'^2) \mu' \sqrt{a'} + \&c.$$

Whence we conclude, as in the No. cited, that the expressions of p, q, p′, q′; &c. contain neither circular arcs nor exponentials, when the bodies μ, μ', μ'', &c. circulate in the same direction: and that therefore the equation in g has all its roots real and unequal.

We may obtain two other integrals of the equations (C). In fact, if we multiply the first of these equations by $\mu \sqrt{a}$, the third by $\mu' \sqrt{a'}$, the fifth by $\mu'' \sqrt{a''}$, &c. we shall have, because of the relations found in No. 522,

$$0 = \frac{d q}{d t} \mu \sqrt{a} + \frac{d q'}{d t} \mu' \sqrt{a'} + \&c.;$$

which by integration gives

$$\text{constant} = q \mu \sqrt{a} + q' \mu' \sqrt{a'} + \&c. \quad . \quad . \quad . \quad . \quad . \quad (1)$$

In the same manner we find

$$\text{constant} = p \mu \sqrt{a} + p' \mu' \sqrt{a'} + \&c. \quad . \quad . \quad . \quad . \quad . \quad (2)$$

Call φ the inclination of the orbit of μ to the fixed plane, and θ the longitude of the ascending node of this orbit upon the same plane; the latitude of μ will be very nearly tan. φ sin. $(n t + \varepsilon - \theta)$: Comparing this value with q sin. $(n t + \varepsilon)$ — p cos. $(n t + \varepsilon)$, we shall have

$$p = \tan. \varphi \sin. \theta; \quad q = \tan. \varphi \cos. \theta;$$

whence we obtain

$$\tan. \varphi = \sqrt{(p^2 + q^2)}; \quad \tan. \theta = \frac{p}{q};$$

We shall, therefore, have the inclination of the orbit of μ, and the position of its node, by means of the values of p and q. By marking successively with one dash, two dashes, &c. relatively to μ', μ'', &c. the values of tan. φ, tan. θ, we shall have the inclinations of the orbits of μ' μ'', &c. and the positions of their nodes by means of p', q', p'', q'', &c.

The quantity $\sqrt{p^2 + q^2}$ is less than the sum $N + N_1 + N_2 +$ &c. of the coefficients of the sines in the expression of q; thus, the coefficients being very small since the orbit is supposed but little inclined to the fixed plane, its inclination will always be inconsiderable; whence it follows, that the system of orbits is also stable, relatively to their inclinations as also to their excentricities. We may therefore consider the inclinations of the orbits, as variable quantities comprised within determinate limits, and the motion of the nodes as not uniform. These variations are very sensible in the satellites of Jupiter, and we shall see hereafter, that they explain the singular phenomena observed in the inclination of the orbit of the fourth satellite.

From the preceding expressions of p and q results this theorem:

Let us imagine a circle whose inclination to a fixed plane is N, *and of which the longitude of the ascending node is* $g t + \beta$; *also let us imagine upon this first circle, a second circle inclined by the angle* N_1, *the longitude of whose intersection with the former circle is* $g_1 t + \beta_1$; *upon this second circle let there be a third inclined to it by the angle* N_2, *the longitude of whose intersection with the second circle is* $g_2 t + \beta_2$, *and so on; the position of the last circle will be that of the orbit of* μ.

Applying the same construction to the expressions of h and l of No. 523, we see that the tangent of the inclination of the last circle upon the fixed plane, is equal to the excentricity of μ's orbit, and that the longitude of the intersection of this circle with the same plane, is equal to that of the perihelion of μ's orbit.

527. It is useful for astronomical purposes, to have the differential variations of the nodes and inclinations of the orbits. For this purpose, let us resume the equations of the preceding No.

$$\tan.\,\varphi = \sqrt{(p^2 + q^2)},\ \tan.\,\theta = \frac{p}{q}.$$

Differentiating these, we shall have

$$d\,\varphi = d\,p\,\sin.\,\theta + d\,q\,\cos.\,\theta;$$

$$d\,\theta = \frac{d\,p\,\cos.\,\theta - d\,q\,\sin.\,\theta}{\tan.\,\varphi}.$$

If we substitute for d p and d q, their values given by the equations (C) of the preceding No. we shall have

$$\frac{d\,\varphi}{d\,t} = (0, 1)\tan.\,\varphi'\sin.\,(\theta - \theta') + (0, 2)\tan\,\varphi''.\sin.\,(\theta - \theta'') + \&c.$$

$$\frac{d\,\theta}{d\,t} = -\{(0, 1) + (0, 2) + \&c.\} + (0, 1)\frac{\tan.\,\varphi'}{\tan.\,\varphi}\cos.\,(\theta - \theta')$$

$$+ (0, 2)\frac{\tan.\,\varphi''}{\tan.\,\varphi}\cos.\,(\theta - \theta'') + \&c.$$

In like manner, we shall have

$$\frac{d\,\varphi'}{d\,t} = (1, 0)\tan.\,\varphi\sin.\,(\theta' - \theta) + (1, 2)\tan.\,\varphi''\sin.\,(\theta' - \theta'') + \&c.$$

$$\frac{d\,\theta'}{d\,t} = -\{(1, 0) + (1, 2) + \&c.\} + (1, 0).\frac{\tan.\,\varphi}{\tan.\,\varphi'}.\cos.\,(\theta' - \theta)$$

$$+ (1, 2).\frac{\tan.\,\varphi''}{\tan.\,\varphi'}.\cos.\,(\theta' - \theta'') + \&c.$$

&c.

Astronomers refer the celestial motions to the moveable orbit of the earth; it is in fact from the plane of this orbit that we observe them; it is therefore important to know the variations of the nodes and the inclinations of the orbits, relatively to the orbit of one of the bodies μ, μ', μ'', &c. for example to the orbit of μ. It is clear that

$$q\sin.\,(n'\,t + \varepsilon') - p\cos.\,(n'\,t + \varepsilon')$$

would be the latitude of μ' above the fixed plane if it were in motion upon the orbit of μ. The latitude of this moveable plane above the same plane is

$$q'\sin.\,(n'\,t + \varepsilon') - p'\cos.\,(n'\,t + \varepsilon');$$

but the difference of these two latitudes is very nearly the latitude of μ' above the orbit of μ; calling therefore $\varphi_{\prime}'$ the inclination, and $\theta_{\prime}'$ the longitude of the node of μ' upon the orbit of μ, we shall have, by what precedes,

$$\tan.\,\varphi_{\prime}' = \sqrt{(p' - p)^2 + (q' - q)^2};\ \tan.\,\theta_{\prime}' = \frac{p' - p}{q' - q}.$$

If we take for the fixed plane, that of μ's orbit at a given epoch; we shall have at that epoch p = 0, q = 0; but the differentials d p and d q will not be zero; thus we shall have.

$$d\,\varphi'_{,} = (d\,p' - d\,p)\sin.\,\theta' + (d\,q' - d\,q)\cos.\,\theta';$$

$$d\,\theta'_{,} = \frac{d\,p' - d\,p)\cos.\,\theta' - (d\,q' - d\,q)\sin.\,\theta'}{\tan.\,\varphi'}.$$

Substituting for d p, d q, d p', d q', &c. their values given by the equations (C) of the preceding No., we shall have

$$\frac{d\,\varphi'_{,}}{d\,t} = \{(1, 2) - (0, 2)\}\tan.\,\varphi''\sin.\,(\theta' - \theta'')$$
$$+ \{(1, 3) - (0, 3)\}\tan.\,\varphi'''\sin.\,(\theta' - \theta''') + \&c.$$
$$\frac{d\,\theta'_{,}}{d\,t} = -\{(1, 0) + (1, 2) + (1, 3) + \&c.\} - (0, 1)$$
$$+ \{(1, 2) - (0, 2)\}\,.\,\frac{\tan.\,\varphi''}{\tan.\,\varphi'}\cos.\,(\theta' - \theta'')$$
$$+ \{(1, 3) - (0, 3)\}\,.\,\frac{\tan.\,\varphi'''}{\tan.\,\varphi'}\cos.\,\theta' - \theta''') + \&c.$$

It is easy to obtain from these expressions the variations of the nodes, and inclinations of the orbits of the other bodies μ'', μ''', &c. upon the moveable orbit of μ.

528. The integrals found above, of the differential equations which determine the variations of the elements of the orbits, are only approximate, and the relations which they give among the elements, only take place on the supposition that the excentricities of the orbits and their inclinations are very small. But the integrals (4), (5), (6), (7), which are given in No. 471, give the same relations, whatever may be the excentricities and inclinations. For this, we shall observe that $\frac{x\,d\,y - y\,d\,x}{d\,t}$ is double the area described during the instant d t, by the projection of the radius-vector of the planet μ upon the plane of x, y. In the elliptic motion, if we neglect the mass of the planet as nothing compared with that of the sun, taken for unity, we shall have, by the Nos. 157, 237, relatively to the plane of μ's orbit,

$$\frac{x\,d\,y - y\,d\,x}{d\,t} = \sqrt{a\,(1 - e^2)}.$$

In order to refer the area upon the orbit to the fixed plane, we must multiply by the cosine of the inclination φ of the orbit to this plane; we shall, therefore, have, with reference to this plane,

$$\frac{x\,d\,y - y\,d\,x}{d\,t} = \cos.\,\varphi\,\sqrt{a\,(1 - e^2)} = \sqrt{\frac{a\,.\,(1 - e^2)}{1 + \tan.^2\,\varphi}}.$$

In like manner

$$\frac{x' d y' - y' d x'}{d t} = \sqrt{\frac{a' (1 - e'^2)}{1 + \tan.^2 \varphi'}};$$

&c.

These values of $x d y - y d x$, $x' d y' - y' d x'$, &c. may be used, abstraction being made of the inequalities of the motion of the planets, provided we consider the elements e, e', &c. φ, φ', &c. as variables, in virtue of the secular inequalities; the equation (4) of No. 471 will therefore give in that case,

$$c = \mu \sqrt{\frac{a (1 - e^2)}{1 + \tan.^2 \varphi}} + \mu' \sqrt{\frac{a' (1 - e'^2)}{1 + \tan.^2 \varphi'}} + \&c.$$

$$+ \Sigma . \mu \mu' \left\{ \frac{(x' - x)(d y' - d y) - (y' - y) d x' - d x)}{d t} \right\}.$$

Neglecting this last term, which always remains of the order $\mu \mu'$, we shall have

$$c = \mu \sqrt{\frac{a . (1 - e^2)}{1 + \tan.^2 \varphi}} + \mu' \sqrt{\frac{a' (1 - e'^2)}{1 + \tan.^2 \varphi'}} + \&c.$$

Thus, whatever may be the changes which the lapse of time produces in the values of e, e', &c. φ, φ', &c. by reason of the secular variations, these values ought always to satisfy the preceding equation.

If we neglect the small quantities of the order e^4, or $e^2 \varphi^2$, this equation will give

$$c = \mu \sqrt{a} + \mu' \sqrt{a'} + \&c. - \tfrac{1}{2} \mu \sqrt{a} \{e^2 + \tan.^2 \varphi\}$$
$$- \tfrac{1}{2} \mu' \sqrt{a'} \{e'^2 + \tan^2 \varphi'\} - \&c.;$$

and consequently, if we neglect the squares of e, e', φ, &c. we shall have $\mu \sqrt{a} + \mu' \sqrt{a'} + \&c.$ constant. We have seen above, that if we only retain the first power of the perturbing force, a, a', &c. will be separately constant; the preceding equation will therefore give, neglecting small quantities of the order e^4 or $e^2 \varphi^2$,

$$\text{const.} = \mu \sqrt{a} \{e^2 + \tan.^2 \varphi\} + \mu' \sqrt{a'} \{e'^2 + \tan.^2 \varphi'\} + \&c.$$

On the supposition that the orbits are nearly circular, and but little inclined to one another, the secular variations are determined (No. 522) by means of differential equations independent of the inclinations, and which consequently are the same as though the orbits were in one plane. But in this hypothesis we have

$$\varphi = 0, \; \varphi' = 0, \; \&c.$$

the preceding equation thus becoming

$$\text{constant} = e^2 \mu \sqrt{a} + e'^2 \mu' \sqrt{a'} + e''^2 \mu'' \sqrt{a''} + \&c.$$

an equation already given in No. 524.

In like manner the secular variations of the inclinations of the orbits, are (No. 526) determined by means of differential equations, independent of excentricities, and which consequently are the same as though the orbits were circular. But in this hypothesis we have $e = 0$, $e' = 0$, &c. Wherefore

$$\text{const.} = \mu \sqrt{a} \,.\, \tan^2 \varphi + \mu' \sqrt{a'} \,.\, \tan.^2 \varphi' + \mu'' \sqrt{a''} \,.\, \tan.^2 \varphi'' + \&c.$$

an equation which has already been given in No. 526.

If we suppose, as in the last No.

$$p = \tan.\ \varphi \sin.\ \theta;\ q = \tan.\ \varphi \cos.\ \theta;$$

it is easy to prove that, the inclination of the orbit of μ to the plane of x, y being φ, and the longitude of its ascending node reckoned from the axis of x being θ, the cosine of the inclination of this orbit to the plane of x, z, will be

$$\frac{q}{\sqrt{(1 + \tan.^2 \varphi)}}.$$

Multiplying this quantity by $\frac{x\,dy - y\,dx}{dt}$, or by its value $\sqrt{a(1 - e^2)}$, we shall have the value of $\frac{x\,dz - z\,dx}{dt}$; the equation (5) of No. 471, will therefore give us, neglecting quantities of the order μ^2,

$$c' = \mu \,.\, q \sqrt{\frac{a(1 - e^2)}{1 + \tan.^2 \varphi}} + \mu' \,.\, q' \sqrt{\frac{a' \,.\, (1 - e'^2)}{1 + \tan.^2 \varphi'}} + \&c.$$

We shall find, in like manner, that the equation (6) of No. 471, gives

$$c'' = \mu \,.\, p \,.\, \sqrt{\frac{a(1 - e^2)}{1 + \tan.^2 \varphi}} + \mu' \,.\, p' \sqrt{\frac{a' \,.\, (1 - e'^2)}{1 + \tan.^2 \varphi'}} + \&c.$$

If in these two equations we neglect quantities of the order e^2 or $e^2 \varphi$; they will become

$$\text{const.} = \mu\, q \,.\, \sqrt{a} + \mu' q' \sqrt{a'} + \&c.$$

$$\text{const.} = \mu\, p \sqrt{a} + \mu' p' \sqrt{a'} + \&c.$$

equations already found in No. 526.

Finally, the equation (7) of No. 471, will give, observing that by 478,

$$\frac{m}{a} = \frac{2\,m}{\varrho} - \frac{dx^2 + dy^2 + dz^2}{dt^2}$$

and neglecting quantities of the order $\mu \cdot \mu'$,

$$\text{const.} = \frac{\mu}{a} + \frac{\mu'}{a'} + \frac{\mu''}{a''} + \&c.$$

These different equations subsist, when we regard inequalities due to very long periods, which affect the elements of the orbits of μ, μ', &c. We have observed in No. 521, that the relation of the mean motions of these bodies may introduce into the expressions of the axis-majors of the orbits considered variable, inequalities whose arguments proportional to the time increase very slowly, and which having for divisors the coefficients of the time t, in these arguments, may become sensible. But it is evident that, retaining the terms only which have like divisors, and considering the orbits as ellipses whose elements vary by reason of those terms, the integrals (4), (5), (6), (7), of No. 471, will always give the relations between these elements already found; because the terms of the order $\mu\,\mu'$ which have been neglected in these integrals, to obtain the relations, have not for divisors the very small coefficients above mentioned, or at least they contain them only when multiplied by a power of the perturbing forces superior to that which we are considering.

529. We have observed already, that in the motion of a system of bodies, there exists an invariable plane, or such as always is of a parallel situation, which it is easy to find at all times by this condition, that the sum of the masses of the system, multiplied respectively by the projections of the areas described by the radius-vectors in a given time is a *maximum*. It is principally in the theory of the solar system, that the research of this plane is important, when viewed with reference to the proper motions of the stars and of the ecliptic, which make it so difficult to astronomers to determine precisely the celestial motions. If we call γ the inclination of this invariable plane to that of x, y, and Π the longitude of its ascending node, it is easily found that

$$\tan.\gamma \sin.\Pi = \frac{c''}{c}; \quad \tan.\gamma \cos.\Pi = \frac{c'}{c};$$

and consequently that

$$\tan.\gamma \sin.\Pi = \frac{\mu\sqrt{a(1-e^2)}\sin.\varphi\sin.\theta + \mu'\sqrt{a'(1-e'^2)}\sin.\varphi'\sin.\theta' + \&c.}{\mu\sqrt{a(1-e^2)}\cos.\varphi + \mu'\sqrt{a'(1-e'^2)}\cos.\varphi' + \&c.}$$

$$\tan.\gamma.\cos.\Pi = \frac{\mu\sqrt{a(1-e^2)}.\sin.\varphi\cos.\theta + \mu'\sqrt{a'(1-e'^2)}\sin.\varphi'\cos.\theta' + \&c.}{\mu\sqrt{a(1-e^2)}.\cos.\varphi + \mu'\sqrt{a'(1-e'^2)}.\cos.\varphi' + \&c.}$$

We shall determine very easily, by means of these values, the angles γ and Π. We see that to determine the invariable plane we ought to know the masses of the comets, and the elements of their orbits; fortunately these masses appear to be so very small that we may, without sensible error, neglect their action upon the planets; but time alone can clear up this point to us. We may observe here, that relatively to this invariable plane the values of p, q, p′, q′, &c. contain no constant terms; for it is evident by the equations (C) of No. 526, that these terms are the same for p, p′, p″, &c. and that they are also the same for q, q′, q″, &c.; and since relatively to the invariable plane, the constants of the first members of the equations (1) and (2) of No. 526 are nothing: the constant terms disappear, by reason of these equations, from the expressions p, p′, &c. q, q′, &c.

Let us consider the motion of the two orbits, supposing them inclined to one another, by any angle whatever: we shall have by No. 528,

$$c' = \sin.\varphi \cos.\theta . \mu \sqrt{a(1-e^2)} + \sin.\varphi' . \cos.\theta' . \mu' \sqrt{a'(1-e'^2)};$$

$$c'' = \sin.\varphi \sin.\theta . \mu \sqrt{a(1-e^2)} + \sin.\varphi' . \sin.\theta' . \mu' \sqrt{a'(1-e'^2)}.$$

Let us suppose that the fixed plane to which we refer the motion of the orbits, is the invariable plane of which we have spoken, and by reference to which the constants of the first members of these equations, are nothing, as may easily be shown. The angles φ and φ' being positive, the preceding equations give the following ones:

$$\mu \sqrt{a(1-e^2)} . \sin.\varphi = \mu' \sqrt{a'(1-e'^2)} . \sin.\varphi';$$

$$\sin.\theta = -\sin.\theta'; \quad \cos.\theta = -\cos.\theta';$$

whence we derive $\theta' = \theta +$ the semi circumference; the nodes of the orbits are consequently upon the same line; but the ascending node of the one coincides with the decending node of the other; so that the mutual inclination of the two orbits is equal to $\varphi + \varphi'$.

We have by No. 528,

$$c = \mu \sqrt{a(1-e^2)} . \cos.\varphi + \mu' \sqrt{a'(1-e'^2)} \cos.\varphi';$$

by combining this equation with the preceding one between sin. φ and sin. φ', we shall have

$$2\mu c . \cos.\varphi . \sqrt{a(1-e^2)} = c^2 + \mu^2 a(1-e^2) - \mu'^2 . a'(1-e'^2).$$

If we suppose the orbits circular, or at least having excentricity so small that we may neglect the squares of their excentricities, the preceding equation will give φ constant: for the same reason φ' will be constant; the inclinations of the planes of the orbits to the fixed plane, and to one another, will therefore be constant, and these three planes will always have a common intersection. It thence results that the mean instantaneous variation of this intersection is always the same; because it can only be a function of these inclinations. When they are very small, we shall easily find by No. 528, and in virtue of the preceding relation between sin. φ and sin. φ', that for the time t, the motion of this intersection is

$$-\{(0,1) + (1,0)\} . t.$$

The position of the invariable plane to which we refer the motion of the orbits, may easily be determined for any instant whatever; for we have only to divide the angle of the mutual inclination of the orbits into two angles, φ and φ', such as that we have in the preceding equation between sin. φ, and sin. φ'. Designating, therefore, this mutual inclination by ϖ, we shall have

$$\tan. \varphi = \frac{\mu' \sqrt{a' (1 - e'^2)} . \sin. \varpi}{\mu \sqrt{a (1 - e^2)} + \mu' \sqrt{a' (1 - e'^2)} . \cos. \varpi}$$

SECOND METHOD OF APPROXIMATION OF THE CELESTIAL MOTIONS.

530. We have already seen that the coordinates of the celestial bodies, referred to the foci of the principal forces which animate them, are determined by differential equations of the second order. We have integrated these equations in retaining only the principal forces, and we have shown that in this case, the orbits are conic sections whose elements are the arbitrary constants introduced by integration.

The perturbing forces adding only small inequalities to the elliptic motion, it is natural to seek to reduce to the laws of this motion the troubled motion of the celestial bodies. If we apply to the differential equations of elliptic motion, augmented by the small terms due to the perturbing forces, the method exposed in No. 512, we can also consider the celestial motions in orbits which turn into themselves, as being elliptic; but the elements of this motion will be variable, and by this method we shall obtain their variations. Hence it results that the equations of motion, being differentials of the second order, not only their finite integrals, but also their infinitely small integrals of the first order, are the same as in the case of invariable ellipses; so that we may differentiate the finite equations of elliptic motion, in treating the elements of this motion as constant. It also results from the same method that the differential equations of the first order may be differentiated, by making vary only the elements of the orbits, and the first differences of the coordinates; provided that instead of the second differences of these coordinates, we substitute only that part of their values which is due to their perturbing forces. These results can be derived immediately from the consideration of elliptic motion.

For that purpose, conceive an ellipse passing through a planet, and through the element of the curve which it describes, and whose focus is occupied by the sun. This ellipse is that which the planet would invariably describe, if the perturbing forces were to cease to act upon it. Its elements are constant during the instant d t; but they vary from one instant to another. Let therefore V = 0, be a finite equation to an invariable ellipse, V being a function of the rectangular coordinates x, y, z and the parameters c, c′, &c. which are functions of the elements of elliptic motion. Since, however, this ellipse belongs to the element of the curve described by the planet during the instant d t; the equation V = 0 will still hold good for the first and last point of this element, by regarding c, c′, &c. as constant. We may, therefore, differentiate this equation once in only supposing x, y, z, to vary, which gives

$$0 = \left(\frac{d\,V}{d\,x}\right) d\,x + \left(\frac{d\,V}{d\,y}\right) d\,y + \left(\frac{d\,V}{d\,z}\right) d\,z; \quad \text{(i)}$$

We also see the reason why the finite equations of the invariable ellipse, may, in the case of the variable ellipse, be differentiated *once* in treating the parameters as constant. For the same reason, every differential equation of the first order relative to the invariable ellipse, equally holds good for the variable ellipse; for let V′ = 0 be an equation of this order, V′ being a function of x, y, z, $\frac{d\,x}{d\,t}$, $\frac{d\,y}{d\,t}$, $\frac{d\,z}{d\,t}$, and the parameters c, c′, &c. It is clear that all these quantities are the same for the variable ellipse as well as for the invariable ellipse, which for the instant d t coincides with it.

Now if we consider the planet, at the end of the instant d t, or at the commencement of the following one; the function V will vary from the ellipse relative to the instant d t to the consecutive ellipse only by the variation of the parameters, since the coordinates x, y, z, relative to the end of the first instant are the same for the two ellipses; thus the function V being nothing, we have

$$0 = \left(\frac{d\,V}{d\,c}\right) . \, d\,c + \left(\frac{d\,V}{d\,c'}\right) d\,c' + \&c. \quad \text{(i')}$$

This equation may be deduced from the equation V = 0, by making x, y, z, c, c′, &c. vary together; for if we take the differential equation (i) from this differential, we shall have the equation (i′).

Differentiating the equation (i), we shall have a new equation in d c, d c′, &c. which with the equation (i′) will serve to determine the parameters c, c′, &c. Thus it is that the geometers, who were first occupied in the theory of celestial perturbations, have determined the variations of the nodes and the inclinations of the orbits: but we may simplify this differentiation in the following manner.

Consider generally the differential equation of the first order V′ = 0, an equation which belongs equally to the variable ellipse, and to the invariable ellipse which, in the instant d t, coincides with it. In the following instant, this equation belongs also to the two ellipses; but with this difference, that c, c′, &c. remain the same in the case of the invariable ellipse, but vary with the variable ellipse. Let V'' be what V' becomes, when the ellipse is supposed invariable, and $V_{/}'$ what this same function becomes in the case of the variable ellipse. It is clear that in order to have V'' we must change in V', the coordinates x, y, z, which are relative to the commencement of the first instant d t, in those which are relative to the commencement of the second instant; we must then augment the first differences d x, d y, d z respectively by the quantities $d^2 x$, $d^2 y$, $d^2 z$, relative to the invariable ellipse, the element d t of the time, being supposed constant.

In like manner, to get $V_{/}'$, we must change in V', the coordinates x, y, z, in those which are relative to the commencement of the second instant, and which are also the same in the two ellipses; we must then augment d x, d y, d z respectively by the quantities $d^2 x$, $d^2 y$, $d^2 z$; finally, we must change the parameters c, c′, &c. into c + d c, c′ + d c′; &c.

The values of $d^2 x$, $d^2 y$, $d^2 z$ are not the same in the two ellipses; they are augmented, in the case of the variable ellipse, by the quantities due to the perturbing forces. We see also that the two functions V'' and $V_{/}'$, differing only in this that in the second the parameters c, c′, &c. increase by d c, d c′, &c.; and the values of $d^2 x$, $d^2 y$, $d^2 z$ relative to the invariable ellipse, are augmented by quantities due to the perturbing forces. We shall, therefore, form $V_{/}' - V''$, by differentiating V' in the supposition that x, y, z are constant, and that d x, d y, d z, c, c′, &c. are variable, provided that in this differential we substitute for $d^2 x$, $d^2 y$, $d^2 z$, &c. the parts of their values due solely to the disturbing forces.

If, however, in the function $V'' - V'$ we substitute for $d^2 x$, $d^2 y$, $d^2 z$ their values relative to elliptic motion, we shall have a function of x, y, z, $\frac{d x}{d t}, \frac{d y}{d t}, \frac{d z}{d t}$, c, c′, &c., which in the case of the invariable ellipse, is nothing; this function is therefore also nothing in the case of the variable ellipse. We evidently have in this last case, $V_{/}' - V' = 0$, since this equation is the differential of the equation $V' = 0$: taking it from the equation $V_{/}' - V' = 0$, we have $V_{/}' - V'' = 0$. Thus, we may, in this case, differentiate the equation $V' = 0$, supposing d x, d y, d z, c, c′, &c. alone to vary, provided that we substitute for $d^2 x$, $d^2 y$, $d^2 z$, the parts of their values relative to the disturbing forces. These results are exactly the same as those which we obtained in No. 512, by considerations purely analytical; but as is due to their importance, we shall here again present them, deduced from the consideration of elliptic motion.

531. Let us resume the equations (P) of No. 513,

$$\left.\begin{aligned} 0 &= \frac{d^2 x}{d t^2} + \frac{m x}{\varrho^3} + \left(\frac{d R}{d x}\right); \\ 0 &= \frac{d^2 y}{d t^2} + \frac{m y}{\varrho^3} + \left(\frac{d R}{d y}\right); \\ 0 &= \frac{d^2 z}{d t^2} + \frac{m z}{\varrho^3} + \left(\frac{d R}{d z}\right). \end{aligned}\right\} \quad (P)$$

If we suppose R = 0, we shall have the equations of elliptic motion, which we have integrated in (478). We have there obtained the seven following integrals.

$$c = \frac{x d y - y d x}{d t};$$

$$c' = \frac{x d z - z d x}{d t};$$

$$c'' = \frac{y d z - z d y}{d t};$$

$$\left.\begin{aligned} 0 &= f + x\left\{\frac{m}{\varrho} - \frac{d y^2 + d z^2}{d t^2}\right\} + \frac{y d y . d x}{d t^2} + \frac{z d z . d x}{d t^2}; \\ 0 &= f' + y\left\{\frac{m}{\varrho} - \frac{d x^2 + d z^2}{d t^2}\right\} + \frac{x d x . d y}{d t^2} + \frac{z d z . d y}{d t^2}; \\ 0 &= f'' + z\left\{\frac{m}{\varrho} - \frac{d x^2 + d y^2}{d t^2}\right\} + \frac{x d x . d z}{d t^2} + \frac{y d y . d z}{d t^2}; \\ 0 &= \frac{m}{a} - \frac{2 m}{\varrho} + \frac{d x^2 + d y^2 + d z^2}{d t^2}. \end{aligned}\right\} \quad (p)$$

These integrals give the arbitraries in functions of their first differences; they are under a very commodious form for determining the variations of these arbitraries. The three first integrals give, by differentiating them, and making vary by the preceding No. the parameters c, c′, c″, and the first differences of the coordinates,

$$d c = \frac{x d^2 y - y d^2 x}{d t}$$

$$d c' = \frac{x d^2 z - z d^2 x}{d t}$$

$$d c'' = \frac{y d^2 z - z d^2 y}{d t}.$$

Substituting for $d^2 x$, $d^2 y$, $d^2 z$, the parts of their values due to the perturbing forces, and which by the differential equations (P) are

$$- d t^2 \left(\frac{d R}{d x}\right), - d t^2 \left(\frac{d R}{d y}\right), - d t^2 \left(\frac{d R}{d z}\right);$$

we shall have

$$d c = d t \left\{ y \left(\frac{d R}{d x}\right) - x \left(\frac{d R}{d y}\right) \right\};$$

$$d c' = d t \left\{ z \left(\frac{d R}{d x}\right) - x \left(\frac{d R}{d z}\right) \right\};$$

$$d c'' = d t \left\{ z \left(\frac{d R}{d y}\right) - y \left(\frac{d R}{d z}\right) \right\}.$$

We know from 478, 479 that the parameters c, c′, c″ determine three elements of the elliptic orbit, viz., the inclination φ of the orbit to the plane of x, y, and the longitude θ of its ascending node, by means of the equations

$$\tan. \varphi = \frac{\sqrt{(c'^2 + c''^2)}}{c}; \ \tan. \theta = \frac{c''}{c'};$$

and the semi-parameter $a(1 - e^2)$ of the ellipse by means of the equation

$$m a (1 - e^2) = c^2 + c'^2 + c''^2.$$

The same equations subsist also in the ease of the variable ellipse, provided we determine c, c′, c″ by means of the preceding differential equations. We shall thus have the parameter of the variable ellipse, its inclination to the fixed plane of x, y and the position of its node.

The three first of the equations (p) have given us in No. (479) the finite integral

$$0 = c'' x - c' y + c z:$$

this equation subsists in the case of the troubled ellipse, as also its first difference

$$0 = c'' d x - c' d y + c d z$$

taken in considering c, c′, c″ constant.

If we differentiate the fourth, the fifth and the sixth of the integrals (p), making only the parameters f, f′, f″, and the differences d x, d y, d z vary; if moreover, we substitute then for $d^2 x$, $d^2 y$, $d^2 z$, the quantities $-d t^2 \left(\frac{d R}{d x}\right)$, $-d t^2 \left(\frac{d R}{d y}\right)$, $-d t^2 \left(\frac{d R}{d z}\right)$, we shall have

$$d f = d y \left\{ y \left(\frac{d R}{d x}\right) - x \left(\frac{d R}{d y}\right) \right\} + d z \left\{ z \left(\frac{d R}{d x}\right) - x \left(\frac{d R}{d z}\right) \right\}$$

$$+ (y d x - x d y) \left(\frac{d R}{d y}\right) + (z d x - x d z) \left(\frac{d R}{d z}\right);$$

$$d f' = d x \left\{ x \left(\frac{d R}{d y}\right) - y \left(\frac{d R}{d x}\right) \right\} + d z \left\{ z \left(\frac{d R}{d y}\right) - y \left(\frac{d R}{d z}\right) \right\}$$

$$+ (x d y - y d x) \left(\frac{d R}{d x}\right) + (z d y - y d z) \left(\frac{d R}{d z}\right);$$

$$d\,f'' = d\,x\left\{x\left(\frac{d\,R}{d\,z}\right) - z\left(\frac{d\,R}{d\,x}\right)\right\} + d\,y\left\{y\left(\frac{d\,R}{d\,z}\right) - z\left(\frac{d\,R}{d\,y}\right)\right\}$$
$$+ (x\,d\,z - z\,d\,x)\left(\frac{d\,R}{d\,x}\right) + (y\,d\,z - z\,d\,y)\left(\frac{d\,R}{d\,y}\right).$$

Finally, the seventh of the integrals (p), differentiated in the same manner, will give the variation of the semi-axis-major a, by means of the equation

$$d.\frac{m}{a} = 2\,d\,R,$$

the differential d R being taken relatively to the coordinates x, y, z alone of the body μ.

The values of f, f′, f″ determine the longitude of the projection of the perihelion of the orbit, upon the fixed plane, and the relation of the excentricity to the semi-axis-major; for I being the longitude of this projection by (479) we have

$$\tan.\,I = \frac{f'}{f};$$

and e being the ratio of the excentricity to the semi-axis-major, we have

$$m\,e = \sqrt{(f^2 + f'^2 + f''^2)}.$$

This ratio may also be determined by dividing the semi-parameter $a\,(1 - e^2)$, by the semi-axis-major a: the quotient taken from unity will give the value of e^2.

The integrals (p) have given by elimination (479) the finite integral

$$0 = m\,\varrho - h^2 + f\,x + f'.y + f''\,z:$$

this equation subsists in the case of the troubled ellipse, and it determines at each instant, the nature of the variable ellipse. We may differentiate it, considering f, f′, f″ as constant; which gives

$$0 = m\,d\,\varrho + f\,d\,x + f'\,d\,y + f''\,d\,z.$$

The semi-axis-major a gives the mean motion of μ, or more exactly, that which in the troubled orbit, corresponds to the mean motion in the invariable orbit; for we have (479) $n = {}^{-\frac{3}{2}}\sqrt{m}$; moreover, if we denote by ς the mean motion of μ, we haveain the invariable elliptic orbit $d\,\zeta = n\,d\,t$: this equation equally holds good in the variable ellipse, since it is a differential of the first order. Differentiating we shall have $d^2\,\zeta = d\,n\,.\,d\,t$; but we have

$$d\,n = \frac{3\,a\,n}{2\,m}\,.\,d.\frac{m}{a} = \frac{3\,a\,n\,d\,R}{m},$$

therefore

$$d^2\,\varsigma = \frac{3\,a\,n\,d\,t\,.\,d\,R}{m};$$

and integrating

$$\zeta = \frac{3}{m} \cdot \int\int a\, n\, d\, t \,.\, d\, R.$$

Finally we have seen in (No. 473) that the integrals (p) are equivalent to but five distinct integrals, and that they give between the seven parameters c, c′, c″, f, f′, f″ e, the two equations of condition

$$0 = f\,c'' - f'\,c' + f''\,c;$$

$$0 = \frac{m}{a} + \frac{f^2 + f'^2 + f''^2 - m^2}{c^2 + c'^2 + c''^2};$$

these equations subsist therefore in the case of the variable ellipse provided that the parameters are determined as above.

We can easily verify these statements a *posteriori.*

We have determined five elements of the variable orbit, viz., its inclination, position of the nodes, its semi-axis-major which gives its mean motion, its excentricity and the position of the perihelion. It remains for us to find the sixth element of elliptic motion,—that which in the invariable ellipse corresponds to the position of μ at a given epoch. For this purpose let us resume the expression of d t (473)

$$\frac{d\,t\,\sqrt{m}}{a^{\frac{3}{2}}} = \frac{d\,v\,(1 - e^2)^{\frac{3}{2}}}{\{1 + e\cos.(v - \varpi)\}^2}.$$

This equation developed into series gives (473)

$$n\,d\,t = d\,v\,\{1 + E^{(1)}\cos.(v - \varpi) + E^{(2)}\cos.2\,(v - \varpi) + \&c.\}.$$

Integrating this equation on the supposition of e and ϖ being constant, we shall have

$$\int n\,d\,t + \varepsilon = v + E^{(1)}\sin.(v - \varpi) + \frac{E^{(1)}}{2}\sin.2.(v - \varpi) + \&c.$$

ε being an arbitrary. This integral is relative to the invariable ellipse: to extend it to the variable ellipse, in making every thing vary even to the arbitraries, ε, e, ϖ which it contains, its differential must coincide with the preceding one; which gives

$$d\,\varepsilon = d\,e\left\{\left(\frac{d\,E^{(1)}}{d\,e}\right)\sin.(v - \varpi) + \tfrac{1}{2}\left(\frac{d\,E^{(2)}}{d\,e}\right)\sin.2\,(v - \varpi) + \&c.\right\}$$

$$- d\,\varpi\,\{E^{(1)}\cos.(v - \varpi) + E^{(2)}\cos.2\,(v - \varpi) + \&c.\}:$$

$v - \varpi$ being the true anomaly of μ measured upon the orbit, and ϖ the longitude of the perihelion also measured upon the orbit. We have determined above, the longitude I of the projection of the perihelion upon a fixed plane. But by (488) we have, in changing v into ϖ and $v_{\prime}$ into I in the expression of $v - \beta$ of this No.

$$\varpi - \beta = I - \theta + \tan.^2\tfrac{1}{2}\,\varphi\,\sin.2\,(I - \theta) + \&c.$$

Supposing next that v, $v_{,}$, are zero in this same expression, we have

$$\beta = \theta + \tan.^2 \tfrac{1}{2} \varphi \sin. 2 \theta + \&c.$$

wherefore,

$$\varpi = I + \tan.^2 \tfrac{1}{2} \varphi . \{\sin. 2 \theta + \sin. 2 (I - \theta) + \&c.\}$$

which gives

$$d \varpi = d I . \{1 + 2 \tan.^2 \tfrac{1}{2} \varphi \cos. 2 (I - \theta) + \&c.\}$$
$$+ 2 d \theta \tan.^2 \tfrac{1}{2} \varphi \{\cos. 2 \theta - \cos. 2 (I - \theta) + \&c.\}$$
$$+ \frac{d \varphi \tan. \frac{1}{2} \varphi}{\cos.^2 \frac{1}{2} \varphi} . \{\sin. 2 \theta + \sin. 2 (I - \theta) + \&c.\}.$$

Thus the values of d I, d θ, and d φ being determined by the above, we shall have that of d ϖ; whence we shall obtain the value of d ε.

It follows from thence that the expressions in series, of the radius-vector, of its projection upon the fixed plane, of the longitude whether referred to the fixed plane or to the orbit, and of the latitude which we have given in (No. 488) for the case of the invariable ellipse, subsist equally in the case of the troubled ellipse, provided we change n t into $\int$ n d t, and we determine the elements of the variable ellipse by the preceding formulas. For since the finite equations between ϱ, v, s, x, y, z, and $\int$ n d t, are the same in the two cases, and because the series of No. 488 result from these equations, by analytical operations entirely independent of the constancy or variability of the elements, it is evident these expressions subsist in the case of variable elements.

When the ellipses are very excentric, as is the case with the orbits of the comets, we must make a slight change in the preceding analysis. The inclination φ of the orbit to the fixed plane, the longitude θ of its ascending node, the semi-axis-major a, the semi-parameter a $(1 - e^2)$, the excentricity e, and the longitude I of the perihelion upon the fixed plane may be determined by what precedes. But the values of ϖ and of d ϖ being given in series ordered according to the powers of tan. $\frac{1}{2} \varphi$, in order to render them convergent, we must choose the fixed plane, so as to make tan. $\frac{1}{2} \varphi$ inconsiderable; and to effect this most simply is to take, for the fixed plane, that of the orbit of μ at a given epoch.

The preceding value of d ε is expressed by a series which is convergent only in the case where the excentricity of the orbit is inconsiderable, we cannot therefore make use of it in this case. Instead, let us resume the equation

$$\frac{d t \sqrt{m}}{a^{\frac{3}{2}}} = \frac{d v (1 - e^2)^{\frac{3}{2}}}{\{1 + e \cos. (v - \varpi)\}^2}.$$

If we make $1 - e = \alpha$, we have by (489) in the case of the invariable ellipse,

$$t + T = \frac{2 a^{\frac{3}{2}} (1 - e^2)^{\frac{3}{2}}}{(2 - \alpha)^2 . \sqrt{m}} . \tan. \tfrac{1}{2} (v - \varpi) \left\{ 1 + \frac{\frac{2}{3} - \alpha}{2 - \alpha} . \tan.^2 \tfrac{1}{2} (v - \varpi) + \&c. \right\}$$

T being an arbitrary. To extend this equation to the variable ellipse, we must differentiate it by making vary T, the semi parameter $a(1 - e^2)$, α, and ϖ. We shall thence obtain a differential equation which will determine T, and the finite equations which subsist in the case of the invariable ellipse, will still hold good in that of the variable ellipse.

532. Let us consider more particularly the variations of the elements of μ's orbit, in the case of the orbits being of small excentricity and but little inclined to one another. We have given in No. 515. the manner of developing R in a series of sines and cosines of the form

$$\mu' k \cos. (i' n' t - i n t + A)$$

k and A being functions of the excentricity and inclinations of the orbits, the positions of their nodes and perihelions, the longitudes of the bodies at a given epoch, and the major-axes. When the ellipses are variable all these quantities must be supposed to vary conformably to what precedes. We must moreover change in the preceding term, the angle $i' n' t - i n t$ into $i' \int n' d t - i \int n d t$, or which is tantamount, into $i' \zeta' - i \zeta$.

However, by the preceding No., we have

$$\frac{m}{a} = 2 \int d R;$$

$$\zeta = \int n d t = \frac{3}{m} . \int\int a n d t . d R.$$

The difference $d R$ being taken relatively to the coordinates x, y, z, of the body μ, we must only make vary, in the term

$$\mu' k \cos. (i' \zeta' - i \zeta + A)$$

of the expression of R developed into a series, what depends upon the motion of this body; moreover, R being a finite function of x, y, z, x', y', z' we may by No. 530, suppose the elements of the orbit constant in the difference $d R$; it suffices therefore to make ζ vary in the preceding term, and since the difference of ζ is n d t, we have

$$i \mu' . k n d t . \sin. (i' \zeta' - i \zeta + A)$$

for the term of $d R$ which corresponds to the preceding term of R. Thus, with respect to this term only, we have

$$\frac{1}{a} = \frac{2 i \mu'}{m} \int k n d t . \sin. (i' \zeta' - i \zeta + A);$$

$$\zeta = \frac{3\,i\,\mu'}{m} \int\!\!\int a\,k\,n^2\,d\,t^2 \sin.\,(i'\,\zeta' - i\,\zeta + A).$$

If we neglect the squares and products of the perturbing masses, we may, in the integrals of the above terms, suppose the elements of elliptic motion constant. Hence ζ becomes n t and ζ', n′ t; whence we get

$$\frac{1}{a} = -\frac{2\,i\,\mu'\,n\,k}{m\,(i'\,n' - i\,n)} \cos.\,(i'\,n'\,t - i\,n\,t + A)$$

$$\zeta = -\frac{3\,i\,\mu'\,a\,n^2\,k}{m\,(i'\,n' - i\,n)^2} \sin.\,(i'\,n'\,t - i\,n\,t + A).$$

Hence we perceive that if i′ n′ — i n is not zero, the quantities a and ζ only contain periodic inequalities, retaining only the first power of the perturbing force; but i and i′ being whole numbers, the equation i n′ — i n = 0 cannot subsist when the mean motions of μ and μ' are incommensurable, which is the case with the planets, and which can be admitted generally, since n and n′ being arbitrary constants susceptible of all possible values, their exact relation of number to number is not at all probable.

We are, therefore, conducted to this remarkable result, viz., *that the principal axes of the planets, and their mean motions, are only subject to periodic inequalities depending on their configuration, and that thus in neglecting these quantities, their principal axes are constant and their mean motions uniform, a result agreeing with what has otherwise been found by* No. 521.

If the mean motions n t and n′ t, without being exactly commensurable, approach very nearly to the ratio i′: i; the divisor i′ n′ — i n is very small, and there may result in ζ and ζ' inequalities, which increasing very slowly, may give reason for observers to suppose that the mean motions of the two bodies μ, μ' are not uniform. We shall see, in the theory of Jupiter and Saturn, that this is actually the case with regard to these two planets: their mean motions are such that twice that of Jupiter is very nearly equal to five times that of Saturn; so that 5 n′ — 2 n is hardly the sixty-fourth part of n. The smallness of this divisor, renders very sensible the term of the expression for ζ, depending upon the angle 5 n′ t — 2 n t, although it is of the order i′ — i, or of the third order, relatively to the excentricities and inclinations of the orbits, as we have seen in No. 515. The preceding analysis gives the most sensible part of these inequalities; for the variation of the mean longitude depends on two integrations, whilst the variations of the other elements of elliptic motion depend only on one integration; only terms of the expression of the mean longitude can therefore have the divisor (i′ n′ — i n)²; consequently with regard only to these terms, which, considering the smallness of the divisor ought to be the more considerable, it will suffice, in the expressions of the radius-vector, the longitude and latitude, to derive from these terms, the mean longitude.

When we have inequalities of this kind, which the action of μ' produces in the mean motion of μ, it is easy thence to get the corresponding inequalities which the action of μ produces in the mean motion of μ' In fact, if we have regard only to the mutual action of three bodies M, μ and μ'; the formula (7) of (471) gives.

$$\text{const.} = \mu \frac{dx^2 + dy^2 + dz^2}{dt^2} + \mu' . \frac{dx'^2 + dy'^2 + dz'^2}{dt^2}$$

$$- \frac{(\mu dx + \mu' dx')^2 + (\mu dy + \mu' dy')^2 + (\mu dz + \mu' dz')^2}{(M + \mu + \mu')^2 dt^2} \quad \ldots (a)$$

$$- \frac{2M\mu}{\sqrt{x^2+y^2+z^2}} - \frac{2M\mu'}{\sqrt{x'^2+y'^2+z'^2}} - \frac{2\mu\mu'}{\sqrt{(x'-x)^2+(y'-y)^2+(z'-z)^2}}.$$

The last of the integrals (p) of the preceding No. gives, by substituting for $\frac{m}{a}$ the integral $2 \int d\,R$,

$$\frac{dx^2 + dy^2 + dz^2}{dt^2} = \frac{2(M + \mu)}{\sqrt{x^2 + y^2 + z^2}} - 2 \int d\,R.$$

If we then call R′, what R becomes when we consider the action of μ upon μ', we shall have

$$R' = \frac{\mu(xx' + yy' + zz')}{(x^2 + y^2 + z^2)^{\frac{3}{2}}} - \frac{\mu}{\sqrt{(x' - x)^2 + (y' - y)^2 + (z' - z)^2}}$$

$$\frac{dx'^2 + dy'^2 + dz'^2}{dt^2} = \frac{2(M + \mu')}{\sqrt{x'^2 + y'^2 + z'^2}} - 2 \int d' R';$$

the differential characteristic d' only belonging to the coordinates of the body μ'. Substituting for $\frac{dx^2 + dy^2 + dz^2}{dt^2}$ and $\frac{dx'^2 + dy'^2 + dz'^2}{dt^2}$ the values in the equation (a), we shall have

$$\mu \int dR + \mu' \int d'R' = \text{const.} - \frac{(\mu dx + \mu' dx')^2 + (\mu dy + \mu' dy')^2 + (\mu dz + \mu' dz')^2}{2(M + \mu + \mu')\, dt^2}$$

$$+ \frac{\mu^2}{\sqrt{x^2 + y^2 + z^2}} + \frac{\mu'^2}{\sqrt{x'^2 + y'^2 + z'^2}}.$$

It is evident that the second member of this equation contains no terms of the order of squares and products of the μ, μ', which have the divisor $i'n' - in$; relative, therefore, only to these terms, we shall have

$$\mu \int d\,R + \mu' \int d'\,R' = 0;$$

thus, by only considering the terms which have the divisor $(i' n' - i n)^2$, we shall have

$$\frac{3\int\int a' n' d t . d' R'}{M + \mu'} = -\frac{\mu (M + \mu). a' n'}{\mu' (M + \mu') a n} \cdot \frac{3\int\int a n d t . d R}{M + \mu};$$

but we have

$$\zeta = \frac{3\int\int a n d t . d R}{M + \mu}; \zeta' = \frac{3\int\int a' n' d t . d' R'}{M + \mu'};$$

we therefore get

$$\mu' (M + \mu') a n \zeta' + \mu (M + \mu) a' n' \zeta = 0.$$

Again, we have

$$n = \frac{\sqrt{(M + \mu)}}{a^{\frac{3}{2}}}; n = \frac{\sqrt{(M + \mu')}}{a'^{\frac{3}{2}}};$$

neglecting therefore μ, μ', in comparison with M, we shall have

$$\mu \sqrt{a} . \zeta + \mu' \sqrt{a'} . \zeta' = 0;$$

or

$$\zeta' = - \frac{\mu \sqrt{a}}{\mu' \sqrt{a'}} . \zeta.$$

Thus the inequalities of ζ, which have the divisor $(i' n' - i n)^2$, give us those of ζ', which have the same divisor. These inequalities are, as we see, affected with the contrary sign, if n and n' have the same sign, or which amounts to the same, if the two bodies μ and μ' circulate in the same direction; they are, moreover, in a constant ratio; whence it follows that if they seem to accelerate the mean motion of μ, they appear to retard that of μ' according to the same law, and the apparent acceleration of μ, will be to the apparent retardation of μ', as $\mu' \sqrt{a'}$ is to $\mu \sqrt{a}$. The acceleration of the mean motion of Jupiter and the retardation of that of Saturn, which the comparison of modern with ancient observations made known to Halley, being very nearly in this ratio; it may be concluded from the preceding theorem, that they are due to the mutual action of the two planets; and, since it is constant, that this action cannot produce in the mean motions any alteration independent of the configuration of the planets, it is very probable that there exists in the theory of Jupiter and Saturn a great periodic inequality, of a very long period. Next, considering that five times the mean motion of Saturn, minus twice that of Jupiter is very nearly equal to nothing, it seems very probable that the phenomenon observed by Halley, was due to an inequality depending upon this argument. The determination of this inequality will verify the conjecture.

The period of the argument i' n' t — i n t being supposed very long, the elements of the orbits of μ, and μ' undergo, in this interval sensible variations, which must be taken into account in the double integral

$$\int\int a\,k\,n^2\,d\,t^2 \sin.(i'\,n'\,t - i\,n\,t + A).$$

For that purpose we shall give to the function $k \sin.(i'\,n'\,t - i\,n\,t + A)$, the form

$$Q \sin.(i'\,n'\,t - i\,n\,t + i'\,\epsilon' - i\,\epsilon) + Q' \cos.(i'\,n'\,t - i\,n\,t + i'\,\epsilon' - i\,\epsilon)$$

Q and Q′ being functions of the elements of the orbits: thus we shall have

$$\int\int a\,k\,n^2\,d\,t^2 \sin.(i'\,n'\,t - i\,n\,t + A) =$$

$$-\frac{n^2 a \sin.(i'n't - int + i'\epsilon - i\epsilon)}{(i'\,n' - i\,n)^2} \cdot \left\{Q - \frac{2\,d\,Q'}{(i'n' - in)dt} - \frac{3\,d^2\,Q}{(i'n' - in)^2dt^2} + \&c.\right\}$$

$$-\frac{n^2 a \cos.(i'n't - i\,n\,t + i'\,\epsilon - i\,\epsilon)}{(i'\,n' - i\,n)^2} \cdot \left\{Q' - \frac{2\,d\,Q}{(i'n' - in)dt} - \frac{3\,d^2\,Q'}{(i'n' - in)^2dt^2} + \&c.\right\}.$$

The terms of these two series decreasing very rapidly, with regard to the slowness of the secular variations of the elliptic elements, we may, in each series, stop at the two first terms. Then substituting for the elements of the orbits their values ordered according to the powers of the time, and only retaining the first power, the double integral above may be transformed in one term to the form

$$(F + E\,t) \sin.(i'\,n'\,t - i\,n\,t + A + H\,t).$$

Relatively to Jupiter and Saturn, this expression may serve for many ages before and after the instant from which we date the given epoch.

The great inequalities above referred to, become sensible amongst the terms depending upon the second power of the perturbing forces. In fact, if in the formula

$$\zeta = \frac{3\,i\,\mu'}{m}\int\int a\,k\,n^2.\,d\,t^2.\,\sin.(i'\,\zeta' - i\,\zeta + A),$$

we substitute for ζ, ζ' their values

$$n\,t - \frac{3\,i\,\mu'\,a\,n^2\,k}{m\,(i'n' - in)^2} \sin.(i'\,n'\,t - i\,n\,t + A);$$

$$n'\,t - \frac{3\,i\,\mu\,a\,n^2\,k}{m\,(i'\,n' - i\,n)^2}\sqrt{\frac{a}{a'}}.\,\sin.(i'\,n'\,t - i\,n\,t + A),$$

there will result among the terms of the order μ^2, the following

$$-\frac{9\,i^2\,\mu'^2\,a^2\,n^4\,k^2}{8\,m^2\,(i'\,n' - i\,n)^4} \cdot \frac{i\,\mu'\,\sqrt{a'} + i'\,\mu\,\sqrt{a}}{\mu'\,\sqrt{a}} \sin.2\,(i'\,n'\,t - i\,n\,t + A).$$

The value of ζ' contains the corresponding term, which is to the one preceding in the ratio $\mu\,\sqrt{a} : -\mu'\,\sqrt{a'}$, viz.

$$\frac{9\,i^2\,\mu'^2\,a^2\,n^4\,k^2}{8\,m^2\,(i'\,n' - i\,n)^4}\{i\,\mu'\,\sqrt{a'} + i'\,\mu\,\sqrt{a}\}.\frac{\mu\,\sqrt{a}}{\mu'^2\,a'} \sin.2\,(i'\,n'\,t - i\,n\,t + A).$$

533. It may happen that the inequalities of the mean motion which are the most sensible, are only to be found among terms of the order of the squares of the perturbing masses. If we consider three bodies, μ, μ', μ'' circulating around M, the expression of $d\,R$ relative to terms of this order, will contain inequalities of the form

$$k \sin. (i\,n\,t - i'\,n'\,t + i''\,n''\,t + A)$$

but if we suppose the mean motions $n\,t$, $n'\,t$, $n''\,t$ such that $i\,n - i'\,n' + i''\,n''$ is an extremely small fraction of n, there will result a very sensible inequality in the value of ζ. This inequality may render rigorously equal to zero, the quantity $i\,n - i'\,n' + i''\,n''$, and thus establish an equation of condition between the mean motions and the mean longitudes of the three bodies μ, μ', μ''. This very singular case exists in the system of Jupiter's satellites. We will give the analysis of it.

If we take M for the mass-unit, and neglect μ, μ', μ'' in comparison with it, we shall have

$$n^2 = \frac{1}{a^3};\ n'^2 = \frac{1}{a'^3},\ n''^2 = \frac{1}{a''^3},$$

we have then

$$d\,\zeta = n\,d\,t;\ d\,\zeta' = n'\,d\,t;\ d\,\zeta'' = n''\,d\,t;$$

wherefore

$$\frac{d^2\,\zeta}{d\,t} = -\frac{3}{2}\,n^{\frac{1}{3}}.\frac{d\,a}{a^2};$$

$$\frac{d^2\,\zeta'}{d\,t} = -\frac{3}{2}\,n'^{\frac{1}{3}}.\frac{d\,a'}{a'^2}$$

$$\frac{d^2\,\zeta''}{d\,t} = -\frac{3}{2}\,n''^{\frac{1}{3}}.\frac{d\,a''}{a''^2}.$$

We have seen in No. 528, that if we neglect the squares of the excentricities and inclinations of the orbits, we have

$$\text{const.} = \mu\,\sqrt{a} + \mu'.\,\sqrt{a'} + \mu''\,\sqrt{a''};$$

which gives

$$0 = \mu.\frac{d\,a'}{\sqrt{a}} + \mu'.\frac{d\,a'}{\sqrt{a'}} + \mu''.\frac{d\,a''}{\sqrt{a''}}.$$

From these several equations, it is easy to get

$$\frac{d^2\,\zeta}{d\,t} = -\frac{3}{2}.\,n^{\frac{1}{3}}.\frac{d\,a}{a^2}$$

$$\frac{d^2\,\zeta}{d\,t} = \frac{3}{2}.\frac{\mu.\,n'^{\frac{4}{3}}}{\mu'.\,n}.\frac{n - n''}{n' - n''}.\frac{d\,a}{a^2};$$

$$\frac{d^2\,\zeta''}{d\,t} = -\frac{3}{2}.\frac{\mu.\,n''^{\frac{4}{3}}}{\mu''.\,n}\,\frac{n - n'}{n' - n''}.\frac{d\,a}{a^2}.$$

Finally the equation

$$\frac{m}{a} = 2\int d\,R$$

of No. 531, gives

$$-\frac{d\,a}{a^2} = 2.\,d\,R.$$

We have therefore only to determine d R.

By No. 513, neglecting the squares and products of the inclinations of the orbits, we have

$$R = \frac{\mu'\,\rho}{\rho'^2}\cos.(v'-v) - \mu'(\rho^2 - 2\,\rho\,\rho'\cos.(v'-v) + \rho'^2)^{-\frac{1}{2}}$$

$$+\frac{\mu''\,\rho}{\rho''^2}\cos.(v''-v) - \mu''(\rho^2 - 2\,\rho\,\rho''\cos.(v''-v) + \rho''^2).^{-\frac{1}{2}}$$

If we develope this function in a series ordered according to the cosines of $v'-v$, $v''-v$ and their multiples; we shall have an expression of this form

$$R = \frac{\mu'}{2}(\rho,\rho')^{(0)} + \mu'(\rho,\rho')^{(1)}\cos.(v'-v) + \mu'(\rho,\rho')^{(2)}\cos.2\,(v'-v)$$

$$+\,\mu'(\rho,\rho')^{(3)}\cdot\cos.3\,(v'-v) + \&c.$$

$$+\frac{\mu'}{2}(\rho,\rho'')^{(0)} + \mu''(\rho,\rho'')^{(1)}\cos.(v''-v) + \mu''(\rho,\rho'')^{(2)}\cos.2\,(v''-v)$$

$$+\,\mu''(\rho,\rho'')^{(3)}\cos.3\,(v''-v) + \&c.;$$

whence we derive

$$dR = \left\{\begin{array}{l} d\rho\left\{\begin{array}{l} \frac{\mu'}{2}\left(\frac{d(\rho,\rho')^{(0)}}{d\rho}\right) + \mu'\left(\frac{d(\rho,\rho')^{(1)}}{d\rho}\right)\cos.(v'-v) + \mu'\left(\frac{d(\rho,\rho')^{(2)}}{d\rho}\right) \\ \cos.2\,(v'-v) + \&c. \\ +\frac{\mu''}{2}\left(\frac{d(\rho,\rho'')^{(0)}}{d\rho}\right) + \mu''\left(\frac{d(\rho,\rho'')^{(1)}}{d\rho}\right)\cos.(v''-v) + \mu''\left(\frac{d(\rho,\rho'')^{(2)}}{d\rho}\right) \\ \cos.2\,(v''-v) + \&c. \end{array}\right\} \\ +\,d\,v\left\{\begin{array}{l} \mu'(\rho,\rho')^{(1)}\sin.(v'-v) + 2\,\mu'(\rho,\rho')^{(2)}\sin.2(v'-v) + \&c. \\ +\mu''(\rho,\rho'')^{(1)}\sin.(v''-v) + 2\mu''(\rho,\rho'')^{(2)}\sin.2(v''-v) + \&c. \end{array}\right\} \end{array}\right\}$$

Suppose, conformably to what observations indicate in the system of the three first satellites of Jupiter, that $n - 2\,n'$ and $n' - 2\,n''$ are very small fractions of n, and that their difference $n - 2\,n' - (n' - 2\,n'')$ or $n - 3\,n' + 2\,n''$ is incomparably smaller than each of them. It results from the expressions of $\frac{\delta\rho}{a}$, and of $\delta\,v$ of No. 517, that the action of μ' produces in the radius-vector and in the longitude of μ, a very sensible inequality depending on the argument $2\,(n'\,t - n\,t + \epsilon' - \epsilon)$. The terms relative to this inequality have the divisor $4\,(n' - n)^2 - n^2$, or (n — 2 n′) (3 n — 2 n′), and this divisor is very small, because of the smallness of the factor n — 2 n′. We also perceive, by the consideration of the same expressions, that the action of μ produces in the radius-vector, and in the longitude of μ', an inequality depending on the argument $(n' t - n t + \varepsilon' - \varepsilon)$, and which having the divisor $(n' - n)^2 - n'^2$, or $n (n - 2 n')$, is very sensible. We see, in like manner, that the action of μ'' upon μ' produces in the same quantities a considerable inequality depending upon the argument $2 (n'' t - n' t + \varepsilon'' - \varepsilon')$. Finally, we perceive that the action of μ' produces in the radius-vector and in the longitude of μ'' a considerable inequality depending upon the argument $n'' t - n' t + \varepsilon'' - \varepsilon$. These inequalities were first recognised by observations; we shall develope them at length in the Theory of Jupiter's Satellites. In the present question we may neglect them, relatively to other inequalities. We shall suppose, therefore,

$$\delta \varrho = \mu' E' \cos. 2 (n' t - n t + \varepsilon' - \varepsilon);$$
$$\delta v = \mu' F' \sin. 2 (n' t - n t + \varepsilon' - \varepsilon);$$
$$\delta \varrho' = \mu'' E'' \cos. 2 (n'' t - n' t + \varepsilon'' - \varepsilon') + \mu G \cos. (n' t - n t + \varepsilon' - \varepsilon)$$
$$\delta v' = \mu'' F'' \sin. 2 (n'' t - n' t + \varepsilon'' - \varepsilon') + \mu H \sin. (n' t - n t + \varepsilon' - \varepsilon)$$
$$\delta \varrho'' = \mu' G' \cos. (n'' t - n' t + \varepsilon'' - \varepsilon')$$
$$d v'' = \mu'' H' \sin. (n'' t - n' t + \varepsilon'' - \varepsilon').$$

We must, however, substitute in the preceding expression of d R for ϱ, v, ϱ', v′, ϱ'', v″, the values of $a \delta \varrho$, $n t + \varepsilon + \delta v$, $a' + \delta \varrho'$, $n' t + \varepsilon' + \delta v'$, $a'' + \delta \varrho''$, $n'' t + \varepsilon'' + \delta v''$, and retain only the terms which depend upon the argument $n t - 3 n' t + 2 n'' t + \varepsilon - 3 \varepsilon' + 2 \varepsilon''$. But it is easy to see that the substitution of the values of $\delta \varrho$, δv, $\delta \varrho''$, $\delta v''$ cannot produce any such term. This is not the case with the substitution of the values of $\delta \varrho'$ and $\delta v'$: the term $\mu' (\varrho, \varrho')^{(1)} d v \sin. (v' - v)$ of the expression of d R, produces the following,

$$-\frac{\mu' \mu'' . n d t}{2} . \left\{E'' \left(\frac{d (a, a')^{(1)}}{d a'}\right) - F'' (a, a')^{(1)}\right\} \times$$
$$\sin. (n t - 3 n' t + 2 n'' t + \varepsilon' - 3 \varepsilon' + 2 \varepsilon'').$$

This is the only expression of the kind which the expression of d R contains. The expressions of $\frac{\delta \varrho}{a}$, and of δv of No. 517, applied to the action of μ'' upon μ', give, retaining only the terms which have the divisor $n' - 2 n''$, and observing that n'' is very nearly equal to $\frac{1}{2} n'$,

$$\frac{E''}{a'} = \frac{1}{4} n'^2 \times \frac{a'^2 \left(\frac{d . (a', a'')^{(2)}}{d a'}\right) + \frac{2 n''}{n' - n''} . a' (a', a'')^{(2)}}{(n' - 2 n'') (3 n' - 2 n'')}$$

$$F' = \frac{2\,E''}{a'};$$

we therefore have

$$d\,R = \frac{\mu'\,\mu''\,n\,d\,t}{2}\,E''.\left\{\frac{2\,(a,a')^{(1)}}{a'} - \left(\frac{d.(a,a')^{(1)}}{d\,a'}\right)\right\}$$

$$\times \sin.(n\,t - 3\,n'\,t + 2\,n''\,t + \epsilon - 3\,\epsilon' + 2\,\epsilon'') = -\tfrac{1}{2}.\frac{d\,a}{a^2}.$$

Substituting this value of $\frac{d\,a}{a^2}$ in the values of $\frac{d^2\,\zeta}{d\,t}$, $\frac{d^2\,\zeta'}{d\,t}$, $\frac{d^2\,\zeta''}{d\,t}$, and making for brevity's sake

$$\beta = \tfrac{3}{2}\,E''\left\{2\,(a,a')^{(1)} - a'\left(\frac{d\,(a,a')^{(1)}}{d\,a'}\right)\right\}.\left\{\frac{a}{a'}\,\mu'\,\mu'' + \frac{9}{4}\,\mu\,\mu'' + \frac{a''}{4\,a'}\,\mu\,\mu'\right\}$$

we shall have, since n is very nearly equal to 2 n′, and n′ to 2 n″,

$$\frac{d^2\zeta}{d\,t^2} - 3.\frac{d^2\,\zeta'}{d\,t^2} + 2.\frac{d^2\,\zeta''}{d\,t^2} = \beta\,n^2\,\sin.(u\,t - 3\,n'\,t + 2\,n''\,t + \epsilon - 3\,\epsilon' + 2\,\epsilon'');$$

or more exactly

$$\frac{d^2\zeta}{d\,t^2} - 3.\frac{d^2\,\zeta'}{d\,t^2} + 2.\frac{d^2\,\zeta''}{d\,t^2} = \beta\,n^2\,\sin.(\zeta - 3\,\zeta' + 2\,\zeta'' + \epsilon - 3\,\epsilon' + 2\,\epsilon'');$$

so that if we suppose

$$V = \zeta - 3\,\zeta' + 2\,\zeta'' + \epsilon - 3\,\epsilon' + 2\,\epsilon'',$$

we shall have

$$\frac{d^2\,V}{d\,t^2} = \beta.\,n^2.\,\sin.\,V.$$

The mean distances a, a′, a″, varying but little as also the quantity n, we may in this equation consider $\beta\,n^2$, as a constant quantity. Integrating, we have

$$d\,t = \frac{\pm\,d\,V}{\sqrt{c - 2\,\beta\,n^2\,\cos.\,V}},$$

c being an arbitrary constant. The different values of which this constant is susceptible, give rise to the three following cases.

If c is positive and greater than $\pm\,2\,\beta\,n^2$, the angle V will increase continually, and this ought to take place, if at the origin of the motion, $(n - 3\,n' + 2\,n'')^2$ is greater than $\pm\,2\,\beta\,n^2\,(1 \mp \cos.\,V)$, the upper or lower signs being taken according as β is positive or negative. It is easy to assure ourselves of this, and we shall see particularly in the theory of the satellites of Jupiter, that β is a positive quantity relatively to the three first satellites. Supposing therefore $\mp\,\varpi = \pi - V$, π being the semi circumference, we shall have

$$d\,t = \frac{d\,\varpi}{\sqrt{c + 2\,\beta\,n^2\,\cos.}}.$$

In the interval from $\varpi = 0$ to $\varpi = \frac{\pi}{2}$, the radical $\sqrt{c + 2\beta n^2 \cos. \varpi}$ is greater than $\sqrt{2\beta n^2}$, when c is equal or greater than $2\beta n^2$; we have therefore in this interval $\varpi > n t \sqrt{2\beta}$. Thus, the time t which the angle ϖ employs in arriving from zero to a right angle is less than $\frac{\pi}{2 n \sqrt{2\beta}}$. The value of β depends upon the masses, μ, μ', μ''; the inequalities observed in the three first satellites of Jupiter, and of which we spoke above, give, between their masses and that of Jupiter, relations from whence it results that $\frac{\pi}{2 n \sqrt{2\beta}}$ is under two years, as we shall see in the theory of these satellites; thus the angle ϖ would employ less than two years to increase from zero to a right angle; but the observations made upon Jupiter's satellites, give since their discovery, ϖ constantly nothing or insensible; the case which we are examining is not therefore that of the three first satellites of Jupiter.

If the constant c is less than $\pm 2\beta n^2$, the angle V will not oscillate; it will never reach two right angles, if β is negative, because then the radical $\sqrt{c - 2\beta n^2 \cos. V}$, becomes imaginary; it will never be nothing if β is positive. In the first case its value will be alternately greater and less than zero; in the second case it will be alternately greater and less than two right angles. All observations of the three first satellites of Jupiter, prove to us that this second case belongs to these stars; thus the value of β ought to be positive relatively to them; and since the theory of gravitation gives β positive, we may regard the phenomenon as a new confirmation of that theory.

Let us resume the equation

$$d t = \frac{d \varpi}{\sqrt{c + 2\beta n^2 \cos. \varpi}}.$$

The angle ϖ being always very small, according to the observations, we may suppose $\cos. \varpi = 1 - \frac{1}{2}\varpi^2$; the preceding equation will give by integration

$$\varpi = \lambda \sin. (n t \sqrt{\beta} + \gamma)$$

λ and γ being two arbitrary constants which observation alone can determine. Hitherto, it has not been recognised, a circumstance which proves it to be very small.

From the preceding analysis result the following consequences. Since the angle $n t + 3 n' t + 2 n'' t + \epsilon - 3\epsilon' + 1\epsilon''$, oscillates being sometimes less and sometimes greater than two right angles, its mean value is equal to two right angles; we shall therefore have, regarding only mean quantities

$$n - 3\,n' + 2\,n'' = 0$$

that is to say, *that the mean motion of the first satellite, minus three times that of the second, plus twice that of the third, is exactly and constantly equal to zero.* It is not necessary that this equality should subsist exactly at the origin, which would not in the least be probable; it is sufficient that it did very nearly so, and that $n - 3\,n' + 2\,n''$ has been less, abstraction being made of the sign, than $\lambda\,n\,\sqrt{\beta}$; and then that the mutual attraction has rendered the equality rigorous.

We have next $\varepsilon - 3\,\varepsilon' + 2\,\varepsilon''$ equal to two right angles; thus *the mean longitude of the first satellite, minus three times that of the second, plus twice that of the third, is exactly and constantly equal to two right angles.*

From this theorem, the preceding values of $\delta\,\varrho'$, and of $\delta\,v'$ are reducible to the two following

$$\delta\,\varrho' = (\mu\,G - \mu''\,E'')\cos.(n'\,t - n\,t + \varepsilon' - \varepsilon)$$

$$\delta\,v' = (\mu\,H - \mu''\,F'')\sin.(n'\,t - n\,t + \varepsilon' - \varepsilon).$$

The two inequalities of the motion of μ' due to the actions of μ and of μ'', merge consequently into one, and constantly remain so.

It also results from this theorem, that the three first satellites can never be eclipsed at the same time. They cannot be seen together from Jupiter neither in opposition nor in conjunction with the sun; for the preceding theorems subsist equally relative to the synodic mean motions, and to the synodic mean longitudes of the three satellites, as we may easily satisfy ourselves. These two theorems subsist, notwithstanding the alterations which the mean motions of the satellites undergo, whether they arise from a cause similar to that which alters the mean motion of the moon, or whether from the resistance of a very rare medium. It is evident that these several causes only require that there should be added to the value of $\frac{d^2\,V}{d\,t^2}$, a quantity of the form of $\frac{d^2\,\psi}{d\,t^2}$, and which shall only become sensible by integrations; supposing therefore $V = \pi - \varpi$, and ϖ very small, the differential equation in V will become

$$0 = \frac{d^2\,\varpi}{d\,t^2} + \beta\,n^2\,\varpi + \frac{d^2\,\psi}{d\,t^2}.$$

The period of the angle $n\,t\,\sqrt{\beta}$ being a very small number of years, whilst the quantities contained in $\frac{d^2\,\psi}{d\,t^2}$ are, either constant, or embrace many ages; by integrating the above equation we shall have

$$\varpi = \lambda \sin. (n t \sqrt{\beta} + \gamma) - \frac{\delta^2 \psi}{\beta n^2 d t^2}.$$

Thus the value of ϖ will always be very small, and the secular equations of the mean motions of the three first satellites will always be ordered by the mutual action of these stars, so, that the secular equation of the first, plus twice that of the third, may be equal to three times that of the second.

The preceding theorems give between the six constants n, n', n'', ϵ, ϵ', ϵ'' two equations of condition which reduce these arbitraries to four; but the two arbitraries λ and γ of the value of ϖ replace them. This value is distributed among the three satellites, so, that calling p, p', p'' the coefficients of sin. $(n t \sqrt{\beta} + \gamma)$ in the expressions of v, v', v'', these coefficients are as the preceding values of $\frac{d^2 \zeta}{d t^2}, \frac{d^2 \zeta'}{d t^2}, \frac{d^2 \zeta''}{d t^2}$; and moreover we have $p - 3 p' + 2 p'' = \lambda$. Hence results, in the mean motions of the three first satellites of Jupiter, an inequality which differs for each only by its coefficients, and which forms in these motions a sort of libration whose extent is arbitrary. Observations show it to be insensible.

534. Let us now consider the variations of the excentricities and of the perihelions of the orbits. For this purpose, resume the expressions of d f, d f', d f'' found in 531: calling ϱ the radius-vector of μ projected upon the plane of x, y; v the angle which this projection makes with the axis of x; and s the tangent of the latitude of μ above the same plane, we shall have

$$x = \varrho \cos. v; \ y = \varrho \sin. v; \ z = \varrho s$$

whence it is easy to obtain

$$x \left(\frac{d R}{d y}\right) - y \left(\frac{d R}{d x}\right) = \left(\frac{d R}{d v}\right)$$

$$x \left(\frac{d R}{d z}\right) - z \left(\frac{d R}{d x}\right) = (1 + s^2) \cos. v \left(\frac{d R}{d s}\right) - \varrho s \cos. v \left(\frac{d R}{d \varrho}\right)$$
$$+ s \sin. v \left(\frac{d R}{d v}\right)$$

$$y \left(\frac{d R}{d z}\right) - z \left(\frac{d R}{d s}\right) = (1 + s^2) \sin. v \left(\frac{d R}{d s}\right) - \varrho s \sin. v \left(\frac{d R}{d \varrho}\right)$$
$$- s \cos. v \left(\frac{d R}{d v}\right)$$

By 531, we also have

$$x d y - y d x = c d t; \ x d z - z d x = c' d t; \ y d z - z d y = c \ d t;$$

the differential equations in f, f′, f″ will thus become

$$\mathrm{d\,f} = -\mathrm{d\,y}\left(\frac{\mathrm{d\,R}}{\mathrm{d\,v}}\right) - \mathrm{d\,z}\left\{(1+s^2)\cos.v\left(\frac{\mathrm{d\,R}}{\mathrm{d\,s}}\right) - \varrho\,s.\cos.v\left(\frac{\mathrm{d\,R}}{\mathrm{d}\,\varrho}\right) + s\sin.v\left(\frac{\mathrm{d\,R}}{\mathrm{d\,v}}\right)\right\}$$

$$-\mathrm{c\,d\,t}\left\{\sin.v\left(\frac{\mathrm{d\,R}}{\mathrm{d}\,\varrho}\right) + \frac{\cos.v}{\varrho}\left(\frac{\mathrm{d\,R}}{\mathrm{d\,v}}\right) - \frac{s.\sin.v}{\varrho}\left(\frac{\mathrm{d\,R}}{\mathrm{d\,s}}\right)\right\} - \frac{\mathrm{c'\,d\,t}}{\varrho}\left(\frac{\mathrm{d\,R}}{\mathrm{d\,s}}\right);$$

$$\mathrm{d\,f'} = \mathrm{d\,x}\left(\frac{\mathrm{d\,R}}{\mathrm{d\,v}}\right) - \mathrm{d\,z}\left\{(1+s^2)\sin.v\left(\frac{\mathrm{d\,R}}{\mathrm{d\,s}}\right) - \varrho\,s.\sin.v\left(\frac{\mathrm{d\,R}}{\mathrm{d}\,\varrho}\right) - s\cos.v\left(\frac{\mathrm{d\,R}}{\mathrm{d\,v}}\right)\right\}$$

$$+\mathrm{c\,d\,t}\left\{\cos.v\left(\frac{\mathrm{d\,R}}{\mathrm{d}\,\varrho}\right) - \frac{\sin.v}{\varrho}\left(\frac{\mathrm{d\,R}}{\mathrm{d\,v}}\right) - \frac{s.\sin.v}{\varrho}\left(\frac{\mathrm{d\,R}}{\mathrm{d\,s}}\right)\right\} - \frac{\mathrm{c''\,d\,t}}{\varrho}\left(\frac{\mathrm{d\,R}}{\mathrm{d\,s}}\right);$$

$$\mathrm{d\,f''} = \mathrm{d\,x}\left\{(1+s^2)\cos.v\left(\frac{\mathrm{d\,R}}{\mathrm{d\,s}}\right) - \varrho\,s\cos.v\left(\frac{\mathrm{d\,R}}{\mathrm{d}\,\varrho}\right) + s\sin.v\left(\frac{\mathrm{d\,R}}{\mathrm{d\,v}}\right)\right\}$$

$$+\mathrm{d\,y}\left\{(1+s^2)\sin.v\left(\frac{\mathrm{d\,R}}{\mathrm{d\,s}}\right) - \varrho\,s.\sin.v\left(\frac{\mathrm{d\,R}}{\mathrm{d}\,\varrho}\right) - s\cos.v\left(\frac{\mathrm{d\,R}}{\mathrm{d\,v}}\right)\right\}$$

$$+\mathrm{c'.d\,t}\left\{\cos.v\left(\frac{\mathrm{d\,R}}{\mathrm{d}\,\varrho}\right) - \frac{\sin.v}{\varrho}\left(\frac{\mathrm{d\,R}}{\mathrm{d\,v}}\right) - \frac{s.\cos.v}{\varrho}.\left(\frac{\mathrm{d\,R}}{\mathrm{d\,s}}\right)\right\}$$

$$+\mathrm{c''.d\,t}\left\{\sin.v\left(\frac{\mathrm{d\,R}}{\mathrm{d}\,\varrho}\right) + \frac{\cos.v}{\varrho}\left(\frac{\mathrm{d\,R}}{\mathrm{d\,v}}\right) - \frac{s.\sin.v}{\varrho}.\left(\frac{\mathrm{d\,R}}{\mathrm{d\,s}}\right)\right\}.$$

The quantities c′, c″ depend, as we have seen in No. 531 upon the inclination of the orbit of μ to the fixed plane, in such a manner that they become zero when the inclination $= 0$; moreover it is easy to see by the nature of R that $\left(\frac{\mathrm{d\,R}}{\mathrm{d\,s}}\right)$ is of the order of the inclinations of the orbits; neglecting therefore the squares and products of these inclinations, the preceding expressions of d f and of d f′, will become

$$\mathrm{d\,f} = -\mathrm{d\,y}\left(\frac{\mathrm{d\,R}}{\mathrm{d\,v}}\right) - \mathrm{c\,d\,t}\left\{\sin.v\left(\frac{\mathrm{d\,R}}{\mathrm{d}\,\varrho}\right) + \frac{\cos.v}{\varrho}.\left(\frac{\mathrm{d\,R}}{\mathrm{d\,v}}\right)\right\};$$

$$\mathrm{d\,f'} = \mathrm{d\,x}\left(\frac{\mathrm{d\,R}}{\mathrm{d\,v}}\right) + \mathrm{c\,d\,t}\left\{\cos.v\left(\frac{\mathrm{d\,R}}{\mathrm{d}\,\varrho}\right) - \frac{\sin.v}{\varrho}\left(\frac{\mathrm{d\,R}}{\mathrm{d\,v}}\right)\right\};$$

but we have

$\mathrm{d\,x} = \mathrm{d}\,(\varrho\cos.v)$; $\mathrm{d\,y} = \mathrm{d}\,(\varrho\sin.v)$; $\mathrm{c\,d\,t} = \mathrm{x\,d\,y} - \mathrm{y\,d\,x} = \varrho^2\,\mathrm{d\,v}$,

we therefore get

$$\mathrm{d\,f} = -\{\mathrm{d}\,\varrho\sin.v + 2\,\varrho.\mathrm{d\,v}\cos.v\}\left(\frac{\mathrm{d\,R}}{\mathrm{d\,v}}\right) - \varrho^2\,\mathrm{d\,v}\sin.v\left(\frac{\mathrm{d\,R}}{\mathrm{d}\,\varrho}\right);$$

$$\mathrm{d\,f'} = \{\mathrm{d}\,\varrho\cos.v - 2\,\varrho\,\mathrm{d\,v}\sin.v\}\left(\frac{\mathrm{d\,R}}{\mathrm{d\,v}}\right) + \varrho^2\,\mathrm{d\,v}\cos.v\left(\frac{\mathrm{d\,R}}{\mathrm{d}\,\varrho}\right).$$

These equations are more exact, if we take for the fixed plane of x, y, that of the orbit of μ, at a given epoch; for then c', c'' and s are of the order of the perturbing forces; thus the quantities which we neglect, are of the order of the squares of the perturbing forces, multiplied by the square of the respective inclination of the two orbits of μ and of μ'.

The values of ϱ, $d\varrho$, dv, $\left(\frac{dR}{d\varrho}\right)$, $\left(\frac{dR}{dv}\right)$, remain clearly the same whatever is the position of the point from which we reckon the longitudes; but in diminishing v by a right angle, sin. v becomes — cos. v, and cos. v becomes sin. v; the expression of d f changes consequently to that of d f'; whence it follows that having developed, into a series of sines and cosines of angles increasing proportionally with the times, the value of d f, we shall have the value of d f', by diminishing in the first the angles ι, ι', ϖ, ϖ', θ and θ' by a right angle.

The quantities f and f' determine the position of the perihelion, and the excentricity of the orbit; in fact we learn from 531, that

$$\tan. I = \frac{f'}{f};$$

I being the longitude of the perihelion referred to the fixed plane. When this plane is that of the primitive orbit of μ, we have up to quantities of the order of the squares of the perturbing forces multiplied by the square of the respective inclinations of the orbits, $I = \varpi$, ϖ being the longitude of the perihelion upon the orbit; we shall therefore then have

$$\tan. \varpi = \frac{f'}{f};$$

which gives

$$\sin. \varpi = \frac{f'}{\sqrt{f^2 + f'^2}}; \ \cos. \varpi = \frac{f}{\sqrt{f^2 + f'^2}}.$$

By 531, we then get

$$m e = \sqrt{f^2 + f'^2 + f''^2}, \ f'' = \frac{f' c' - f c''}{c};$$

thus c' and c'' being in the preceding supposition of the order of the perturbing forces, f'' is of the same order, and neglecting the terms of the square of these forces, we have

$$m e = \sqrt{f^2 + f'^2}.$$

If we substitute for $\sqrt{f^2 + f'^2}$, its value m e, in the expressions of sin. ϖ, and of cos. ϖ, we shall have

$$m e \sin. \varpi = f'; \ m e \cos. \varpi = f;$$

these two equations will determine the excentricity and the position of the perihelion, and we thence easily obtain

$$m^2 . e d e = f d f + f' d f'; \ m^2 e^2 d \varpi = f d f' - f' d f.$$

Taking for the plane of x, y that of the orbit of μ; we have for the cases of the invariable ellipses,

$$\varrho = \frac{a(1-e^2)}{1+e\cos.(v-\varpi)};\ d\varrho = \frac{\varrho^2 dv.e.\sin.(v-\varpi)}{a(1-e^2)};$$

$$\varrho^2 dv = a^2 n dt \sqrt{1-e^2};$$

and by No. 530, these equations also subsist in the case of the variable ellipses; the expressions of d f and of d f' will thus become

$$df = -\frac{a n d t}{\sqrt{1-e^2}}.\{2\cos.v + \tfrac{3}{2}e\cos.\varpi + \tfrac{1}{2}e\cos.(2v-\varpi)\}.\left(\frac{dR}{dv}\right)$$

$$-a^2 n dt\sqrt{1-e^2}.\sin.v.\left(\frac{dR}{d\varrho}\right)$$

$$df' = -\frac{a n d t}{\sqrt{1-e^2}}.\{2\sin.v + \tfrac{3}{2}e\sin.\varpi + \tfrac{1}{2}e\sin.(2v-\varpi)\}\left(\frac{dR}{dv}\right)$$

$$+a^2 n dt\sqrt{1-e^2}.\cos._v\left(\frac{dR}{d\varrho}\right);$$

wherefore

$$e\,d\varpi = -\frac{a n d t}{m\sqrt{1-e^2}}\sin.(v-\varpi)\{2+e\cos.(v-\varpi)\}\left(\frac{dR}{dv}\right)$$

$$+\frac{a^2.n dt\sqrt{1-e^2}}{m}\cos.(v-\varpi).\left(\frac{dR}{d\varrho}\right)$$

$$de = -\frac{a n d t}{m\sqrt{1-e^2}}.\{2\cos.(v-\varpi)+e+e\cos.^2(v-\varpi)\}\left(\frac{dR}{dv}\right)$$

$$-\frac{a^2 n d t}{m}\sqrt{1-e^2}.\sin.(v-\varpi).\left(\frac{dR}{d\varrho}\right).$$

This expression of d e may be put into a more commodious form in some circumstances. For that purpose, we shall observe that

$$d\varrho\left(\frac{dR}{d\varrho}\right) = dR - dv\left(\frac{dR}{dv}\right)$$

substituting for ϱ and d ϱ their preceding values, we shall have

$$\varrho^2.dv.e.\sin.(v-\varpi)\left(\frac{dR}{d\varrho}\right) = a(1-e^2).dR - a(1-e^2)dv.\left(\frac{dR}{dv}\right);$$

but we have

$$\varrho^2 dv = a^2 n dt\sqrt{1-e^2};$$

$$dv = \frac{n dt\{1+e\cos.(v-\varpi)\}^2}{(1-e^2)^{\frac{3}{2}}};$$

wherefore

$$a^2 n dt\sqrt{1-e^2}.\sin.(v-\varpi).\left(\frac{dR}{d\varrho}\right)$$

$$= \frac{a(1-e^2)}{e}dR - \frac{a n d t}{e.\sqrt{1-e^2}}.\{1+e\cos.(v-\varpi)\}^2.\left(\frac{dR}{dv}\right);$$

the preceding expression of d e, will thus give

$$e\,d\,e = \frac{a\,n\,d\,t\,\sqrt{1-e^2}}{m}\cdot\left(\frac{d\,R}{d\,v}\right) - \frac{a\,(1-e^2)}{m}\,d\,R.$$

We can arrive very simply at this formula, in the following manner. We have by No. 531,

$$\frac{d\,c}{d\,t} = y\left(\frac{d\,R}{d\,x}\right) - x\left(\frac{d\,R}{d\,y}\right) = -\left(\frac{d\,R}{d\,v}\right);$$

but by the same No. $c = \sqrt{m\,a\,(1-e^2)}$ which gives

$$d\,c = \frac{d\,a\,\sqrt{m\,a\,(1-e^2)}}{2\,a} - \frac{e\,d\,e\,\sqrt{m\,a}}{\sqrt{1-e^2}};$$

therefore

$$e\,d\,e = \frac{a\,n\,d\,t\,\sqrt{1-e^2}}{m}\left(\frac{d\,R}{d\,v}\right) + a\,(1-e^2)\frac{d\,a}{2\,a^2}:$$

and then we have by No. 531

$$\frac{m\,d\,a}{2\,a^2} = -\,d\,R.$$

We thus obtain for e d e the same expression as before.

535. We have seen in 532, that if we neglect the squares of the perturbing forces, the variations of the principal axis and of the mean motion contain only periodic quantities, depending on the configuration of the bodies μ, μ', μ'', &c. This is not the case with respect to the variations of the excentricities and inclinations: their differential expressions contain terms independent of this configuration and which, if they were rigorously constant, would produce by integration, terms proportional to the time, which at length would render the orbits very excentric and greatly inclined to one another; thus the preceding approximations, founded upon the smallness of the excentricity and inclination of the orbits, would become insufficient and even faulty. But the terms apparently constant, which enter the differential expressions of the excentricities and inclinations, are functions of the elements of the orbits; so that they vary with an extreme slowness, because of the changes they there introduce. We conceive there ought to result in these elements, considerable inequalities independent of the mutual configuration of the bodies of the system, and whose periods depend upon the ratios of the masses μ, μ', &c. to the mass M. These inequalities are those which we have named *secular* inequalities, and which have been considered in (520). To determine them by this method we resume the value of d f of the preceding No.

$$d\,f = -\frac{a\,n\,d\,t}{\sqrt{1-e^2}}\{2\cos v + \tfrac{3}{2}\,e\cos.\varpi + \tfrac{1}{2}\,e\cos.(2\,v-\varpi)\}\left(\frac{d\,R}{d\,v}\right)$$

$$- a^2 n d t \sqrt{1 - e^2}. \sin v. \left(\frac{d R}{d \varrho}\right).$$

We shall neglect in the developement of this equation the square and products of the excentricities and inclinations of the orbits; and amongst the terms depending upon the excentricities and inclinations, we shall retain those only which are constant: we shall then suppose, as in No. 515.

$$\varrho = a (1 + u_{,}); \quad \varrho' = a' (1 + u_{,}');$$
$$v = n t + \varepsilon + v_{,}; \quad v' = n' t + \varepsilon' + v_{,}'.$$

Again, if we substitute for R, its value found in 515; if we next consider that by the same No. we have,

$$\left(\frac{d R}{d \varrho}\right) = \frac{a}{\varrho} \left(\frac{d R}{d a}\right) = (1 - u_{,}) \left(\frac{d R}{d a}\right),$$

and lastly if we substitute for $u_{,}$, $u_{,}'$, $v_{,}$, $v_{,}'$ their values $- e \cos. (n t + \varepsilon - \varpi)$, $- e' \cos. (n' t + \varepsilon' - \varpi')$, $2 e \sin. (n t + \varepsilon - \varpi)$, $2 e' \sin. (n' t + \varepsilon' - \varpi')$ given in No. 484, &c. by retaining only the constant terms of those which depend upon the first power of the excentricities of the orbits, and neglecting the squares of the excentricities and inclinations, we shall find that

$$d f = \frac{a \mu' n d t}{2} . e . \sin. \varpi \left\{ a \left(\frac{d A^{(0)}}{d a}\right) + \tfrac{1}{2} a^2 \left(\frac{d^2 A^{(0)}}{d a^2}\right) \right\}$$
$$+ a \mu' n d t . e' . \sin. \varpi' \left\{ A^{(1)} + \tfrac{1}{2} \left(\frac{d A^{(1)}}{d a}\right) + \tfrac{1}{2} a' \left(\frac{d A^{(1)}}{d a'}\right) + \tfrac{1}{4} a a' \left(\frac{d^2 A^{(1)}}{d a d a'}\right) \right\}$$
$$- a \mu' n d t . \Sigma \left\{ i A^{(i)} + \tfrac{1}{2} a \left(\frac{d A^{(i)}}{d a}\right) \right\} \sin. \{ i (n' t - n t + \varepsilon' - \varepsilon) + n t + \varepsilon \};$$

the integral sign belonging as in the value of R of 515, to all the whole positive and negative values of i, including also the value of $i = 0$.

We shall have by the preceding No. the value of $d f'$, by diminishing in that of $d f$ the angles ε, ε', ϖ, ϖ' by a right angle; whence we get

$$d f' = - \frac{a \mu' n d t}{2} . e . \cos. \varpi \left\{ a \left(\frac{d A^{(0)}}{d a}\right) + \tfrac{1}{2} a^2 \left(\frac{d^2 A^{(0)}}{d a^2}\right) \right\}$$
$$- a \mu' n d t . e' . \cos. \varpi' \left\{ A^{(1)} + \tfrac{1}{2} a \left(\frac{d A^{(1)}}{d a}\right) + \tfrac{1}{2} a' \left(\frac{d A^{(1)}}{d a'}\right) + \tfrac{1}{4} a a' \left(\frac{d^2 A^{(1)}}{d a d a'}\right) \right\}$$
$$+ a \mu' n d t . \Sigma \left\{ i A^{(i)} + \tfrac{1}{2} a \left(\frac{d A^{(i)}}{d a}\right) \right\} \cos. \{ i (n' t - n t + \varepsilon' - \varepsilon) + n t + \varepsilon \}.$$

Let X, for the greater brevity, denote that part of $d f$, which is contained under the sign Σ, and Y the corresponding part of $d f'$. Make also, as in No. 522,

$$(0, 1) = - \frac{\mu' n}{2} \left\{ a^2 . \left(\frac{d A^{(0)}}{d a}\right) + \tfrac{1}{2} a^3 \left(\frac{d^2 A^{(0)}}{d a^2}\right) \right\};$$

$$\boxed{0, 1} = \frac{\mu' n}{2}\left\{a A^{(1)} - a^2\left(\frac{d A^{(1)}}{d a}\right) - \tfrac{1}{2} a^3\left(\frac{d^2 A^{(1)}}{d a^2}\right)\right\};$$

then observe that the coefficient of e′ d t sin. ϖ', in the expression of d f, is reducible to $\boxed{0, 1}$ when we substitute for the partial differences in a′, their values in partial differences relative to a; finally suppose, as in 517, that

$$e \sin. \varpi = h;\ e' \sin. \varpi' = h'$$
$$e \cos. \varpi = l;\ e' \cos. \varpi' = l'$$

which gives by the preceding No. f = m l, f′ = m h or simply f = l, f′ = h, by taking M for the mass-unit, and neglecting μ with regard to M; we shall obtain

$$\frac{d h}{d t} = (0, 1).\, l - \boxed{0, 1}.\, l' + a \mu' n Y;$$

$$\frac{d l}{d t} = -(0, 1).\, h + \boxed{0, 1}.\, h' - a \mu' n.\, X.$$

Hence, it is easy to conclude that if we name (Y) the sum of the terms analogous to a μ' n Y, due to the motion of each of the bodies μ', μ'', &c. upon μ; that if we name in like manner (X) the sum of the terms analogous to — a μ' n X due to the same actions, and finally if we mark successively with one dash, two dashes, &c. what the quantities (X), (Y), h, and l become relatively to the bodies μ', μ'', &c.; we shall have the following differential equations;

$$\frac{d h}{d t} = \{(0, 1) + (0, 2) + \&c.\}\, l - \boxed{0, 1}\, l' - \boxed{0, 2}\, l'' - \&c. + (Y);$$

$$\frac{d l}{d t} = -\{(0, 1) + (0, 2) + \&c.\}\, h + \boxed{0, 1}\, h' + \boxed{0, 2}\, h'' + \&c + (X);$$

$$\frac{d h'}{d t} = \{(1, 0) + (1, 2) + \&c.\}\, l' - \boxed{1, 0}\, l - \boxed{1, 2}\, l'' - \&c. + (Y')\cdot$$

$$\frac{d l'}{d t} = -\{(1, 0) + (1, 2) + \&c.\}\, h' + \boxed{1, 0}\, h + \boxed{1, 2}\, h'' + \&c. + (X')$$

&c.

To integrate these equations, we shall observe that each of the quantities h, l, h′, l′, &c. consists of two parts; the one depending upon the mutual configuration of the bodies μ, μ', &c.; the other independent of this configuration, and which contains the secular variations of these quantities. We shall obtain the first part by considering that if we regard hat alone, h, l, h′, l′, &c. are of the order of the perturbing masses, and consequently, (0, 1). h, (0, 1). l, &c. are of the order of the squares of these masses. By neglecting therefore quantities of this order, we shall have

$$\frac{d\,h}{d\,t} = (Y); \quad \frac{d\,l}{d\,t} = (X);$$

$$\frac{d\,h'}{d\,t} = (Y'); \quad \frac{d\,l'}{d\,t} = (X');$$

wherefore,

$$h = \int (Y)\,d\,t; \quad l = \int (X)\,d\,t; \quad h' = \int (Y')\,d\,t; \ \&c.$$

If we take these integrals, not considering the variability of the elements of the orbits and name Q what $\int (Y)\,d\,t$ becomes; by calling $\delta\,Q$ the variation of Q due to that of the elements we shall have

$$\int (Y)\,d\,t = Q - \int \delta\,Q;$$

but Q being of the order of the perturbing masses, and the variations of the elements of the orbits being of the same order, $\delta\,Q$ is of the order of the squares of the masses; thus, neglecting quantities of this order, we shall have

$$\int (Y)\,d\,t = Q.$$

We may, therefore, take the integrals $\int (Y)\,d\,t$, $\int (X)\,d\,t$, $\int (Y')\,d\,t$, &c. by supposing the elements of the orbits constant, and afterwards consider the elements variable in the integrals; we shall after a very simple method, obtain the periodic portions of the expressions of h, l, h', &c.

To get those parts of the expressions which contain the secular inequalities, we observe that they are given by the integration of the preceding differential equations deprived of their last terms, (Y), (X), &c.; for it is clear that the substitution of the periodic parts of h, l, h', &c. will cause these terms to disappear. But in taking away from the equations their last terms, they will become the same as those of (A) of No. 522, which we have already considered at great length.

536. We have observed in No. 532, that if the mean motions n t and n' t of the two bodies μ and μ', are very nearly in the ratio of i' to i so that i' n' — i n may be a very small quantity; there may result in the mean motions of these bodies very sensible inequalities. This relation of the mean motions may also produce sensible variations in the excentricities of the orbits, and in the positions of their perihelions. To determine them, we shall resume the equation found in 534,

$$e\,d\,e = \frac{a\,n\,d\,t.\sqrt{1 - e^2}}{m}.\left(\frac{d\,R}{d\,v}\right) - \frac{a\,(1 - e^2)}{m}\,d\,R.$$

It results from what has been asserted in 515, that if we take for the fixed plane, that of the orbit of μ, at a given epoch, which allows us to neglect in R the inclination φ of the orbit of μ to this plane; all the terms of the expression of R depending upon the angle $i' n' t - i n t$, will be comprised in the following form,

$$\mu' k \cos. (i' n' t - i n t + i' \epsilon' - i \epsilon - g \varpi - g' \varpi' - g'' \theta'),$$

i, i', g, g', g'' being whole numbers and such that we have $0 = i' - i - g - g' - g''$. The coefficient k has the factor $e^{g}. e'^{g'} (\tan. \frac{1}{2} \varphi')^{g''}$; g, g', g'' being taken positively in the exponents: moreover, if we suppose i and i' positive, and i' greater than i; we have seen in No. 515, that the terms of R which depend upon the angle $i' n' t - i n t$ are of the order $i' - i$, or of a superior order of two, of four, &c. units; taking into account therefore only terms of the order $i' - i$, k will be of the form $e^{g}. e'^{g'} (\tan. \frac{1}{2} \varphi')^{g''}. Q$, Q being a function independent of the excentricities and the inclination of the orbits. The numbers g, g', g'' comprehended under the symbol *cos.*, are then positive; for if one of them, g for instance, be negative and equal to $- f$, k will be of the order $f + g' + g''$; but the equation $0 = i' - i - g - g' - g''$ gives $f + g' + g'' = i' - i + 2 f$; thus k will be of an order superior to $i' - i$, which is contrary to the supposition. Hence by No. 515, we have $\left(\frac{d R}{d v}\right) = \left(\frac{d R}{d \epsilon}\right)$ provided that in this last partial difference, we make $\epsilon - \varpi$ constant; the term of $\left(\frac{d R}{d v}\right)$ corresponding to the preceding term of R, is therefore

$$\mu' (i + g) k \sin. (i' n' t - i n t + i' \epsilon' - i \epsilon - g \varpi - g' \varpi' - g'' \theta').$$

The corresponding term of d R is

$$\mu' \beta i n k d t \sin. (i' n' t - i n t + i' \epsilon' - i \epsilon - g \varpi - g' \varpi' - g'' \theta').$$

Hence only regarding these terms and neglecting e^2 in comparison with unity, the preceding expression of e d e, will give

$$d e = \frac{\mu' a n d t}{m} . \frac{g k}{e} \sin. (i' n' t - i n t + i' \epsilon' - i \epsilon - g \varpi - g' \varpi' - g'' \theta'),$$

but we have

$$\frac{g k}{e} = g e^{g-1}. e'^{g'}. (\tan. \tfrac{1}{2} \varphi')^{g''}. Q = \left(\frac{d k}{d e}\right);$$

integrating therefore we get

$$e = - \frac{\mu' a n}{m (i' n - i n)} \left(\frac{d k}{d e}\right) \cos. (i' n' t - i n t + i' \epsilon' - i \epsilon - g \varpi - g' \varpi' - g'' \theta').$$

The sum of all the terms of R, however, which depend on the angle $i' n' t - i n t$, being represented by the following quantity

$$\mu'. P \sin. (i' n' t - i n t + i' \epsilon' - i \epsilon) + \mu' P' \cos. (i' n' t - i n t + i' \epsilon' - i \epsilon)$$

the corresponding part of e will be

$$\frac{-\mu' a n}{m(i' n' - i n)} \left\{ \left(\frac{d P}{d e}\right) \sin.(i' n' t - i n t + i' \epsilon' - i \epsilon) + \left(\frac{d P'}{d e}\right) \cos.(i' n' t - i n t + i' \epsilon' - i \epsilon) \right\}.$$

This inequality may become very sensible, if the coefficient $i' n' - i n$ is very small, for it actually takes place in the theory of Jupiter and Saturn. In fact, it has for a divisor only the first power of $i' n' - i n$, whilst the corresponding inequality of the mean motion, has for a divisor the second power of this quantity, as we see in No. 532; but $\left(\frac{d P}{d e}\right)$ and $\left(\frac{d P'}{d e}\right)$ being of an order inferior to P and P', the inequality of the excentricity may be considerable, and even surpass that of the mean motion, if the excentricities e and e' are very small; this will be exemplified in the theory of Jupiter's satellites.

Let us now determine this corresponding inequality of the motion of the perihelion. For that purpose, resume the two equations

$$e\,d\,e = \frac{f\,d\,f + f'\,d\,f'}{m^2}, \quad e^2\,d\,\varpi = \frac{f\,d\,f' - f'\,d\,f}{m^2};$$

which we found in No. 534. These equations give

$$d\,f = m\,d\,e\cos.\varpi - m\,e\,d\,\varpi.\sin.\varpi;$$

thus with regard only to the angle

$$i'n't - i\,n\,t + i'\epsilon' - i\,\epsilon - g\,\varpi - g\,\varpi' - g''\theta',$$

we shall have

$$d\,f = \mu'.\,a\,n\,d\,t\left(\frac{d\,k}{d\,e}\right)\cos.\varpi\sin.(i'n't - i\,n\,t + i'\epsilon' - i\,\epsilon - g\,\varpi - g'\varpi' - g''\theta')$$

$$- m\,e\,d\,\varpi.\sin.\varpi.$$

Representing by

$$-\mu'.\,a\,n\,d\,t\left\{\left(\frac{d\,k}{d\,e}\right) + k'\right\}\cos.(i'n't - i\,n\,t + i'\epsilon' - i\,\epsilon - g\,\varpi - g'\varpi' - g''\theta'),$$

the part of $m\,e\,d\,\varpi$, which depends upon the same angle, we shall have

$$d\,f = \mu'.\,a\,n\,d\,t\left\{\left(\frac{d\,k}{d\,e}\right) + \tfrac{1}{2}k'\right\}\sin.(i'n't - i\,n\,t + i'\epsilon' - i\,\epsilon - (g-1)\varpi - g'\varpi' - g''\theta')$$

$$-\frac{\mu'.\,a\,n\,d\,t}{2}\,k'\sin.(i'n't - i\,n\,t + i'\epsilon' - i\,\epsilon - (g+1)\,\varpi - g'\varpi' - g''\theta').$$

It is easy to see by the last of the expressions of $d\,f$, given in the No. 534, that the coefficient of this last sine has the factor $e^{g+1}.\,e'^{g'}(\tan.\frac{1}{2}\varphi)^{g''}$; k' is therefore of an order superior to that of $\left(\frac{d\,k}{d\,e}\right)$ by two units; thus, in neglecting it in comparison to $\left(\frac{d\,k}{d\,e}\right)$, we shall have

$$-\frac{\mu'.\,a\,n\,d\,t}{m}.\left(\frac{d\,k}{d\,e}\right)\cos.(i'n't - i\,n\,t + i'\epsilon' - i\,\epsilon - g\,\varpi - g'\varpi' - g''\theta')$$

for the term of $e\,d\,\varpi$, which corresponds to the term

$$\mu'\,k\cos.(i'n't - i\,n\,t + i'\epsilon' - i\,\epsilon - g\,\varpi - g'\varpi' - g''\theta'),$$

of the expression of R. Hence it follows that the part of ϖ, which corresponds to the part of R expressed by

μ' P sin. (i' n' t — i n t + i' ε' — i ε) + μ' P' cos. (i' n' t — i n t + i' ε' — i ε),

is equal to

$$\frac{\mu'.\,a\,n}{m(i'n'-in)e}.\left\{\left(\frac{d\,P}{d\,e}\right)\cos.(i'n't-int+i'\varepsilon'-i\varepsilon)-\left(\frac{d\,P'}{d\,e}\right)\sin.(i'n't-int+i'\varepsilon'-i\varepsilon)\right\};$$

we shall therefore, thus, after a very simple manner, find the variations of the excentricity and of the perihelion, depending upon the angle i' n' t — i n t + i' e' — i e. They are connected with the variation ζ of the corresponding mean motion, in such a way that the variation of the excentricity is

$$\frac{1}{3\,i\,n}.\left(\frac{d^2\,\zeta}{d\,e.\,d\,t}\right);$$

and the variation of the longitude of the perihelion is

$$\frac{i'\,n'-i\,n}{3\,i\,n\,e}.\left(\frac{d\,\zeta}{d\,e}\right).$$

The corresponding variation of the excentricity of the orbit of μ', due to the action of μ, will be

$$-\frac{1}{3\,i'\,n'.\,e'}.\left(\frac{d^2\,\zeta'}{d\,e'.\,d\,t}\right),$$

and the variation of the longitude of its perihelion, will be,

$$-\frac{i'\,n'-i\,n}{3\,i'\,n'\,e'}\left(\frac{d\,\zeta'}{d\,e'}\right),$$

and since by No. 532, $\zeta' = -\frac{\mu\,\sqrt{a}}{\mu'\,\sqrt{a'}}.\,\zeta$, the variations will be

$$\frac{\mu\,\sqrt{a}}{3\,i'.\,n'.\,\mu'\,\sqrt{a'}}\left(\frac{d^2\,\zeta}{d\,e'.\,d\,t}\right),\ \text{and}\ \frac{(i'\,n'-i\,n)\,\mu\,\sqrt{a}}{3\,i'\,n'.\,e'\,\mu'\,\sqrt{a'}}.\frac{d\,\zeta}{d\,e'}.$$

When the quantity i' n' — i n is very small, the inequality depending upon the angle i' n' t — i n t, produces a sensible one in the expression of the mean motion, amongst the terms depending on the squares of the perturbing masses; we have given the analysis of this in No. 532. This same inequality produces in the expression of d e and of d ϖ, terms of the order of the squares of the masses, and which, being only functions of the elements of the orbits, have a sensible influence upon the secular variations of these elements. Let us consider, in fact, the expression of d e, depending on the angle i' n' t — i n t.

By what precedes, we have

$$d\,e = -\frac{\mu'.\,a\,n.\,d\,t}{m}\left\{\left(\frac{d\,P}{d\,e}\right)\cos.\,(i'\,n'\,t-i\,n\,t+i'\,\varepsilon'-i\,\varepsilon)\right.$$

$$\left.-\left(\frac{d\,P'}{d\,e}\right).\sin.\,(i'\,n'\,t-i\,n\,t+i'\,\varepsilon'-i\,\varepsilon)\right\}.$$

By No. 532 the mean motion n t, ought to be augmented by

$$\frac{3\mu' a n^2 i}{(i'n'-in)^2.m}.\left\{P\cos.(i'n't-int+i'\varepsilon'-i\varepsilon)-P'\sin.(i'n't-int+i'\varepsilon'-i\varepsilon)\right\}$$

and the mean motion n′ t, ought to be augmented by

$$-\frac{3\mu' a n^2.i}{(i'n'-in)^2.m}.\frac{\mu\sqrt{a}}{\mu'\sqrt{a'}}.\{P\cos.(i'n't-int+i'\varepsilon'-i\varepsilon)-$$
$$P'\sin.(i'n't-int+i'\varepsilon'-i\varepsilon)\}.$$

In virtue of these augments, the value of d e will be augmented by the function

$$-\frac{3\mu' a^2.in^3.dt}{2m^2\sqrt{a'}.(i'n'-in)^2}\{i.\mu'\sqrt{a'}+i'\mu\sqrt{a}\}\left\{P.\left(\frac{dP}{de}\right)+P'\left(\frac{dP}{de}\right)\right\};$$

and the value of d ϖ will be augmented by the function

$$\frac{3\mu' a^2.in^3.dt}{2m^2\sqrt{a'}(i'n'-in)^2.e}.\{i\mu'\sqrt{a'}+i'\mu\sqrt{a}\}.\left\{P.\left(\frac{dP}{de}\right)+P'\left(\frac{dP'}{de}\right)\right\}.$$

In like manner we find that the value of d e′ will be augmented by the function

$$-\frac{3\mu a^2.\sqrt{a}.in^3.dt}{2m^2.a'.(i'n'-in)^2}.\{i.\mu'.\sqrt{a'}+i'.\mu.\sqrt{a}\}\left\{P.\left(\frac{dP'}{de'}\right)-P'.\left(\frac{dP}{de}\right)\right\};$$

and that the value of d ϖ' will be augmented by the function

$$\frac{3\mu a^2.\sqrt{a}.in^3.dt}{2m^2a'.(i'n'-in)^2.e'}.\{i\mu'\sqrt{a'}+i'\mu\sqrt{a}\}\left\{P.\left(\frac{dP}{de'}\right)+P'\left(\frac{dP'}{de'}\right)\right\}.$$

These different terms are sensible in the theory of Jupiter and Saturn, and in that of Jupiter's satellites. The variations of e, e′, ϖ, ϖ' relative to the angle $i'n't-int$ may also introduce some constant terms of the order of the square of the perturbing masses in the differentials d e, d e′, dϖ, and dϖ', and depending on the variations of e, e′, ϖ, ϖ' relative to the same angle. This may easily be discussed by the preceding analysis. Finally it will be easy, by our analysis, to determine the terms of the expressions of e, ϖ, e′, ϖ' which depending upon the angle $i'n't-int+i'\varepsilon'-i\varepsilon$ have not $i'n'-in$ for a divisor, and those which, depending on the same angle and the double of this angle, are of the order of the square of the perturbing forces. These different terms are sufficiently considerable in the theory of Jupiter and Saturn, for us to notice them: we shall develope them to the extent they merit when we come to that theory.

537. Let us determine the variations of the nodes and inclinations of the orbits, and for that purpose resume the equations of 531,

$$dc=dt\left\{y.\left(\frac{dR}{dx}\right)-x\left(\frac{dR}{dy}\right)\right\};$$

$$d\,c' = d\,t \left\{ z \left(\frac{d\,R}{d\,x}\right) - x \left(\frac{d\,R}{d\,z}\right) \right\};$$

$$d\,c'' = d\,t \left\{ z \left(\frac{d\,R}{d\,y}\right) - y \left(\frac{d\,R}{d\,z}\right) \right\}.$$

If we only notice the action of μ', the value of R of No. 513, gives

$$y \left(\frac{d\,R}{d\,x}\right) - x \left(\frac{d\,R}{d\,y}\right) = \mu' \,(x'\,y - x\,y') \times$$

$$\left\{ \frac{1}{(x'^2 + y'^2 + z'^2)^{\frac{3}{2}}} - \frac{1}{\{(x' - x)^2 + (y' - y)^2 + (z' - z)^2\}^{\frac{3}{2}}} \right\};$$

$$z \left(\frac{d\,R}{d\,x}\right) - x \left(\frac{d\,R}{d\,z}\right) = \mu' \,(x'\,z - x\,z') \times$$

$$\left\{ \frac{1}{(x'^2 + y'^2 + z'^2)^{\frac{3}{2}}} - \frac{1}{\{(x' - x)^2 + (y' - y)^2 + (z' - z)^2\}^{\frac{3}{2}}} \right\};$$

$$z \left(\frac{d\,R}{d\,y}\right) - y \left(\frac{d\,R}{d\,z}\right) = \mu' \,(y'\,z - y\,z') \times$$

$$\left\{ \frac{1}{(x'^2 + y'^2 + z'^2)^2} - \frac{1}{\{(x' - x)^2 + (y' - y)^2 + (z' - z)^2\}^{\frac{3}{2}}} \right\}.$$

Let however,

$$\frac{c''}{c} = p; \quad \frac{c'}{c} = q;$$

the two variables p and q will determine, by No. **531**, the tangent of the inclination φ of the orbit of μ, and the longitude θ of its node by means of the equations

$$\tan.\,\varphi = \sqrt{p^2 + q^2}; \quad \tan.\,\theta = \frac{p}{q}.$$

Call p', q', p'', q'', &c. what p and q become relatively to the bodies μ', μ'', &c.: we shall have by **531**,

$$z = q\,y - p\,x; \quad z' = q'\,y' - p'\,x', \text{ \&c.}$$

The preceding value of p differentiated gives

$$\frac{d\,p}{d\,t} = \frac{1}{c} \cdot \frac{d\,c'' - p\,d\,c}{d\,t};$$

substituting for d c, and d c'' their values we get

$$\frac{d\,p}{d\,t} = \frac{\mu'}{c} \,\{(q - q')\,y\,y' + (p' - p)\,x'\,y\} \times$$

$$\left\{ \frac{1}{(x'^2 + y'^2 + z'^2)^{\frac{3}{2}}} - \frac{1}{\{(x' - x)^2 + (y' - y)^2 + (z' - z)^2\}^{\frac{3}{2}}} \right\}.$$

In like manner we find

$$\frac{d\,q}{d\,t} = \frac{\mu'}{c} \,\{(p' - p)\,x\,x' + (q - q')\,x\,y'\} \times$$

$$\left\{\frac{1}{(x'^2 + y'^2 + z'^2)^{\frac{3}{2}}} - \frac{1}{\{(x'-x)^2+(y'-y)^2+(z'-z)^2\}^{\frac{3}{2}}}\right\}.$$

If we substitute for x, y, x', y' their values ϱ cos. v, ϱ sin. v, ϱ' cos. v', ϱ' sin. v', we shall have

$$(q-q')\,y\,y' + (p'-p)\,x'\,y = \frac{q'-q}{2}.\varrho\,\varrho'.\{\cos.(v'+v) - \cos.(v'-v)\}$$

$$+ \frac{p'-p}{2}.\varrho\,\varrho'.\{\sin.(v'+v) - \sin.(v'-v)\};$$

$$(p'-p)\,x\,x' + (q-q')\,x\,y' = \frac{p'-p}{2}.\varrho\,\varrho'.\{\cos.(v'+v) + \cos.(v'-v)\}$$

$$+ \frac{q-q'}{2}.\varrho\,\varrho'.\{\sin.(v'+v) + \sin.(v'-v)\}.$$

Neglecting the excentricities and inclinations of the orbits, we have

$$\varrho = a;\ v = n\,t + \varepsilon;\ \varrho' = a';\ v' = n'\,t + \varepsilon';$$

which give

$$\frac{1}{(x'^2 + y'^2 + z'^2)^{\frac{3}{2}}} - \frac{1}{\{(x'-x)^2 + (y'-y)^2 + (z'-z)^2\}^{\frac{3}{2}}} = \frac{1}{a'^3} - \frac{1}{\{a^2 - 2\,a\,a'\cos.(n'\,t - n\,t + \varepsilon' - \varepsilon) + a'^2\}^{\frac{3}{2}}};$$

moreover by No. 516,

$$\frac{1}{\{a^2-2\,a\,a'\cos.(n'\,t-n\,t+\varepsilon'-\varepsilon)+a'^2\}^{\frac{3}{2}}} = \tfrac{1}{2}\,\Sigma.\,B^{(i)}.\cos.i\,(n'\,t-n\,t+\varepsilon'-\varepsilon)$$

the integral sign Σ belonging to all whole positive and negative values of i, including the value i = 0; we shall thus have, neglecting terms of the order of the squares and products of the excentricities and inclinations of the orbits,

$$\frac{d\,p}{d\,t} = \frac{q'-q}{2\,c}.\frac{\mu'\,a}{a'^2}.\{\cos.(n'\,t+n\,t+\varepsilon'+\varepsilon) - \cos.(n'\,t-n\,t+\varepsilon'-\varepsilon)\}$$

$$+ \frac{p'-p}{2\,c}.\frac{\mu'\,a}{a'^2}.\{\sin.(n'\,t+n\,t+\varepsilon'+\varepsilon) - \sin.(n'\,t-n\,t+\varepsilon'-\varepsilon)\}$$

$$+ \frac{q'-q}{4\,c}.\mu'.a\,a'.\Sigma.B^{(i)}\{\cos.[(i+1)(n'\,t-n\,t+\varepsilon'-\varepsilon)]$$

$$- \cos.[(i+1)(n'\,t-n\,t+\varepsilon'-\varepsilon) + 2\,n\,t+2\,\varepsilon]\}$$

$$+ \frac{p'-p}{4\,c}.\mu'.a\,a'.\Sigma.B^{(i)}\{\sin.[(i+1)(n'\,t-n\,t+\varepsilon'-\varepsilon)]$$

$$- \sin.[(i+1)(n'\,t-n\,t+\varepsilon'-\varepsilon) + 2\,n\,t+2\,\varepsilon]\}.$$

$$\frac{d\,q}{d\,t} = \frac{p'-p}{2\,c}.\frac{\mu'\,a}{a'^2}.\{\cos.(n'\,t+n\,t+\varepsilon'+\varepsilon) + \cos.(n'\,t-n\,t+\varepsilon'-\varepsilon)\}$$

$$+ \frac{q'-q}{2c} \cdot \frac{\mu' a}{a'^2} \cdot \{\sin. (n' t + n t + \epsilon' + \epsilon) + \sin. (n' t - n t + \epsilon' - \epsilon)\}$$

$$+ \frac{p-p'}{4c} \cdot \mu'. a a'. \Sigma. B^{(i)}. \{\cos. [(i+1)(n' t - n t + \epsilon' - \epsilon)]$$

$$+ \cos. [(i+1)(n' t - n t + \epsilon' - \epsilon) + 2 n t + 2 \epsilon]\}$$

$$+ \frac{q'-q}{4c} \cdot \mu'. a a'. \Sigma. B^{(i)}. \{\sin. [(i+1)(n' t - n t + \epsilon' - \epsilon)]$$

$$+ \sin. [(i+1)(n' t - n t + \epsilon' - \epsilon) + 2 n t + 2 \epsilon]\}.$$

The value $i = -1$ gives in the expression of $\frac{dp}{dt}$, the constant quantity $\frac{q'-q}{4c} \cdot \mu'. a a' B^{(-1)}$; all the other terms of the expression of $\frac{dp}{dt}$ are periodic: denoting their sum by P, and observing that $B^{(-1)} = B^{(1)}$ by 516, we shall have

$$\frac{dp}{dt} = \frac{q'-q}{4c} \cdot \mu'. a a'. B^{(1)} + P.$$

By the same process we shall find, that if we denote by Q the sum of all the periodic terms of the expression of $\frac{dq}{dt}$, we shall have

$$\frac{dq}{dt} = \frac{p-p'}{4c} \cdot \mu'. a a'. B^{(1)} + Q.$$

If we neglect the squares of the excentricities and inclinations of the orbits, by 531, we have $c = \sqrt{m a}$, and then supposing $m = 1$, we have $n^2 a^3 = 1$ which gives $c = \frac{1}{a n}$; the quantity $\frac{\mu'. a a'. B^{(1)}}{4c}$ thus becomes $\frac{\mu'. a^2 a'. n B^{(1)}}{4}$ which by 526, is equal to (0, 1); hence we get

$$\frac{dp}{dt} = (0, 1). (q' - q) + P;$$

$$\frac{dq}{dt} = (0, 1). (p - p') + Q.$$

Hence it follows that, if we denote by (P) and (Q) the sum of all the functions P and Q relative to the action of the different bodies μ', μ'', &c. upon μ; if in like manner we denote by (P′), (Q′), (P″), (Q″), &c. what (P) and (Q) become when we change successively the quantities relative to μ into those which are relative to μ', μ'', &c. and reciprocally; we shall have for determining the variables p, q, p′, q′, p″, q″, &c. the following system of differential equations,

$$\frac{dp}{dt} = -\{(0, 1) + (0, 2) + \&c.\} q + (0, 1). q' + (0, 2) q'' + \&c. + (P);$$

$$\frac{dq}{dt} = \{(0,1) + (0,2) + \&c.\}\, p - (0,1)\, p' - (0,2)\, p'' - \&c. + (Q);$$

$$\frac{dp'}{dt} = -\{(1,0) + (1,2) + \&c.\}\, q' + (1,0)\, q + (1,2)\, q'' + \&c. + (P');$$

$$\frac{dq'}{dt} = \{(1,0) + (1,2) + \&c.\}.\, p' - (1,0)\, p - (1,2)\, p'' - \&c. + (Q');$$

&c.

The analysis of 535, gives for the periodic parts of p, q, p′, q′, &c.

$$p = \int (P).\, dt; \quad q = \int (Q).\, dt;$$
$$p' = \int (P').\, dt; \quad q' = \int (Q').\, dt;$$
&c.

We shall then have the secular parts of the same quantities, by integrating the preceding differential equations deprived of their last terms (P), (Q), (P′), &c.; and then we shall again hit upon the equations (C) of No. 526, which have been sufficiently treated of already to render it unnecessary again to discuss them.

538. Let us resume the equations of No. 531,

$$\tan.\varphi = \frac{\sqrt{c'^2 + c''^2}}{c}; \quad \tan.\theta = \frac{c''}{c}$$

whence result these

$$\frac{c'}{c} = \tan.\varphi \cos.\theta; \quad \frac{c''}{c} = \tan.\varphi \sin.\theta.$$

Differentiating, we shall have

$$d\tan.\varphi = \frac{1}{c}\{dc' \cos.\theta + dc'' \sin.\theta - dc \tan.\varphi\}$$

$$d\theta \tan.\varphi = \frac{1}{c}\{dc'' \cos.\theta - dc' \sin.\theta\}.$$

If we substitute in these equations for $\frac{dc}{dt}$, $\frac{dc'}{dt}$, $\frac{dc''}{dt}$, their values $y\left(\frac{dR}{dx}\right) - x\left(\frac{dR}{dy}\right)$, $z\left(\frac{dR}{dx}\right) - x\left(\frac{dR}{dz}\right)$, $z\left(\frac{dR}{dy}\right) - y\left(\frac{dR}{dz}\right)$, and for these last quantities their values given in 534; if moreover we observe that $s = \tan.\varphi \sin.(v - \theta)$, we shall have

$$d.\tan.\varphi = \frac{dt \tan.\varphi \cos.(v-\theta)}{c}.\left\{\varrho\left(\frac{dR}{d\varrho}\right)\sin.(v-\theta) + \left(\frac{dR}{dv}\right)\cos.(v-\theta)\right\}$$
$$-\frac{(1+s^2)\, dt}{c}\cos.(v-\theta).\left(\frac{dR}{ds}\right);$$

$$d\theta.\tan.\varphi = \frac{dt \tan.\varphi \sin.(v-\theta)}{c}\left\{\varrho.\left(\frac{dR}{d\varrho}\right)\sin.(v-\theta) + \left(\frac{dR}{dv}\right)\cos.(v\theta-)\right\}$$
$$-\frac{(1+s^2)\, dt}{c}\sin.(v-\theta)\left(\frac{dR}{ds}\right).$$

These two differential equations will determine directly the inclination of the orbit and the motion of the nodes.

They give

$$\sin.(v - \theta)\, d \tan.\varphi - d\theta \cos.(v - \theta) \tan.\varphi = 0;$$

an equation which may be deduced from this

$$s = \tan.\varphi \sin.(v - \theta);$$

in fact, this last equation being finite, we may (530) differentiate it whether we consider φ and θ constant or variable; so that its differential, taken by only making φ and θ vary, is nothing; whence results the preceding differential equation.

Suppose, however, that the fixed plane is inclined extremely little to the orbit of μ, so that we may neglect the squares of s and tan. φ, we shall have

$$d.\tan.\varphi = -\frac{dt}{c}\cos.(v - \theta).\left(\frac{dR}{ds}\right);$$

$$d\theta \tan.\varphi = -\frac{dt}{c}\sin.(v - \theta)\left(\frac{dR}{ds}\right);$$

by making therefore as before

$$p = \tan.\varphi \sin.\theta; \quad q = \tan.\varphi \cos.\theta;$$

we shall have, instead of the preceding differential equations, the following ones,

$$dq = -\frac{dt}{c}\cos.v.\left(\frac{dR}{ds}\right);$$

$$dp = -\frac{dt}{c}\sin.v.\left(\frac{dR}{ds}\right);$$

But we have also

$$s = q\sin.v - p\cos.v$$

which gives

$$\left(\frac{dR}{ds}\right) = \frac{1}{\sin.v}.\left(\frac{dR}{dq}\right), \quad \left(\frac{dR}{ds}\right) = -\frac{1}{\cos.v}.\left(\frac{dR}{dp}\right),$$

wherefore

$$dq = \frac{dt}{c}\left(\frac{dR}{dp}\right);$$

$$dp = \frac{dt}{c}\left(\frac{dR}{dq}\right).$$

We have seen in 515 that the function R is independent of the position of the fixed plane of x, y; supposing, therefore, all the angles of that function referred to the orbit of μ, it is evident that R will be a function of these angles and the respective inclination of two orbits, an inclination we denote by $\varphi_{,}'$. Let $\theta_{,}'$ be the longitude of the node of the orbit of μ' upon the orbit of μ; and supposing that

$$\mu' k (\tan. \varphi_{,}')^{g} \cos. (i' n' t - i n t + A - g \theta_{,}')$$

is a term of R depending on the angle i′ n′ t — i n t, we shall have, by 527,

$$\tan. \varphi_{,}' . \sin. \theta_{,}' = p' - p; \quad \tan. \varphi_{,}' \cos. \theta_{,}' = q' - q;$$

whence we get

$$(\tan. \varphi_{,}')^{g} \sin. g \theta_{,}' = \frac{\{q' - q + (p' - p)\sqrt{-1}\}^{g} - \{q' - q - (p' - p)\sqrt{-1}\}^{g}}{2\sqrt{-1}}$$

$$(\tan. \varphi_{,}')^{g} \cos. g \theta_{,}' = \frac{\{q' - q + (p' - p)\sqrt{-1}\}^{g} + \{q' - q - (p' - p)\sqrt{-1}\}^{g}}{2}.$$

With respect to the preceding term of R, we shall have

$$\left(\frac{d R}{d p}\right) = - g (\tan. \varphi_{,}')^{g-1} \mu' k . \sin. \{i' n' t - i n t + A - (g-1) \theta_{,}'\};$$

$$\left(\frac{d R}{d q}\right) = - g (\tan. \varphi_{,}')^{g-1} \mu' k \cos. \{i' n' t - i n t + A - (g-1) \theta_{,}'\}.$$

If we substitute these values in the preceding expressions of d p and d q, and observe that very nearly $c = \frac{m}{a n}$, we shall have

$$p = \frac{g \mu' k . a n}{m (i' n' - i n)} . (\tan. \varphi_{,}')^{g-1} . \sin. \{i' n' t - i n t + A - (g-1) \theta_{,}'\}$$

$$q = \frac{g \mu' k . a n}{\mu (i' n' - i n)} . (\tan. \varphi_{,}')^{g-1} \cos. \{i' n' t - i n t + A - (g-1) \theta_{,}'\}.$$

Substituting these values in the equation

$$s = q \sin. v - p \cos. v$$

we shall have

$$s = - \frac{g . \mu' k . a n}{m (i' n' - i n)} (\tan. \varphi_{,}')^{g-1} \sin. \{i' n' t - i n t - v + A - (g-1) \theta_{,}'\}.$$

This expression of s is the variation of the latitude corresponding to the preceding term of R: it is evident that it is the same whatever may be the fixed plane to which we refer the motions of μ and μ', provided that it is but little inclined to the plane of the orbits; we shall therefore thus have that part of the expression of the latitude, which the smallness of the divisor i′ n′ — i n may make sensible. Indeed the inequality of the latitude, containing only the first power of this divisor, is in that degree less sensible than the corresponding inequality of the mean longitude, which contains the square of the same divisor; but, on the other hand, tan. $\varphi_{,}'$ is then raised to a power less by one; a remark analogous to that which was made in No. 536, upon the corresponding inequality of the excentricities of the orbits. We thus see that all these inequalities are connected with one another, and with the corresponding part of R, by very simple relations.

If we differentiate the preceding expressions of p and q, and if in the values of $\frac{d\,p}{d\,t}$ and $\frac{d\,q}{d\,t}$ we augment the angles n t and n′ t by the inequalities of the mean motions, depending on the angle i′ n′ t — i n t, there will result in these differentials, quantities which are functions only of the elements of the orbits, and which may influence, in a sensible manner, the secular variations of the inclinations and nodes although of the order of the squares of the masses. This is analogous to what was advanced in No. 536 upon the secular variations of the excentricities and aphelions.

539. It remains to consider the variation of the longitude ε of the epoch. By No. 531 we have

$$d\,\varepsilon = d\,e \left\{ \left(\frac{d\,E^{(1)}}{d\,e}\right) \sin.(v - \varpi) + \tfrac{1}{2}\left(\frac{d\,E^{(2)}}{d\,e}\right)_{\sin} . \, 2\,(v - \varpi) + \&c. \right\}$$
$$- d\,\varpi \{E^{(1)} \cos.(v - \varpi) + E^{(2)} \cos. 2\,(v - \varpi) + \&c.\};$$

substituting for $E^{(1)}$, $E^{(2)}$, &c. their values in series ordered according to the powers of e, series which it is easy to form from the general expression of $E^{(i)}$ (473) we shall have

$$d\,\varepsilon = -2\,d\,e \sin.(v - \varpi) + 2\,e\,d\,\varpi \cos.(v - \varpi)$$
$$+ e\,d\,e \{\tfrac{3}{2} + \tfrac{1}{2}\,e^2 + \&c.\} \sin. 2\,(v - \varpi) - e^2\,d\,\varpi \{\tfrac{3}{2} + \tfrac{1}{4}\,e^2 + \&c.\} \cos. 2\,(v - \varpi)$$
$$- e^2\,d\,e \{1 + \&c.\} \sin. 3\,(v - \varpi) + e^3\,d\,\varpi \{1 + \&c.\} \cos. 3\,(v - \varpi)$$
$$+ \&c.$$

If we substitute for d e and e d ϖ their values given in 534, we shall find, carrying the approximation to quantities of the order e^2 inclusively,

$$d\,\varepsilon = \frac{a^2 . n\,d\,t}{m} \sqrt{1 - e^2}\, \{2 - \tfrac{3}{2}\,e \cos.(v - \varpi) + e^2 \cos. 2\,(v - \varpi)\} . \left(\frac{d\,R}{d\,\varrho}\right)$$
$$- \frac{a\,n\,d\,t}{m \sqrt{1 - e^2}} . \, e . \sin.(v - \varpi)\, \{1 + \tfrac{1}{2}\,e \cos.(v - \varpi)\} \left(\frac{d\,R}{d\,\varrho}\right).$$

The general expression of d ε contains terms of the form

$$\mu' \, k . n\,d\,t . \cos.(i'\,n'\,t - i\,n\,t + A)$$

and consequently the expression of ε contains terms of the form

$$\frac{\mu' \, k\,n}{i'\,n' - i\,n} \sin.(i'\,n'\,t - i\,n\,t + A);$$

but it is easy to be convinced that the coefficient k in these terms is of the order i′ — i, and that therefore these terms are of the same order as those of the mean longitude, which depend upon the same angle. These having the divisor $(i'\,n' - i\,n)^2$, we see that we may neglect the corresponding terms of ε, when i′ n′ — i n is a very small quantity.

If in the terms of the expression of d ϵ, which are solely functions of the elements of the orbits, we substitute for these elements the secular parts of their values; it is evident that there will result constant terms, and others affected with the sines and cosines of angles, upon which depend the secular variations of the excentricities and inclinations of the orbits. The constant terms will produce, in the expression of ϵ, terms proportional to the time, and which will merge into the mean motion μ. As to the terms affected with sines and cosines, they will acquire by integration, in the expression of ϵ, very small divisors of the same order as the perturbing forces; so that these terms being at the same time multiplied and divided by the forces, may become sensible, although of the order of the squares and products of the excentricities and inclinations. We shall see in the theory of the planets, that these terms are there insensible; but in the theory of the moon and of the satellites of Jupiter, they are very sensible, and upon them depend the secular equations.

We have seen in No. 532, that the mean motion of μ, is expressed by

$$\frac{3}{m}\int\int a\, n\, d\, t\, .\, d\, R,$$

and that if we retain only the first power of the perturbing masses, d R will contain none but periodic quantities. But if we consider the squares and products of the masses, this differential may contain terms which are functions only of the elements of the orbits. Substituting for the elements the secular parts of their values, there will thence result terms affected with sines and cosines of angles depending upon the secular variations of the orbits. These terms will acquire, by the double integration, in the expression of the mean motion, small divisors, which will be of the order of the squares and products of the perturbing masses; so that being both multiplied and divided by the squares and products of the masses, they become sensible, although of the order of the squares and products of the excentricities and inclinations of the orbits. We shall see that these terms are insensible in the theory of the planets.

540. The elements of μ's orbit being determined by what precedes, by substituting them in the expressions of the radius-vector, of the longitude and latitude which we have given in 484, we shall get the values of these three variables, by means of which astronomers determine the position of the celestial bodies. Then reducing them into series of sines and cosines, we shall have a series of inequalities, whence tables being formed, we may easily calculate the position of μ at any given instant.

This method, founded on the variation of the parameters, is very useful in the research of inequalities, which, by the relations of the mean motions of the bodies of the system, will acquire great divisors, and thence become very sensible. This sort of inequality principally affects the elliptic elements of the orbits; determining, therefore, the variations which result in these elements, and substituting them in the expression of elliptic motion, we shall obtain, in the simplest manner, all the inequalities made sensible by these divisors.

The preceding method is moreover useful in the theory of the comets. We perceive these stars in but a very small part of their courses, and observations only give that part of the ellipse which coincides with the arc of the orbit described during their apparitions; thus, in determining the nature of the orbit considered a variable ellipse, we shall see the changes undergone by this ellipse in the interval between two consecutive apparitions of the same comet. We may therefore announce its return, and when it reappears, compare theory with observation.

Having given the methods and formulas for determining, by successive approximations, the motions of the centers of gravity of the celestial bodies, we have yet to apply them to the different bodies of the solar system: but the ellipticity of these bodies having a sensible influence upon the motions of many of them, before we come to numerical applications, we must treat of the figure of the celestial bodies, the consideration of which is as interesting in itself as that of their motions.

SUPPLEMENT

TO

SECTIONS XII. AND XIII.

ON ATTRACTIONS AND THE FIGURE OF THE CELESTIAL BODIES.

541. The figure of the celestial bodies depends upon the law of gravitation at their surface, and the gravitation itself being the result of the attractions of all their parts, depends upon their figure; the law of gravity at the surface of the celestial bodies, and their figure have, therefore, a reciprocal connexion, which renders the knowledge of the one necessary to the determination of the other. The research is thus very intricate,

and seems to require a very particular sort of analysis. If the planets were entirely solid, they might have any figure whatever; but if, like the earth, they are covered with a fluid, all the parts of this fluid ought to be disposed so as to be in equilibrium, and the figure of its exterior surface depends upon that of the fluid which covers it, and the forces which act upon it. We shall suppose generally that the celestial bodies are covered with a fluid, and on that hypothesis, which subsists in the case of the earth, and which it seems natural to extend to the other bodies of the system of the world, we shall determine their figure and the law of gravity at their surface. The analysis which we propose to use is a singular application of the Calculus of Partial Differences, which by simple differentiation, will conduct us to very extensive results, and which with difficulty we should obtain by the method of integrations.

THE ATTRACTIONS OF HOMOGENEOUS SPHEROIDS BOUNDED BY SURFACES OF THE SECOND ORDER.

542. The different bodies of the solar system may be considered as formed of shells very nearly spherical, of a density varying according to any law whatever; and we shall show that the action of a spherical shell upon a body exterior to it, is the same as if its mass were collected at its center. For that purpose we shall establish upon the attractions of spheroids, some general propositions which will be of great use hereafter.

Let x, y, z be the three coordinates of the point attracted which we call μ; let also d M be the element or molecule of the spheroid, and x′, y′, z′ the coordinates of this element; if we call ϱ its density, ϱ being a function of x′, y′, z′ independent of x, y, z, we shall have

$$d M = \varrho . d x' . d y' . d z'.$$

The action of d M upon μ decomposed parallel to the axis of x and directed towards their origin, will be

$$\frac{\varrho d x' . d y' . d z' (x - x')}{\{(x - x')^2 + (y - y')^2 + (z - z')^2\}^{\frac{3}{2}}}$$

and consequently it will be equal to

$$-\left(\frac{d . \frac{\varrho d x' . d y' . d z'}{\sqrt{(x - x')^2 + (y - y')^2 + (z - z')^2}}}{d x}\right);$$

calling therefore V the integral

$$\int \frac{\varrho d x' . d y' . d z'}{\sqrt{(x - x')^2 + (y - y')^2 + (z - z')^2}}$$

extended to the entire mass of the spheroid, we shall have $-\left(\frac{d V}{d x}\right)$

for the total action of the spheroid upon the point μ, resolved parallel to the axis of x and directed towards its origin.

V is the sum of the elements of the spheroid, divided by their respective distances from the point attracted; to get the attraction of the spheroid upon this point, parallel to any straight line, we must consider V as a function of three rectangular coordinates, one of which is parallel to this straight line, and differentiate this function relatively to this coordinate; the coefficient of this differential taken with a contrary sign, will be the expression of the attraction of the spheroid, parallel to the given straight line, and directed towards the origin of the coordinate which is parallel to it.

If we represent by β, the function $\{(x-x')^2+(y-y')^2+(z-z')^2\}^{-\frac{1}{2}}$; we shall have

$$V = \int \beta \,.\, \varrho \,.\, dx'\, dy'\, dz'.$$

The integration being only relative to the variables x', y', z', it is evident that we shall have

$$\left(\frac{d^2 V}{d x^2}\right) + \left(\frac{d^2 V}{d y^2}\right) + \left(\frac{d^2 V}{d z^2}\right) = \int \varrho\, dx'\, dy'\, dz' \left\{ \left(\frac{d^2 \beta}{d x^2}\right) + \left(\frac{d^2 \beta}{d y^2}\right) + \left(\frac{d^2 \beta}{d z^2}\right) \right\};$$

But we have

$$0 = \left(\frac{d^2 \beta}{d x^2}\right) + \left(\frac{d^2 \beta}{d y^2}\right) + \left(\frac{d^2 \beta}{d z^2}\right);$$

in like manner we get

$$0 = \left(\frac{d^2 V}{d x^2}\right) + \left(\frac{d^2 V}{d y^2}\right) + \left(\frac{d^2 V}{d z^2}\right); \quad \ldots\ldots \quad (A)$$

This remarkable equation will be of the greatest use in the theory of the figure of the celestial bodies. We may present it under more commodious forms in different circumstances; conceive, for example, from the origin of coordinates we draw to the point attracted a radius which we call ϱ; let θ be the angle which this radius makes with the axis of x, and ϖ the angle which the plane formed by ϱ and this axis makes with the plane of x, y; we shall have

$$x = \varrho \cos.\theta; \quad y = \varrho \sin.\theta \cos.\varpi; \quad z = \varrho \sin.\theta \sin.\varpi;$$

whence we derive

$$\varrho = \sqrt{x^2+y^2+z^2}; \quad \cos.\theta = \frac{x}{\sqrt{x^2+y^2+z^2}}; \quad \tan.\varpi = \frac{z}{y};$$

thus we can obtain the partial differences of ϱ, θ, ϖ, relative to the variables x, y, z, and thence get the values of $\left(\frac{d^2 V}{d x^2}\right)$, $\left(\frac{d^2 V}{d y^2}\right)$, $\left(\frac{d^2 V}{d z^2}\right)$

in partial differences of V relative to the variables ϱ, θ, ϖ. Since we shall often use these transformations of partial differences, it is useful here to lay down the principle of it. Considering V as a function of the variables x, y, z, and then of the variables ϱ, θ, ϖ, we have

$$\left(\frac{dV}{dx}\right) = \left(\frac{dV}{d\varrho}\right)\left(\frac{d\varrho}{dx}\right) + \left(\frac{dV}{d\theta}\right)\left(\frac{d\theta}{dx}\right) + \left(\frac{dV}{d\varpi}\right)\left(\frac{d\varpi}{dx}\right).$$

To get the partial differences $\left(\frac{d\varrho}{dx}\right)$, $\left(\frac{d\theta}{dx}\right)$, $\left(\frac{d\varpi}{dx}\right)$, we must make x alone vary in the preceding expressions of ϱ, cos. θ, tan. ϖ; differentiating therefore these expressions, we shall have

$$\left(\frac{d\varrho}{dx}\right) = \cos.\theta; \quad \left(\frac{\theta}{dx}\right) = -\frac{\sin.\theta}{\varrho}; \quad \left(\frac{d\varpi}{dx}\right) = 0;$$

which gives

$$\left(\frac{dV}{dx}\right) = \cos.\theta\left(\frac{dV}{d\varrho}\right) - \frac{\sin.\theta}{\varrho}\left(\frac{dV}{d\theta}\right).$$

Thus we therefore get the partial difference $\left(\frac{dV}{dx}\right)$, in partial differences of the function V, taken relatively to the variables ϱ, θ, ϖ. Differentiating again this value of $\left(\frac{dV}{dx}\right)$, we shall have the partial difference $\left(\frac{d^2V}{dx^2}\right)$ in partial differences of V taken relatively to the variables ϱ, θ, ϖ. By the same process the values of $\left(\frac{d^2V}{dy^2}\right)$ and $\left(\frac{d^2V}{dz^2}\right)$ may be found.

In this way we shall transform equation (A) into the following one:

$$0 = \left(\frac{d^2V}{d\theta^2}\right) + \frac{\cos.\theta.}{\sin.\theta}\left(\frac{dV}{d\theta}\right) + \frac{\left(\frac{d^2V}{d\varpi^2}\right)}{\sin.^2\theta} + \varrho\left(\frac{d^2.\varrho V}{d\varrho^2}\right); \quad . \quad . \quad (B)$$

And if we make cos. θ = m, this last equation will become

$$0 = \left(\frac{d.\left\{(1-m^2)\left(\frac{dV}{d\mu}\right)\right\}}{dm}\right) + \frac{\left(\frac{d^2V}{d\varpi^2}\right)}{1-m^2} + \varrho\left(\frac{d^2.\varrho V}{d\varrho^2}\right). \quad (C)$$

543. Suppose, however, that the spheroid is a spherical shell whose origin of coordinates is at the center; it is evident that V will only depend upon ϱ, and contain neither m nor ϖ; the equation (C) will therefore give

$$0 = \left(\frac{d^2.\varrho V}{d\varrho^2}\right);$$

whence by integration we get

$$V = A + \frac{B}{\varrho},$$

A and B being two arbitrary constants. We therefore have

$$-\left(\frac{d\,V}{d\,\varrho}\right) = \frac{B}{\varrho^2}.$$

$-\frac{d\,V}{d\,\varrho}$ expresses, by what precedes, the action of the spherical shell upon the point μ, decomposed along the radius ϱ and directed towards the center of the shell; but it is evident that the total action of the shell ought to be directed along this radius; $-\left(\frac{d\,V}{d\,\varrho}\right)$ expresses therefore the total action of the spherical shell upon the point μ.

First suppose this point placed within the shell. If it were at the center itself, the action of the shell would be nothing; we have therefore,

$$-\left(\frac{d\,V}{d\,\varrho}\right) = 0, \text{ or } \frac{B}{\varrho^2} = 0,$$

when $\varrho = 0$, which gives B = 0, and consequently $-\left(\frac{d\,V}{d\,\varrho}\right) = 0$, whatever ϱ may be; whence it follows that a point placed in the interior of the shell, suffers no action, or which comes to the same thing, it is equally attracted on all sides.

If the point μ is situated without the spherical shell, it is evident, supposing it infinitely distant from the center, that the action of the shell upon the point will be the same, as if all the mass of the shell were condensed at this center; calling, therefore M the mass of the shell, $-\left(\frac{d\,V}{d\,\varrho}\right)$ or $\frac{B}{\varrho^2}$ will become in this case equal to $\frac{M}{\varrho^2}$, which gives B = M; we have therefore generally relatively to exterior points,

$$-\left(\frac{d\,V}{d\,\varrho}\right) = \frac{M}{\varrho^2}.$$

that is to say, the shell attracts them as if all its mass were collected at its center.

A sphere being a spherical shell, the radius of whose interior surface is nothing, we see that its attraction, upon a point placed at or above its surface, is the same as if its mass were collected at its center.

This result obtains for globes formed of concentric shells, varying in density from the center to the circumference according to any law whatever, for it is true for each of the shells: thus since the sun, the planets, and satellites may be considered nearly as globes of this nature, they attract exterior bodies very nearly as if their masses were collected into their centers of gravity. This is conformable with what has been found by observations. Indeed the figure of the celestial bodies departs a little from the sphere, but the difference is very little, and the error which results from the preceding supposition is of the same order as this supposition relatively to points near the surface; and relatively to distant points, the error is of the same order as the product of this difference by the square of the ratio of the radii of the attracting bodies to their distances from the points attracted; for we know that the consideration alone of the distance of the points attracted, renders the error of the preceding supposition of the same order as tne square of this ratio. The celestial bodies, therefore, attract one another very nearly as if their masses were collected at their centers of gravity, not only because they are very distant from one another relatively to their respective dimensions, but also because their figures differ very little from the sphere.

The property of spheres, by the law of Nature, of attracting as if their masses were condensed into their centers, is very remarkable, and we may be curious to learn whether it also obtains in other laws of attraction. For that purpose we shall observe, that if the law of gravity is such, that a homogeneous sphere attracts a point placed without it as if all its mass were collected at its center, the same result ought to obtain for a spherical shell of a constant thickness; for if we take from a sphere a spherical shell of a constant thickness, we form a new sphere of a smaller radius with the remainder, but which, like the former, shall have the property of attracting as if all its mass were collected at its center; but it is evident, that these two spheres can only have this common property, unless it also belongs to the spherical shell which forms their difference. The problem, therefore, is reduced to determine the laws of attraction according to which a spherical shell, of an infinitely small and constant thickness, attracts an exterior point as if all its mass were condensed into its center.

Let ϱ be the distance of the point attracted to the center of the spherical shell, u the radius of the shell, and d u its thickness. Let θ be the angle which the radius u makes with the straight line ϱ, ϖ the angle which the plane passing through the straight lines ϱ, u, makes with a fixed plane passing through ϱ, the element of the spherical shell will be $u^2 d u . d \varpi . d \theta \sin. \theta$. If we then call f the distance of this element from the point attracted, we shall have

$$f^2 = \varrho^2 - 2 \varrho u \cos. \theta + u^2.$$

Represent by φ (f) the law of attraction to the distance f; the action of the shell's element upon the point attracted, decomposed parallel to ϱ and directed towards the center of the shell, will be

$$u^2 \, d\,u \, . \, d\,\varpi \sin.\theta \frac{\varrho - u \cos.\theta}{f} \varphi(f);$$

but we have

$$\frac{\varrho - u \cos.\theta}{f} = \left(\frac{d\,f}{d\,\varrho}\right);$$

which gives to the preceding quantity this form

$$u^2 \, d\,u \, . \, d\,\varpi \sin.\theta \left(\frac{d\,f}{d\,\varrho}\right) \varphi(f);$$

wherefore if we denote $\int d\,f\,\varphi(f)$ by $\varphi_{,}(f)$ we shall have the whole action of the spherical shell upon the point attracted, by means of the integral $u^2 \, d\,u \int d\,\varpi \, d\,\theta \sin.\theta \, . \, \varphi_{,}(f)$, differentiated relatively to ϱ, and divided by $d\,\varrho$.

This integral ought to be taken relatively to ϖ, from $\varpi = 0$ to ϖ equal to the circumference, and after this integration it becomes

$$2\,\pi\,u^2 \int d\,\theta \sin.\theta \, \varphi_{,}(f);$$

If we differentiate the value of f relatively to θ, we shall have

$$d\,\theta \sin.\theta = \frac{f\,d\,f}{\varrho\,u};$$

and consequently

$$2\,\pi \, . \, u^2 d\,u \int d\,\theta \sin.\theta \, . \, \varphi_{,}(f) = 2\,\pi \, . \, \frac{u\,d\,u}{\varrho} \int f\,d\,f \, . \, \varphi_{,}(f).$$

The integral relative to θ ought to be taken from $\theta = 0$ to $\theta = \pi$, and at these two limits we have $f = \varrho - u$, and $f = \varrho + u$; thus the integral relative to f must be taken from $f = \varrho - u$ to $f = \varrho + u$; let therefore $\int f\,d\,f \, . \, \varphi_{,}(f) = \psi(f)$, we shall have

$$\frac{2\,\pi \, . \, u\,d\,u}{\varrho} \int f\,d\,f\,\varphi_{,}(f) = \frac{2\,\pi \, . \, u\,d\,u}{\varrho} \{\psi(\varrho + u) - \psi(\varrho - u)\}.$$

The coefficient of $d\,\varrho$, in the differential of the second member of this equation, taken relatively to ϱ, will give the attraction of the spherical shell upon the point attracted; and it is easy thence to conclude that in nature where $\varphi(f) = \frac{1}{f^2}$ this attraction is equal to

$$\frac{4\,\pi \, . \, u^2\,d\,u}{\varrho^2}.$$

That is to say, that it is the same as if all the mass of the spherical shell were collected at its center. This furnishes a new demonstration of the property already established of the attraction of spheres.

Let us determine $\varphi(f)$ on the condition that the attraction of the shell is the same as if its mass were condensed into its center. This mass is equal to $4\,\pi \, . \, u^2\,d\,u$, and if it were condensed into its center, its action upon the point attracted would be $4\pi \,.\, u^2 \, du \,.\, \varphi(\rho)$; we shall therefore have

$$2\pi \,.\, u\, du \left(\frac{d \,.\, \left\{ \frac{1}{\rho} \,.\, (\psi[\rho+u] - \psi[\rho-u]) \right\}}{d\rho} \right) = 4\pi u^2 \, du \, \varphi(\rho); \,.\, (D)$$

integrating relatively to ρ, we shall get

$$\psi(\rho + u) - \psi(\rho - u) = 2\rho u \int d\rho \,.\, \varphi(\rho) + \rho U,$$

U being a function of u and constants, added to the integral $2u\int d\rho\, \varphi(\rho)$. If we represent $\psi(\rho + u) - \psi(\rho - u)$ by R, we shall have by differentiating the preceding equation

$$\left(\frac{d^2 R}{d\rho^2}\right) = 4u\varphi(\rho) + 2\rho u \,.\, \frac{d\varphi(\rho)}{d\rho}$$

$$\left(\frac{d^2 R}{du^2}\right) = \rho \left(\frac{d^2 U}{du^2}\right);$$

But we have, by the nature of the function R,

$$\left(\frac{d^2 R}{d\rho^2}\right) = \frac{d^2 R}{du^2};$$

wherefore

$$2u \left\{ 2\varphi(\rho) + \rho \frac{d\varphi(\rho)}{d\rho} \right\} = \rho \left(\frac{d^2 U}{du^2}\right);$$

or

$$\frac{2\varphi(\rho)}{\rho} + \frac{d \,.\, \varphi(\rho)}{d\rho} = \frac{1}{2u}\left(\frac{d^2 U}{du^2}\right).$$

Thus the first member of this equation being independent of u and the functions of ρ, each of its members must be equal to an arbitrary which we shall designate by 3 A; we therefore have

$$\frac{2\varphi\rho}{\rho} + \frac{d\varphi\rho}{d\rho} = 3A;$$

whence in integrating we derive

$$\varphi\rho = A\rho + \frac{B}{\rho^2}$$

B being a new arbitrary constant. All the laws of attraction in which a sphere acts upon an exterior point placed at the distance ρ from its center, as if all the mass were condensed into its center, are therefore comprised in the general formula

$$A\rho + \frac{B}{\rho^2};$$

it is easy to see in fact that this value satisfies equation (D) whatever may be A and B.

If we suppose $A = 0$, we shall have the law of nature, and we see that in the infinity of laws which render attraction very small at great distances, that of nature is the only one in which spheres have the property of acting as if their masses were condensed into their centers.

This law is also the only one in which a body placed within a spherical shell, every where of an equal thickness, is equally attracted on all sides. It results from the preceding analysis that the attraction of the spherical shell, whose thickness is d u, upon a point placed in its interior, has the expression

$$2\pi u^2 d u \left(\frac{d . \frac{1}{\varrho} \{\psi(u+\varrho) - \psi(u-\varrho)\}}{d \varrho} \right).$$

To make this function nothing, we must have

$$\psi(u+\varrho) - \psi(u-\varrho) = \varrho . U,$$

U being a function of u independent of ϱ, and it is easy to see that this obtains in the law of nature, where $\varphi(f) = \frac{B}{\varrho^2}$. But to show that it takes place only in this law, we shall denote by $\psi'(f)$ the difference of $\psi(f)$ divided by d f, we shall also denote by $\psi''(f)$ the difference of $\psi'(f)$ divided by d f, and so on; thus we shall get, by differentiating twice successively, the preceding equation relatively to ϱ,

$$\psi''(u+\varrho) - \psi''(u-\varrho) = 0.$$

This equation obtaining whatever may be u and ϱ, it thence results that $\psi''(f)$ ought to be equal to a constant whatever f may be, and that therefore $\psi'''(f) = 0$. But, by what precedes,

$$\psi(f) = f . \varphi_{,}(f),$$

whence we get

$$\psi'''(f) = 2\varphi(f) + f\varphi'(f);$$

we therefore have

$$0 = 2\varphi(f) + f\varphi'(f);$$

which gives by integration

$$\varphi(f) = \frac{b}{f^2},$$

and consequently the law of nature.

554. Let us resume the equation (C) of No. 541. If this equation could generally be integrated, we should have an expression of V, which would contain two arbitrary functions, which we should determine by finding the attraction of a spheroid, upon a point situated so as to facilitate this research, and by comparing this attraction with its general expression. But the integration of the equation (C) is possible only in some particular cases, such as that where the attracting spheroid is a sphere, which reduces this equation to ordinary differences; it is also possible in

the case where the attracting body is a cylinder whose base is an oval or curve returning into itself, and whose length is infinite. This particular case contains the theory of Saturn's ring.

Fix the origin of ϱ upon the same axis of the cylinder, which we shall suppose of an infinite length on each side of the origin. Naming ϱ' the distance of the point attracted from the axis; we shall have

$$\varrho' = \varrho\sqrt{1 - m^2}.$$

It is evident that V only depends on ϱ' and ϖ, since it is the same for all the points relatively to which these two variations are the same; it contains therefore only m inasmuch as ϱ' is a function of this variable. This gives

$$\left(\frac{dV}{dm}\right) = \left(\frac{dV}{d\varrho'}\right) \cdot \left(\frac{d\varrho'}{dm}\right) = -\frac{\varrho m}{\sqrt{1 - m^2}} \left(\frac{dV}{d\varrho'}\right);$$

$$\left(\frac{d^2 V}{dm^2}\right) = \frac{\varrho^2 m^2}{1 - m^2} \cdot \left(\frac{d^2 V}{d\varrho'^2}\right) - \frac{\varrho}{(1-m^2)^{\frac{3}{2}}} \cdot \left(\frac{dV}{d\varrho'}\right);$$

the equation (C) hence becomes

$$0 = \varrho'^2 \left(\frac{d^2 V}{d\varrho'^2}\right) + \frac{d^2 V}{d\varpi^2} + \varrho' \left(\frac{dV}{d\varrho'}\right);$$

whence by integrating we get

$$V = \varphi\{\varrho' \cos.\varpi + \varrho'\sqrt{-1}\sin.\varpi\} + \psi\{\varrho' \cos.\varpi - \varrho'\sqrt{-1}\sin.\varpi\};$$

$\varphi(\varrho')$ and $\psi(\varrho')$ being arbitrary functions of ϱ', which we can determine by seeking the attraction of the cylinder when ϖ is nothing and when it is a right angle.

If the base of the cylinder is a circle, V will be evidently a function of ϱ' independent of ϖ; the preceding equation of partial differences will thus become

$$0 = \varrho'^2 \left(\frac{d^2 V}{d\varrho'^2}\right) + \varrho' \left(\frac{dV}{d\varrho'}\right);$$

which gives by integrating,

$$-\left(\frac{dV}{d\varrho'}\right) = \frac{H}{\varrho'},$$

H being a constant. To determine it, we shall suppose ϱ' relatively to the radius of the base of the cylinder extremely great, which supposition permits us to consider the cylinder as an infinite straight line. Let A be this base, and z the distance of any point whatever of the axis of the cylinder, to the point where this axis is met by ϱ'; the action of the cylinder considered as concentrated or condensed upon its axis, will be, parallel to ϱ', equal to

$$\int \frac{A\varrho' . dz}{(\varrho'^2 + z^2)^{\frac{3}{2}}},$$

the integral being taken from $z = -\infty$ to $z = \infty$; this reduces the integral to $\frac{2A}{\varrho'}$; which is the expression of $-\left(\frac{dV}{d\varrho'}\right)$ when ϱ' is very considerable. Comparing this with the preceding one we have $H = 2A$, and we see that whatever is ϱ', the action of the cylinder upon an exterior point, is $\frac{2A}{\varrho'}$.

If the attracted point is within a circular cylindrical shell, of a constant thickness, and infinite length, we shall have $-\left(\frac{dV}{d\varrho}\right) = \frac{H}{\varrho'}$; and since the attraction is nothing when the point attracted is upon the axis of the shell, we have $H = 0$, and consequently, a point placed in the interior of the shell is equally attracted on all sides.

545. We have thus determined the attraction of a sphere and of a spherical shell: let us now consider the attraction of spheroids terminated by surfaces of the second order.

Let x, y, z be the three rectangular coordinates of an element of the spheroid; designating d M this element, and taking for unity the density of the spheroid which we shall suppose homogeneous, we shall have

$$dM = dx \cdot dy \cdot dz.$$

Let a, b, c be the rectangular coordinates of the point attracted by the spheroid, and denote by A, B, C the attractions of the spheroid upon this point resolved parallel to the axes of x, y, z and directed to the origin of the coordinates.

It is easy to show that we have

$$A = \iiint \frac{(a - x)\, dx \cdot dy \cdot dz}{\{(a - x)^2 + (b - y)^2 + (c - z)^2\}^{\frac{3}{2}}};$$

$$B = \iiint \frac{(b - y)\, dx \cdot dy \cdot dz}{\{(a - x)^2 + (b - y)^2 + (c - z)^2\}^{\frac{3}{2}}};$$

$$C = \iiint \frac{(c - z)\, dx \cdot dy \cdot dz}{(a - x)^2 + (b - y)^2 + (c - z)^2\}^{\frac{3}{2}}}.$$

All these triple integrals ought to be extended to the entire mass of the spheroid. The integrations under this form present great difficulties, which we can often in part remove by transforming the differentials into others more convenient. This is the general principle of such transformations.

Let us consider the differential function $P\, dx \cdot dy \cdot dz$, P being any function whatever of x, y, z. We may suppose x a function of y and z and of a new variable p: let $\varphi(y, z, p)$ denote this function; in this case, we shall have, making y and z constant, $dx = \beta . dp$, β being a function of y, z and p. The preceding differential will thus become $\beta . P . dp . dy . dz$; and to integrate it, we must substitute in P, for x, its value $\varphi(y, z, p)$.

In like manner we may suppose in this new differential, $y = \varphi'(z, p, q)$, q being a new variable, and $\varphi'(z, p, q)$ being any function of the three variables z, p and q. We shall have, considering z and p constant, $dy = \beta' dq$, β' being a function of z, p, q; the preceding differential will thus take this new form $\beta \beta' P . dp . dq . dz$, and to integrate it, we must substitute in βP for y its value $\varphi'(z, p, q)$.

Lastly we may suppose z equal to $\varphi''(p, q, r)$, r being a new variable, and $\varphi''(p, q, r)$ being any function whatever of p, q, r. We shall have, considering p and q constant, $dz = \beta'' dr$, β'' being a function of p, q, r; the preceding differential will thus become $\beta . \beta' . \beta'' . P . dp . dq . dr$ and to integrate it, we must substitute in $\beta . \beta' . P$ for z its value $\varphi''(p, q, r)$. The proposed differential function is thence transformed to another relative to the three new variables p, q, r, which are connected with the preceding by the equations

$$x = \varphi(y, z, p);\ y = \varphi'(z, p, q);\ z = \varphi''(p, q, r).$$

It only remains to derive from these equations the values of β, β', β''. For that purpose we shall observe that they give x, y, z, in functions of the variables p, q and r; let us consider therefore the three first variables as functions of the three last. Since β'' is the coefficient of d r in the differential of z, taken by considering p and q constant, we have

$$\beta'' = \left(\frac{dz}{d\varrho}\right).$$

β' is the coefficient of d q, in the differential of y taken on the supposition that p and z are constant; we shall therefore have β', by differentiating y on the supposition that p is constant, and by eliminating d r by means of the differential of z taken on the supposition that p is constant, and equating it to zero. Thus we shall have the two equations

$$dy = \left(\frac{dy}{dq}\right) dq + \left(\frac{dy}{dr}\right) dr$$

$$0 = \left(\frac{dz}{dq}\right) dq + \left(\frac{dz}{dr}\right) dr;$$

which give

$$dy = dq \times \frac{\left(\frac{dy}{dq}\right)\left(\frac{dz}{dr}\right) - \left(\frac{dy}{dr}\right)\left(\frac{dz}{dq}\right)}{\left(\frac{dz}{dr}\right)};$$

wherefore

$$\beta' = \frac{\left(\frac{d\,y}{d\,q}\right)\left(\frac{d\,z}{d\,r}\right) - \left(\frac{d\,y}{d\,r}\right)\left(\frac{d\,z}{d\,q}\right)}{\left(\frac{d\,z}{d\,r}\right)}.$$

Finally, β is the coefficient of d p, in the differential of x taken on the supposition that y and z are constant. This gives the three following equations

$$d\,x = \left(\frac{d\,x}{d\,p}\right) d\,p + \left(\frac{d\,x}{d\,q}\right) d\,q + \left(\frac{d\,x}{d\,r}\right) d\,r;$$

$$0 = \left(\frac{d\,y}{d\,p}\right) d\,p + \left(\frac{d\,y}{d\,q}\right) d\,q + \left(\frac{d\,y}{d\,r}\right) d\,r;$$

$$0 = \left(\frac{d\,z}{d\,p}\right) d\,p + \left(\frac{d\,z}{d\,q}\right) d\,q + \left(\frac{d\,z}{d\,r}\right) d\,r.$$

If we make

$$\epsilon = \left(\frac{d\,x}{d\,p}\right)\left(\frac{d\,y}{d\,q}\right)\left(\frac{d\,z}{d\,r}\right) - \left(\frac{d\,x}{d\,p}\right)\left(\frac{d\,y}{d\,r}\right)\left(\frac{d\,z}{d\,q}\right)$$
$$+ \left(\frac{d\,x}{d\,q}\right)\left(\frac{d\,y}{d\,r}\right)\left(\frac{d\,z}{d\,p}\right) - \left(\frac{d\,x}{d\,q}\right)\left(\frac{d\,y}{d\,p}\right)\left(\frac{d\,z}{d\,r}\right)$$
$$+ \left(\frac{d\,x}{d\,r}\right)\left(\frac{d\,y}{d\,p}\right)\left(\frac{d\,z}{d\,q}\right) - \left(\frac{d\,x}{d\,r}\right)\left(\frac{d\,y}{d\,q}\right)\left(\frac{d\,z}{d\,p}\right);$$

we shall have

$$d\,x = \frac{\epsilon\,d\,p}{\left(\frac{d\,y}{d\,q}\right)\left(\frac{d\,z}{d\,r}\right) - \left(\frac{d\,y}{d\,r}\right)\left(\frac{d\,z}{d\,q}\right)};$$

which gives

$$\beta = \frac{\epsilon}{\left(\frac{d\,y}{d\,q}\right)\left(\frac{d\,z}{d\,r}\right) - \left(\frac{d\,y}{d\,r}\right)\left(\frac{d\,z}{d\,q}\right)};$$

wherefore $\beta . \beta' . \beta'' = \epsilon$ and the differential P. d x . d y . d z is transformed into ϵ. P. d p . d q . d r; P being here what P becomes when we substitute for x, y, z their values in p, q, r. The whole is therefore reduced to finding the variables p, q, r such that the integrations may become possible.

Let us transform the coordinates x, y, z into the radius drawn from the point attracted to the molecule, and into the angles which this radius makes with given straight lines or with given planes. Let r be this radius, p the angle which it forms with a straight line drawn through the attracted point parallel to the axis of x, and let q be the angle which its projection makes on the plane of y, z with the axis of y; we shall have

$x = a - r \cos. p$; $y = b - r \sin. p \cos. q$; $z = c - r \sin. p \sin. q$.

We shall then find $\varepsilon = - r^2 \sin. p$, and the differential $dx . dy . dz$ will thus be transformed into $- r^2 \sin. p . dp . dq . dr$: this is the expression of the element d M, and since this expression ought to be positive in considering sin. p, d p, d q, d r as positive, we must change its sign, which amounts to changing that of ε, and to making $\varepsilon = r^2 \sin. p$.

The expressions of A, B, C will thus become

$$A = \iiint dr\, dp\, dq . \sin. p \cos. p;$$
$$B = \iiint dr\, dp\, dq . \sin.^2 p \cos. p;$$
$$C = \iiint dr\, dp\, dq . \sin.^2 p . \sin. q.$$

It is easy to arrive by another way at these expressions, by observing that the element d M may be supposed equal to a rectangular parallelopiped, whose dimensions are d r, r d p and r d q sin. p, and by then observing that the attraction of the element, parallel to the three axes of x, y, z is

$$\frac{dM}{r^2} \cos. p; \quad \frac{dM}{r^2} \sin. p \cos. q; \quad \frac{dM}{r^2} \sin. p \sin. q.$$

The triple integrals of the expressions of A, B, C must extend to the entire mass of the spheroid: the integrations relative to r are easy, but they are different according as the point attracted is within or without the spheroid; in the first case, the straight line which passing through the point attracted, traverses the spheroid, is divided into two parts by this point; and if we call r and r′ these parts, we shall have

$$A = \iint (r + r')\, dp\, dq . \sin. p \cos. p;$$
$$B = \iint (r + r')\, dp\, dq . \sin.^2 p \cos. p;$$
$$C = \iint (r + r')\, dp\, dq . \sin.^2 p \sin. q;$$

the integrals relative to p and q ought to be taken from p and q equal to zero, to p and q equal to two right angles.

In the second case, if we call r, the radius at its entering the spheroid, and r′ the radius at its farther surface, we shall have

$$A = \iint (r' - r)\, dp\, dq . \sin. p \cos. p;$$
$$B = \iint (r' - r)\, dp\, dq . \sin.^2 p \cos. q;$$
$$C = \iint (r' - r)\, dp\, dq . \sin.^2 p \sin. q.$$

The limits of the integrals relative to p and to q, must be fixed at the points where $r' - r = 0$, that is to say, where the radius r is a tangent to the surface of the spheroid.

546. Let us apply these results to spheroids bounded by surfaces of the

second order. The general equation of these surfaces, referred to the three orthogonal coordinates x, y, z is

$$0 = A + B.x + C.y + E.z + F.x^2 + H.xy + L.y^2 + M.xz + N.yz + O.z^2.$$

The change of the origin of coordinates introduces three arbitraries, since the position of this new origin relating to the first depends upon three arbitrary coordinates. The changing the position of the coordinates around their origin introduces three arbitrary angles; supposing, therefore, the coordinates of the origin and position in the preceding equation to change at the same time, we shall have a new equation of the second degree whose coefficients will be functions of the preceding coefficients and of the six arbitraries. If we then equate to zero the first powers of the coordinates, and their products two and two, we shall determine these arbitraries, and the general equation of the surfaces of the second order, will take this very simple form

$$x^2 + m y^2 + n z^2 = k^2;$$

it is under this form that we shall discuss it.

In these researches we shall only consider solids terminated by finite surfaces, which supposes m and n positive. In this case, the solid is an ellipsoid whose three semi-axes are what the variables x, y, z become when we suppose two of them equal to zero; we shall thus have k, $\frac{k}{\sqrt{m}}$, $\frac{k}{\sqrt{n}}$ for the three semi-axes respectively parallel to x, to y and to z. The solid content of the ellipsoid will be $\frac{4\pi . k^3}{3\sqrt{mn}}$.

If, however, in the preceding equation we substitute for x, y, z their values in p, q, r given by the preceding No., we shall have

$$r^2 (\cos.^2 p + m \sin.^2 p \cos.^2 q + n \sin.^2 p \sin.^2 q)$$
$$- 2r (a \cos. p + m b \sin. p \cos. q + n c \sin. p \sin. q) = k^2 - a^2 - m b^2 - n c^2;$$

so that if we suppose

$$I = a \cos. p + m b \sin. p \cos. q + n c \sin. p \sin. q;$$
$$L = \cos.^2 p + m \sin.^2 p \cos.^2 q + n \sin.^2 p \sin.^2 q;$$
$$R = I^2 + (k^2 - a^2 - m b^2 - n c^2). L$$

we shall have

$$r = \frac{I \pm \sqrt{R}}{L};$$

whence we obtain r′ by taking +, and r by taking —; we shall therefore have

$$r + r' = \frac{2I}{L}; \quad r' - r = \frac{2\sqrt{R}}{L}.$$

Hence relatively to the interior points of the spheroid, we get

$$A = 2\iint \frac{dp.dq.I.\sin.p.\cos.p}{L};$$

$$B = 2\iint \frac{dp.dq.I.\sin.^2 p.\cos.q}{L};$$

$$C = 2\iint \frac{dp.dq.I.\sin.^2 p.\sin.q}{L};$$

and relatively to the exterior points

$$A = 2\iint \frac{dp.dq.\sin.p.\cos.p\sqrt{R}}{L};$$

$$B = 2\iint \frac{dp.dq.\sin.^2 p\cos.q\sqrt{R}}{L};$$

$$C = 2\iint \frac{dp.dq.\sin.^2 p\sin.q\sqrt{R}}{L};$$

the three last integrals being to be taken between the two limits which correspond to $R = 0$.

547. The expressions relative to the interior points being the most simple, we shall begin with them. First, we shall observe that the semi-axis k of the spheroid does not enter the values of I and L; the values of A, B, C are consequently independent; whence it follows that we may augment at pleasure, the shells of the spheroid which are above the point attracted, without changing the attraction of the spheroid upon this point, provided the values of m and n are constant. Thence results the following theorem.

A point placed within an elliptic shell whose interior and exterior surfaces are similar and similarly situated, is equally attracted on all sides.

This theorem is an extension of that which we have demonstrated in 542, relative to a spherical shell.

Let us resume the value of A. If we substitute for I and L their values, it will become

$$A = 2\iint \frac{dp.dq.\sin.p.\cos.p.(a\cos.p + mb\sin.p\cos.q + nc\sin.p\sin.q)}{\cos.^2 p + m\sin.^2 p\cos.^2 q + n\sin.^2 p\sin.^2 q}.$$

Since the integrals relative to p and q, must be taken from p and q equal to zero, to p and q equal to two right angles, it is clear we have generally $\int P\,dp.\cos.p = 0$, P being a rational function of sin. p and of $\cos.^2 p$; because the value of p being taken at equal distances greater and less than the right angle, the corresponding values of P. cos. p are equal and have contrary signs; thus we have

$$A = 2a\iint \frac{dp.dq.\sin.p\cos.^2 p}{\cos.^2 p + m\sin.^2 p\cos^2 q + n\sin^2 p\sin.^2 q}.$$

If we integrate relatively to q from q = 0 to q = two right angles, we shall find

$$A = \frac{2 a \pi}{\sqrt{m n}} \int \frac{d p . \sin. p \cos.^2 p}{\sqrt{\left(1 + \frac{1-m}{m} \cos.^2 p\right). \left(1 + \frac{1-n}{n} \cos.^2 p\right)}};$$

an integral which must be taken from cos. p = 1 to cos. p = — 1. Let cos. p = x, and call M the entire mass of the spheroid; we shall have by 545, $M = \frac{4 \pi . k^3}{\sqrt{m n}}$ and consequently $\frac{4 \pi}{\sqrt{m n}} = \frac{3 M}{k^3}$; we shall therefore have

$$A = \frac{3 a M}{k^3} \int \frac{x^2 d x}{\sqrt{\left(1 + \frac{1-m}{m} . x^2\right). \left(1 + \frac{1-n}{n} . x^2\right)}};$$

which must be taken from x = 0, to x = 1.

Integrating in the same manner the expressions of B, C we shall reduce them to simple integrals; but it is easier to get these integrals from the preceding expression of A. For that purpose, we shall observe that this expression may be considered as a function of a and of the squares k^2, $\frac{k^2}{m}$, $\frac{k^2}{n}$ of the semi-axes of the spheroid, parallel to the coordinates a, b, c of the point attracted; calling therefore k'^2 the square of the semi-axis parallel to b, and consequently $k'^2 . m$, and $k'^2 n$ the squares of the two other semi-axes, B will be a similar function of b, k'^2, $k'^2 m$, $k'^2 \frac{m}{n}$; thus to get B we must change in the expression of A, a into b, k into k' or $\frac{k}{\sqrt{m}}$, m into $\frac{1}{m}$, and n into $\frac{n}{m}$, which gives

$$B = \frac{3 b M}{k^3} \int \frac{m^{\frac{3}{2}} . x^2 d x}{\sqrt{\{1 + (m - 1) x^2\}. \left\{1 + \frac{m-n}{n} . x^2\right\}}}.$$

Let

$$x = \frac{t}{\sqrt{m + (1-m). t^2}};$$

we shall have

$$B = \frac{3 b M}{k^3} \int \frac{t^2 d t}{\left(1 + \frac{1-m}{m} . t^2\right)^{\frac{3}{2}} \left(1 + \frac{1-n}{n} . t^2\right)^{\frac{1}{2}}};$$

an integral relative to t which must be taken, like the integral relative to x from t = 0 to t = 1, because x = 0 gives t = 0 and x = 1, gives t = 1

Hence it follows that if we suppose

$$\frac{1-m}{m} = \lambda^2; \ \frac{1-n}{n} = \lambda'^2; \ F = \int \frac{x^2 \, dx}{\sqrt{(1+\lambda^2 x^2).(1+\lambda'^2 x^2)}};$$

we shall have

$$B = \frac{3\,b\,M}{k^3}\left(\frac{d.\lambda F}{d\lambda}\right).$$

If we change in this expression, b into c, λ into λ′ and reciprocally, we shall have the value of C. The attractions A, B, C of the spheroid, parallel to its three axes are thus given by the following formulas

$$A = \frac{3\,a\,M}{k^3}.F; \ B = \frac{3\,b\,M}{k^3}\left(\frac{d.\lambda F}{d\lambda}\right); \ C = \frac{3\,c\,M}{k^3}\left(\frac{d.\lambda' F}{d\lambda'}\right).$$

We may observe that these expressions obtaining for all the interior points, and consequently for those infinitely near to the surface, they also hold good for the points of the surface.

The determination of the attractions of a spheroid thus depends only on the value of F; but although this value is only a definite integral, it has, however, all the difficulty of indefinite integrals when λ and λ′ are indeterminate, for if we represent this definite integral, taken from x = 0 to x = 1, by $\varphi(\lambda^2, \lambda'^2)$, it is easy to see that the indefinite integral will be $x^3 \varphi(\lambda x^2, \lambda'^2 x^2)$, so that the first being given, the second is likewise given. The indefinite integral is only possible in itself when one of the quantities λ, λ′ is nothing, or when they are equal: in these two cases, the spheroid is an ellipsoid of revolution, and k will be its semi-axis of revolution if λ and λ′ are equal. In this last case we have

$$F = \int \frac{x^2 \, dx}{1+\lambda^2 x^2} = \frac{1}{\lambda^3}\{\lambda - \tan.^{-1}\lambda\}.$$

To get the partial differences $\left(\frac{d.\lambda F}{d\lambda}\right)$, $\left(\frac{d.\lambda' F}{d\lambda'}\right)$, which enter the expressions of B, C, we shall observe that

$$dF = \frac{d\lambda}{\lambda}\left(\frac{d.\lambda F}{d\lambda}\right) + \frac{d\lambda'}{\lambda'}\left(\frac{d.\lambda' F}{d\lambda'}\right) - F\left(\frac{d\lambda}{\lambda} + \frac{d\lambda'}{\lambda'}\right).$$

but when λ = λ′, we have

$$\left(\frac{d.\lambda F}{d\lambda}\right) = \left(\frac{d.\lambda' F}{d\lambda'}\right); \ \frac{d\lambda}{\lambda} = \frac{d\lambda'}{\lambda'};$$

wherefore

$$\left(\frac{d.\lambda F}{d\lambda}\right).d\lambda = \tfrac{1}{2}.\lambda\, dF + F\, d\lambda = \frac{1}{2\lambda} d.\lambda^2 F.$$

Substituting for F its value, we shall have

$$\left(\frac{d.\lambda F}{d\lambda}\right) = \frac{1}{2\lambda^3}\left(\tan.^{-1}\lambda - \frac{\lambda}{1+\lambda^2}\right).$$

we shall therefore have relatively to ellipsoids of revolution, whose semi-axis of revolution is k,

$$A = \frac{3\,a.\,M}{k^3.\,\lambda^3}\,(\lambda - \tan.^{-1}\lambda);$$

$$B = \frac{3\,b.\,M}{2\,k^3.\,\lambda^3}\left(\tan.^{-1}\lambda - \frac{\lambda}{1+\lambda^2}\right);$$

$$C = \frac{3\,c.\,M}{2\,k^3.\,\lambda^3}\left(\tan.^{-1}\lambda - \frac{\lambda}{1+\lambda^2}\right).$$

548. Now let us consider the attraction of spheroids upon an exterior point. This research presents greater difficulties than the preceding because of the radical $\sqrt{.R}$ which enters the differential expressions, and which under this form renders the integrations impossible. We may render them possible by a suitable transformation of the variables of which they are functions; but instead of that method, let us use the following one, founded solely upon the differentiation of functions.

If we designate by V the sum of all the elements of the spheroid divided by their respective distances from the point attracted, and x, y, z the coordinates of the element d M of the spheroid, and a, b, c those of the point attracted, we shall have

$$V = \int \frac{d\,M}{\sqrt{(a-x)^2 + (b-y)^2 + (c-z)^2}}.$$

Then designating, as above, by A, B, C the attractions of the spheroid parallel to the axes of x, y, z, and directed towards their origin, we shall have

$$A = \int \frac{(a-x).\,d\,M}{\{(a-x)^2 + (b-y)^2 + (c-z)^2\}^{\frac{3}{2}}} = -\left(\frac{d\,V}{d\,a}\right).$$

In like manner we get

$$B = -\left(\frac{d\,V}{d\,b}\right),\; C = -\left(\frac{d\,V}{d\,c}\right);$$

whence it follows that if we know V, it will be easy thence to obtain by differentiation alone, the attraction of a spheroid parallel to any straight line whatever, by considering this straight line as one of the rectangular coordinates of the point attracted; a remark we have already made in 541.

The preceding value of V, reduced into a series, becomes

$$V = \int \frac{d\,M}{\sqrt{a^2+b^2+c^2}} \left\{ \begin{array}{l} 1 + \frac{1}{2}.\,\frac{2\,a\,x + 2\,b\,y + 2\,c\,z - x^2 - y^2 - z^2}{a^2+b^2+c^2} \\ \quad + \frac{3}{8}.\,\frac{(2\,a\,x + 2\,b\,y + 2\,c\,z - x^2 - y^2 - z^2)^2}{(a^2+b^2+c^2)^2} + \&c. \end{array} \right\}.$$

This series is ascending relatively to the dimensions of the spheroid.

and descending relatively to the coordinates of the point attracted. If we only retain the first term, which is sufficient when the attracted point is at a very great distance, we shall have

$$V = \frac{M}{\sqrt{a^2 + b^2 + c^2}};$$

M being the entire mass of the spheroid. This expression will be still more exact, if we place the origin of coordinates at the center of gravity of the sphere; for by the property of this center we have

$$\int x . d M = 0; \int y . d M = 0; \int z . d M = 0;$$

so that if we consider a very small quantity of the first order, the ratio of the dimensions of the spheroid to its distance from the point attracted, the equation

$$V = \frac{M}{\sqrt{a^2 + b^2 + c^2}}$$

will be exact to quantities nearly of the third order.

We shall now investigate a rigorous expression of V relatively to elliptic spheroids.

549. If we adopt the denominations of 544, we shall have

$$V = \int \frac{d M}{r} = \iiint r \, d r \, d p \, d q \sin. p = \tfrac{1}{2} \iint (r'^2 - r^2) \, d p \, d q . \sin. p.$$

Substituting for r and r′ their values found in 544, we shall have

$$V = 2 \iint \frac{d p . d q \sin. p . I . \sqrt{R}}{L^2}.$$

Let us resume the values of A B, C relative to the exterior points, and given in 546,

$$A = 2 \iint \frac{d p . d q \sin. p \cos. p \sqrt{R}}{L};$$

$$B = 2 \iint \frac{d p . d q \sin.^2 p \cos. q \sqrt{R}}{L};$$

$$C = 2 \iint \frac{d p . d q \sin.^2 p \sin. q \sqrt{R}}{L}.$$

Since at the limits of the integrals, we have $\sqrt{R} = 0$, it is easy to see that by taking the first differences of V, A, B, C relatively to any of the six quantities a, b, c, k, m, n, we may dispense with regarding the variations of the limits; so that we have, for example,

$$\left(\frac{d V}{d a}\right) = 2 \iint d p \, d q \sin. p \left(\frac{d . \frac{I \sqrt{R}}{L^2}}{d a} \right);$$

for the integral

$$\int \frac{d p \sin. p \, I \sqrt{R}}{L^2}$$

is towards these limits, very nearly proportional to $R^{\frac{3}{2}}$, which renders equal to zero, its differential at these limits. Hence it is easy to see by differentiation that if for brevity we make

$$a\,A + b\,B + c\,C = F;$$

we shall have between the four quantities B, C, F, and V the following equation of partial differences,

$$0 = \frac{a^2 + b^2 + c^2 - k^2}{2}.k.\left\{\left(\frac{d\,V}{d\,k}\right) - \left(\frac{d\,F}{d\,k}\right)\right\} + k^2(V - F)$$

$$+ k^2\frac{m-1}{m}.b.\left\{\left(\frac{d\,F}{d\,b}\right) - \tfrac{1}{2}\left(\frac{d\,V}{d\,b}\right) - B\right\}$$

$$+ k^2\frac{n-1}{n}.c\left\{\left(\frac{d\,F}{d\,c}\right) - \tfrac{1}{2}\left(\frac{d\,V}{d\,c}\right) - C\right\}$$

$$- k^2(m-1).\left(\frac{d\,F}{d\,m}\right) - k^2(n-1).\left(\frac{d\,F}{d\,n}\right).(1)$$

We may eliminate from this equation, the quantities B, C, F by means of their values

$$-\left(\frac{d\,V}{d\,b}\right), - \left(\frac{d\,V}{d\,c}\right) \text{and} - a\left(\frac{d\,V}{d\,a}\right) - b\left(\frac{d\,V}{d\,a}\right) - c\left(\frac{d\,V}{d\,c}\right).$$

We shall thus get an equation of partial differences in V alone. Let therefore

$$V = \frac{4\,\pi.k^3}{3\sqrt{m\,n}}.v = M.v,$$

M being by 545, the mass of the elliptic spheroid; and for the variables m and n let us here introduce θ and ϖ which shall be such that we have

$$\theta = \frac{1-m}{m}.k^2; \quad \varpi = \frac{1-n}{n}.k^2;$$

θ will be the difference of the square of the axis of the spheroid parallel to y and the square of the axis parallel to x; ϖ will be the difference of the square of the axis of z and the square of the axis of x; so that if we take for the axis of x, the smallest of the three axes of the spheroid, $\sqrt{\theta}$ and $\sqrt{\varpi}$ will be its two excentricities. Thus we shall have

$$k\left(\frac{d\,V}{d\,k}\right) = M.\left\{2\,\theta\left(\frac{d\,v}{d\,\theta}\right) + 2\,\varpi\left(\frac{d\,v}{d\,\varpi}\right) + k\left(\frac{d\,v}{d\,k}\right) + 3\,v\right\};$$

$$\left(\frac{d\,V}{d\,m}\right) = -M\left\{\frac{k^2}{m^2}.\left(\frac{d\,v}{d\,\theta}\right) + \frac{v}{2\,m}\right\};$$

$$\left(\frac{d\,V}{d\,m}\right) = -M\left\{\frac{k^2}{n^2}\left(\frac{d\,v}{d\,\varpi}\right) + \frac{v}{2\,n}\right\};$$

V being considered in the first members of those equations as a function of a, b, c, k, m, n; and v being considered in their second members as a function of a, b, c, θ, ϖ, k.

If we make

$$Q = a\left(\frac{d\,v}{d\,a}\right) + b\left(\frac{d\,v}{d\,b}\right) + c\left(\frac{d\,v}{d\,c}\right);$$

we shall have $F = -M\,Q$, and we shall get the values of $k\left(\frac{d\,F}{d\,k}\right)$, $\left(\frac{d\,F}{d\,m}\right)$, $\left(\frac{d\,F}{d\,n}\right)$ by changing in the preceding values of $k\left(\frac{d\,V}{d\,k}\right)$, $\left(\frac{d\,V}{d\,m}\right)$, $\left(\frac{d\,V}{d\,n}\right)$, v into $-$ Q. Moreover V and F are homogeneous functions in a, b, c, k, $\sqrt{\theta}$, $\sqrt{\varpi}$ of the second dimension, for V being the sum of the elements of the spheroid, divided by their distances from the point attracted, and each element being of three dimensions, V is necessarily of two dimensions, as also F which has the same number of dimensions as V; v and Q are therefore homogeneous functions of the same quantities and of the dimension -1; thus we shall have by the nature of homogeneous functions,

$$a.\left(\frac{d\,v}{d\,a}\right) + b.\left(\frac{d\,v}{d\,b}\right) + c.\left(\frac{d\,v}{d\,c}\right) + 2\,\theta.\left(\frac{d\,v}{d\,\theta}\right) + 2\,\varpi.\left(\frac{d\,v}{d\,\varpi}\right) + k\left(\frac{d\,v}{d\,k}\right) = -v;$$

an equation which may be put under this form

$$2\,\theta\left(\frac{d\,v}{d\,\theta}\right) + 2\,\varpi\left(\frac{v}{d\,\varpi}\right) + k\left(\frac{d\,v}{d\,k}\right) = -v - Q.$$

We shall have in like manner

$$a.\left(\frac{d\,Q}{d\,a}\right) + b\left(\frac{d\,Q}{d\,b}\right) + c\left(\frac{d\,Q}{d\,c}\right) + 2\theta\left(\frac{d\,Q}{d\,\theta}\right) + 2\varpi\left(\frac{d\,Q}{d\,\varpi}\right) + k\left(\frac{d\,Q}{d\,k}\right) = -Q;$$

then, if in equation (1) we substitute for V, F and their partial differences; if moreover we substitute $\frac{k^2}{k^2+\theta}$ for m and $\frac{k^2}{k^2+\varpi}$ for n, we shall have

$$0 = (a^2 + b^2 + c^2)\left[v + \tfrac{1}{2}\,Q - \tfrac{1}{2}\left\{a\left(\frac{d\,Q}{d\,a}\right) + b\left(\frac{d\,Q}{d\,b}\right) + c\left(\frac{d\,Q}{d\,c}\right)\right\}\right]$$

$$+ \theta^2\left(\frac{d\,Q}{d\,\theta}\right) + \varpi^2\left(\frac{d\,Q}{d\,\varpi}\right) - \frac{k^3}{2}\left(\frac{d\,Q}{d\,k}\right) + \tfrac{1}{2}(\theta + \varpi).\,Q \qquad (2)$$

$$b\,\theta.\left(\frac{d\,Q}{d\,b}\right) + c\,\varpi\left(\frac{d\,Q}{d\,c}\right) - \tfrac{1}{2}\,b\,\theta\left(\frac{d\,v}{d\,b}\right) - \tfrac{1}{2}\,c\,\varpi\left(\frac{d\,v}{d}\right).$$

550. Conceive the function v expanded into a series ascending relatively to the dimensions k, $\sqrt{\theta}$, $\sqrt{\varpi}$ of the spheroid, and consequently descending relatively to the quantities a, b, c: this series will be of the following form:

$$v = U^{(0)} + U^{(1)} + U^{(2)} + U^{(3)} + \&c.;$$

$U^{(0)}$, $U^{(1)}$, &c. being homogeneous functions of a, b, c, k, $\sqrt{\theta}$, $\sqrt{\varpi}$, and separately homogeneous relatively to the three first and to the three last of these six quantities; the dimensions relative to the three first always decreasing, and the dimensions relative to the three last increasing continually. These functions being of the same dimension as v, are all of the dimension — 1.

If we substitute in equation (2) for v its preceding expanded value; if we call s the dimension of $U^{(i)}$ in k, $\sqrt{\theta}$, $\sqrt{\varpi}$, and consequently — s — 1 its dimension in a, b, c; if in like manner we name s′ the dimension of $U^{(i+1)}$ in k, $\sqrt{\theta}$, $\sqrt{\varpi}$, and consequently — s′ — 1 its dimension in a, b, c; if we then consider that by the nature of homogeneous functions we have

$$a\left(\frac{d\,U^{(i)}}{d\,a}\right) + b\left(\frac{d\,U^{(i)}}{d\,b}\right) + c\left(\frac{d\,U^{(i)}}{d\,c}\right) = -(s+1)\,U^{(i)};$$

$$a\left(\frac{d\,U^{(i+1)}}{d\,a}\right) + b\left(\frac{d\,U^{(i+1)}}{d\,b}\right) + c\left(\frac{d\,U^{(i+1)}}{d}\right) = -(s'+1)\,U^{(i+1)};$$

we shall have, by rejecting the terms of a dimension superior in k, $\sqrt{\theta}$, $\sqrt{\varpi}$ to that of the terms which we retain,

$$U^{(i+1)} = \frac{\left\{\begin{array}{l} \frac{1}{2}(s+1).k^3.\left(\frac{d\,U^{(i)}}{d\,k}\right) - (s+1).\theta^2\left(\frac{d\,U^{(i)}}{d\,\theta}\right) \\ -(s+1).\varpi^2.\left(\frac{d\,U^{(i)}}{d\,\varpi}\right) - \frac{s+1}{2}.(\theta+\varpi).U^{(i)} \\ -(s+\frac{3}{2})\,b\,\theta.\left(\frac{d\,U^{(i)}}{d\,b}\right) - (s+\frac{3}{2}).\,c\,\varpi.\left(\frac{d\,U^{(i)}}{d\,c}\right) \end{array}\right\}}{s'.\frac{s'+3}{2}.(a^2+b^2+c^2)} \quad . \; . \; . \; (3)$$

This equation gives the value of $U^{(i+1)}$, by means of $U^{(i)}$ and of its partial differences; but we have

$$U^{(0)} = \frac{1}{(a^2+b^2+c^2)^{\frac{1}{2}}};$$

since, retaining only the first term of the series, we have found in 548, that

$$V = \frac{M}{(a^2+b^2+c^2)^{\frac{1}{2}}}.$$

Substituting therefore this value of $U^{(0)}$ in the preceding formula, we shall get that of $U^{(1)}$; by means of that of $U^{(1)}$ we shall have that of $U^{(2)}$, and so on. But it is remarkable that none of these quantities contains k: for it is evident by the formula (3) that $U^{(0)}$, not containing $U^{(1)}$, does not contain it; that $U^{(1)}$ not containing it, $U^{(2)}$ will not contain it, and so on; so that the entire series $U^{(0)} + U^{(1)} + \&c.$ is independent of k, or which is the same thing $\left(\frac{d\,v}{dk}\right) = 0$. The values of v, $-\left(\frac{d\,v}{d\,a}\right)$, $-\left(\frac{d\,v}{d\,b}\right)$,

$-\left(\frac{d\,v}{d\,c}\right)$, are therefore the same for all elliptic spheroids similarly situated, and which have the same excentricities $\sqrt{\theta}$, $\sqrt{\varpi}$; but $-M\left(\frac{d\,v}{d\,a}\right)$, $-M\left(\frac{d\,v}{d\,b}\right)$, $-M\left(\frac{d\,v}{d\,c}\right)$, express by 548, the attractions of the spheroid parallel to its three axes; therefore the attractions of different elliptic spheroids which have the same center, the same position of the axes and the same excentricities, upon an exterior point, are to one another as their masses.

It is easy to see by formula (3) that the dimensions of $U^{(0)}$, $U^{(1)}$, $U^{(2)}$, &c. in $\sqrt{\theta}$ and $\sqrt{\varpi}$, increase two units at a time, so that $s=2i$, $s'=2i+2$; moreover we have by the nature of homogeneous functions

$$\varpi\left(\frac{d\,U^{(i)}}{d\,\varpi}\right)=i\,.\,U^{(i)}-\theta\left(\frac{d\,U^{(i)}}{d\,\theta}\right);$$

this formula will therefore become

$$U^{(i+1)}=\frac{\left\{\begin{array}{l}(2i+1).\theta.(\varpi-\theta)\left(\frac{d\,U^{(i)}}{d\,\theta}\right)-(2i+\frac{3}{2}).b.\theta\left(\frac{d\,U^{(i)}}{d\,b}\right)\\-(2i+\frac{3}{2})c\,\varpi\left(\frac{d\,U^{(i)}}{d\,c}\right)-\frac{1}{2}(2i+1).\{\theta+(2i+1).\varpi\}\,U^{(i)}\end{array}\right\}}{(i+1)\,(2i+5)\,(a^2+b^2+c^2)}. \quad (4)$$

By means of this equation, we shall have the value of v in a series very convergent, whenever the excentricities $\sqrt{\theta}$, $\sqrt{\varpi}$ are very small, or when the distance $\sqrt{a^2+b^2+c^2}$ of the point attracted from the center of the spheroid is very great relatively to the dimensions of the spheroid.

If the spheroid is a sphere, we shall have $\theta=0$, and $\varpi=0$, which give $U^{(1)}=0$, $U^{(2)}=0$, &c.; wherefore

$$v=U^{(0)}=\frac{1}{\sqrt{a^2+b^2+e^2}};$$

and

$$V=\frac{M}{\sqrt{a^2+b^2+c^2}};$$

whence it follows that the value of V is the same as if all the mass of the sphere were condensed into its center, and that thus, a sphere attracts any exterior point, as if all its whole mass were condensed into its center; a result already obtained in 542.

551. The property of the function of v being independent of k, furnishes the means of reducing its value to the most simple form of which it is susceptible; for since we can make k vary at pleasure without changing this value, provided the spheroid retain the same excentricities, $\sqrt{\theta}$ and

$\sqrt{\varpi}$, we may suppose k such that the spheroid shall bè infinitely flattened, or so contrived that its surface pass through the point attracted. In these two cases, the research of the attractions of the spheroid is rendered more simple; but since we have already determined the attractions of elliptic spheroids, upon points at the surface, we shall now suppose k such that the surface of the spheroid passes through the point of attraction.

If we call k′, m′, n′ relatively to this new spheroid what in 545, we named k, m, n relatively to the spheroid we there considered; the condition that the point attracted is at the surface, and that also a, b, c are the coordinates of a point of the surface, will give

$$a^2 + m' b^2 + n' c^2 = k^2;$$

and since we suppose the excentricities $\sqrt{\theta}$ and $\sqrt{\varpi}$ to remain the same, we shall have

$$\frac{1 - m'}{m'} . k'^2 = \theta; \quad \frac{1 - n'}{n'} . k'^2 = \varpi;$$

whence we obtain

$$m' = \frac{k'^2}{k'^2 + \theta}; \quad n' = \frac{k'^2}{k'^2 + \varpi};$$

we shall therefore have to determine k′, the equation

$$a^2 + \frac{k'^2}{k'^2 + \theta} b^2 + \frac{k'^2}{k'^2 + \varpi} . c^2 = k'^2. \quad . \quad . \quad . \qquad (5)$$

It is easy hence to conclude that there is only one spheroid whose surface passes through the point attracted, θ and ϖ remaining the same. For if we suppose, which we always may do, that θ and ϖ are positive, it is clear that augmenting in the preceding equation, k'^2 by any quantity which we may consider an aliquot part of k'^2, each of the terms of the first member of this equation, will increase in a less ratio than k'^2; therefore if in the first state of k'^2, there subsist an equality between the two members of this equation, this equality will no longer obtain in the second state; whence it follows that k'^2 is only susceptible of one real and positive value.

Let M′ be the mass of the new spheroid, and A′, B′, C′ its attractions parallel to the axes of a, b, c; if we make

$$\frac{1 - m'}{m'} = \lambda^2; \quad \frac{1 - n'}{n} = \lambda'^2$$

$$F = \int \frac{x^2 \, d x}{\sqrt{(1 + \lambda^2 . x^2) . (1 + \lambda'^2 . x^2)}};$$

by 547, we shall have

$$A' = \frac{3 a M' F}{k'^3}; \; B' = \frac{3 b M'}{k'^3} \left(\frac{d.}{q} \frac{\lambda F}{\lambda}\right); \; C' = \frac{3 c M'}{k'^3} \left(\frac{d . \lambda' F}{d \lambda'}\right).$$

Changing in these values of A′, B′, C′, M′ into M, we shall have by the preceding No., the values of A, B, C relatively to the first spheroid but the equations

$$\frac{1-m'}{m'} \cdot k'^2 = \theta; \quad \frac{1-n'}{n'} \cdot k'^2 = \varpi,$$

give

$$\lambda^2 = \frac{\theta}{k'^2}; \quad \lambda'^2 = \frac{\varpi}{k'^2};$$

k'^2 being given by equation (5) which we may put under this form

$$0 = k'^6 - (a^2+b^2+c^2-\theta-\varpi)\,k'^4 - \{(a^2+c^2)\,\theta + (a^2+b^2)\,\varpi - \theta\,\varpi\}\,k'^2 - a^2.\theta\,\varpi;$$

we shall therefore have

$$A = \frac{3\,a\,M}{k'^3} \cdot F; \quad B = \frac{3\,b\,M}{k'^3}\left(\frac{d.\lambda\,F}{d\,\lambda}\right); \quad C = \frac{3\,c\,M}{k'^3}\left(\frac{d.\lambda'\,F}{d\,\lambda'}\right).$$

These values obtain relatively to all points exterior to the spheroid, and to extend them to those of the surface, and even to the interior points we have only to change k′ to k.

If the spheroid is one of revolution, so that $\theta = \varpi$, the formula (5) will give

$$2\,k'^2 = a^2+b^2+c^2-\theta+\sqrt{(a^2+b^2+c^2-\theta)^2+4\,a^2.\theta};$$

and by 547, we shall have

$$A = \frac{3\,a\,M}{k'^3.\lambda^3}\left(\lambda - \tan.^{-1}\lambda\right)$$

$$B = \frac{3\,b\,M}{2\,k'^3.\lambda^3}\left(\tan.^{-1}\lambda - \frac{\lambda}{1+\lambda^2}\right)$$

$$C = \frac{3\,c\,M}{2\,k'^3.\lambda^3}\left(\tan.^{-1}\lambda - \frac{\lambda}{1+\lambda^2}\right).$$

Thus we have terminated the complete theory of the attractions of elliptic spheroids; for all that remains to be done is the integration of the differential expression of F, and this integration in the general sense is impossible, not only by known methods, but also in itself. The value of F cannot be expressed in finite terms by algebraic, logarithmic or circular quantities; or which is tantamount, by any algebraic function of quantities whose exponents are constant, nothing or variable. Functions of this kind being the only ones which can be expressed independently of the symbol ∫, all the integrals which cannot be reduced to such functions, are impossible in finite terms.

If the elliptic spheroid is not homogeneous, and if it is composed of elliptic shells varying in position, excentricity and density according to any law whatever, we shall have the attraction of one of its shells, by determining as above the difference of the attractions of two homogeneous elliptic spheroids, having the same density as the shell, one of which shall have for its surface the exterior surface of the shell, and the other the interior surface of the shell. Then summing this differential attraction, we shall have the attraction of the whole spheroid.

THE DEVELOPEMENT INTO SERIES, OF THE ATTRACTIONS OF ANY SPHEROIDS WHATEVER.

552. Let us consider generally the attractions of any spheroids whatever. We have seen in No. 547, that the expression V of the sum of the elements of the spheroid, divided by their distances from the attracted points, possesses the advantage of giving by its differentiation, the attraction of this spheroid parallel to any straight line whatever. We shall see moreover, when treating of the figure of the planets, that the attraction of their elements presents itself under this form in the equation of their equilibrium; thus we proceed particularly to investigate V.

Let us resume the equation of No. 548,

$$V = \int \frac{d\,M}{\sqrt{(a-x)^2 + (b-y)^2 + (c-z)^2}};$$

a, b, c being the coordinates of the point attracted; x, y, z those of the element d M of the spheroid; the origin of coordinates being in the interior of the spheroid. This integral must be taken relatively to the variables x, y, z, and its limits are independent of a, b, c; hence we shall find by differentiation,

$$0 = \left(\frac{d^2 V}{d\,a^2}\right) + \left(\frac{d^2 V}{d\,b^2}\right) + \left(\frac{d^2 V}{d\,c^2}\right); \quad \ldots\ldots\ldots \quad (1)$$

an equation already obtained in 541,

Let us transform the coordinates to others more commodious. For that purpose, let r be the distance of the point attracted from the origin of coordinates; θ the angle which the radius r makes with the axis of a; ϖ the angle which the plane formed by the radius and this axis, makes with the plane of the axis of a, and of b; we shall have

$$a = r\cos.\theta;\ b = r\sin.\theta\cos.\theta;\ c = r\sin.\theta\sin.\varpi.$$

If in like manner we name R, θ', ϖ' what r, θ, ϖ become relatively to the element d M of the spheroid; we shall have

$$x = R\cos.\theta';\ y = R\sin.\theta'\cos.\varpi';\ z = R\sin.\theta'.\sin.\varpi'.$$

Moreover, the element d M of the spheroid is equal to a rectangular parallelopiped whose dimensions are d R, R d θ', R d ϖ' sin. θ', and consequently it is equal to $\rho . R^2 . dR . d\theta' . d\varpi' . \sin.\theta'$, ρ being its density; we shall thus have

$$V = \iiint \frac{\rho R^2 . dR . d\theta' . d\varpi' \sin.\theta'}{\sqrt{r^2 - 2rR\{\cos.\theta . \cos.\theta' + \sin.\theta \sin.\theta' \cos.(\varpi' - \varpi)\} + R^2}};$$

the integral relative to R must be taken from R = 0 to the value of R at the surface of the spheroid; the integral relative to ϖ' must be taken from $\varpi' = 0$ to ϖ' equal to the circumference; and the integral relative to θ' must be taken from $\theta' = 0$ to θ' equal to the semi-circumference. Differentiating this expression of V, we shall find

$$0 = \left(\frac{d^2 V}{d\theta^2}\right) + \frac{\cos.\theta}{\sin.\theta} . \left(\frac{dV}{d\theta}\right) + \frac{\left(\frac{d^2 V}{d\varpi^2}\right)}{\sin.^2\theta} + r\left(\frac{d^2 . rV}{dr^2}\right); \quad . . . (2)$$

an equation which is only equation (1) in another form.

If we make $\cos.\theta = m$, we may give it this form

$$0 = \left(\frac{d . \left\{(1 - m^2)\left(\frac{dV}{dm}\right)\right\}}{dm}\right) + \frac{\left(\frac{d^2 V}{d\varpi}\right)}{1 - m^2} + r\left(\frac{d^2 . rV}{dr^2}\right) . . (3)$$

We have already arrived at these several equations in 541.

553. First, let us suppose the point attracted to be exterior to the spheroid. If we wish to expand V into a series, it ought in this case, to descend relatively to powers of r, and consequently to be of this form

$$V = \frac{U^{(0)}}{r} + \frac{U^{(1)}}{r^2} + \frac{U^{(2)}}{r^3} + \&c.$$

Substituting this value of V in equation (3) of the preceding No., the comparison of the same powers of r will give, whatever i may be

$$0 = \left(\frac{d . \left\{(1 - m^2)\left(\frac{dU^{(i)}}{dm}\right)\right\}}{dm}\right) + \frac{\left(\frac{d^2 U^{(i)}}{d\varpi}\right)}{1 - m^2} + i(i+1)U^{(i)}.$$

It is evident from the integral expression alone of V that $U^{(i)}$ is a rational and entire function of m, $\sqrt{1 - m^2} . \sin.\varpi$, and $\sqrt{1 - m^2} . \cos.\varpi$, depending upon the nature of the spheroid. When i = 0, this function becomes a constant; and in the case of i = 1, it assumes the form

$$H m + H' \sqrt{1 - m^2} . \sin.\varpi + H'' \sqrt{1 - m^2} . \cos.\varpi;$$

H, H′, H″ being constants.

To determine generally $U^{(i)}$ call T the radical

$$\frac{1}{\sqrt{r^2 - 2Rr\{\cos.\theta \cos.\theta' + \sin.\theta \sin.\theta' \cos.(\varpi' - \varpi)\} + R^2}},$$

we shall have

$$0=\left(\frac{d\left\{(1-m^2)\left(\frac{dT}{dm}\right)\right\}}{dm}\right)+\frac{\left(\frac{d^2T}{d\varpi^2}\right)}{1-m^2}+r\left(\frac{d^2.rT}{dr^2}\right).$$

This equation will still subsist if we change θ into θ', ϖ into ϖ', and reciprocally; because T is a similar function of θ', ϖ' and of θ, ϖ.

If we expand T, in a series descending relatively to r, we shall have

$$T=\frac{Q^{(0)}}{r}+Q^{(1)}.\frac{R}{r^2}+Q^{(2)}.\frac{R^2}{r^3}+\&c.$$

$Q^{(i)}$ being, whatever i may be, subject to the condition that

$$0=\left(\frac{d\left\{(1-m^2)\left(\frac{dQ^{(i)}}{dm}\right)\right\}}{dm}\right)+\frac{\left(\frac{d^2Q^{(i)}}{d\varpi^2}\right)}{1-m^2}+i\,(i+1).Q^{(i)};$$

and moreover it is evident, that $Q^{(i)}$ is a rational and entire function of m, and $\sqrt{1-m^2}.\cos.(\varpi'-\varpi)$: $Q^{(i)}$ being known, we shall have $U^{(i)}$ by means of the equation

$$U^{(i)}=\int\rho\,R^{(i+2)}.dR.d\varpi'.d\theta'.\sin.\theta'.Q^{(i)}.$$

Now suppose the point attracted in the interior of the spheroid: we must then develope the integral expression of V, in a series ascending relatively to r, which gives for V a series of the form

$$V=v^{(0)}+r.v^{(1)}+r^2.v^{(2)}+r^3.v^{(3)}+\&c.$$

$v^{(i)}$ being a rational and whole function of m, $\sqrt{1-m^2}.\sin.\varpi$ and $\sqrt{1-m^2}\cos.\varpi$, which satisfies the same equation of partial differences that $U^{(i)}$ does; so that we have

$$0=\left(\frac{d\left\{(1-m^2)\left(\frac{dv^{(i)}}{dm}\right)\right\}}{dm}\right)+\frac{\left(\frac{d^2v^{(i)}}{d\varpi^2}\right)}{1-m^2}+i\,(i+1).v^{(i)}.$$

To determine $v^{(i)}$, we shall expand the radical T into a series ascending according to r, and we shall have

$$T=\frac{Q^{(0)}}{R}+Q^{(1)}.\frac{r}{R^2}+Q^{(2)}.\frac{r^2}{R^3}+\&c.$$

the quantities $Q^{(0)}$, $Q^{(1)}$, $Q^{(2)}$, &c. being the same as above; we shall therefore get

$$v^{(i)}=\int\frac{\rho.dR.d\varpi'.d\theta'.\sin.\theta'.Q^{(i)}}{R^{i-1}}.$$

But since the preceding expression of T is only convergent so long as R is equal to or greater than r, the preceding value of V only relates to the shells of the spheroid, which envelope the point attracted. This point being exterior, relatively to the other shells, we shall determine that part of V which is relative to them by the first series of V.

554. First let us consider those spheroids which differ but very little from the sphere, and determine the functions $U^{(0)}$, $U^{(1)}$, $U^{(2)}$, &c. $v^{(0)}$, $v^{(1)}$, $v^{(2)}$, &c. relatively to these spheroids. There exists a differential equation in V, which holds good at their surface, and which is remarkable because it gives the means of determining those functions without any integration.

Let us suppose generally, that gravity is proportional to a power n of the distance; let d M be an element of the spheroid, and f its distance from the point attracted; call V the integral $\int f^{n+1}\, dM$, which shall extend to the entire mass of the spheroid. In nature we have $n = -2$, it becomes $\int \frac{dM}{f}$, and we have expressed it in like manner by V in the preceding Nos. The function V possesses the advantage of giving, by its differentiation, the attraction of the spheroid, parallel to any straight line whatever; for considering f as a function of the three coordinates of the point attracted perpendicular to one another, and one of which is parallel to this straight line. Call r this coordinate, the attraction of the spheroid along r and directed towards its origin, will be $\int . f^n . \left(\frac{df}{dr}\right) . dM$. Consequently it will be equal to $\frac{1}{n+1}\left(\frac{dV}{dr}\right)$, which, in the case of nature, becomes $-\left(\frac{dV}{dr}\right)$, conformably with what has been already shown.

Suppose, however, that the spheroid differs very little from a sphere of the radius a, whose center is upon the radius r perpendicular to the surface of the spheroid, the origin of the radius being supposed to be arbitrary, but very near to the center of gravity of the spheroid; suppose, moreover, that the sphere touches the spheroid, and that the point attracted is at the point of contact of the two surfaces. The spheroid is equal to the sphere plus the excess of the spheroid above the sphere; but we may conceive this excess as being formed of an infinite number of molecules spread over the surface of the sphere, these molecules being supposed negative wherever the sphere exceeds the spheroid; we shall therefore have the value of V by determining this value, 1st, relatively to the sphere; 2dly, relatively to the different molecules.

Relatively to the sphere, V is a function of a, which we denote by A; if we name d m one of the molecules of the excess of the spheroid above the sphere, and f its distance from the point attracted; the value of V relative to this excess will be $\int . f^{n+1} . dm$; we shall therefore have, for the entire value of V, relative to the spheroid,

$$V = A + \int . f^{n+1} . dm.$$

Conceive that the point attracted is elevated by an infinitely small quantity d r, above the surface of the spheroid and the sphere upon r or a produced; the value of V, relative to this new position of the attracted point, will become

$$V + \left(\frac{dV}{dr}\right) dr;$$

A will increase by a quantity proportional to d r, and which we shall represent by A′. d r. Moreover, if we name γ the angle formed by the two radii drawn from the center of the sphere to the point attracted, and to the molecule d m, the distance f of this element or molecule from the point attracted, will be in the first position of the point, equal to

$$\sqrt{2 a^2 (1 - \cos. \gamma)};$$

in the second position it will be

$$\sqrt{(a + dr)^2 - 2 a (a + dr) \cos. \gamma + a^2},$$

or

$$f \left(1 + \frac{dr}{2a}\right);$$

the integral $\int . f^{n+1} dm$, will thus become

$$\left\{1 + \frac{n+1}{2} . \frac{dr}{a}\right\} \int . f^{n+1} dm;$$

we shall therefore have

$$\left(\frac{dV}{dr}\right) . dr = A' dr + \frac{n+1}{2} . \frac{dr}{a} \int . f^{n+1} . dm;$$

substituting for $\int . f^{n+1} . dm$, its value V — A, we shall have

$$\left(\frac{dV}{dr}\right) = A' - \frac{(n+1) A}{2a} + \frac{n+1}{2a} . V; \quad . \; . \; . \; (1)$$

In the case of nature, the equation (1) becomes

$$-a\left(\frac{dA}{dr}\right) = -a A' - \tfrac{1}{2} A + \tfrac{1}{2} V.$$

The value of V relative to the sphere of radius a, is, by 550, equal to $\frac{4 \pi a^3}{3 r}$, which gives $A = \frac{4 \pi a^3}{3}$; $A' = -\frac{4 \pi a}{3}$; we shall therefore get

$$-a\left(\frac{dV}{dr}\right) = \frac{2 \pi . a^2}{3} + \tfrac{1}{2} V; \quad . \; . \; . \; . \; . \; . \; . \; (2)$$

We must here observe that this equation obtains, whatever may be the position of the straight line r, and even in the case where it is not perpendicular to the surface of the spheroid, provided that it passes very near its center of gravity, for it is easy to see that the attraction of the spheroid, resolved parallel to these straight lines, and which, as we have seen, is equal to $-\left(\frac{dV}{dr}\right)$, is, whatever may be their position, always the same, to quantities nearly of the order of the square of the excentricity of the spheroid.

555. Let us resume the general expression of V of 553, relative to a point attracted exterior to the spheroid,

$$V = \frac{U^{(0)}}{r} + \frac{U^{(1)}}{r^2} + \frac{U^{(2)}}{r^3} + \&c.$$

the function $U^{(i)}$ being, whatever i may be, subject to the equation of partial differences

$$0 = \left(\frac{d\left\{(1-m^2)\cdot\left(\frac{dU^{(i)}}{dm}\right)\right\}}{dm}\right) + \frac{\left(\frac{d^2U^{(i)}}{d\varpi^2}\right)}{1-m^2} + i(i+1)\cdot U^{(i)}.$$

By differentiating the value of V relatively to r, we have

$$-\left(\frac{dV}{dr}\right) = \frac{U^{(0)}}{r^2} + \frac{2U^{(1)}}{r^3} + \frac{3U^{(2)}}{r^4} + \&c.$$

Let us represent by $a(1+\alpha y)$ the radius drawn from the origin of r to the surface of the spheroid, α being a very small constant coefficient, whose square and higher powers we shall neglect, and y being a function of m and ϖ depending on the nature of the spheroid. We shall have to quantities nearly of the order α, $V = \frac{4\pi a^3}{3r}$; whence it follows that in the preceding expression of V, 1st, the quantity $U^{(0)}$ is equal to $\frac{4\pi a^3}{3}$ *plus* a very small quantity of the order α, and which we shall denote by $U'^{(0)}$; 2dly, that the quantities $U^{(1)}$, $U^{(2)}$, &c. are small quantities of the order α. Substituting $a(1+\alpha y)$ for r in the preceding expressions of V and of $-\left(\frac{dV}{dr}\right)$, and neglecting quantities of the order α^2, we shall have relatively to an attracted point placed at the surface

$$\tfrac{1}{2}V = \tfrac{2}{3}\pi\cdot a^2(1-\alpha y) + \frac{U'^{(0)}}{2a} + \frac{U^{(1)}}{2a^2} + \frac{U^{(2)}}{2a^3} + \&c.;$$

$$-a\left(\frac{dV}{dr}\right) = \tfrac{4}{3}\pi a^2(1-2\alpha y) + \frac{U'^{(0)}}{a} + \frac{2\cdot U^{(1)}}{a^2} + \frac{3U^{(2)}}{a^3} + \&c.$$

If we substitute these values in equation (2) of the preceding No. we shall have

$$4\alpha\pi a^2\cdot y = \frac{U'^{(0)}}{a} + \frac{3\cdot U^{(1)}}{a^2} + \frac{5U^{(2)}}{a^3} + \frac{7U^{(3)}}{a^4} + \&c.$$

It thence follows that the function y is of this form

$$y = Y^{(0)} + Y^{(1)} + Y^{(2)} + \&c.$$

the quantities $Y^{(0)}$, $Y^{(1)}$, $Y^{(2)}$, &c. as well as $U^{(0)}$, $U^{(1)}$, &c. being subject to the equation of partial differences

$$0 = \left(\frac{d \left\{ (1 - m^2) . \left(\frac{d Y^{(i)}}{d m} \right) \right\}}{d m} \right) + \frac{\left(\frac{d^2 Y^{(i)}}{d \varpi^2} \right)}{1 - m^2} + i (i + 1) . Y^{(i)};$$

this expression of y is not therefore arbitrary, but it is derived from the developement of the attractions of spheroids. We shall see in the following No. that y cannot be thus developed except in one manner only; we shall therefore have generally, by comparing similar functions,

$$U^{(i)} = \frac{4 \alpha \pi}{2 i + 1} a^{i+3} . Y^{(i)};$$

whence, whatever r may be, we derive

$$V = \frac{4 \pi a^3}{3 r} + \frac{4 \alpha \pi . a^3}{r} \left\{ Y^{(0)} + \frac{a}{3 r} . Y^{(1)} + \frac{a^2}{5 r^2} Y^{(2)} + \&c. \right\}; \quad . \; . \; (3)$$

To get V, therefore, it remains only to reduce y to the form above described; for which object we shall give, in what follows, a very simple method.

If we had $y = Y^{(i)}$, the part of V relative to the excess of the spheroid above the sphere whose radius is a, or which is the same thing, relative to a spherical shell whose radius is a, and thickness α a y, would be $\frac{4 \alpha \pi a^{i+3} . Y^{(i)}}{(2 i + 1) r^{i+1}}$; this value would consequently be proportional to y, and it is evident that it is only in this case that the proportionality can subsist.

556. We may simplify the expression $Y^{(0)} + Y^{(1)} + Y^{(2)} + \&c.$ of y, and cause to disappear the two first terms, by taking for a, the radius of a sphere equal in solidity to the spheroid, and by fixing the arbitrary origin of r at the center of gravity of the spheroid. To show this, we shall observe that the mass M of the spheroid supposed homogeneous, and of a density represented by unity, is by 552, equal to $\int R^2 \, d R \, d m \, d \varpi$, or to $\frac{1}{3} \int R'^3 \, d m \, d \varpi$, R′ being the radius R produced to the surface of the spheroid. Substituting for R′ its value a $(1 + \alpha y)$ we shall have

$$M = \frac{4 \pi a^3}{3} + \alpha a^3 \int y \, d m \, d \varpi.$$

All that remains to be done, therefore, is to substitute for y its value $Y^{(0)} + Y^{(1)} + \&c.$ and then to make the integrations. For this purpose here is a general theorem, highly useful also in this analysis.

"If $Y^{(i)}$ and $Z^{(i)}$ be rational and entire functions of m, $\sqrt{1-m^2}\,.\sin.\varpi$
"and $\sqrt{1-m^2}\,.\cos.\varpi$, which satisfy the following equations:

$$\text{"}\,0=\left(\frac{d.\left\{(1-m^2).\left(\frac{dY^{(i)}}{dm}\right)\right\}}{dm}\right)+\frac{\left(\frac{d^2Y^{(i)}}{d\varpi^2}\right)}{1-m^2}+i\,(i+1)\,.\,Y^{(i)};$$

$$\text{"}\,0=\left(\frac{d\left\{(1-m^2).\left(\frac{dZ^{(i)}}{dm}\right)\right\}}{dm}\right)+\frac{\left(\frac{d^2Z^{(i')}}{d\varpi^2}\right)}{1-m^2}+i'\,(i'+1)\,.\,Z^{(i')};$$

"we shall have generally

$$\text{"}\int Y^{(i)}\,.\,Z^{(i')}\,.\,dm\,d\varpi=0,$$

"whilst i and i′ are whole positive numbers differing from one another;
"the integrals being taken from m = −1 to m = 1, and from $\varpi = 0$
"to $\varpi = 2\pi$."

To demonstrate this theorem, we shall observe that in virtue of the first of the two preceding equations of partial differences, we have

$$\int Y^{(i)}.\,Z^{(i')}.\,dm.\,d\varpi=-\frac{1}{i\,(i+1)}.\int Z^{(i')}\left(\frac{d.\left\{(1-m^2).\left(\frac{dY^{(i)}}{dm}\right)\right\}}{dm}\right).dm.d\varpi$$

$$-\frac{1}{i\,(i+1)}.\int\frac{Z^{(i')}.\left(\frac{d^2Y^{(i)}}{d\varpi^2}\right)}{1-m^2}.\,dm\,.\,d\varpi;$$

But integrating by parts relatively to m we have

$$\int Z^{(i')}.\left(\frac{d\left\{(1-m^2)\left(\frac{dY^{(i)}}{dm}\right)\right\}}{dm}\right).dm=(1-m^2)\left(\frac{dY^{(i)}}{dm}\right).\,Z^{(i')}$$

$$-(1-m^2)\,Y^{(i)}.\left(\frac{dZ^{(i')}}{dm}\right)$$

$$+\int Y^{(i)}\left(\frac{d\left\{(1-m^2)\left(\frac{dZ^{(i')}}{dm}\right)\right\}}{dm}\right)dm;$$

and it is clear that if we take the integral from m = − 1 to m = 1, the second member of this equation will be reduced to its last term. In like manner, integrating by parts relatively to ϖ, we get

$$\int Z^{(i')}.\left(\frac{d^2Y^{(i)}}{d\varpi^2}\right).\,d\varpi=\text{const.}+Z^{(i')}.\left(\frac{Y^{(i)}}{dd\varpi}\right)$$

$$-Y^{(i)}\left(\frac{dZ^{(i')}}{d\varpi}\right)+\int Y^{(i)}\left(\frac{d^2Z^{(i')}}{d\varpi^2}\right).\,d\varpi;$$

and this second member also reduces to its last term, when the integral is taken from $\varpi = 0$ to $\varpi = 2\pi$, because the values of $Y^{(i)}$, $\left(\frac{d\,Y^{(i)}}{d\,\varpi}\right)$, $Z^{(i')}$, $\left(\frac{d\,Z^{(i')}}{d\,\varpi}\right)$ are the same at these two limits; thus we shall have

$$\int Y^{(i)} \,.\, Z^{(i')} \,.\, d\,m \,.\, d\,\varpi =$$

$$-\frac{1}{i\,(i+1)} \,.\int Y^{(i)} . d\,m .\, d\,\varpi \left\{ \left(\frac{d \,.\, \left\{ (1-m^2) \,.\, \left(\frac{d\,Z^{(i')}}{d\,m}\right) \right\}}{d\,m} \right) + \frac{\left(\frac{d^2\,Z^{(i')}}{d\,\varpi^2}\right)}{1 - m^2} \right\};$$

whence we derive, in virtue of the second of the two preceding equations of partial differences,

$$\int Y^{(i)} .\, Z^{(i')} .\, d\,m \,.\, d\,\varpi = \frac{i'\,(i'+1)}{i\,(i+1)} \,.\int Y^{(i)} .\, Z^{(i')} .\, d\,m \,.\, d\,\varpi,$$

we therefore have

$$0 = \int Y^{(i)} .\, Z^{(i')}\, d\,m \,.\, d\,\varpi,$$

when i is different from i′.

Hence it is easy to conclude that y can be developed into a series of the form $Y^{(0)} + Y^{(1)} + Y^{(2)} +$ &c. in one way only; for we have generally

$$\int y \,.\, Z^{(i)}\, d\,m\, d\,\varpi = \int Y^{(i)} .\, Z^{(i)}\, d\,m \,.\, d\,\varpi;$$

If we could develope y into another series of the same form, $Y_{,}^{(0)} + Y_{,}^{(1)} + Y_{,}^{(2)} +$ &c. we should have

$$\int y \,.\, Z^{(i)} = \int Y_{,}^{(i)} .\, Z^{(i)}\, d\,m \,.\, d\,\varpi;$$

wherefore

$$\int Y_{,}^{(i)} .\, Z^{(i)} .\, d\,m\, d\,\varpi = \int Y^{(i)} .\, Z^{(i)}\, d\,m \,.\, d\,\varpi.$$

But it is easy to perceive that if we take for $Z^{(i)}$ the most general function of its kind, the preceding equation can only subsist in the case wherein $Y_{,}^{(i)} = Y^{(i)}$; the function y can therefore be developed thus in only one manner.

If in the integral $\int y\, d\,m \,.\, d\,\varpi$, we substitute for y its value $Y^{(0)} + Y^{(1)} + Y^{(2)} +$ &c., we shall have generally $0 = \int Y^{(i)}\, d\,m \,.\, d\,\varpi$, i being equal to or greater than unity; for the unity which multiplies $d\,m \,.\, d\,\varpi$ is comprised in the form $Z^{(0)}$, which extends to every constant and quantity independent of m and ϖ. The integral $\int y\, d\,m \,.\, d\,\varpi$ reduces therefore to $\int Y^{(0)}\, d\,m \,.\, d\,\varpi$, and consequently to $4\,\pi\, Y^{(0)}$; we have therefore

$$M = \tfrac{4}{3}\,\pi\, a^3 + 4\,\alpha\,\pi\, a^3 \,.\, Y^{(0)};$$

thus, by taking for a, the radius of the sphere equal in solidity to the spheroid, we shall have $Y^{(0)} = 0$, and the term $Y^{(0)}$ will disappear from the expression of y.

The distance of the element d M, or $R^2 . dR\,dm . d\varpi$, from the plane of the meridian from whence we measure the angle ϖ, is equal to $R\sqrt{1-m^2} . \sin. \varpi$; the distance of the center of gravity of the spheroid from this plane, will be therefore $\int R^3\,dR\,dm . d\varpi\sqrt{1-m^2} . \sin. \varpi$, and integrating relatively to R, it will be $\frac{1}{4}\int R'^4\,dm . d\varpi\sqrt{1-m^2}\sin. \varpi$, R' being the radius R produced to the surface of the spheroid. In like manner the distance of the element d M from the plane of the meridian perpendicular to the preceding, being $R\sqrt{1-m^2} . \cos. \varpi$, the distance of the center of gravity of the spheroid from this plane will be $\frac{1}{4}\int R'^4\,dm . d\varpi . \sqrt{1-m^2} . \cos. \varpi$. Finally, the distance of the element d M from the plane of the equator being m, the distance of the center of gravity of the spheroid from this plane will be $\frac{1}{4}\int R'^4\,m . dm . d\varpi$. These functions m, $\sqrt{1-m^2} . \sin. \varpi$, $\sqrt{1-m^2} . \cos. \varpi$, are of the form $Z^{(1)}$, $Z^{(1)}$ being subject to the equation of partial differences

$$0=\left(\frac{d\left\{(1-m^2) . \left(\frac{dZ^{(1)}}{dm}\right)\right\}}{dm}\right)+\frac{\left(\frac{d^2Z^{(1)}}{d\varpi^2}\right)}{1-m^2}+2Z^{(1)}.$$

If we conceive R'^4 developed into the series $N^{(0)}+N^{(1)}+N^{(2)}+$ &c. $N^{(i)}$ being a rational and entire function of m, $\sqrt{1-m^2} . \sin. \varpi$, $\sqrt{1-m^2} . \cos. \varpi$, subject to the equation of partial differences.

$$0=\left(\frac{d\left\{(1-m^2) . \left(\frac{dN^{(i)}}{dm}\right)\right\}}{dm}\right)+\frac{\left(\frac{d^2N^{(i)}}{d\varpi^2}\right)}{1-m^2}+i\,(i+1) . N^{(i)};$$

the distances of the center of gravity of the spheroid, from the three preceding planes, will be, in virtue of the general theorem above demonstrated,

$$\tfrac{1}{4}\int N^{(1)} . dm . d\varpi . \sqrt{1-m^2} . \sin. \varpi,$$
$$\tfrac{1}{4}\int N^{(1)} . dm . d\varpi . \sqrt{1-m^2} . \cos. \varpi;$$
$$\tfrac{1}{4}\int N^{(1)}\,m . dm . d\varpi.$$

$N^{(1)}$ is, by No. 553, of the form $A\,m+B\sqrt{1-m^2} . \sin. \varpi+C\sqrt{1-m^2} . \cos. \varpi$, A, B, C being constants; the preceding distances will thus become $\frac{\pi}{3} . B$, $\frac{\pi}{3} . C$, $\frac{\pi}{3} . A$. The position of the center of gravity of the spheroid, thus depends only on the function $N^{(1)}$. This gives a very simple way of determining it. If the origin of the radius R' is at the center; this origin being upon the three preceding planes, the distances of the center of gravity from these planes will be nothing. This gives $A=0$, $B=0$, $C=0$; therefore $N^{(1)}=0$.

These results obtain whatever may be the spheroid: when it is very little different from a sphere, we have $R' = a(1 + \alpha y)$, and $R'^4 = a^4(1 + 4\alpha y)$; thus, y being equal to $Y^{(0)} + Y^{(1)} + Y^{(2)} + \&c.$, we have $N^{(1)} = 4\alpha a^4 Y^{(1)}$, the function $Y^{(1)}$ disappears, therefore, from the expression of y, when we fix the origin of R' at the center of gravity of the spheroid.

557. Now let the point attracted be in the interior of the spheroid, we shall have by 553

$$V = v^{(0)} + r \cdot v^{(1)} + r^2 \cdot v^{(2)} + r^3 v^{(3)} + \&c.$$

$$v^{(i)} = \int \frac{dR \cdot d\varpi' \cdot d\theta' \cdot \sin.\theta' \cdot Q^{(i)}}{R^{i-1}}.$$

Suppose that this value of V is relative to a shell whose interior surface is spherical and of the radius a, and the radius of whose exterior surface is $a(1 - \alpha y)$; the thickness of the shell is $\alpha a y$. If we denote by y' what y becomes when we change θ, ϖ into θ', ϖ', we may, neglecting quantities of the order α^2, change r into a, and dR into $\alpha a y'$, in the integral expression of $v^{(i)}$; thus we shall have

$$v^{(i)} = \frac{\alpha}{a^{i-2}} \int y' \, d\varpi' \cdot d\theta' \cdot \sin.\theta' \cdot Q^{(i)}.$$

Relatively to a point placed without the spheroid, we have, by 553,

$$V = \frac{U^{(0)}}{r} + \frac{U^{(1)}}{r^2} + \&c.;$$

$$U^{(i)} = \int R^{i+2} \cdot dR \cdot d\varpi' \cdot d\theta' \cdot \sin.\theta' \cdot Q^{(i)}.$$

If we suppose this value of V relative to a shell, whose interior and exterior radii are respectively a, $a(1 + \alpha y)$, we shall have

$$U^{(i)} = \alpha \cdot a^{i+3} \cdot \int y' \cdot d\varpi' \cdot d\theta' \cdot \sin.\theta' \cdot Q^{(i)};$$

wherefore

$$v^{(i)} = \frac{U^{(i)}}{a^{2i+1}}.$$

We have by 555

$$U^{(i)} = \frac{4\alpha\pi a^{i+3} \cdot Y^{(i)}}{2i+1};$$

therefore

$$v^{(i)} = \frac{4 a \pi Y^{(i)}}{(2i+1) a^{i-2}};$$

which gives

$$V = 4\alpha\pi a^2 \left\{ Y^{(0)} + \frac{r}{3a} \cdot Y^{(1)} + \frac{r^2}{5a^2} \cdot Y^{(2)} + \&c. \right\}.$$

To this value of V we must add that which is relative to the spherical shell of the thickness a — r which envelopes the attracted point *plus* that which is relative to the sphere of radius r, and which is below the same point. If we make $\cos.\ \theta' = m'$, we shall have, relatively to the first of the two parts of V,

$$v^{(i)} = \int \frac{d\,R \,.\, d\,\varpi' \,.\, d\,m' \,.\, Q^{(i)}}{R^{i-1}};$$

an integral which, relative to m′, must be taken from $m' = -1$ to $m = 1$ Integrating relative to R, from $R = r$ to $R = a$, we shall have

$$v^{(i)} = \frac{1}{2-i}\left(\frac{1}{a^{i-2}} - \frac{1}{r^{i-2}}\right)\int d\,\varpi' \,.\, d\,m' \,.\, Q^{(i)};$$

But we have generally, by the theorem of the preceding No., $\int d\,\varpi' \,.\, d\,m' \,.\, Q^{(i)} = 0$ when i is equal to or greater than unity; when $i = 0$, we have, by 553, $Q^{(i)} = 1$; moreover the integration relative to ϖ' must be taken from $\varpi' = 0$ to $\varpi' = 2\pi$; we shall therefore have

$$v^{(0)} = 2\pi\,(a^2 - r^2).$$

This value of $v^{(0)}$ is that part of V which is relative to the spherical shell whose thickness is $a - r$.

The part of V which is relative to the sphere whose radius is r is equal to the mass of this sphere, divided by the distance of the attracted point from its center: it is consequently equal to $\frac{4\,\pi\,r^2}{3}$. Collecting the different parts of V, we shall have its whole value

$$V = 2\pi a^2 - \tfrac{2}{3}\pi r^2 + 4\alpha\pi a^2\left\{Y^{(0)} + \frac{r}{3a}Y^{(1)} + \frac{r^2}{5a^2}Y^{(2)} + \&c.\right\}; \quad (4)$$

Suppose the point attracted, placed within a shell very nearly spherical, whose interior radius is

$$a + \alpha a\,\{Y^{(0)} + Y^{(1)} + Y^{(2)} + \&c.\}$$

and whose exterior radius is

$$a' + \alpha a'\,\{Y'^{(0)} + Y'^{(1)} + Y'^{(2)} + \&c.\}$$

The quantities $\alpha\,a\,Y^{(0)}$ and $\alpha\,a'\,Y'^{(0)}$ may be comprised in the quantities a, a′. Moreover, by fixing the origin of coordinates at the center of gravity of the spheroid whose radius is

$$a + \alpha a\,\{Y^{(0)} + Y^{(1)} + \&c.\},$$

we may cause $Y^{(1)}$ to disappear from the expression of this radius; and then the interior radius of the shell will be of this form,

$$a + \alpha a\,\{Y^{(2)} + Y^{(3)} + \&c.\},$$

and the exterior radius will be of the form,

$$a' + \alpha a'\,\{Y'^{(1)} + Y'^{(2)} + \&c.\}.$$

We shall have the value of V relative to this shell, by taking the difference of the values of V relative to two spheroids, the smaller of which shall have for the radius of its surface the first quantity, and the greater the second quantity for the radius of its surface; calling therefore $\triangle . V$, what V becomes relatively to this shell, we shall have

$$\triangle V = 2\,\pi(a'^2 - a^2) + 4\,\alpha\,\pi \left\{ \frac{r\,a'}{3} Y'^{(1)} + \frac{r^2}{5}\{Y'^{(2)} - Y^{(2)}\} + \frac{r^3}{7}\left(\frac{Y'^{(3)}}{a'} - \frac{Y^{(3)}}{a}\right) + \&c.\right\}$$

If we wish that the point placed in the interior of the shell, should be equally attracted on all sides, $\triangle . V$ must be reduced to a constant independent of r, θ, ϖ; for we have seen that the partial differences of $\triangle . V$, taken relatively to these variables, express the partial attractions of the shell upon the point attracted; we therefore, in this case have $Y'^{(1)} = 0$, and generally

$$Y'^{(i)} = \left(\frac{a'}{a}\right)^{i-2} . \ Y^{(i)};$$

so that the radius of the interior surface being given, that of the exterior surface will be found.

When the interior surface is elliptic, we have $Y^{(3)} = 0$, $Y^{(4)} = 0$, &c. and consequently $Y'^{(3)} = 0$, $Y'^{(4)} = 0$; the radii of the two surfaces, interior and exterior, are therefore

$$a\,\{I + \alpha\,Y^{(2)}\}; \quad a'\,\{1 + \alpha\,Y^{(2)}\};$$

thus we see that these two surfaces are similar and similarly situated, which agrees with what we found in 547.

558. The formulas (3), (4) of Nos. 555, and 557, comprehend all the theory of the attractions of *homogeneous* spheroids, differing but little from the sphere; whence it is easy to obtain that of heterogeneous spheroids, whatever may be the law of the variation of the figure and density of their shells. For that purpose let a $(1 + \alpha\,y)$ be the radius of one of the shells of a heterogeneous spheroid, and suppose y to be of this form

$$Y^{(0)} + Y^{(1)} + Y^{(2)} + \&c.$$

the coefficients which enter the quantities $Y^{(0)}$, $Y^{(1)}$, &c. being functions of a, and consequently variable from one shell to another. If we differentiate relatively to a, the value of V given by the form (3) of No. 555; and call ϱ the density of the shell whose radius is a $(1 + \alpha\,y)$, ϱ being a function of a only; the value of V corresponding to this shell will be, for an exterior attracted point,

$$\frac{4\,\pi}{3\,r}\varrho\,d\,a^3 + \frac{4\,\alpha\,\pi.\varrho}{r} d\left\{a^3\,Y^{(0)} + \frac{a^4}{3\,r}.\,Y^{(1)} + \frac{a_5}{5\,r^2}.\,Y^{(2)} + \&c.\right\};$$

this value will be, therefore, relatively to the whole spheroid,

$$V = \frac{4\,\pi}{3\,r}\!\int\!\varrho\,d\,a^3 + \frac{4\,\alpha\,\pi}{r}\!\int\!\varrho\,d\left\{a^3\,Y^{(0)} + \frac{a^4}{3r} Y^{(1)} + \frac{a^5}{5\,r^2} Y^{(2)} + \&c.\right\}; \ . \ . \ (5)$$

the integrals being taken from a $= 0$ to that value of a which subsists at the surface of the spheroid, and which we denote by *a*.

To get the part of V relative to an attracted point in the interior of the spheroid, we shall determine first the part of this value relative to all the shells to which this point is exterior. This first part is given by formula (5) by taking the integral from a = 0 to a = a, a being relative to the shell in which is the point attracted. We shall find the second part of V relative to all the shells in the interior of which is placed the point attracted, by differentiating the formula (4) of the preceding No. relatively to a; then multiplying this differential by ρ, and taking the integral from a = a, to a = a, the sum of the two parts of V will be its entire value relative to an interior point, which sum will be

$$V = \frac{4\pi}{3}\int \rho\, d\, a^3 + \frac{4\alpha\pi}{r}\int \rho\, d\,.\left\{a^3 Y^{(0)} + \frac{a^4}{3r} Y^{(1)} + \frac{a^5}{5r^2} Y^{(2)} + \&c.\right\}$$

$$+ 2\pi \int \rho\, d\, a^2 + 4\alpha\pi \int \rho\, d\,.\left\{a^2 Y^{(0)} + \frac{a\,r}{3} Y^{(1)} + \frac{r^2}{5} Y^{(2)} + \&c.\right\}. \quad (6)$$

the two first integrals being taken from a = 0 to a = a, and the two last being taken from a = a to a = a; after the integrations, moreover, we must substitute a for r in the terms multiplied by α, and $\frac{1-\alpha y}{\alpha}$ for $\frac{1}{r}$ in the term $\frac{4\pi}{3r}\int \rho\, d\,.\,a^3$.

559. Now let us consider any spheroids whatever. The research of their attraction is reduced, by 553, to forming the quantities $U^{(i)}$ and $v^{(i)}$, by that No. we have

$$U^{(i)} = \int \rho\, R^{i+2}\,.\,d\,R\,d\,m'\,d\,\varpi'\,.\,Q^{(i)};$$

in which the integrals must be taken from R = 0 to its value at the surface, from m' = — 1 to m' = 1, and from $\varpi' = 0$ to $\varpi' = 2\pi$.

To determine this integral, $Q^{(i)}$ must be known. This quantity may be developed into a finite function of cosines of the angle $\varpi - \varpi'$, and of its multiples. Let β cos. n $(\varpi - \varpi')$ be the term of $Q^{(i)}$ depending on cos. n $(\varpi - \varpi')$, β being a function m, m'. If we substitute for $Q^{(i)}$ its value in the equation of partial differences in $Q^{(i)}$ of No. 553, we shall have, by comparing the terms multiplied by cos. n $(\varpi - \varpi')$, this equation of ordinary differences,

$$0 = \frac{d\left\{(1-m^2)\,.\left(\frac{d\beta}{dm}\right)\right\}}{d\,m} - \frac{n^2\beta}{1-m^2} + i\,(i+1)\,.\,\beta;$$

$Q^{(i)}$ being the coefficient of $\frac{R^{(i)}}{r^{i+1}}$, in the developement of the radical

$$\frac{1}{\sqrt{r^2 - 2Rr\{m m' + \sqrt{1-m'^2}\,.\,\sqrt{1-m^2}\,.\,\cos.(\varpi - \varpi') + R^2\}}}$$

The term depending on cos. n $(\varpi - \varpi')$, in the developement of this radical, can only result from the powers of cos. $(\varpi - \varpi')$, equal to n, n+2, n + 4, &c.; thus cos. $(\varpi - \varpi')$ having the factor $\sqrt{1 - m^2}$, β must have the factor $(1 - m^2)^{\frac{n}{2}}$. It is easy to see, by the consideration of the developement of the radical, that β is of this form

$$(1 - m^2)^{\frac{n}{2}} \cdot \{A \cdot m^{i-n} + A^{(1)} \cdot m^{i-n-2} + A^{(2)} \cdot m^{i-n-4} + \&c.\}.$$

If we substitute this value in the differential equation in β, the comparison of like powers of m will give

$$A^{(s)} = -\frac{(i - n - 2s + 2) \cdot (i - n - 2s + 1)}{2s(2i - 2s + 1)} \cdot A^{(s-1)};$$

whence we derive, by successively putting s = 1, s = 2, &c. the values of $A^{(1)}$, $A^{(2)}$, and consequently,

$$\beta = A(1-m^2)^{\frac{n}{2}} \left\{ \begin{array}{l} m^{i-n} - \frac{(i-n)(i-n-1)}{2(2i-1)} m^{i-n-2} + \frac{(i-n)(i-n-1)(i-n-2)(i-n-3)}{2.4.(2i-1)(2i-3)} m^{i-n-4} \\ - \frac{(i-n)(i-n-1)(i-n-2)(i-n-3)(i-n-4)(i-n-5)}{2.4.6(2i-1)(2i-3)(2i-5)} m^{i-n-6} + \&c. \end{array} \right\}.$$

A is a function of m′ independent of m; but m and m′ entering alike into the preceding radical, they ought to enter similarly into the expression of β; we have therefore

$$A = \gamma (1 - m'^2)^{\frac{n}{2}} \left\{ m'^{i-n} - \frac{(i-n)(i-n-1)}{2(2i-1)} \cdot m'^{i-n-2} + \&c. \right\};$$

γ being a coefficient independent of m and m′; therefore

$$\beta = \gamma (1 - m'^2)^{\frac{n}{2}} \left\{ m'^{i-n} - \frac{(i-n)(i-n-1)}{2(2i-1)} m'^{i-n-2} + \&c. \right\} \cdot (1 - m^2)^{\frac{n}{2}} \times$$

$$\left\{ m^{i-n} - \frac{(i-n)(i-n-1)}{2(2i-1)} m^{i-n-2} + \&c. \right\}.$$

Thus we see that β is split into three factors, the first independent of m and m′; the second a function of m′ alone; and the third a like function of m. We have only now to determine γ.

For that purpose, we shall observe, that if i — n be even; we have, supposing m = 0, and m′ = 0,

$$\beta = \frac{\gamma \cdot \{1.2.\ldots i - n\}^2}{\{2.4.\ldots(i-n).(2i-1).(2i-3)\ldots(i+n+1)\}^2}$$

$$= \frac{\gamma \cdot \{1.3.5.\ldots(i-n-1).1.3.5.\ldots(i+n-1)\}^2}{\{1.3.5.\ldots(2i-1)\}^2}$$

If i — n is odd, we shall have, in retaining only the first power of m, and m′,

$$\beta = \frac{\gamma . m . m' \{1. 2. \ldots (i - n)\}^2}{\{2. 4. \ldots (i - n - 1)(2i - 1)(2i - 3) \ldots (i + n + 2)\}^2}$$
$$= \frac{\gamma . m . m' \{1. 3. 5. \ldots (i - n) . 1. 3. 5. \ldots (i + n)\}^2}{\{1. 3. 5. \ldots (2i - 1)\}^2}.$$

The preceding radical becomes, neglecting the squares of m, m′,

$$\{r^2 - 2 R r \cos.(\varpi - \varpi') + R^2\}^{-\frac{1}{2}} + R r . m m' \{r^2 - 2 r R \cos.(\varpi - \varpi') + R^2\}^{-\frac{3}{2}}; \quad (f)$$

If we substitute for cos. $(\varpi - \varpi')$, its value in imaginary exponentials, and if we call c the number whose hyperbolic logarithm is unity, the part independent of m m′, becomes

$$\{r - R . c^{(\varpi - \varpi')\sqrt{-1}}\}^{-\frac{1}{2}} . \{r - R . c^{-(\varpi - \varpi')\sqrt{-1}}\}^{-\frac{1}{2}}.$$

The coefficient of

$$\frac{R^i}{r^{i+1}} . \frac{c^{n(\varpi - \varpi')\sqrt{-1}} + c^{-n(\varpi - \varpi')\sqrt{-1}}}{2}, \text{ or of } \frac{R^i}{r^{i+1}} . \cos. n(\varpi - \varpi')$$

in the developement of this function is

$$\frac{2. 1. 3. 5. \ldots (i + n - 1). 1. 3. 5. \ldots (i - n - 1)}{2. 4. 6. \ldots (i + n)\ 2. 4. 6. \ldots (i - n)}.$$

This is the value of β when i — n is even. Comparing it with that which in the same case we have already found, we shall have

$$\gamma = 2\left(\frac{1. 3. 5. \ldots (2i - 1)}{1. 2. 3. \ldots i}\right)^2 \times \frac{i(i - 1) \ldots (i - n + 1)}{(i + 1)(i + 2) \ldots (i + n)}$$

When n = 0, we must take only half this coefficient, and then we have

$$\gamma = \left(\frac{1. 3. 5. \ldots 2i - 1}{1. 2. 3. \ldots i}\right)^2.$$

In like manner, the coefficient of $\frac{R^i}{r^{i+1}} m . m' \cos. n(\varpi - \varpi')$ in the function (f) is

$$\frac{2. 1. 3. 5. \ldots (i + n) . 1. 3. 5. \ldots (i - n)}{2. 4. 6. (i + n - 1) . 2. 4. 6. \ldots (i - n - 1)};$$

this is the coefficient of m m′ in the value of β, when we neglect the squares of m, m′, and when i — n is odd. Comparing this with the value already found, we shall have

$$\gamma = 2\left(\frac{1. 3. 5. \ldots (2i - 1)}{1. 2. 3. \ldots i}\right)^2 . \frac{i(i - 1) \ldots (i - n + 1)}{(i + 1)(i + 2) \ldots (i + n)};$$

an expression which is the same as in the case of i — n being even.

If n = 0, we also have

$$\gamma = \left(\frac{1. 3. 5. \ldots (2i - 1)}{1. 2. 3. \ldots i}\right)^2.$$

560. From what precedes, we may obtain the general form of functions $Y^{(i)}$ of m, $\sqrt{1-m^2}.\sin.\varpi$, and $\sqrt{1-m^2}.\cos.\varpi$, which satisfy the equation of partial differences

$$0=\left(\frac{d\left\{(1-m^2).\left(\frac{dY}{dm}\right)\right\}}{dm}\right)+\frac{\left(\frac{d^2Y^{(i)}}{d\varpi^2}\right)}{1-m^2}+i(i+1)Y^{(i)}.$$

Designating by β, the coefficient of sin. n ϖ, or of cos. n ϖ, in the function $Y^{(i)}$, we shall have

$$0=\frac{d\left\{(1-m^2)\left(\frac{d\beta}{dm}\right)\right\}}{dm}--\frac{n^2\beta}{1-m^2}+i(i+1).\beta.$$

β is equal to $(1-m^2)^{\frac{n}{2}}$ multiplied by a rational and entire function of m, and in this case, by the preceding No., we have

$$\beta=A^{(n)}(1-m^2)^{\frac{n}{2}}\left\{m^{i-n}-\frac{(i-n)(i-n-1)}{2(2i-1)}m^{i-n-2}+\&c.\right\},$$

$A^{(n)}$ being an arbitrary constant; thus the part of $Y^{(i)}$ depending on the angle n ϖ, is

$$(1-m^2)^{\frac{n}{2}}\left\{m^{i-n}-\frac{(i-n)(i-n-1)}{2(2i-1)}.m^{i-n-2}+\&c.\right\}.\{A^{(n)}\sin.n\varpi$$
$$+B^{(n)}\cos.n\varpi\};$$

$A^{(n)}$ and $B^{(n)}$ being two arbitraries. If we make successively in this function, n = 0, n = 1, n = 2 . . . n = i; the sum of all the functions which thence result, will be the general expression of $Y^{(i)}$, and this expression will contain 2 i + 1 arbitraries $B^{(0)}$, $A^{(1)}$, $B^{(1)}$, $A^{(2)}$, $B^{(2)}$, &c.

Let us now consider a rational and entire function S of the order s, of the three rectangular coordinates x, y, z. If we represent by R the distance of the point determined by these coordinates from their origin; by θ the angle formed by R and the axis of x; and by ϖ the angle which the plane of x, y forms with the plane passing through R and the axis of x; we shall have

$$x=Rm;\ y=R.\sqrt{1-m^2}.\cos.\varpi;\ z=R\sqrt{1-m^2}.\sin.\varpi.$$

Substituting these values in S, and developing this function into sines and cosines of the angle ϖ and its multiples, if S is the most general function of the order s, then sin. n ϖ, and cos. n ϖ, will be multiplied by functions of the form

$$(1-m^2)^{\frac{n}{2}}\{A.m^{s-n}+B.m^{s-n-1}+C.m^{s-n-2}+\&c.\};$$

thus the part of S, depending on the angle n ϖ, will contain 2 (s — n + 1) indeterminate constants. The part of S depending on the angle ϖ and its multiples will contain therefore s (s + I) indeterminates; the part independent of ϖ will contain s + 1, and S will therefore contain $(s+1)^2$ indeterminate constants.

The function $Y^{(0)} + Y^{(1)} + \&c.\ Y^{(s)}$ contains in like manner $(s+1)^2$ indeterminate constants, since the function $Y^{(i)}$ contains $2i+1$; we may therefore put S into a function of this form, and this may be effected as follows:

From what precedes we shall learn the most general expression of $Y^{(s)}$, we shall take it from S and determine the arbitraries of $Y^{(s)}$ so that the powers and products of m and $\sqrt{1-m^2}$ of the order s shall disappear from the difference $S - Y^{(s)}$; this difference will thus become a function of the order s − 1 which we shall denote by S′. We shall take the most general expression of $Y^{(s-1)}$; we shall subtract it from S′, and determine the arbitraries of $Y^{(i-1)}$ so that the powers and products of m and $\sqrt{1-m^2}$ of the order s − 1 may disappear from the difference $S' - Y^{(s-1)}$. Thus proceeding we shall determine the functions $Y^{(s)}$, $Y^{(s-1)}$, $Y^{(s-2)}$, &c. of which the sum is S.

561. Resume now, the equation of No. 559,

$$U^{(i)} = \int \varrho \,.\, R^{i+2}\, dR \,.\, dm' \,.\, d\varpi' \,.\, Q^{(i)}.$$

Suppose R a function of m′, ϖ' and of a parameter a, constant for all shells of the same density, and variable from one shell to another. The difference d R being taken on the supposition that m′, ϖ' are constant we shall have

$$dR = \left(\frac{dR}{da}\right) da;$$

therefore

$$U^{(i)} = \frac{1}{i+3} \cdot \int \varrho \left(\frac{d\,.\,R^{i+3}}{da}\right) da \,.\, dm' \,.\, d\varpi' \,.\, Q^{(i)}.$$

Let R^{i+3} be developed into a series of the form

$$Z'^{(0)} + Z'^{(1)} + Z'^{(2)} + \&c,$$

$Z'^{(i)}$ being whatever i may be, a rational and entire function of m′, $\sqrt{1-m'^2}$. sin. ϖ', and $\sqrt{1-m'^2}$. cos. ϖ', which satisfies the equation of partial differences

$$0 = \left(\frac{d\left\{(1-m^2).\left(\frac{dZ'^{(i)}}{dm'}\right)\right\}}{dm'}\right) + \frac{\left(\frac{d^2 Z'^{(i)}}{d\varpi'}\right)}{1-m'^2} + i(i+1)\,Z'^{(i)}.$$

The difference of $Z'^{(i)}$ taken relatively to a, satisfies also this equation, and consequently it is of the same form; by the general theorem of 556, we ought therefore only to consider the term $Z'^{(i)}$ in the developement of R^{i+3}, and then we have

$$U^{(i)} = \frac{1}{i+3} \cdot \int \varrho \left(\frac{dZ'^{(i)}}{da}\right) . \, da \,.\, dm' \,.\, d\varpi' \,.\, Q^{(i)}.$$

When the spheroid is homogeneous and differing but little from a sphere, we may suppose $\rho = 1$, and $R = a(1 + \alpha y')$; then we have, by integrating relatively to a

$$U^{(i)} = \frac{1}{i+3} \int Z'^{(i)} . d m' . d \varpi' . Q^{(i)}.$$

Moreover, if we suppose y' developed into a series of the form

$$Y'^{(0)} + Y'^{(1)} + Y'^{(2)} + \&c.,$$

$Y^{(i)}$ satisfying the same equation of partial difference as $Z'^{(i)}$; we shall have, neglecting quantities of the order α^2, $Z'^{(i)} = (i+3) . \alpha . a^{i+3} Y'^{(i)}$; we shall therefore have

$$U^{(i)} = \alpha . a^{i+3} . \int Y'^{(i)} . d m' . d \varpi' . Q^{(i)}.$$

If we denote by $Y^{(i)}$ what $Y'^{(i)}$ becomes when we change m' and ϖ' into m and ϖ; we shall have by No. 554,

$$U^{(i)} = \frac{4 \alpha \pi . a^{i+3}}{2i+1} . Y^{(i)};$$

we therefore have this remarkable result,

$$\int Y'^{(i)} . d m' . d \varpi' . Q^{(i)} = \frac{4 \pi Y^{(i)}}{2i+1} \quad . \quad . \quad . \quad . \quad . \quad . \quad (1)$$

This equation subsisting whatever may be $Y'^{(i)}$ we may conclude generally that the double integration of the function $\int Z'^{(i)} d m' . d \varpi' . Q^{(i)}$ taken from $m' = -1$ to $m' = 1$, and from $\varpi' = 0$ to $\varpi' = 2\pi$, only transforms $Z'^{(i)}$ into $\frac{4 \pi Z^{(i)}}{2i+1}$; $Z^{(i)}$ being what $Z'^{(i)}$ becomes when we change m' and ϖ' into m and ϖ; we therefore have

$$U^{(i)} = \frac{4 \pi}{(i+3)(2i+1)} \int \rho \left(\frac{d Z^{(i)}}{d a}\right) . d a;$$

and the triple integration upon which $U^{(i)}$ depends, reduces to one integration only taken relatively to a, from $a = 0$ to its value at the surface of the spheroid.

The equation (1) presents a very simply method of integrating the function $\int Y^{(i)} . Z^{(i)} . d m . d \varpi$, from $m = -1$ to $m = 1$, and from $\varpi = 0$ to $\varpi = 2\pi$. In fact, the part of $Y^{(i)}$ depending on the angle $n \varpi$, is by what precedes, of the form $\lambda \{A^{(n)} \sin. n \varpi + B^{(n)} \cos. n \varpi\}$, λ being equal to

$$(1 - m^2)^{\frac{n}{2}} . \left\{ m^{i-n} - \frac{(i-n)(i-n-1)}{3(2i-1)} . m^{i-n-2} + \&c. \right\};$$

we shall have therefore

$$Y'^{(i)} = \lambda' \{A^{(n)} \sin. n \varpi' + B^{(n)} \cos. n \varpi'\};$$

λ' being what λ becomes when m is changed into m'. The part of $Q^{(i)}$ depending on the angle $n \varpi$, is by the preceding No., $\gamma \lambda \lambda' \cos. n (\varpi - \varpi')$, or $\gamma\,\lambda'.\,\lambda.\{\cos.\,n\,\varpi.\,\cos.\,n\,\varpi' + \sin.\,n\,\varpi.\,\sin.\,n\,\varpi'\}$; thus that part of the integral $\int Y^{(i)}.\,d\,m\,.\,d\,\varpi'.\,Q^{(i)}$ which depends on the angle $n\,\varpi$, will be

$$\gamma\,\lambda.\,\sin.\,n\,\varpi.\int\lambda'^2.\,d\,m'.\,d\,\varpi'.\,\sin.\,n\,\varpi'\,\{A^{(n)}\sin.\,n\,\varpi' + B^{(n)}\cos.\,n\,\varpi'\}$$

$$\gamma\,\lambda.\,\cos.\,n\,\varpi\int\lambda'^2.\,d\,m'.\,d\,\varpi'.\,\cos.\,n\,\varpi'\,\{A^{(n)}\sin.\,n\,\varpi' + B^{(n)}\cos.\,n\,\varpi'\}.$$

Executing the integrations relative to ϖ', that part becomes

$$\gamma\,\lambda\,\varpi\,\{A^{(n)}\sin.\,n\,\varpi + B^{(n)}\cos.\,n\,\varpi\}.\int\lambda'^2.\,d\,m';$$

but in virtue of equation (1), the same part is equal to

$$\frac{4\,\pi}{2\,i+1}.\,\lambda.\,\{A^{(n)}\sin.\,n\,\varpi + B^{(n)}\cos.\,n\,\varpi\}$$

we therefore have

$$\int\lambda'^2.\,d\,m' = \frac{4}{(2\,i+1)\,\gamma}.$$

Now represent by $\lambda\,\{A'^{(n)}\sin.\,n\,\varpi + B'^{(n)}\cos.\,n\,\varpi\}$ that part of $Z^{(i)}$ which depends on the angle $n\,\varpi$. This part ought to be combined with the corresponding part of $Y^{(i)}$; because the terms depending on the sines and cosines of the angle ϖ and its multiples, disappear by integration, in the function $\int Y^{(i)}\,Z^{(i)}\,d\,m\,.\,d\,\varpi$, integrated from $\varpi = 0$ to $\varpi = 2\,\pi$; we shall thus obtain, in regarding only that part of $Y^{(i)}$ which depends on the angle $n\,\varpi$,

$$\int Y^{(i)}.\,Z^{(i)}\,d\,m\,d\,\varpi =$$

$$\int\lambda^2.\,d\,m\,.\,d\,\varpi\{A^{(n)}\sin.\,n\,\varpi + B^{(n)}\cos.\,n\,\varpi\}\{A'^{(n)}\sin.\,n\,\varpi + B'^{(n)}\cos.\,n\,\varpi\}$$

$$= \pi\{A^{(n)}.\,A'^{(n)} + B^{(n)}.\,B'^{(n)}\}.\int\lambda^2\,d\,m = \frac{4\,\pi}{(2\,i+1)\,\gamma}.\{A^{(n)}.\,A'^{(n)} + B^{(n)}.B'^{(n)}\}.$$

Supposing therefore successively in the last member $n = 0$, $n = 1$; $n = 2 \ldots n = i$; the sum of all the terms, will be the value of the integral $\int Y^{(i)}\,Z^{(i)}\,d\,m\,.\,d\,\varpi$.

If the spheroid is one of revolution, so that the axis with which the radius R forms the angle ϖ, may be the axis of revolution; the angle ϖ will disappear from the expression of $Z^{(i)}$, which then takes the following form:

$$\frac{1.3.5\ldots(2\,i-1)}{1.2.3\ldots i}\,A^{(i)}\left\{m^{(i)} - \frac{i.(i-1)}{2.(2\,i-1)}\,m^{i-2} + \frac{i.(i-1)(i-2)(i-3)}{2.4.(2i-1)(2i-3)}\,m^{i-4} - \&c.\right\};$$

$A^{(i)}$ being a function of a. Call $\lambda^{(i)}$ the coefficient of $A^{(i)}$, in this function: the product

$$\left(\frac{1.3.5\ldots(2\,i-1)}{1.2.3\ldots i}\right)^2.\left\{1 - \frac{i.(i-1)}{2.(2\,i-1)} + \&c.\right\}^2,$$

is by the preceding No., the coefficient of $\frac{R^i}{r^{i+1}}$ in the developement of the radical

$$\{r^2 - 2\,R\,r\,\{m\,m' + \sqrt{1-m^2}.\,\sqrt{1-m'^2}\cos.\,(\varpi - \varpi')\} + R^2\}^{-\frac{1}{2}};$$

when we therein suppose m and m' equal to unity. This coefficient is then equal to 1; we have therefore

$$\frac{1.3.5\ldots(2i-1)}{1.2.3\ldots i}\left\{1-\frac{i(i-1)}{2(2i-1)}+\&c.\right\}=1,$$

that is to say, $\lambda^{(i)}$ reduces to unity, when m = 1. We have then

$$U^{(i)}=\frac{4\pi\lambda^{(i)}}{(i+3).(2i+1)}.\int\rho\left(\frac{d\,A^{(i)}}{d\,a}\right)d\,a.$$

Relatively to the axis of revolution, m = 1, and consequently,

$$U^{(i)}=\frac{4\pi}{(1+3)(2i+1)}.\int\rho\left(\frac{d\,A^{(i)}}{d\,a}\right)d\,a;$$

therefore if we suppose that relatively to a point placed upon this axis produced, we have

$$V=\frac{B^{(0)}}{r}+\frac{B^{(1)}}{{}_2}+\frac{B^{(2)}}{{}_3}+\&c.;$$

we shall have the value of V relative to another point placed at the mean distance from the origin of coordinates, but upon a radius which makes with the axis of revolution, an angle whose cosine is m; by multiplying the terms of this value respectively by $\lambda^{(0)}$, $\lambda^{(1)}$, $\lambda^{(2)}$, &c.

In the case when the spheroid is not of revolution, this method will give the part of V independent of the angle ϖ: we shall determine the other part in this manner. Suppose for the sake of simplicity, the spheroid such that it is divided into two equal and similar parts by the equator, whether by the meridian where we fix the origin of the angle ϖ, or by the meridian which is perpendicular to the former. Then V will be a function of m^2, $\sin.^2\varpi$, and $\cos.^2\varpi$, or which comes to the same, it will be a function of m^2, and of the cosine of the angle 2ϖ and its multiples; $U^{(i)}$ will therefore be nothing, when i is odd, and in the case when it is even, the term which depends on the angle $2n\varpi$, will be of the form

$$C^{(i)}.(1-m^2)^n\left\{m^{i-2n}-\frac{(i-2n)(i-2n-1)}{2(2i-1)}m^{i-2n-2}+\&c.\right\}\cos.2n\varpi.$$

Relatively to an attracted point placed in the plane of the equator, where m = 0, that part of V which depends on this term becomes

$$\pm\frac{C^{(i)}}{r^{i+1}}.\frac{1.3.5\ldots(i-2n-1)}{2(i+2n+1)(i+2n+2)\ldots(2i-1)}.\cos.2n\varpi;$$

whence it follows that having developed V into a series ordered according to the cosines of the angle 2ϖ and its multiples, when the point attracted is situated in the plane of the equator; to extend this value to any attracted point whatever, it will be sufficient to multiply the terms which depend on $\frac{\cos.2n\varpi}{r^{i+1}}$ by the function

$$\pm\frac{2(i+2n+1)\ldots(2i-1)}{1.3.5\ldots(i-2n-1)}.(1-m^2)^n\left\{m^{i-2n}-\frac{(i-2n)(i-2n-1)}{2(2i-1)}m^{i-2n-2}+\&c.\right\};$$

we shall hence obtain, therefore, the entire value of V, when this value shall be determined in a series, for the two cases where the part attracted is situated upon the polar axis produced, and where it is situated in the plane of the equator; this greatly simplifies the research of this value.

The spheroid which we are considering comprehends the ellipsoid. Relatively to an attracted point situated upon the polar axis, which we shall suppose to be the axis of x, by 546, we have $b = 0$, $c = 0$, and then the expression of V of No. 549, is integrable relatively to p. Relatively to a point situated in the plane of the equator, we have $a = 0$, and the same expression of V still becomes, by known methods, integrable relatively to q, by making $\tan. q = t$. In the two cases, the integral being taken relatively to one of these variables in its limits, it then becomes possible relatively to the other, and we find that M being the mass of the spheroid, the value of $\frac{V}{M}$ is independent of the semi-axis k of the spheroid perpendicular to the equator, and depends only on the excentricities of the ellipsoid. Multiplying therefore the different terms of the values of $\frac{V}{M}$ relative to these two cases, and reduced into series proceeding according to the powers of $\frac{1}{r}$, by the factors above mentioned, to get the value of $\frac{V}{M}$ relative to any attracted point whatever; the function which thence results will be independent of k, and only depend on the excentricities; this furnishes a new demonstration of the theorem already proved in 550.

If the point attracted is placed in the interior of the spheroid, the attraction which it undergoes, depends, as we have seen in No. 553, on the function $v^{(i)}$, and by the No. cited, we have

$$v^{(i)} = \int\frac{\rho\, d R\, d m'\, d\varpi'.\, Q^{(i)}}{R^{i-1}};$$

an equation which we can put under this form

$$v^{(i)} = \frac{1}{2-i}.\int\rho\left(\frac{d R^{2-i}}{d a}\right) d a.d m'.d\varpi'.\,Q^{(i)}.$$

Suppose R^{2-i} developed into a series of the form

$$z'^{(0)} + z'^{(1)} + z'^{(2)} + \&c.$$

$z'^{(i)}$ satisfying the equation of partial differences,

$$0 = \left(\frac{d \left\{ (1 - m'^2) \left(\frac{d z'^{(i)}}{d m'} \right) \right\}}{d m'} \right) + \frac{\left(\frac{d^2 z'^{(i)}}{d \varpi'^2} \right)}{1 - m'^2} + i (i + 1) z'^{(i)};$$

if moreover we call $z^{(i)}$ what $z'^{(i)}$ becomes when we change m' into m, and ϖ' into ϖ, we shall have by what precedes,

$$v^{(i)} = \frac{4 \pi}{(2 i + 1)(2 - i)} \int \rho \left(\frac{d z^{(i)}}{d a} \right) d a;$$

thus therefore we shall get the expression of V relative to all the shells of the spheroid which envelope the point attracted. The value of V relative to shells to which it is interior, we have already shown how to determine.

ON THE FIGURE OF A FLUID HOMOGENEOUS MASS IN EQUILIBRIUM, AND ENDOWED WITH A ROTATORY MOTION.

562. Having exposed in the preceding Nos. the theory of the attractions of spheroids, we now proceed to consider the figure which they must assume in virtue of the mutual action of their parts, and the other forces which act upon them. We shall first seek the figure which satisfies the equilibrium of a fluid homogeneous mass endowed with a rotatory motion, and of that problem we shall give a rigorous solution.

Let a, b, c be the rectangular coordinates of any point of the surface of the mass, and P, Q, R the forces which solicit it parallel to the coordinates, the forces being supposed as tending to diminish them. We know that when the mass is in equilibrium, we have

$$0 = P . d a + Q . d b + R . d c;$$

provided that in estimating the forces P, Q, R, we reckon the centrifugal force due to the motion of rotation.

To estimate these forces, we shall suppose that the figure of the fluid mass, is that of the ellipsoid of revolution, whose axis of rotation, is the axis itself of revolution. If the forces P, Q, R which result from this hypothesis, substituted in the preceding equation of equilibrium give the differential equation of the surface of the ellipsoid; the preceding hypothesis is legitimate, and the elliptic figure satisfies the equilibrium of the fluid mass.

Suppose that the axis of a is that also of revolution; the equation of the surface of the ellipsoid will be of this form

$$a^2 + m (b^2 + c^2) = k^2;$$

the origin of the coordinates a, b, c being at the center of the ellipsoid, k will be the semi-axis of revolution, and if we call M the mass of the ellipsoid, by 546, we shall have

$$M = \frac{4\,\pi\,\rho\,k^3}{3\,m}$$

ρ being the density of the fluid. If we make as in 547, $\frac{1-m}{m} = \lambda^2$, we shall have $m = \frac{1}{1+\lambda^2}$, and consequently

$$M = \frac{4\,\pi}{3}\,k^3.\,(1+\lambda^2);$$

an equation which will give the semi-axis k, when λ is known.

Let

$$A' = \frac{4\,\pi\,\rho\,(1+\lambda^2)}{\lambda^3}\,(\lambda - \tan.^{-1}\lambda)$$

$$B' = \frac{4\,\pi\,\rho}{2\,\lambda^3}\,\{(1+\lambda^2)\,\tan.^{-1}\lambda - \lambda)\};$$

we shall have by 547, regarding only the attraction of the fluid mass

$$P = A'\,a;\ Q = B'\,b;\ R = B'\,c.$$

If we call g, the centrifugal force at the distance 1, from the axis of rotation; this force at the distance $\sqrt{b^2+c^2}$ from the same axis, will be $g\sqrt{b^2+c^2}$: resolving this parallel to the coordinates b, c there will result in Q the term $-g\,b$, and in R the term $-g\,c$; thus we shall have, reckoning all the forces which animate the molecules of the surface,

$$P = A'\,a;\ Q = (B'-g)\,b;\ R = (B'-g).\,c;$$

the preceding equation of equilibrium, will therefore become

$$0 = a\,d\,a + \frac{B'-g}{A'}\,(b\,d\,b + c\,d\,c).$$

The differential equation of the surface of the ellipsoid is by substituting for m its value $\frac{1}{1+\lambda^2}$,

$$0 = a\,d\,a + \frac{b\,d\,b + c\,d\,c}{1+\lambda^2};$$

by comparing this with the preceding one, we shall have

$$(1+\lambda^2)\,(B'-g) = A'; \quad \ldots \quad (1)$$

if we substitute for A', B' their values, and if we make $\frac{g}{\frac{4}{3}\pi\,\rho} = q$; we shall have

$$0 = \frac{9.\lambda + 2\,q.\lambda^3}{9 + 3.\lambda^2} - \tan.^{-1}\lambda; \quad \ldots \quad (2)$$

determining therefore λ by this equation which is independent of the coordinates a, b, c, the equation of equilibrium will coincide with the equation of the surface of the ellipsoid; whence it follows, that the elliptic figure satisfies the equilibrium, at least, when the motion of rotation is such that the value of λ^2 is not imaginary, or when being negative, it is neither equal to nor greater than unity. The case where λ^2 is imaginary would give an imaginary solid; that where $\lambda^2 = -1$, would give a paraboloid, and that where λ^2 is negative and greater than unity, would give a hyperboloid.

563. If we call p the gravity at the surface of the ellipsoid, we shall have

$$p = \sqrt{P^2 + Q^2 + R^2}.$$

In the interior of the ellipsoid, the forces P, Q, R, are proportional to the coordinates a, b, c; for we have seen in No. 547, that the attractions of the ellipsoid, parallel to these coordinates, are respectively proportional to them, which equally takes place for the centrifugal force resolved parallel to the same coordinates. Hence it follows, that the gravities at different points of a radius drawn from the center of the ellipsoid to its surface, have parallel directions, and are proportional to the distances from the center; so that if we know the gravity at its surface, we shall have the gravity in the interior of the spheroid.

If in the expression of p, we substitute for P, Q, R, their values given in the preceding No., we shall have

$$p = \sqrt{A'^2 a^2 + (B' - g)^2 . (b^2 + c^2)};$$

whence we derive, in virtue of equation (1) of the preceding No.

$$p = A' \sqrt{a^2 + \frac{b^2 + c^2}{(1 + \lambda^2)^2}};$$

but the equation of the surface of the ellipsoid gives $\frac{b^2 + c^2}{1 + \lambda^2} = k^2 - a^2$;

we shall therefore have

$$p = A' . \sqrt{\frac{k^2 + \lambda^2 a^2}{1 + \lambda^2}},$$

a is equal to k at the pole, and it is nothing at the equator; whence it follows, that the gravity at the pole is to the gravity at the equator, as $\sqrt{1 + \lambda^2}$ is to unity, and consequently, as the diameter of the equator is to the polar axis.

Call t the perpendicular at the surface of the ellipsoid, produced to meet the axis of revolution, we shall have

$$t = \sqrt{(1 + \lambda^2)(k^2 + \lambda^2 a^2)};$$

wherefore

$$p = \frac{A' t}{1 + \lambda^2};$$

thus gravity is proportional to t.

Let ψ be the complement of the angle which t makes with the axis of revolution; ψ will be the latitude of the point of the surface, which we are considering, and by the nature of the ellipse, we shall have

$$t = \frac{(1 + \lambda^2) k}{\sqrt{1 + \lambda^2 \cos.^2 \psi}};$$

we therefore shall have

$$p = \frac{A' k}{\sqrt{1 + \lambda^2. \cos.^2 \psi}};$$

and substituting for A' its value, we shall get

$$p = \frac{4 \pi \rho . k . (1 + \lambda^2). (\lambda - \tan.^{-1} \lambda)}{\lambda^3 \sqrt{1 + \lambda^2. \cos.^2 \psi}}; \quad . \quad . \quad . \quad . \quad . \quad (3)$$

this equation gives the relation between gravity and the latitude; but we must determine the constants which it contains.

Let T be the number of seconds in which the fluid mass will effect a revolution; the centrifugal force at the distance 1 from the axis of revolution, will be equal to $\frac{4 \pi^2}{T^2}$; we therefore have

$$q = \frac{g}{\frac{4}{3} \pi . \rho} = \frac{12 \pi^2}{4 \pi \rho T^2};$$

which gives

$$4 \pi \rho = \frac{12 . \pi^2}{q . T^2}.$$

The radius of curvature of the elliptic meridian is

$$\frac{(1 + \lambda^2) k}{(1 + \lambda^2 \cos.^2 \psi)^{\frac{3}{2}}};$$

calling therefore c the length of a degree at the latitude ψ, we shall have

$$\frac{(1 + \lambda^2) \pi k}{(1 + \lambda^2 \cos.^2 \psi)^{\frac{3}{2}}} = 200 \text{ c}.$$

This equation combined with the preceding one, gives

$$\frac{4 \pi \rho (1 + \lambda^2). k}{\sqrt{1 + \lambda^2 \cos.^2 \psi}} = 200 \text{ c} (1 + \lambda^2 \cos.^2 \psi). \frac{12 \pi}{q T^2};$$

thus we shall have

$$p = 200 \text{ c} (1 + \lambda^2 \cos.^2 \psi) \frac{\lambda - \tan.^{-1} \lambda}{\lambda^3} . \frac{12 \pi}{q T^2}.$$

Let I be the length of the simple pendulum which oscillates seconds;

from dynamics it results that $p = \pi^2 l$ (see § X.); comparing these two expressions of p, we get

$$q = \frac{2400\, c\, (\lambda - \tan.^{-1} \lambda)\, (1 + \lambda^2 \cos.^2 \psi)}{\pi\, l\, T^2\, \lambda^3}; \quad . \quad . \quad . \quad . \quad (4)$$

this equation and equation (2) of the preceding No. will give the values of q and λ by means of the length l of the seconds' pendulum, and the length c of the degree of the meridian, both being observed at the latitude ψ.

Suppose $\psi = 50^\circ$, these equations will give

$$q = \frac{800\, c}{\pi\, l\, T^2} - \frac{1}{4}\left(\frac{800\, c}{\pi\, l\, T^2}\right)^2 + \&c.;$$

$$\lambda^2 = \frac{5}{2} q + \frac{75}{14} q^2 + \&c.;$$

observations give, as we shall see hereafter,

$$c = 100000;\ l = 0.741608;$$

moreover we have $T = 99727$; we shall thus obtain

$$q = 0.00344957;\ \lambda^2 = 0.00868767.$$

The ratio of the axis of the equator to the polar axis, being $\sqrt{1 + \lambda^2}$, it becomes in this case 1.00433441; these two axes are very nearly in the ratio of 231.7 to 230.7, and by what precedes, the gravities at the pole and at the equator are in the same ratio.

We shall have the semi polar axis k, by means of the equation

$$k = \frac{200\, c\, (1 + \frac{1}{2} \lambda^2)^{\frac{3}{2}}}{\pi\, (1 + \lambda^2)} = \frac{200\, c}{\pi} \{1 - \frac{1}{4} \lambda^2 + \&c.\};$$

which gives

$$k = 6352534.$$

To get the attraction of a sphere, whose radius is k, and density any whatever; we shall observe that a sphere, having the radius k and density ϱ, acts upon a point placed at its surface, with a force equal to $\frac{4}{3} \pi \varrho . k$, and consequently, in virtue of equation (3) equal to $\frac{\lambda^3 p \sqrt{1 + \frac{1}{2} \lambda^2}}{3\, (1 + \lambda^2)\, (\lambda - \tan.^{-1} \lambda)}$, or to $p . \left(1 - \frac{3}{20} \lambda^2 + \&c.\right)$, or finally to 0.998697. p, p being the gravity upon the parallel of 50°. Hence it is easy to obtain the attractive force of a sphere of any radius and density whatever, upon a point placed within or without it.

564. If the equation (2) of No. 562, were susceptible of many real roots, many figures of equilibrium might result from the same motion of rotation; let us examine therefore whether this equation has several real

roots. For that purpose, call φ the function $\frac{9\lambda + 2q\lambda^3}{9 + 3\lambda^2} - \tan.^{-1}\lambda$, which being equated to zero, produces the equation (2). It is easy to see, that by making λ increase from zero to infinity, the expression of φ begins and ends by being positive; thus, by imagining a curve whose abscissa is λ and ordinate φ, this curve will cut its axis when $\lambda = 0$; the ordinates will afterwards be positive and increasing; when arrived at their maximum, they will decrease; the curve will cut the axis a second time at a point which will determine the value of λ corresponding to the state of equilibrium of the fluid mass; the ordinates will then be negative, and since they are positive when $\lambda = \infty$; the curve necessarily cuts the axis a third time, which determines a second value of λ which satisfies the equilibrium. Thus we see, that for one and the same value of q, or for one given motion of rotation, there are several figures for which the equilibrium may subsist.

To determine the number of these figures, we shall observe, that we have

$$d\varphi = \frac{6\lambda^2 d\lambda\{q\lambda^4 + (10q - 6)\lambda^2 + 9q\}}{(3\lambda^2 + 9)^2.(1 + \lambda^2)}.$$

The supposition of $d\varphi = 0$, gives

$$0 = q\lambda^4 + (10q - 6)\lambda^2 + 9q;$$

whence we derive, considering only the positive values of λ

$$\lambda = \sqrt{\frac{3}{q} - 5 \pm \sqrt{\left(\frac{3}{q} - 5\right)^2 - 9}}.$$

These values of λ determine the *maxima* and *minima* of the ordinate φ; there being only two similar ordinates on the side of positive abscissas, on that side the curve cuts its axis in three points, one of them being the origin; thus, the number of figures which satisfy the equilibrium is reduced to two.

The curve on the side of negative abscissas being exactly the same as on the side of positive abscissas; it cuts its axis on each side in corresponding points equidistant from the origin of coordinates; the negative values of λ which satisfy the equilibrium, are therefore, as to the sign taken, the same as the positive values; which gives the same elliptic figures, since the square of λ only enters the determination of these figures; it is useless therefore to consider the curve on the side of negative abscissas.

If we suppose q very small, which takes place for the earth, we may satisfy equation (2) of 562, in the two hypotheses of λ^2 being very small, and of λ^2 being very great. In the first, by the preceding No., we have

$$\lambda^2 = \frac{5}{2} q + \frac{75}{14} q^2 + \&c.$$

To get the value of λ^2 in the second hypothesis, we shall observe that then $\tan.^{-1} \lambda$ differs very little from $\frac{1}{2} \pi$, so that if we suppose $\lambda = \frac{\pi}{2} - \alpha$, α will be a very small angle of which the tangent is $\frac{1}{\lambda}$; we shall therefore have, p. 27. Vol. I.

$$\alpha = \frac{1}{\lambda} - \frac{1}{3\lambda^3} + \frac{1}{5\lambda^5} - \&c.$$

and consequently

$$\tan.^{-1} \lambda = \frac{\pi}{2} - \frac{1}{\lambda} + \frac{1}{3\lambda^3} - \frac{1}{5\lambda^5} + \&c.;$$

equation (2) of No. 562, will thus become

$$\frac{9\lambda + 2q.\lambda^3}{9 + 3\lambda^2} = \frac{\pi}{2} - \frac{1}{\lambda} + \frac{1}{3\lambda^3} - \&c.;$$

whence by the reversion of series we get

$$\lambda = \frac{3\pi}{4q} - \frac{8}{\pi} + \frac{4q}{\pi}\left(1 - \frac{64}{3\pi^2}\right) + \&c.$$

$$= 2.356195.\frac{1}{q} - 2.546479 - 1.478885\, q + \&c.$$

We have seen in the preceding No., that relatively to the earth, $q = 0.00344957$; this value of q substituted in the preceding expression, gives $\lambda = 680.49$. Thus the ratio of the two axes equatorial and polar, a ratio which is equal to $\sqrt{1 + \lambda^2}$, is in the case of a very thin spheroid, equal to 680.49.

The value of q has a limit beyond which the equilibrium is impossible, the figure being elliptic. Suppose, in fact, that the curve cuts its axis only at its origin, and that in the other points it only touches; at the point of contact we shall have $\varphi = 0$, and $d\varphi = 0$; the value of φ will never therefore be negative on the side of positive abscissas, which are the only ones we shall here consider. The value of q determined by the two equations $\varphi = 0$, $d\varphi = 0$, will therefore be the limit of those with which the equilibrium can take place, so that a greater value will render the equilibrium impossible; for q being supposed to increase by f, the function φ increases by the term $\frac{2f\lambda^3}{9 + 3\lambda^2}$; thus, the value of φ corresponding to q, being never negative, whatever λ may be, the same function corresponding to q + f, is constantly positive, and can never become nothing; the equilibrium is then therefore impossible. It results also from this analysis, that there is only one real and positive value of q, which would satisfy the two equations $\varphi = 0$, and $d\,\varphi = 0$. These equations give

$$q = \frac{6\,\lambda^2}{(1 + \lambda^2)(9 + \lambda^2)}$$

$$0 = \frac{7\,\lambda^5 + 30\,\lambda^3 + 27\,\lambda}{(1 + \lambda^2)(3 + \lambda^2)(9 + \lambda^2)} - \tan.^{-1}\lambda.$$

The value of λ which satisfies this last equation is $\lambda = 2.5292$; whence we get $q = 0.337007$; the quantity $\sqrt{1 + \lambda^2}$, which expresses the ratio of the equatorial axis to the polar axis, is in this case equal to 2.7197.

The value of q relatively to the earth is equal to 0.00344957. This value corresponds to a time of rotation of 0.99727 days; but we have generally $q = \frac{g}{\frac{4}{3}\pi\rho}$, so that relatively to masses of the same density, q is proportional to the centrifugal force g of the rotatory motion, and consequently in the inverse ratio of the square of the time of rotation; whence it follows, that relatively to a mass of the same density as the earth, the time of rotation which answers to $q = 0.337007$, is 0.10090 days. Whence result these two theorems.

"Every homogeneous fluid mass of a density equal to the mean density of the earth, cannot be in equilibrium having an elliptic figure, if the time of its rotation is less than 0.10090 days. If this time be greater, there will be always two elliptic figures and no more which satisfy the equilibrium."

"If the density of the fluid mass is different from that of the earth; we shall have the time of rotation in which the equilibrium ceases to be possible under an elliptic figure, by multiplying 0.10090 days by the square root of the ratio of the mean density of the earth to that of the fluid mass."

This relatively to a fluid mass, whose density is only a fourth part of that of the earth, which nearly is the case with the sun, this time will be 0.20184 days; and if the density of the earth supposed fluid and homogeneous were about 98 times less than its actual density, the figure which it ought to take to satisfy its actual motion of rotation, would be the limit of all the elliptic figures with which the equilibrium can subsist. The density of Jupiter being about five times less than that of the earth, and the time of its rotation being 0.41377 days; we see that this duration is in the limits of those of equilibrium.

It may be thought that the limit of q, is that where the fluid would begin to fly off by reason of a too rapid motion of rotation; but it is easy to be convinced of the contrary, by observing that by 563, the gravity at the equator of the ellipsoid is to that at the pole in the ratio of the polar axis to that of the equator, a ratio which in this case, is that of 1 to 2.7197; the equilibrium ceases therefore to be possible, because with a motion of rotation more rapid, it is impossible to give to the fluid mass, an elliptic figure such that the resultant of its attraction and of the centrifugal force, may be perpendicular to the surface.

Hitherto we have supposed λ^2 positive, which gives the spheroids flattened towards the poles; let us now examine whether the equilibrium can subsist with a figure lengthened towards the poles, or with a prolate spheroid. Let $\lambda^2 = -\lambda'^2$; λ'^2 must be positive and less than unity, otherwise, the ellipsoid will be changed into a hyperboloid. The preceding value of d φ gives

$$\varphi = \int \frac{\lambda^2 d\lambda \{q\lambda^4 + (10q - 6)\lambda^2 + 9q\}}{(1+\lambda^2)(9+3\lambda^2)^2};$$

the integral being taken from $\lambda = 0$. Substituting for λ its value $\pm \lambda' \sqrt{-1}$, we shall have

$$\varphi = \pm \sqrt{-1} . \int \frac{\lambda'^2 d\lambda' \{q.(1-\lambda'^2).(9-\lambda'^2)+6\lambda'^2\}}{(1-\lambda'^2)(9-3\lambda'^2)};$$

but it is evident that the elements of this last integral are all of the same sign from $\lambda'^2 = 0$, to $\lambda'^2 = 1$; the function φ can therefore never become nothing in this interval. Thus then the equilibrium cannot subsist in the case of the prolate spheroid.

565. If the motion of rotation primitively impressed upon the fluid mass, is more rapid than that which belongs to the limit of q, we must not thence infer that it cannot be in equilibrium with an elliptic figure; for we may conceive, that by flattening it more and more, it will take a rotatory motion less and less rapid; supposing therefore that there exists, as in the case of all known fluids, a force of tenacity between its molecules, this mass, after a great number of oscillations, may at length arrive at a rotatory motion, comprised within the limits of equilibrium, and may continue in that state. But this possibility it would also be interesting to verify; and it would be equally interesting to know whether there would not be many possible states of equilibrium; for what we have already demonstrated upon the possibility of two states of equilibrium, corresponding to one motion of rotation, does not infer the possibility of two states of equilibrium corresponding to one primitive force; because the two

states of equilibrium relative to one motion of rotation, require two primitive forces, either different in quantity or differently applied.

Consider therefore a fluid mass agitated primitively by any forces whatever, and then left to itself, and to the mutual attractions of all its parts. If through the center of gravity of this mass supposed immoveable, we conceive a plane relatively to which the sum of the areas described upon this plane, by each molecule, multiplied respectively by the corresponding molecules, is a *maximum* at the origin of motion; this plane will always have this property, whatever may be the manner in which the molecules act upon one another, whether by their tenacity, by their attraction, and their mutual collision, even in the very case where there is finite loss of motion in an instant of time; thus, when after a great number of oscillations, the fluid mass shall take a uniform rotatory motion about a fixed axis, this axis shall be perpendicular to the plane above-mentioned, which will be that of the equator, and the motion of rotation will be such that the sum of the areas described during the instant $d\,t$, by the molecules projected upon this plane, will be the same as at the origin of motion; we shall denote by $E\,d\,t$ this last sum.

We shall here observe, that the axis in question, is that relatively to which the sum of the moments of the primitive forces of the system was a *maximum*. It retains this property during the motion of the system, and finally becomes the axis of rotation; for what is above asserted as to the *plane of the maximum of projected areas*, equally applies to the *axis of the greatest moment of forces*; since the elementary area described by the projection of the radius-vector of a body upon a plane, and multiplied by its mass, is evidently proportional to the moment of the finite force of this body relatively to the axis perpendicular to this plane.

Let, as above, g be the centrifugal force due to the rotatory motion at the distance 1 from the axis; $\sqrt{g}$ will be the angular velocity of rotation (p. 166. Vol. I.); then call k the semi-axis of rotation of the fluid mass, and $k\sqrt{1+\lambda^2}$ the semi-axis of its equator. It is easy to show that the sum of the areas described during the instant $d\,t$, by all the molecules projected upon the plane of the equator and multiplied respectively by the corresponding molecules, is

$$\frac{4\pi\rho}{15}(1+\lambda^2)^2 \,.\, k^5\,d\,t\,\sqrt{g}$$

we shall therefore have

$$\frac{4\pi\rho}{15}(1+\lambda^2)^2 \,.\, k^5\,\sqrt{g} = E.$$

Then calling M, the fluid mass, we shall have

$$\tfrac{4}{3}\pi k^3 \varrho (1 + \lambda^2) = M;$$

the quantity $\frac{g}{\frac{4}{3}\pi . \varrho}$, which we have called q, in No. 562, thus becomes $q' (1 + \lambda^2)^{-\frac{2}{3}}$, denoting by q′ the function $\frac{25 E^2 (\frac{4}{3}\pi \varrho)^{\frac{1}{3}}}{M^{\frac{1}{3}}}$ The equation of the same No. becomes

$$0 = \frac{9\lambda + 2 q' \lambda^3 (1 + \lambda^2)^{-\frac{2}{3}}}{9 + 3\lambda^2} - \tan.^{-1}\lambda.$$

This equation will determine λ; we shall then have k by means of the preceding expression of M.

Call φ the function

$$\frac{9\lambda + 2 q' \lambda^3 (1 + \lambda^2)^{-\frac{2}{3}}}{9 + 3\lambda^2} - \tan.^{-1}\lambda,$$

which, by the condition of equilibrium, ought to be equal to zero: this equation begins by being positive, when λ is very small, and ends by being negative, when λ is infinite; there exists therefore between $\lambda = 0$, and $\lambda =$ infinity, a value of λ which renders this function nothing, and consequently, there is always, whatever q′ may be, an elliptic figure, with which the fluid mass may be in equilibrium.

The value of φ may be put under this integral form

$$\varphi = 2 \int \frac{\lambda^4 d\lambda \left\{\frac{27 q'}{\lambda^2} + 18 q' - \{q' \lambda^2 + 18 (1 + \lambda^2)^{\frac{2}{3}}\}\right\}}{(9 + 3\lambda^2)^2 (1 + \lambda^2)^{\frac{5}{3}}}.$$

When it becomes nothing the function

$$\frac{27 q'}{\lambda^2} + 18 q' - \{q' \lambda^2 + 18 (1 + \lambda^2)^{\frac{2}{3}}\},$$

has already passed through zero to become negative; but from the instant when this function begins to be negative, it continues to be so as λ increases, because the positive part $\frac{27 q'}{\lambda^2} + 18 q'$ decreases whilst the negative part $- \{q' \lambda^2 + 18 (1 + \lambda^2)^{\frac{2}{3}}\}$ increases; the function φ cannot therefore twice become nothing; whence it follows, that there is but one real and positive value of λ which satisfies the equation of equilibrium, and consequently, the fluid can be in equilibrium with one elliptic figure only.

ON THE FIGURE OF A SPHEROID DIFFERING VERY LITTLE FROM A SPHERE, AND COVERED WITH A SHELL OF FLUID IN EQUILIBRIUM.

566. We have already discussed the equilibrium of a homogeneous fluid mass, and we have found that the elliptic figure satisfies this equilibrium; but in order to get a complete solution of the problem, we must determine *a priori* all the figures of equilibrium, or we must be certified that the elliptic is the only figure which will fulfil these conditions; besides, it is very probable that the celestial bodies have not homogeneous masses, and that they are denser towards the center than at the surface. In the research, therefore, of their figure, we must not rest satisfied with the case of homogeneity; but then this research presents great difficulties. Happily it is simplified by the consideration of the little difference which exists between the spherical figure and those of the planets and satellites; by which we are permitted to neglect the square of this difference, and of the quantities depending on it. Notwithstanding, the research of the figure of the planets is still very complex. To treat it with the greatest generality, we proceed to consider the equilibrium of a fluid mass which covers a body formed of shells of variable density, endowed with a rotatory motion, and sollicited by the attraction of other bodies. For that purpose, we proceed to recapitulate the laws of equilibrium of fluids, as laid down in works upon hydrostatics.

If we name ϱ the density of a fluid molecule, Π the pressure it sustains, F, F', F'', &c. the forces which act upon it, and $d f$, $d f'$, $d f''$ the elements of the directions of these forces; then the general equation of the equilibrium of the fluid mass will be

$$\frac{d\Pi}{\varrho} = F\, d f + F'\, d f' + F''\, d f'' + \&c.$$

Suppose that the second member of this equation is an exact difference; designating by $d\varphi$ this difference, ϱ will necessarily be a function of Π and of φ: the integral of this equation will give φ in a function of Π; we may therefore reduce to a function of Π only, from which we can obtain Π in a function of ϱ; thus, relatively to shells of a given constant density, we shall have $d\Pi = 0$, and consequently

$$0 = F\, d f + F'\, d f' + F''\, d f'' + \&c.;$$

an equation which indicates the tangential force at the surface of those shells is nothing, and consequently, that the resultant of all the forces F, F', F'', &c. is perpendicular to this surface; so that the shells are spherical.

The pressure Π being nothing at the exterior surface, ϱ must there be constant, and the resultant of all the forces which animate each molecule of the surface is perpendicular to it. This resultant is what we call gravity. The conditions of equilibrium of a fluid mass, are therefore 1st, that the direction of gravity be perpendicular to each point of the exterior surface: 2dly, that in the interior of the mass the directions of the gravity of each molecule be perpendicular to the surface of the shells of a constant density. Since we may take, in the interior of a homogeneous mass, such shells as we wish for shells of a constant density, the second of two preceding conditions of equilibrium, is always satisfied, and it is sufficient for the equilibrium that the first should be fulfilled; that is to say, that the resultant of all the forces which animate each molecule of the exterior surface should be perpendicular to the surface.

567. In the theory of the figure of the celestial bodies, the forces F, F′, F″, &c. are produced by the attraction of their molecules, by the centrifugal force due to their motion of rotation, and by the attraction of distant bodies. It is easy to be certified that the difference $F\, d\, f + F'\, d\, f' +$ &c. is there exact; but we shall clearly perceive that, by the analysis which we are about to make of these different forces, in determining that part of the integral $\int (F\, d\, f + F'\, d\, f' +$ &c.) which is relative to each of them.

If we call d M any molecule of the spheroid, and f its distance from the point attracted, its action upon this latter will be $\frac{d\, M}{f^2}$. Multiplying this action by the element of its direction, which is $- d\, f$, since it tends to diminish f, we shall have, relatively to the action of the molecule d M, $\int F\, d\, f = \frac{d\, M}{f}$; whence it follows that that part of the integral $\int (F\, d\, f + F'\, d\, f' +$ &c.), which depends on the attraction of the molecules of the spheroid, is equal to the sum of all these molecules divided by their respective distances from the molecule attracted. We shall represent this sum by V, as we have already done.

We propose, in the theory of the figure of the planets, to determine the laws of the equilibrium of all their parts, about their common center of gravity; we must, therefore, transfer into a contrary direction to the molecule attracted, all the forces by which this center is animated in virtue of the reciprocal action of all the parts of the spheroid; but we know that, by the property of this center, the resultant of all the actions upon this point is nothing. To get, therefore, the total effect of the attraction of the spheroid upon the molecules attracted, we have nothing to add to V.

To determine the effect of the centrifugal force, we shall suppose the position of the molecule determined by the three rectangular coordinates x', y', z', whose origin we fix at the center of gravity of the spheroid. We shall then suppose that the axis of x' is the axis of rotation, and that g expresses the centrifugal force due to the velocity of rotation at the distance 1 from the axis. This force will be nothing in the direction of x' and equal to $g\,y'$ and $g\,z'$ in the direction of y' and of z'; multiplying, therefore, these two last forces respectively by the elements $d\,y'$, $d\,z'$ of their directions, we shall have $\frac{1}{2}\,g\,(y'^2 + z'^2)$ for that part of the integral $\int(F\,d\,f + F'\,d\,f' + \&c.)$, which is due to the centrifugal force of the rotatory motion.

If we call, as above, r the distance of the molecule attracted from the center of gravity of the spheroid, θ the angle which the radius r forms with the axis of x', and ϖ the angle the plane which passes through the axis of x', and through the molecule, forms with the plane of x', y'; finally, if we make $\cos.\,\theta = m$, we shall have

$$x' = r\,m; \quad y' = r\,\sqrt{1 - m^2}.\cos.\,\varpi; \quad z' = r\,\sqrt{1 - m^2}.\sin.\,\varpi;$$

whence we get

$$\tfrac{1}{2}\,g\,(y'^2 + z'^2) = \tfrac{1}{2}\,g\,r^2\,(1 - m^2).$$

We shall put this last quantity under the following form:

$$\tfrac{1}{3}\,g\,r^2 - \tfrac{1}{2}\,g\,r^2\,(m^2 - \tfrac{1}{3})$$

to assimilate its terms to those of the expression V which are given in No. 559; that is to say, to give them the property of satisfying the equation of partial differences

$$0 = \left(\frac{d\,(1 - m^2).\left(\frac{d\,Y^{(i)}}{d\,m}\right)}{d\,m}\right) + \frac{\left(\frac{d^2\,Y^{(i)}}{d\,\varpi^2}\right)}{1 - m^2} + i\,(i + 1)\,Y^{(i)};$$

in which $Y^{(i)}$ is a rational and entire function of m, $\sqrt{1 - m^2}.\cos.\,\varpi$ and $\sqrt{1 - m^2}\,\sin.\,\varpi$ of the degree i; for it is clear that each of the two terms $\frac{1}{3}\,g\,r^2$ and $-\frac{1}{2}\,g\,r^2\,(m^2 - \frac{1}{3})$ satisfies for $Y^{(i)}$, the preceding equation.

It remains now for us to determine that part of the integral $\int(F\,d\,f + F'\,d\,f' + \&c.)$ which results from the action of distant bodies. Let S be the mass of one of these bodies, f its distance from the molecule attracted, and s its distance from the center of gravity of the spheroid. Multiplying its action by the element $-\,d\,f$ of its direction, and then integrating we shall have $\frac{S}{f}$. This is not the entire part of the integral $\int(F\,d\,f + F'\,d\,f' + \&c.)$ which is due to the action of S; we have still to transfer, in a contrary direction to the molecule, the action of this body upon the center of gravity of the spheroid. For that purpose, call v the angle which s forms with the axis of x', and ψ the angle which the plane passing through this star and through the body S, makes with the plane of x', y'. The action of $\frac{S}{s^2}$ of this body upon the center of gravity of the spheroid, resolved parallel to the axes of x', y', z', will produce the three following forms:

$$\frac{S}{s^2}\cos. v; \quad \frac{S}{s^2}\sin. v \cos. \psi; \quad \frac{S}{s^2}\sin. v \sin. \psi.$$

Transferring them in a contrary direction to the molecule attracted, which amounts to prefixing to them the sign —, then multiplying them by the elements d x', d y', d z', of their directions, and integrating them, the sum of the integrals will be

$$-\frac{S}{s^2}.\{x' \cos. v + y' \sin. v. \cos. \psi + z' \sin. v \sin. \psi\} + \text{const.};$$

the entire part of the integral $\int(F\,d\,f + F'\,d\,f' + \&c.)$, due to the action of the body S, will therefore be

$$\frac{S}{f} - \frac{S}{s^2}\{x' \cos. v + y' \sin. v \cos. \psi + z' \sin. v \text{ siu. } \psi\} + \text{const.};$$

and since this quantity ought to be nothing relatively to the center of gravity of the spheroid, which we suppose immoveable, and that relatively to this point, f becomes s, and x', y', z', are nothing, we shall have

$$\text{const.} = -\frac{S}{s}.$$

However, f is equal to

$$\{(s \cos. v - x')^2 + (s. \sin. v \cos. \psi - y')^2 + (s \sin. v \sin. \psi - z')^2\}^{\frac{1}{2}};$$

which gives, by substituting for x', y', z', their preceding values

$$\frac{S}{f} = \frac{S}{\sqrt{s^2 - 2\,s\,r\{\cos. v \cos. \theta + \sin. v \sin. \theta \cos. (\varpi - \psi) + r^2\}}}.$$

If we reduce this function into a series descending relatively to powers of s, and if we thus represent the series,

$$\frac{S}{s}\left\{P^{(0)} + \frac{r}{s}P^{(1)} + \frac{r^2}{s^2}P^{(2)} + \&c.\right\};$$

we shall have generally by 561 and 562,

$$P^{(i)} = \frac{1.3.5..(2i-1)}{1.2.3.\,....\,i}\left\{\delta^i - \frac{i\,(i-1)}{2(2\,i-1)}\delta^{i-2} + \frac{i(i-1)(i-2)(1-3)}{2.\,4(2\,i-1)(2\,i-3)}\delta^{i-4} - \&c.\right\};$$

δ being equal to cos. v cos. θ + sin. v sin. θ . cos. $(\varpi - \psi)$; it is evident that by 553, we have

$$0 = \left(\frac{d \left\{ (1-m^2) \left(\frac{d P^{(i)}}{d m} \right) \right\}}{d m} \right) + \frac{\left(\frac{d^2 P^{(i)}}{d \varpi^2} \right)}{1-m^2} + i(i+1) P^{(i)};$$

so that the terms of the preceding have this property, common with those of V. This being shown, we have

$$\frac{S}{f} - \frac{S}{s} - \frac{S}{s^2}(x' \cos. v + y' \sin. v \cos. \psi + z' \sin. v \sin. \psi)$$

$$= \frac{S r^2}{s^3} \left\{ P^{(2)} + \frac{r}{s} P^{(3)} + \frac{r^2}{s^2} P^{(4)} + \&c. \right\}.$$

If there were other bodies S', S'', &c.; denoting by s', v', ψ', $P'^{(i)}$; s'', v'', ψ'', $P''^{(i)}$, &c. what we have called s, v, ψ, $P^{(i)}$, relatively to the body S, we shall have the parts of the integral $\int (F\, d f + F' d f' + \&c.)$ due to their action, by marking with one, two, &c. dashes, the letters s, v, ψ, and P in the preceding expression of that part of this integral, which is due to the action of S.

If we collect all the parts of this integral, and make

$$\frac{g}{3} = \alpha Z^{(0)};$$

$$\frac{S}{s^3} P^{(2)} + \frac{S'}{s'^3} P'^{(2)} + \&c. - \frac{g}{2}\left(m^2 - \frac{1}{3}\right) = \alpha Z^{(2)};$$

$$\frac{S}{s^4} P^{(3)} + \frac{S'}{s'^4} P'^{(3)} + \&c. = \alpha Z^{(3)}$$

&c.

α being a very small coefficient, because the condition that the spheroid is very little different from a sphere, requires that the forces which produce this difference should themselves be very small; we shall have

$$\int (F\, d f + F' d f' + \&c.) = V + \alpha r^2 \{Z^{(0)} + Z^{(2)} + r Z^{(3)} + r^2 Z^{(4)} + \&c.\}$$

$Z^{(i)}$ satisfying, whatever i may be, in the equation of partial differences

$$0 = \left(\frac{d \left\{ (1-m^2) \frac{d Z^{(i)}}{d m} \right\}}{d m} \right) + \frac{\left(\frac{d^2 Z^{(i)}}{d \varpi^2} \right)}{1-m^2} + i(i+1) Z^{(i)}.$$

The general equation of equilibrium will therefore be

$$\int \frac{d \Pi}{\rho} = V + \alpha r^2 \{Z^{(0)} + Z^{(2)} + r Z^{(3)} r^2 Z^{(4)} + \&c.\} \quad (1)$$

If the extraneous bodies are very distant from the spheroid, we may neglect the quantities $r^3 Z^{(3)}$, $r^4 Z^{(4)}$, &c., because the different terms of these quantities being divided respectively by s^4, s^5, &c. s'^4, s'^5, &c. these terms become very small when s, s', &c. are very great compared with r. This case subsists for the planets and satellites with the exception of Saturn, whose ring is too near his surface for us to neglect the preceding terms. In the theory of the figure of that planet, we must therefore prolong the second member of equation (1), which possesses the advantage of forming a series always convergent; and since then the number of corpuscles exterior to the spheroid is infinite, the values of $Z^{(0)}$, $Z^{(2)}$, &c. are given in definite integrals, depending on the figure and interior constitution of the ring of Saturn.

568. The spheroid may be entirely fluid; it may be formed of a solid nucleus covered by a fluid. In both cases the equation (1) of the preceding No. will determine the figure of the shells of the fluid part, by considering, that since Π must be a function of ϱ, the second member of this equation must be constant for the exterior surface, and for that of the shells in equilibrium, and can only vary from one shell to another.

The two preceding cases reduce to one when the spheroid is homogeneous; for it is indifferent as to the equilibrium whether it is entirely fluid, or contains an interior solid nucleus. It is sufficient by No. 556, that at the exterior surface we have

$$\text{constant} = V + \alpha r^2 \{Z^{(0)} + Z^{(2)} + r Z^{(3)} + \&c.\}.$$

If we substitute in this equation for V its value given by formula (3) of No. 555, and if we observe that by No. 556, $Y^{(0)}$ disappears by taking for a the radius of a sphere of the same volume as the spheroid, and that $Y^{(1)}$ is nothing when we fix the origin of coordinates at the center of the spheroid; we shall have

$$\text{constant} = \frac{4\pi a^3}{3r} + \frac{4\pi\alpha a^5}{r^3}\left\{\frac{1}{5}Y^{(2)} + \frac{a}{7r}\cdot Y^{(3)} + \frac{a^2}{9r^2}Y^{(4)} + \&c.\right\}$$
$$+ \alpha r^2 \{Z^{(0)} + Z^{(2)} + r Z^{(3)} + r^2 Z^{(4)} + \&c.\}$$

Substituting in the equation of the surface of the spheroid for r its value at the surface $1 + \alpha y$, or

$$a + \alpha a \{Y^{(2)} + Y^{(3)} + Y^{(4)} + \&c.\}$$

which gives

$$\text{const.} = \frac{4\pi}{3}a^2 - \frac{8\alpha\pi a^2}{3}\left\{\frac{1}{5}Y^{(2)} + \frac{2}{7}Y^{(3)} + \frac{3}{9}Y^{(4)} + \&c.\right\}$$
$$+ \alpha a^2 \{Z^{(0)} + Z^{(2)} + a Z^{(3)} + a^2 Z^{(4)} + \&c.\}$$

We shall determine the arbitrary constant of the first member of this equation, by means of this equation,

$$\text{const.} = \frac{4}{3}\pi a^2 + \alpha a^2 Z^{(0)};$$

we shall then have by comparing like functions, that is to say, such as are subject to the same equation of partial differences,

$$Y^{(i)} = \frac{3\,(2\,i + 1)}{8\,(i - 1)\,\pi} \cdot a^{i-2}\, Z^{(i)};$$

i being greater than unity. The preceding equation may be put under the form

$$Y^{(i)} = \frac{3}{4\,\pi} \cdot a^{i-2} \cdot Z^{(i)} + \frac{9}{8\,a\,\pi} \int r^{i-2}\, d\,r\, Z^{(i)}$$

the integral being taken from $r = 0$ to $r = a$. The radius $a\,(I - \alpha\,y)$ of the surface of the spheroid will hence become

$$a\,(1 + \alpha\,y) = a \left\{ \begin{array}{l} 1 + \frac{3\,\alpha}{4\,\pi} \{Z^{(2)} + a\,Z^{(3)} + a^2\,Z^{(4)} + \&c.\} \\ + \frac{9\,\alpha}{8\,a\,\pi} \int d\,r\, \{Z^{(2)} + r\,Z^{(3)} + r^2\,Z^{(4)} + \&c.\} \end{array} \right\}, \quad (2)$$

We may put this equation under a finite form, by considering that we have by the preceding No.

$$\alpha\,\{Z^{(2)} + r\,Z^{(3)} + r^2\,Z^{(4)} + \&c.\} = -\frac{g}{2}(m^2 - \tfrac{1}{3}) - \frac{S}{s\,r^2} - \frac{S\,.\,\delta}{s^2\,r}$$
$$+ \frac{S}{r^2 \sqrt{s^2 - 2\,s\,r\,\delta + r^2}} - \frac{S'}{s'\,r^2} - \&c.,$$

so that the integral $\int d\,r\,\{Z^{(2)} + r\,Z^{(3)} + \&c.\}$ is easily found by known methods.

569. The equation (1) of 567 not only has the advantage of showing the figure of the spheroid, but also that of giving by differentiation the law of gravity at its surface; for it is evident that the second member of this equation being the integral of the sum of all the forces with which each molecule is animated, multiplied by the elements of their respective directions, we shall have that part of the resultant which acts along the radius r, by differentiating the second member relatively to r; thus calling p the force by which a molecule of the surface is sollicited towards the center of gravity of the spheroid, we shall have

$$p = -\left(\frac{d\,V}{d\,r}\right) - \frac{\alpha}{d\,r}\, d\,\{r^2\,Z^{(0)} + r^2\,Z^{(2)} + r^3\,Z^{(3)} + r^4\,Z^{(4)} + \&c.\}.$$

If we substitute in this equation for $-\left(\frac{d\,V}{d\,r}\right)$, its value at the surface $\frac{2}{3}\pi\,a + \frac{V}{2\,a}$, given by equation (2) of No. 554, and for V, its value given by equation (1) of No. 567; we shall have

$$p = \frac{4}{3}\pi\,a - \frac{1}{2}\alpha\,a\,\{Z^{(2)} + a\,Z^{(3)}\,a^2\,Z^{(4)} + \&c.\}$$
$$- \frac{\alpha}{d\,r} \cdot d \cdot \{r^2\,Z^{(0)} + r^2\,Z^{(2)} + r^3\,Z^{(3)} + r^4\,Z^{(4)} + \&c.\} \quad (3)$$

r must be changed into a after the differentiations in the second member of this equation, which by the preceding No. may always be reduced to a finite function.

p does not represent exactly gravity, but only that part of it which is directed towards the center of gravity of the spheroid, by supposing it resolved into two forces, one of which is perpendicular to the radius r, and the other p is directed along this radius. The first of these two forces is evidently a small quantity of the order α; denoting it therefore by $\alpha\gamma$, gravity will be equal to $\sqrt{p^2 + \alpha^2 \gamma^2}$, a quantity which, neglecting the terms of the order α^2, reduces to p. We may thus consider p as expressing gravity at the surface of the spheroid, so that the equations (2) and (3) of the preceding No. and of this, determine both the figure of homogeneous spheroids in equilibrium, and the law of gravity at their surfaces; they contain the complete theory of the equilibrium of these spheroids, on the supposition that they differ very little from the sphere.

If the extraneous bodies S, S′, &c. are nothing, and therefore the spheroid is only sollicited by the attraction of its molecules, and the centrifugal force of its rotatory motion, which is the case of the Earth and primary planets with the exception of Saturn, when we only regard the permanent state of their figures; then designating by $\alpha\varphi$, the ratio of the centrifugal force to gravity at the equator, a ratio which is very nearly equal to $\frac{g}{\frac{4}{3}\pi}$, the density of the spheroid being taken for unity; we shall find,

$$a(1 + \alpha y) = a\{1 - \frac{5\alpha\varphi}{4}(m^2 - \tfrac{1}{3})\};$$

$$p = \tfrac{4}{3}\pi a\{1 - \tfrac{2}{3}\alpha\varphi + \frac{5\alpha\varphi}{4}(m^2 - \tfrac{1}{3})\};$$

the spheroid is then therefore an ellipsoid of revolution, upon which increments of gravity, and decrements of the radii, from the equator to the poles, are very nearly proportional to the square of the sine of the latitude, m being to quantities of the order α, equal to this sine.

a, by what precedes, is the radius of a sphere, equal in solidity to the spheroid; gravity at the surface of this sphere will be $\frac{4}{3}\pi a$; thus we shall have the point of the surface of the spheroid, where gravity is the same as at the surface of the sphere, by determining m by the equation

$$0 = -\tfrac{2}{3} + \tfrac{5}{4}(m^2 - \tfrac{1}{3});$$

which gives

$$m = \sqrt{\frac{13}{15}}.$$

570. The preceding analysis conducts us to the figure of a homogeneons fluid mass in equilibrium, without employing other hypotheses than that of a figure differing very little from the sphere: it also shows that the elliptic figure which satisfies this equilibrium, is the only figure which does it. But as the expansion of the radius of the spheroid into a series of the form $a\{1 + \alpha Y^{(0)} + \alpha Y^{(1)} + \&c.\}$ may give rise to some difficulties, we proceed to demonstrate directly, and independently of this expansion, that the elliptic figure is the only figure of the equilibrium of a homogeneous fluid mass endowed with a rotatory motion; which by confirming the results of the preceding analysis, will at the same time serve to remove any doubts we may entertain against the generality of this analysis.

First suppose the spheroid one of revolution, and that its radius is $a(1 + \alpha y)$, y being a function of m, or of the cosine of the angle θ which this radius makes with the axis of revolution. If we call f any straight line drawn from the extremity of this radius in the interior of the spheroid; p the complement of the angle which this straight line makes with the plane which passes through the radius $a(1 + \alpha y)$ and through the axis of revolution; q the angle made by the projection of f upon this plane and by the radius; finally, if we call V the sum of all the molecules of the spheroid, divided by their distances from the molecules placed at the extremity of the radius $a(1 + \alpha y)$; each molecule being equal to $f^2\, df . dp . dq . \sin. p$, we shall have

$$V = \tfrac{1}{2}\int f'^2\, dp . dq . \sin. p,$$

f′ being what f becomes at its quitting the spheroid. We must now determine f′ in terms of p and q.

For that purpose, we shall observe that if we call θ', the value of θ relative to this point of exit, and $a(1 + \alpha y')$, the corresponding radius of the spheroid, y′ being a similar function of cos. θ' or of m′ that y is of m; it is easily seen that the cosine of the angle formed by the two straight lines f′ and $a(1 + \alpha y)$ is equal to sin. p . cos. q; and therefore that in the triangle formed by the three straight lines f′, $a(1 + \alpha y)$ and $a(1 + \alpha y')$ we have

$$a^2(1 + \alpha y')^2 = f'^2 - 2af'(1 + \alpha y)\sin. p . \cos. q + a^2(1 + \alpha y)^2.$$

This equation gives for f'^2 two values; but one of them being of the order α^2 is nothing when we neglect the quantities of that order; the other becomes

$$f'^2 = 4a^2 \sin.^2 p \cos.^2 q\,(1 + 2\alpha y) + 4\alpha a^2 (y' - y);$$

which gives

$$V = 2\,a^2 \int d\,p\,d\,q\,\sin.\,p\,\{(1 + 2\,\alpha\,y)\sin.^2 p\cos.^2 q + \alpha\,(y' - y)\}.$$

It is evident that the integrals must be taken from $p = 0$, to $p = \pi$, and from $q = -\frac{1}{2}\pi$ to $q = \frac{1}{2}\pi$; we shall therefore have

$$V = \tfrac{4}{3}\pi\,a^2 - \tfrac{4}{3}\,\alpha\,\pi\,a^2\,y + 2\,\alpha\,a^2 \int d\,p\,.\,d\,q\,.\,y'\sin.\,p\,.$$

y' being a function of $\cos.\,\theta'$, we must determine this cosine in a function of p and q; we may therefore in this determination neglect the quantities of the order α, since y' is already multiplied by α; hence we easily find

$$a\cos.\,\theta' = (a - f'\sin.\,p\cos.\,q)\cos.\,\theta + f'\sin.\,p\,.\,\sin.\,q\,.\,\sin.\,\theta\,;$$

whence we derive, substituting for f' its value $2\,a\sin.\,p\cos.\,q$,

$$m' = m\cos.^2 p - \sin.^2 p\cos.\,(2\,q + \theta).$$

Here we must observe, relatively to the integral $\int y'\,d\,p\,.\,d\,q\,.\,\sin.\,p$, taken relatively to q from $2\,q = -\pi$ to $2\,q = \pi$, that the result would be the same, if this integral were taken from $2\,q = -\theta$ to $2\,q = 2\pi - \theta$, because the values of m', and consequently of y' are the same from $2\,q = -\pi$ to $2\,q = -\theta$ as from $2\,q = \pi$ to $2\,q = 2\pi - \theta$; supposing therefore $2\,q + \theta = q'$, which gives

$$m' = m\cos.^2 p - \sin.^2 p\cos.\,q'\,;$$

we shall have

$$V = \tfrac{4}{3}\pi\,a^2 - \tfrac{4}{3}\,\alpha\,\pi\,a^2\,y + \alpha\,a^2 \int y'\,d\,p\,d\,q'\,\sin.\,p\,;$$

the integrals being taken from $p = 0$ to $p = \pi$ and from $q' = 0$ to $q' = 2\,\pi$.

Now if we denote by $a^2\,N$ the integral of all the forces extrinsic to the attraction of the spheroid, and multiplied by the elements of their directions; by 568 we shall have in the case of equilibrium

$$\text{constant} = V + a^2\,N,$$

and substituting for V its value, we shall have

$$\text{const.} = \tfrac{4}{3}\,\alpha\,\pi\,.\,y - \alpha \int y'\,d\,p\,.\,d\,q'\,\sin.\,p - N\,;$$

an equation which is evidently but the equation of equilibrium of No. 568, presented under another form. This equation being linear, it thence results that if any number i of radii $a\,(1 + \alpha\,y)$, $a\,(1 + \alpha\,v)$, and satisfy it; the radius $a\,\{1 + \frac{\alpha}{i}\,(y + v + \&c.)\}$ will also satisfy it.

Suppose that the extraneous forces are reduced to the centrifugal force due to the rotation, and call g this force at the distance 1 from the axis of rotation; we shall have, by 567, $N = \frac{1}{2}\,g\,(1 - m^2)$; the equation of equilibrium will therefore be

$$\text{const.} = \tfrac{4}{3}\,\alpha\,\pi\,y - \alpha \int y'\,d\,p\,d\,q'\,\sin.\,p - \tfrac{1}{2}\,g\,(1 - m^2).$$

Differentiating three times successively, relatively to m, and observing that $\left(\frac{d\,m'}{d\,m}\right) = \cos.^2 p$, in virtue of the equation

$$m' = m \cos.^2 p - \sin.^2 p \cos. q';$$

we shall have

$$0 = \tfrac{4}{3}\pi \left(\frac{d^3 y}{d m^3}\right) - \int d p\, d q' \sin. p \cos.^6 p \left(\frac{d^3 y'}{d m'^3}\right),$$

but we have $\int d p\, d q' \sin. p \cos.^6 p = \frac{4\pi}{7}$; we may therefore put the preceding equation under this form,

$$0 = \int d p\, d q' \sin. p \cos.^6 p \left\{\tfrac{7}{3}\left(\frac{d^3 y}{d m^3}\right) - \left(\frac{d^3 y'}{d m'^3}\right)\right\}.$$

This equation subsists, whatever m may be; but it is evident, that amongst all the values between $m = -1$ and $m = 1$, there is one which we shall designate by h, and which is such that, abstraction being made of the sign, each of the values of $\left(\frac{d^3 y}{d m^3}\right)$ will not exceed that which is relative to h; denoting therefore by H, this latter value, we shall have

$$0 = \int d p\, d q' \sin. p \cos.^6 p \left\{\tfrac{7}{3} . H - \left(\frac{d^3 y'}{d m'^3}\right)\right\}.$$

The quantity $\tfrac{7}{3} H - \left(\frac{d^3 y'}{d m'^3}\right)$ has evidently the same sign as H, and the factor $\sin. p . \cos.^6 p$, is constantly positive in the whole extent of the integral; the elements of this integral have, therefore, all of them the same sign as H; whence it follows that the entire integral cannot be nothing, at least H cannot be so, which requires that we have generally $0 = \left(\frac{d^3 y}{d m^3}\right)$, whence by integrating we get

$$y = l + m . m + n . m^2;$$

l, *m*, n, being arbitrary constants.

If we fix the origin of the radii in the middle of the axis of revolution, and take for a the half of this axis, y will be nothing when $m = 1$ and when $m = -1$, which gives $m = 0$ and $n = -l$; the value of y thus becomes, $l(1 - m^2)$; substituting in the equation of equilibrium,

$$\text{const.} = \tfrac{4}{3}\alpha\pi y - \alpha \int y' \, d p\, d q' \sin. p - \tfrac{1}{2} g (1 - m^2);$$

we shall find $\alpha l = \frac{15 g}{16 \pi} = \frac{5}{4}\alpha\varphi$, $\alpha\varphi$ being the ratio of the centrifugal force to the equatorial gravity, a ratio which is very nearly equal to $\frac{3 g}{4 \pi}$; the radius of the spheroid will therefore be

$$a \left\{1 + \frac{5\alpha\varphi}{4}(1 - m^2)\right\};$$

whence it follows that the spheroid is an ellipsoid of revolution, which is conformable to what precedes.

Thus we have determined directly and independently of series, the figure of a homogeneous spheroid of revolution, which turns round its axis, and we have shown that it can only be that of an ellipsoid which becomes a sphere when $\varphi = 0$; so that the sphere is the only figure of revolution which would satisfy the equilibrium of an immoveable homogeneous fluid mass.

Hence we may conclude generally, that if the fluid mass is sollicited by any very small forces, there is only one possible figure of equilibrium; or, which comes to the same, there is only one radius $a(1 + \alpha y)$ which can satisfy the equation of equilibrium,

$$\text{const.} = \tfrac{4}{3}\alpha\pi . y - \alpha\int y' \, d p . d q' \sin. p - N;$$

y being a function of θ and of the longitude ϖ, and y' being what y becomes when we change θ and ϖ into θ' and ϖ'. Suppose, in fact, that there are two different rays $a(1 + \alpha y)$ and $a(1 + \alpha y + \alpha v)$ which satisfy this equation; we shall have

$$\text{const.} = \tfrac{4}{3}\alpha\pi(y + v) - \alpha\int(y' + v') \, d p \, d q' \sin. p - N.$$

Taking the preceding equation from this, we shall have

$$\text{const.} = \tfrac{4}{3}\pi v - \int v' \, d p \, d q' \sin. p.$$

This equation is evidently that of a homogeneous spheroid in equilibrium, whose radius is $a(1 + \alpha v)$, and which is not sollicited by any force extraneous to the attraction of its molecules. The angle ϖ disappearing in this equation, the radius $a(1 + \alpha v)$ will still satisfy it if ϖ be successively changed to $\varpi + d\varpi$, $\varpi + 2 d\varpi$, &c., whence it follows, that if we call v_1, v_2, &c. what v becomes in virtue of these changes; the radius

$$a\{1 + \alpha v \, d\varpi + \alpha v_1 \, d\varpi + \alpha v_2 \, d\varpi + \&c.\},$$

or

$$a(1 + \alpha\int v \, d\varpi),$$

will satisfy the preceding equation. If we take the integral $\int v \, d\varpi$ from $\varpi = 0$ to $\varpi = 2\pi$, the radius $a(1 + \alpha\int v \, d\varpi)$ becomes that of a spheroid of revolution, which, by what precedes, can only be a sphere: see the condition which results for v.

Suppose that a is the shortest distance of the center of gravity of the spheroid whose radius is $a(1 + \alpha v)$, to the surface, and fix the pole or origin of the angle θ at the extremity of a; v will be nothing at the pole, and positive every where else; it will be the same for the integral $\int v \, d\varpi$. But, since the center of gravity of the spheroid whose radius is $a(1+\alpha v)$, is at the center of the sphere whose radius is a, this point will, in like manner, be the center of gravity of the spheroid whose radius is

$a(1 + \alpha \int v . d \varpi)$; the different radii drawn from this center to the surface of this last spheroid are therefore unequal to one another, if v is not nothing; there can only therefore be a sphere in the case of $v = 0$; thus we learn for a certainty, that a homogeneous spheroid, sollicited by any small forces whatever, can only be in equilibrium in one manner.

571. We have supposed that N is independent of the figure of the spheroid; which is what very nearly takes place when the forces, extraneous to the action of the fluid molecules, are due to the centrifugal force of rotatory motion, and to the attraction of bodies exterior to the spheroid. But if we conceive at the center of the spheroid a finite force depending on the distance r, its action upon the molecules placed at the surface of the fluid, will depend on the nature of this surface, and consequently N will depend upon y. This is the case of a homogeneous fluid mass which covers a sphere of a density different from that of the fluid; for we may consider this sphere as of the same density as the fluid, and may place at its center a force reciprocal to the square of the distances; so that, if we call c the radius of the sphere, and ϱ its density, that of the fluid being taken for unity, this force at the distance r will be equal to $\frac{4}{3}\pi . \frac{c^3(\varrho - 1)}{\varrho^2}$. Multiplying by the element $- d r$ of its direction the integral of the product will be $\frac{4}{3}\pi . \frac{c^3(\varrho - 1)}{\varrho}$, a quantity which we must add to $a^2 N$; and since at the surface we have $r = a(1 + \alpha y)$, in the equation of equilibrium of the preceding No., we must add to N,

$$\frac{4}{3}\pi . \frac{(\varrho - 1) c^3}{a^3} . (1 - \alpha y).$$

This equation will become

$$\text{const.} = \frac{4 \alpha \pi}{3} \left\{1 + (\varrho - 1) . \frac{c^3}{a^3}\right\} y - \alpha \int y' d p . d q \sin. p - N.$$

If we denote by $a(1 + \alpha y + \alpha . v)$, a new expression of the radius of the spheroid in equilibrium, we shall have to determine v, the equation

$$\text{const.} = \frac{4}{3}\pi \left\{1 + (\varrho - 1) \frac{c^3}{a^3}\right\} - \int v' d p d q' \sin. p;$$

an equation which is that of the equilibrium of the spheroid, supposing it immoveable, and abstracting every external force.

If the spheroid is of revolution, v will be a function of cos. θ or m only; but in this case we may determine it by the analysis of the preceding No.; for if we differentiate this equation i + 1 times successively relatively to m, we shall have

$$0 = \frac{4}{3}\pi \left\{1 + (\varrho - 1) \frac{c^3}{a^3}\right\} \left(\frac{d^{i+1} v}{d m^{i+1}}\right) - \int \left(\frac{d^{i+1} v'}{d m'^{i+1}}\right) d p d q' \sin. p \cos.^{2i+2} p.$$

but we have

$$\int d\,p\,d\,q'\,\sin.\,p\cos.^{2i+2}p = \frac{4\,\pi}{2\,i+3};$$

the preceding equation may therefore be put under this form,

$$0 = \int d\,p\,d\,q'\sin.p\cos.^{2i+2}p\left\{\frac{2i+3}{3}\left(1+\overline{\varrho-1}\,.\,\frac{c^3}{a^3}\right)\left(\frac{d^{i+1}v}{d\,m^{i+1}}\right)-\left(\frac{d^{i+1}v'}{dm'^{i+1}}\right)\right\}$$

We may take i such that, abstraction being made of the sign, we have

$$\frac{2\,i+3}{3}\left\{1+(\varrho-1)\,\frac{c^3}{a^3}\right\} > 1;$$

Supposing, therefore, that i is the smallest positive whole number which renders this quantity greater than unity, we may see, as in the preceding No., that this equation cannot be satisfied unless we suppose $\left(\frac{d^{i+1}v'}{d\,m^{i+1}}\right) = 0$, which gives

$$v = m^i + A\,m^{i-1} + B\,m^{i-2} + \&c.$$

Substituting in the preceding equation of equilibrium for v, this value, and for v′

$$m'^i + A\,m'^{i-1} + B\,m'^{i-2} + \&c.$$

m′ being by the preceding No. equal to m cos.^{2} p — sin.^{2} p cos. q′, first we shall find

$$1 + (\varrho - 1)\,\frac{c^3}{a^3} = \frac{3}{2\,i+1};$$

which supposes ϱ equal to or less than unity; thus, whenever a, c, and ϱ are not such as to satisfy this equation, i being a positive whole number, the fluid can be in equilibrium only in one manner. Then we shall have

$$A = 0; \quad B = -\frac{i\,(i-1)}{2\,(2\,i-1)}; \ \&c.$$

so that

$$v = m^i - \frac{i\,(i-1)}{2\,(2\,i-1)}\,.\,m^{i-2} + \frac{i\,(i-1)\,(i-2)\,(i-3)}{2.\,4.\,(2\,i-1)\,(2\,i-3)}\,m^{i-4} - \&c.;$$

there are, therefore, generally two figures of equilibrium, since α v is susceptible of two values, one of which is given by the supposition of $\alpha = 0$, and the other is given by the supposition of v being equal to the preceding function of m.

If the spheroid has no rotatory motion, and is not sollicited by any extraneous force, the first of these two figures is a sphere, and the second has for its meridian a curve of the order i. These two curves coincide in the case of i = 1, because the radius a (1 + α m) is that of a sphere in which the origin of the radii is at the distance α from its center; but then it is easy to see that $\varrho = 1$, that is, the spheroid is homogeneous, a result agreeing with that of the preceding No.

572. When we have figures of revolution which satisfy the equilibrium, it is easy to obtain those which are not of revolution by the following method. Instead of fixing the origin of the angle θ at the extremity of the axis of revolution, suppose it at the distance γ from this extremity, and call θ' the distance from this same extremity of the point of the surface whose distance from the new origin of the angle θ is θ. Call, moreover, $\varpi - \beta$ the angle comprised between the two arcs θ and γ; we shall have

$$\cos.\theta' = \cos.\gamma\cos.\theta + \sin.\gamma\sin.\theta.\cos.(\varpi - \beta);$$

designating therefore by $\Gamma.(\cos.\theta')$ the function

$$\cos.^{i}\theta' - \frac{i(i-1)}{2.(2i+1)}.\cos.^{i-2}\theta' + \&c.;$$

the radius of the immoveable spheroid in equilibrium, which we have seen is equal to $a\{1 + \alpha\,\Gamma.(\cos.\theta')\}$, will be

$$a + \alpha a\,\Gamma.\{\cos.\gamma.\cos.\theta + \sin.\gamma.\sin.\theta\cos.(\varpi - \beta)\};$$

and although it is a function of the angle ϖ, it belongs to a solid of revolution, in which the angle θ is not at the extremity of the axis of revolution.

Since this radius satisfies the equation of equilibrium, whatever may be α, β, and γ, it will also satisfy in changing these quantities into α', β', γ', α'', β'', γ'', &c. whence it follows that this equation being linear, the radius

$$\begin{aligned} &a + \alpha\, a\,\Gamma.\{\cos.\gamma\ \cos.\theta + \sin.\gamma\ \sin.\theta\cos.(\varpi - \beta)\} \\ &+ \alpha' a\,\Gamma.\{\cos.\gamma'\ \cos.\theta + \sin.\gamma'\sin.\theta\cos.(\varpi - \beta')\} \\ &+ \&c. \end{aligned}$$

will likewise satisfy it. The spheroid to which this radius belongs is no longer one of revolution; it is formed of a sphere of the radius a, and of any number of shells similar to the excess of the spheroid of revolution whose radius is $a + \alpha a\,\Gamma.(m)$ above the sphere whose radius is a, these shells being placed arbitrarily one over another.

If we compare the expression of $\Gamma.(\cos.\theta')$ with that of $P^{(i)}$ of No. 567, we shall see that these two functions are similar, and that they differ only by the quantities γ and β, which in $P^{(i)}$ are v and ψ, and by a factor independent of m and ϖ; we have, therefore,

$$0 = \left(\frac{d.(1-m^2).\left(\frac{d.\Gamma(\cos.\theta')}{dm}\right)}{dm}\right) + \frac{\left(\frac{d^2\Gamma.(\cos.\theta')}{d\varpi^2}\right)}{1-m^2} + i(i+1)\,\Gamma.(\cos.\theta')$$

It is easy hence to conclude, that if we represent by $\alpha\,Y^{(i)}$ the function

$$\begin{aligned} &\alpha\ .\Gamma.\{\cos.\gamma\ \cos.\theta + \sin.\gamma\ \sin.\theta.\cos.(\varpi - \beta)\} \\ &+ \alpha'.\Gamma.\{\cos.\gamma'\cos.\theta + \sin.\gamma'\sin.\theta.\cos.(\varpi - \beta')\} \\ &+ \&c. \end{aligned}$$

$Y^{(i)}$ will be a rational and entire function of m, $\sqrt{1 - m^2}\cos.\varpi$, $\sqrt{1 - m^2}\sin.\varpi$, which will satisfy the equation of partial differences,

$$0 = \left(\frac{d\left\{(1-m^2).\left(\frac{d\,Y^{(i)}}{d\,m}\right)\right\}}{d\,m}\right) + \frac{\left(\frac{d^2\,Y^{(i)}}{d\,\varpi^2}\right)}{1-m^2} + i\,(i+1)\,Y^{(i)};$$

choosing for $Y^{(i)}$, therefore, the most general function of that nature, the function $a\,(1 + \alpha\,Y^{(i)})$ will be the most general expression of the equilibrium of an immoveable spheroid.

We may arrive at the same result by means of the series for V in 555; for the equation of equilibrium being, by the preceding No.,

$$\text{const.} = V + a^2\,N;$$

if we suppose that all the forces extraneous to the reciprocal action of the fluid molecules, are reducible to a single attractive force equal to $\frac{4}{3}\pi.\frac{(\rho - 1)\,c^3}{r^2}$, placed at the center of the spheroid, by multiplying this force by the element $-\,d\,r$ of its direction, and then integrating, we shall have

$$\tfrac{4}{3}\pi.\frac{(\rho - 1)\,c^3}{r} = a^2\,N;$$

and since at the surface $r = a\,(1 + \alpha\,y)$ the preceding equation of equilibrium will become

$$\text{const.} = V + \tfrac{4}{3}\alpha\,\pi.\frac{c^3}{a}\,(1 - \rho)\,y.$$

Substituting in this equation for V its value given by formula (3) of No. 555, in which we shall put for r its value $a\,(1 + \alpha\,y)$, and by substituting for y its value

$$Y^{(0)} + Y^{(1)} + Y^{(2)} + \&c.;$$

we shall have

$$0 = \left\{(1-\rho)\frac{c^3}{a^3} + 2\right\}Y^{(0)} + (1-\rho)\frac{c^3}{a^3}Y^{(1)} + \left\{(1-\rho)\frac{c^3}{a^3} - \tfrac{2}{5}\right\}Y^{(2)}$$

$$\ldots\ldots + \left\{(1-\rho)\frac{c^3}{a^3} - \frac{2\,i - 2}{2\,i + 1}\right\}Y^{(i)} + \&c.;$$

the constant a being supposed such, that const. $= \frac{4}{3}\pi\,a^2$. This equation gives $Y^{(0)} = 0$, $Y^{(1)} = 0$, $Y^{(2)} = 0$, &c. unless the coefficient of one of these quantities, of $Y^{(i)}$ for example, is nothing, which gives

$$(1 - \rho)\frac{c^3}{a^3} = \frac{2\,i - 2}{2\,i + 1},$$

i being a positive whole number, and in this case all these quantities except $Y^{(i)}$ are nothing; we shall therefore have $y = Y^{(i)}$, which agrees with what is found above.

Thus we see, that the results obtained by the expansion of V into a series, have all possible generality, and that no figure of equilibrium has escaped the analysis founded upon this expansion; which confirms what we have seen *à priori*, by the analysis of 555, in which we have proved that the form which we have given to the radius of spheroids, is not arbitrary but depends upon the nature itself of their attractions.

573. Let us now resume equation (1) of No. 567. If we therein substitute for V its value given by formula (6) of No. 558, we shall have relatively to the different fluid shells

$$\int \frac{d\Pi}{\rho} = 2\pi \int \rho\, d\, a^2 + 4\alpha\pi \int \rho\, d \left\{ a^2 Y^{(0)} + \frac{a r}{3} Y^{(1)} + \frac{r^2}{5} Y^{(2)} + \frac{r^3}{7a} Y^{(3)} + \&c. \right\}$$
$$+ \frac{4\pi}{3r} \int \rho\, d\, a^3 + \frac{4\alpha\pi}{r} \int \rho\, d \left\{ a^3 Y^{(0)} + \frac{a^4}{3r} Y^{(1)} + \frac{a^5}{5r^2} Y^{(2)} + \frac{a^6}{7r^3} Y^{(3)} + \&c. \right\}$$
$$+ \alpha r^2 \{ Z^{(0)} + Z^{(2)} + r Z^{(3)} + r^2 Z^{(4)} + \&c. \}; \quad . \quad . \quad . \quad . \quad (1)$$

the differentials and integrals being relative to the variable a; the two first integrals of the second member of this equation must be taken from a = a to a = 1, a being the value of a, relative to the *leveled* fluid shell, which we are considering, and this value at the surface being taken for unity: the two last integrals ought to be taken from a = 0 to a = a: finally, the radius r ought to be changed into a $(1 + \alpha y)$ after all the differentiations and integrations. In the terms multiplied by α it will suffice to change r into a; but in the term $\frac{4\pi}{3r} \int \rho\, d \, . \, a^3$ we must substitute a $(1 + \alpha y)$ for r; which changes it into this

$$\frac{4\pi}{3a} (1 - \alpha y) \int \rho\, d\, a^3,$$

and consequently, into the following

$$\frac{4\pi}{3a} \{ 1 - \alpha Y^{(0)} - \alpha Y^{(1)} - \alpha Y^{(2)} - \&c. \} . \int \rho\, d\, a^3.$$

Hence if in equation (1) we compare like functions, we shall have

$$\int \frac{d\Pi}{\rho} = 2\pi \int \rho\, d\, a^2 + 4\alpha\pi \int \rho\, d\, (a^2 Y^{(0)}) + \frac{4\pi}{3a} \int \rho\, d\, a^3$$
$$- \frac{4\alpha\pi}{3a} Y^{(0)} \int \rho\, d\, a^3 + \frac{4\alpha\pi}{a} \int \rho\, d\, (a^3 Y^{(0)}) + \alpha a^2 Z^{(0)};$$

the two first integrals of the second member of this equation being taken from a = a to a = 1, the three other integrals must be taken from a = 0 to a = a. This equation determining neither a nor $Y^{(0)}$, but only a relation between them, we see that the value of $Y^{(0)}$ is arbitrary, and may be determined at pleasure. We shall have then, i being equal to, or greater than unity,

$$0 = \frac{4\pi a^i}{2i+1} \int \varrho \, d. \left(\frac{Y^{(i)}}{a^{i-2}}\right) - \frac{4\pi}{3a} Y^{(i)} \int \varrho \, d\, a^3$$

$$+ \frac{4\pi}{(2i+1)\, a^{i+1}} \int \varrho \, d\, (a^{i+3}\, Y^{(i)}) + a^i\, Z^{(i)}; \quad . \quad . \quad (2)$$

the first integral being taken from a = a, to a = 1, and the two others being taken from a = 0 to a = a. This equation will give the value of $Y^{(i)}$ relative to each fluid shell, when the law of the densities ϱ shall be known.

To reduce these different integrals within the same limits, let

$$\frac{4\pi}{2i+1} \int \varrho \, d \left(\frac{Y^{(i)}}{a^{i-2}}\right) + Z^{(i)} = \frac{4\pi}{2i+1} Z'^{(i)},$$

the integral being taken from a = 0 to a = 1; $Z'^{(i)}$ will be a quantity independent of a, and the equation (2) will become

$$0 = (2i+1)\, a^i\, Y^{(i)} \int \varrho \, d\, a^3 + 3\, a^{2i+1}. \int \varrho \, d \left(\frac{Y^{(i)}}{a^{i-2}}\right)$$

$$- 3 \int \varrho \, d\, (a^{i+3}\, Y^{(i)}) - 3\, a^{2i+1}\, Z'^{(i)};$$

all the integrals being taken from a = 0 to a = a.

We may make the signs of integration disappear by differentiating relatively to a, and we shall have the differential equation of the second order,

$$\left(\frac{d^2\, Y^{(i)}}{d\, a^2}\right) = \left\{\frac{i\,(i+1)}{a^2} - \frac{6\, \varrho\, a}{\int \varrho \, d\, a^3}\right\} Y^{(i)} - \frac{6\, \varrho\, a^2}{\int \varrho \, d.\, a^3} \left(\frac{d\, Y^{(i)}}{d\, a}\right).$$

The integral of this equation will give the value of $Y^{(i)}$ with two arbitrary constants; these constants are rational and entire functions of the order i, of m, $\sqrt{1-m^2}.\sin.\,\varpi$, and $\sqrt{1-m^2}.\cos.\,\varpi$, such, that representing them by $U^{(i)}$, they satisfy the equation of partial differences,

$$0 = \left(\frac{d\left\{(1-m^2).\left(\frac{d\, U^{(i)}}{d m}\right)\right\}}{d m}\right) + \frac{\left(\frac{d^2\, U^{(i)}}{d\, \varpi^2}\right)}{1-m^2} + i\,(i+1).\, U^{(i)}.$$

One of these functions will be determined by means of the function $Z'^{(i)}$ which disappears by differentiation, and it is evident that it will be a multiple of this function. As to the other function, if we suppose that the fluid covers a solid nucleus, it will be determined by means of the equation of the surface of the nucleus, by observing that the value of $Y^{(i)}$ relative to the fluid shell contiguous to this surface, is the same as that of the surface. Thus the figure of the spheroid depends upon the figure of the internal nucleus, and upon the forces which sollicit the fluid.

574. If the mass is entirely fluid, nothing then determining one of the arbitrary constants, it would seem that there ought to be an infinity of

figures of equilibrium. Let us examine this case particularly, which is the more interesting inasmuch as it appears to have subsisted primitively for the celestial bodies.

First, we shall observe that the shells of the spheroid ought to decrease in density from the center to the surface; for it is clear that if a denser shell were placed above a shell of less density, its molecules would penetrate into the other in the same manner that a ponderous body sinks into a fluid of less density; the spheroid will not therefore be in equilibrium. But whatever may be its density at the center, it can only be finite; reducing therefore the expression of ϱ into a series ascending relatively to the powers of a, this series will be of the form $\beta - \gamma . a^n$ — &c. β, γ and n being positive; we shall thus have

$$\frac{a^3 . \varrho}{\int \varrho d . a^3} = 1 - \frac{n \gamma . a^n}{(n+3) \beta} + \&c.;$$

and the differential equation in $Y^{(i)}$ will become

$$\left(\frac{d^2 Y^{(i)}}{d a^2}\right) = \left\{(i-2)(i+3) + \frac{6 n \gamma . a^n}{(n+3) . \beta} - \&c.\right\} . \frac{Y^{(i)}}{a^2}$$
$$- \frac{6}{a}\left\{1 - \frac{n \gamma . a^n}{(n+3)\beta} + \&c.\right\} . \left(\frac{d Y^{(i)}}{d a}\right).$$

To integrate this equation, suppose that $Y^{(i)}$ is developed into a series ascending according to the powers of a, of this form

$$Y^{(i)} = a^s . U^{(i)} + a^{s'} . U'^{(i)} + \&c.;$$

the preceding differential equation will give

$$(s+i+3)(s-i+2) a^{s-2} U^{(i)} + (s'+i+3)(s'-i+2) a^{s'-2} U'^{(i)} + \&c.$$
$$= \frac{6 n \gamma a^n}{(n+3)\beta} \{(s+1) a^{s-2} . U^{(i)} + (s'+1) a^{s'-2} U'^{(i)} + \&c.\} \quad (e)$$

Comparing like powers of a, we have $(s+i+3)(s-i+2) = 0$, which gives $s = i - 2$, and $s = -i - 3$. To each of these values of s, belongs a particular series, which, being multiplied by an arbitrary, will be an integral of the differential equation in $Y^{(i)}$; the sum of these two integrals will be its complete integral. In the present case, the series which answers to $s = -i - 3$ must be rejected; for there thence results for a $Y^{(i)}$, an infinite value, when a shall be infinitely small, which would render infinite the radii of the shells which are infinitely near to the center. Thus of the two particular integrals of the expression of $Y^{(i)}$, that which answers to $s = i - 2$ ought alone to be admitted. This expression then, contains no more than one arbitrary which will be determined by the function $Z^{(i)}$.

$Z^{(1)}$ being nothing by No. 567, $Y^{(1)}$ is likewise nothing, so that the center of gravity of each shell, is at the center of gravity of the entire.

spheroid. In fact the differential equation in $Y^{(1)}$ of the preceding No. gives

$$\left(\frac{d^2 Y^{(1)}}{d a^2}\right) = \left(\frac{2}{a^2}\right) - \frac{6 \rho a^2}{\int \rho d . a^3} . Y^{(1)} - \frac{6 \rho a^2}{\int \rho d . a^3} . \left(\frac{d Y^{(1)}}{d a}\right).$$

We satisfy this equation by making $Y^{(1)} = \frac{U^{(1)}}{a}$, $U^{(1)}$ being independent of a. This value of $Y^{(1)}$ is that which answers to the equation $s = i - 2$; it is, consequently, the only one which we ought to admit. Substituting it in the equation (2) of the preceding No., and supposing $Z^{(1)} = 0$, the function $U^{(1)}$ disappears, and consequently remains arbitrary; but the condition that the origin of the radius r is at the center of gravity of the terrestrial spheroid, renders it nothing; for we shall see in the following No. that then $Y^{(1)}$ is nothing at the surface of every spheroid covered over with a shell of fluid in equilibrium; we shall have, therefore, in the present case $U^{(1)} = 0$; thus, $Y^{(1)}$ is nothing relatively to all the fluid shells which form the spheroid.

Now consider the general equation,

$$Y^{(i)} = a^s . U^{(i)} + a^{s'} . U'^{(i)} + \&c.;$$

s being, as we have seen, equal to $i - 2$, s is nothing or positive, when i is equal to or greater than 2; moreover, the functions $U'^{(i)}$, $U''^{(i)}$, &c. are given in $U^{(i)}$, by the equation (e) of this No.; so that we have

$$Y^{(i)} = h . U^{(i)};$$

h being a function of a, and $U^{(i)}$ being independent of it. If we substitute this value of $Y^{(i)}$ in the differential equation in $Y^{(i)}$, we shall have

$$\frac{d^2 h}{d a^2} = \left\{ i (i + 1) - \frac{6 \rho a^3}{\int \rho d . a^3} \right\} . \frac{h}{a^2} - \frac{6 \rho a^2}{\int \rho d . a^3} . \frac{d h}{d a}.$$

The product $i (i + 1)$ is greater than $\frac{6 \rho a^3}{\int \rho d . a^3}$, when i is equal to or greater than 2, for the fraction $\frac{\rho a^3}{\int \rho d . a^3}$ is less than unity; in fact its denominator $\int \rho d . a^3$ is equal to $\rho a^3 - \int a^3 d \rho$; and the quantity $-\int a^3 d \rho$ is positive, since ρ decreases from the center to the surface.

Hence it follows that h and $\frac{d h}{d a}$ are constantly positive, from the center to the surface. To show this, suppose that both these quantities are positive in going from the center; d h ought to become negative before h, and it is clear that in order to do this it must pass through zero; but from the instant it is nothing, $d^2 h$ becomes positive in virtue of the preceding equation, and consequently d h begins to increase; it can never therefore become negative. Whence it follows that h and d h always preserve the same sign from the center to the surface. Now both of these quantities are positive in going from the center; for we have in virtue of equation (e), $s' - 2 = s + n - 2$, which gives $s' = i + n - 2$; hence we have

$$(s' + i + 3)\ (s' - i + 2)\ U'^{(i)} = \frac{6\,n\,(s + 1)\,\gamma\,U^{(i)}}{(n + 3)\,\beta};$$

whence we derive

$$U^{(i)} = \frac{6\,(i - 1)\,\gamma\,.\,U^{(i)}}{(n + 3)\,(2\,i + n + 1)\,.\,\beta};$$

we shall therefore get

$$h = a^{i-2} + \frac{6\,(i - 1)\,\gamma\,.\,a^{i+n-2}}{(n + 3)\,(2\,i + n + 1)\,\beta} + \&c.;$$

$$\frac{d\,h}{d\,a} = (i - 2)\,a^{i-3} + \frac{6\,(i - 1)\,(i + n - 2)\,\gamma\,.\,a^{i+n-3}}{(n + 3)\,(2\,i + n + 1)\,\beta} + \&c.$$

γ, β, n, being positive, we see that at the center h and $d\,h$ are positive, when i is equal to or greater than 2; they are therefore constantly positive from the center to the surface.

Relatively to the Earth, to the Moon, to Jupiter, &c. $Z^{(i)}$ is nothing or insensible, when i is equal to or greater than 3; the equation (2) of the preceding No. then becomes

$$0 = \{3\,a^{2i+1}\textstyle\int \rho\,d\left(\frac{h}{a^{i-2}}\right) - (2\,i + 1)\,a^{i}\,h\int \rho\,d\,a^{3} + 3\int \rho\,d\,(a^{i+3}\,h)\}\,.\,U^{(i)};$$

the first integral being taken from $a = a$, to $a = 1$, and the two others being taken from $a = 0$, to $a = a$. At the surface where $a = 1$, this equation becomes

$$0 = \{-(2\,i + 1)\,h\textstyle\int \rho\,d\,.\,a^{3} + 3\int \rho\,d\,(a^{i+3}\,h)\}\,.\,U^{(i)};$$

an equation which we can put under this form

$$0 = \{-(2\,i - 2)\,\rho\,h + (2\,i + 1)\,h\textstyle\int a^{3}\,d\,\rho - 3\int a^{i+3}\,h\,.\,d\,\rho\}\ U^{(i)}.$$

$d\,\rho$ is negative from the center to the surface, and h increases in the same interval; the function $(2\,i + 1)\,h\int a^{3}\,d\,\rho - 3\int a^{i+3}\,h\,d\,\rho$ is therefore negative in the same interval; thus in the preceding equation the coefficient of $U^{(i)}$ is negative and cannot be nothing at the surface; $U^{(i)}$ ought therefore to be nothing, which gives $Y^{(i)} = 0$; the expression of the radius of the spheroid thus reduces to $a + \alpha\,a\,\{Y^{(0)} + Y^{(2)}\}$; that is to say, that the surface of each *leveled* shell of the spheroid is elliptic, and cousequently its exterior surface is elliptic.

$Z^{(2)}$, relatively to the Earth is, by No. 567, equal to $-\frac{g}{2\,\alpha}\,(m^{2} - \frac{1}{3})$; the equation (2) of the preceding No. gives therefore

$$0 = \{\tfrac{4}{5}\pi a^5 \int \rho\, dh - \tfrac{4}{3}\pi a^2 h . \int \rho\, d a^3 + \tfrac{4}{5}\pi \int \rho\, d(a^5 h)\} U^{(2)} - \frac{g}{2\alpha} a^5 (m^2 - \tfrac{1}{3}).$$

At the surface, the first integral $\int \rho\, dh$ is nothing; we have therefore at this surface where $a = 1$,

$$U^{(2)} = \frac{-\frac{1}{2\alpha} . g . (m^2 - \frac{1}{3})}{\frac{4}{3}\pi . h . \int \rho\, d . a^3 - \frac{4}{5}\pi \int \rho\, d(a^5 h)}.$$

Let $\alpha \varphi$, be the ratio of the centrifugal force to the equatorial gravity; the expression of gravity to quantities of the order α, being equal to $\frac{4}{3}\pi \int \rho\, d . a^3$; we shall have $g = \frac{4}{3}\pi\alpha\varphi \int \rho\, d . a^3$; wherefore

$$U^{(2)} = \frac{-\varphi (m^2 - \frac{1}{3})}{2h - \frac{2}{5} . \frac{\int \rho . d(a^5 h)}{\int \rho . a^2 d a}};$$

comprising therefore in the arbitrary constant a, what we have taken for unity, the function

$$\alpha Y^{(0)} - \frac{\alpha h \varphi}{3h - \frac{3}{5} . \frac{\int \rho\, d(a^5 h)}{\int \rho . a^2 d a}},$$

the radius of the terrestrial spheroid at the surface will be

$$1 + \frac{\alpha h \varphi (1 - m^2)}{2h - \frac{2}{5} . \frac{\int \rho\, d(a^5 h)}{\int \rho . a^2 d a}}.$$

The figure of the earth supposed fluid, can therefore only be that of an ellipsoid of revolution; all of whose shells of constant density are elliptic, and of revolution, and in which the ellipticities increase, and the densities decrease from the center to the surface. The relation between the ellipticities and densities is given by the differential equation of the second order,

$$\frac{d^2 h}{d a^2} = \frac{6h}{a^2}\left(1 - \frac{\rho a^3}{3\int \rho\, a^2 d a}\right) - \frac{2\rho a^2}{\int \rho . a^2 d a} . \frac{d h}{d a}.$$

This equation is not integrable by known methods except in some particular suppositions of the densities ρ; but if the law of the ellipticities were given, we should easily obtain that of the corresponding densities. We have seen that the expression of h given by the integral of this equation contains, in the present question, only one arbitrary, which disappears from the preceding value of the radius of the spheroid; there is therefore only one figure of equilibrium differing but little from a sphere, which is possible, and it is easy to see that the limits of the flattening of this figure are $\frac{\alpha\varphi}{2}$ and $\frac{5}{4}\alpha\varphi$, the former of which corresponds to the case where all the mass of the spheroid is collected at its center, and the second to the case where this mass is homogeneous.

The directions of gravity from any point of the surface to the center do not form a straight line, but a curve whose elements are perpendicular to the *leveled* shells which they traverse: this curve is the orthogonal trajectory of all the ellipses which by their revolution form these shells. To determine its nature, take for the axis, the radius drawn from the center to a point of the surface, θ being the angle which this radius forms with the axis of revolution. We have just seen that the general expression of any shell of the spheroid is $a + \alpha k . a h . (1 - m^2)$, k being independent of a: whence it is easy to conclude that if we call $\alpha y'$, the ordinate let fall from any point of the curve upon its axis, we shall have

$$\alpha y' = \alpha a k . \sin. 2 \theta \left\{ c - \int \frac{h d a}{a} \right\},$$

c being the entire value of the integral $\int \frac{h d a}{a}$, taken from the center to the surface.

575. Now consider the general case in which the spheroid always fluid at its surface, may contain a solid nucleus of any figure whatever, but differing but little from the sphere. The radius drawn from the center of gravity of the spheroid to its surface, and the law of gravity at this surface have some general properties, which it is the more essential to consider, inasmuch as these properties are independent of every hypothesis.

The first of these properties is, that in the state of equilibrium the fluid part of the spheroid must always be disposed so, that the function $Y^{(1)}$ may disappear from the expression of the radius drawn from the center of gravity of the whole spheroid to its surface; so that the center of gravity of this surface coincides with that of the spheroid.

To show this, we shall observe that R being supposed to represent the radius drawn from the center of gravity of the spheroid to any one of its molecules, the expression of this molecule will be $\rho R^2 . d R . d m . d \varpi$, and we shall have by 556, in virtue of the properties of the center of gravity,

$$0 = \int \rho R^3 . d R . d m . d \varpi . m;$$

$$0 = \int \rho R^3 . d R . d m . d \varpi . \sqrt{1 - m^2} . \sin. \varpi;$$

$$0 = \int \rho R^3 . d R . d m . d \varpi . \sqrt{1 - m^2} . \cos. \varpi.$$

Conceive the integral $\int \rho R^3 . d R$ taken relatively to R from the origin of R to the surface of the spheroid, and then developed into a series of the form

$$N^{(0)} + N^{(1)} + N^{(2)} + \&c.;$$

$N^{(i)}$ being whatever i may be, subject to the equation of partial differences,

$$0 = \left(\frac{d \left\{ (1 - m^2) \left(\frac{d N^{(i)}}{d m} \right) \right\}}{d m} \right) + \frac{\left(\frac{d^2 N^{(i)}}{d \varpi^2} \right)}{1 - m^2} + i (i + 1) N^{(i)};$$

we shall have by No. 556, when i is different from unity,

$$0 = \int N^{(i)} . m \, d m . d \varpi; \quad 0 = \int N^{(i)} . d m . d \varpi . \sqrt{1 - m^2} . \sin. \varpi;$$

and

$$0 = \int . N^{(i)} . d m . d \varpi . \sqrt{1 - m^2} . \cos. \varpi.$$

The three preceding equations given by the nature of the center of gravity, will become

$$0 = \int N^{(1)} m \, d m . d \varpi; \quad 0 = \int N^{(1)} d m . d \varpi . \sqrt{1 - m^2} . \sin. \varpi;$$

$$0 = \int N^{(1)} d m . d \varpi . \sqrt{1 - m^2} . \cos. \varpi.$$

$N^{(1)}$ is of the form

$$H m + H' . \sqrt{1 - m^2} . \sin. \varpi + H'' . \sqrt{1 - m^2} . \cos. \varpi.$$

Substituting this value, in these three equations, we shall have

$$H = 0; \quad H' = 0; \quad H'' = 0;$$

where $N^{(1)} = 0$; this is the condition necessary that the origin of R is at the center of gravity of the spheroid.

Now let us see, what $N^{(1)}$ becomes relatively to the spheroids differing little from the sphere, and covered over with a fluid in equilibrium. In this case we have $R = a (1 + \alpha y)$, and the integral $\int \rho . R^3 . d R$, becomes $\frac{1}{4} \int \rho \, d . \{a^4 (1 + 4 \alpha y)\}$, the differential and integral being relative to the variable a, of which ρ is a function. Substituting for y its value $Y^{(0)} + Y^{(1)} + Y^{(2)} + \&c.$, we shall have

$$N^{(1)} = \alpha \int \rho \, d (a^4 Y^{(1)}).$$

The equation (2) of No. 573 gives, at the surface where $a = 1$, and observing that $Z^{(1)}$ is nothing

$$\int \rho \, d (a^4 Y^{(1)}) = Y^{(1)} \int \rho \, d . a^5,$$

the value of $Y^{(1)}$ in the second member of this equation, being relative to the surface; thus, $N^{(1)}$ being nothing, when the origin of R is at the center of gravity of the spheroid, we have in like manner $Y^{(1)} = 0$.

576. The permanent state of equilibrium of the celestial bodies, makes known also some properties of their radii. If the planets did not turn exactly, or at least if they turned not nearly, round one of their three principal axes of rotation, there would result in the position of their axes of rotation, changes which for the earth above all would be sensible; and since the most exact observations have not led to the discovery of any, we may conclude that long since, all the parts of the celestial bodies, and princi-

pally the fluid parts of their surfaces, are so disposed as to render stable their state of equilibrium, and consequently their axes of rotation. It is in fact very natural to suppose that after a great number of oscillations, they must settle in this state, in virtue of the resistances which they suffer. Let us see, however, the conditions which thence result in the expression of the radii of the celestial bodies.

If we name x, y, z the rectangular coordinates of a molecule d M of the spheroid, referred to three principal axes, the axis of x being the axis of rotation of the spheroid; by the properties of these axes as shown in dynamics, we have

$$0 = \int x y . d M; \; 0 = \int x z . d M; \; 0 = \int y z . d M;$$

the integrals ought to be extended to the entire mass of the spheroid, R being the radius drawn from the origin of coordinates to the molecule d M; θ being the angle formed by R and by the axis of rotation; and ϖ being the angle which the plane formed by this axis and by R, makes with the plane formed by this axis and by that of the principal axes, which is the axis of y; we shall have

$$x = R m; \; y = R \sqrt{1 - m^2}. \cos. \varpi; \; z = R \sqrt{1 - m^2}. \sin. \varpi;$$
$$d M = \rho R^2 d R d m . d \varpi.$$

The three equations given by the nature of the principal axes of rotation, will thus become

$$0 = \int \rho . R^4. d R . d m . d \varpi . m \sqrt{1 - m^2}. \cos. \varpi;$$
$$0 = \int \rho . R^4. d R . d m . d \varpi . m \sqrt{1 - m^2}. \sin. \varpi;$$
$$0 = \int \rho . R^4. d R . d m . d \varpi . (1 - m^2) \sin. 2 \varpi.$$

Conceive the integral $\int \rho R^4 d R$ taken relatively to R, from R = 0, to the value of R at the surface of the spheroid, and developed into a series of the form $U^{(0)} + U^{(1)} + U^{(2)} + \&c.$; $U^{(i)}$ being, whatever i may be, subject to the equation of partial differences,

$$0 = \left(\frac{d \left\{ (1 - m^2). \left(\frac{d U^{(i)}}{d m} \right) \right\}}{d m} \right) + \frac{\left(\frac{d^2 U^{(i)}}{d \varpi^2} \right)}{1 - m^2} + i (i + 1). U^{(i)}.$$

We shall have by the theorem of No. 556, where i is different from 2, and by observing that the functions $m \sqrt{1 - m^2}. \cos. \varpi$, $m \sqrt{1 - m^2}. \sin. \varpi$, and $(1 - m^2) \sin. 2 \varpi$, are comprised in the form $U^{(2)}$;

$$0 = \int U^{(i)}. d m . d \varpi . m . \sqrt{1 - m^2}. \cos. \varpi;$$
$$0 = \int U^{(i)}. d m . d \varpi . m . \sqrt{1 - m^2}. \sin. \varpi;$$
$$0 = \int U^{(i)}. d m . d \varpi . (1 - m^2) \sin. 2 \varpi.$$

The three equations relative to the nature of the axes of rotation, will thus become

$$0 = \int U^{(2)}.\, d\, m\,.\, d\, \varpi\,.\, m\,.\, \sqrt{1 - m^2}.\, \cos.\, \varpi\,;$$
$$0 = \int U^{(2)}.\, d\, m\,.\, d\, \varpi\,.\, m\,.\, \sqrt{1 - m^2}.\, \sin.\, \varpi\,;$$
$$0 = \int U^{(2)}.\, d\, m\,.\, d\, \varpi\,.\, (1 - m^2) \sin.\, 2\, \varpi.$$

These equations therefore depend only on the value of $U^{(2)}$: this value is of the form

$$H\,(m^2 - \tfrac{1}{3}) + H'\, m\, \sqrt{1 - m^2}.\, \sin.\, \varpi + H''\, m\, \sqrt{1 - m^2}.\, \cos.\, \varpi + H'''\,(1 - m^2) \sin.\, 2\, \varpi + H''''\,(1 - m^2) \cos.\, 2\, \varpi:$$

substituting it in the three preceding equations, we shall have

$$H' = 0;\quad H'' = 0;\quad H''' = 0.$$

It is to these three conditions that the conditions necessary to make the three axes of x, y, z the true axes of rotation are reduced, and then $U^{(2)}$ will be of the form

$$H\,(m^2 - \tfrac{1}{3}) + H''''\,(1 - m^2) \cos.\, 2\, \varpi.$$

When the spheroid is a solid differing but little from the sphere, and covered with a fluid in equilibrium, we have $R = a\,(1 + \alpha\, y)$, and consequently

$$\int \varrho\, R^4.\, d\, R = \tfrac{1}{5} \int \varrho\, d\,.\, \{a^5.\, (1 + 5\, \alpha\, y)\}.$$

If we substitute for y, its value $Y^{(0)} + Y^{(1)} + Y^{(2)} +$ &c.; we shall have

$$U^{(2)} = \alpha \int \varrho\, d\, (a^5\, Y^{(2)}).$$

The equation (2) of No. 573, gives for the surface of the spheroid,

$$\frac{4\,\pi}{5} \int \varrho\, d\, (a^5\, Y^{(2)}) = \tfrac{4}{3}\, \pi\, Y^{(2)} \int \varrho\, d\,.\, a^3 - Z^{(2)};$$

$Y^{(2)}$ and $Z^{(2)}$ in the second member of this equation being relative to the surface; we have therefore,

$$U^{(2)} = \tfrac{5}{3}\, \alpha\, Y^{(2)} \int \varrho\, d\,.\, a^3 - \frac{5\, \alpha\, Z^{(2)}}{4\, \pi}.$$

The value of $Z^{(2)}$ is of the form

$$-\frac{g}{2}\,(m^2 - \tfrac{1}{3}) + g'\, m\, \sqrt{1 - m^2}.\, \sin.\, \varpi + g''\, m\, \sqrt{1 - m^2}.\, \cos.\, \varpi\,; + g'''\,(1 - m^2) \sin.\, 2\, \varpi + g''''\,(1 - m^2) \cos.\, 2\, \varpi\,;$$

and that of $Y^{(2)}$ is of the form

$$-h\,(m^2 - \tfrac{1}{3}) + h'\, m\, \sqrt{1 - m^2}.\, \sin.\, \varpi + h''\, m\, \sqrt{1 - m^2}.\, \cos.\, \varpi + h'''.\,(1 - m^2) \sin.\, 2\, \varpi + h''''\,(1 - m^2) \cos.\, 2\, \varpi.$$

Substituting in the preceding equation, these values, and $H\,(m^2 - \tfrac{1}{3}) + H''''\,(1 - m^2) \cos 2\, \varpi$, for $U^{(2)}$; we shall have

$$h' = \frac{g'}{4\, \pi \int \varrho\,.\, a^2\, d\, a};\quad h'' = \frac{g''}{4\, \pi \int \varrho\,.\, a^2\, d\, a};\quad h''' = \frac{g'''}{4\, \pi \int \varrho\,.\, a^2\, d\, a}.$$

Such are the conditions which result from the supposition that the spheroid turns round one of its principal axes of rotation. This supposition determines the constants h′, h″, h‴ by means of the values g′, g″, g‴; but it leaves indeterminate the quantities h and h⁗ as also the functions $Y^{(3)}$, $Y^{(4)}$, &c.

If the forces extraneous to the attraction of the molecules of the spheroid are reduced to the centrifugal force due to its rotatory motion; we shall have $g' = 0$, $g'' = 0$, $g''' = 0$; wherefore $h' = 0$, $h'' = 0$, $h''' = 0$, and the expression of $Y^{(2)}$, will be of the form

$$-h(m^2 - \tfrac{1}{3}) + h''''(1 - m^2)\cos. 2\varpi.$$

577. Let us consider the expression of gravity at the surface of the spheroid. Call p this force; it is easy to see by No. 569, that we shall have its value by differentiating the second member of the equation (1) of 573 relatively to r, and by dividing its differential by $-dr$; which gives at the surface

$$p = \frac{4\pi}{3r^2}\int\rho\, d.a^3 + \frac{4\alpha\pi}{r^2}\int\rho\, d.\left\{a^3 Y^{(0)} + \frac{2a^4}{3r}Y^{(1)} + \frac{3a^5}{5r^2}Y^{(2)} + \frac{4a^6}{7r^3}Y^{(3)} + \&c\right\}$$

$$-\alpha r\{2Z^{(0)} + 2Z^{(2)} + 3r.Z^{(3)} + 4r^2.Z^{(4)} + \&c.\};$$

these integrals being taken from $a = 0$, to $a = 1$. The radius r at the surface is equal to $1 + \alpha y$, or equal to

$$1 + \alpha\{Y^{(0)} + Y^{(1)} + Y^{(2)} + \&c.\};$$

we shall hence obtain

$$p = \frac{4\pi}{3}\int\rho\, d.a^3 - \frac{8\alpha\pi}{3}\{Y^{(0)} + Y^{(1)} + Y^{(2)} + \&c.\}\int\rho\, d.a^3$$

$$+ 4\alpha\pi\int\rho\, d.\{a^3 Y^{(0)} + \frac{2a^4}{3}Y^{(1)} + \frac{3a^3}{5}Y^{(2)} + \&c.\}$$

$$-\alpha\{2Z^{(0)} + 2Z^{(2)} + 3Z^{(3)} + 4Z^{(4)} + \&c.\}.$$

The integrals of this expression may be made to disappear by means of equation (2) of No. 573, which becomes at the surface,

$$\frac{4\pi}{2i+1}.\int\rho\, d.(a^{i+3}Y^{(i)}) = \tfrac{4}{3}\pi Y^{(i)}.\int\rho\, d.a^3 - Z^{(i)};$$

supposing therefore

$$P = \tfrac{4}{3}\pi\int\rho\, d.a^3 - \frac{8\alpha\pi}{3}Y^{(0)} + 4\alpha\pi\int\rho\, d.(a^3 Y^{(0)}) - 2\alpha Z^{(0)};$$

we shall have

$$p = P + \alpha P.\{Y^{(2)} + 2Y^{(3)} + 3Y^{(4)} + \ldots + (i-1)Y^{(i)} + \&c.\}$$

$$-\alpha\{5Z^{(2)} + 7Z^{(3)} + 9Z^{(4)} + \ldots + (2i+1)Z^{(i)} + \&c.\}.$$

By observations of the lengths of the seconds' pendulum, has been recognised the variation of gravity at the surface of the earth. By dynamics it appears that these lengths are proportional to gravity; let therefore l, L be the lengths of the pendulum corresponding to the gravities p, P; the preceding equation will give

$$l = L + \alpha L \{Y^{(2)} + 2 Y^{(3)} + 3 Y^{(4)} + \ldots + (i-1) Y^{(i)}\}$$
$$- \frac{\alpha L}{P} . \{5 Z^{(2)} + 7 Z^{(3)} + \ldots\ldots\ldots\ldots + (2i+1) Z^{(i)}\}.$$

Relatively to the earth $\alpha Z^{(2)}$ reduces by 567, to $-\frac{g}{2}(m^2 - \frac{1}{3})$, or, which comes to the same, to $-\frac{\alpha \varphi}{2} . P. (m^2 - \frac{1}{3})$, $\alpha \varphi$ being the ratio of the centrifugal force to the equatorial gravity; moreover, $Z^{(3)}$, $Z^{(4)}$, &c. are nothing; we have therefore

$$l = L + \alpha L . \{Y^{(2)} + 2 Y^{(3)} + 3 Y^{(4)} + \ldots + (i-1) Y^{(i)}\}$$
$$+ \tfrac{5}{2} \alpha \varphi . L . (m^2 - \tfrac{1}{3}).$$

The radius of curvature of the meridian of a spheroid which has for its radius $1 + \alpha y$, is

$$1 + \alpha \left(\frac{d . m y}{d m}\right) + \alpha \left(\frac{d . \left\{(1 - m^2)\left(\frac{d y}{d m}\right)\right\}}{d m}\right)$$

designating therefore by c, the magnitude of the degree of a circle whose radius is what we have taken for unity; the expression of the degree of the spheroid's meridian, will be

$$c \left\{1 + \alpha \left(\frac{d . m y}{d m}\right) + \alpha \left(\frac{d . \left\{(1 - m^2)\left(\frac{d y}{d m}\right)\right\}}{d m}\right)\right\}.$$

y is equal to $Y^{(0)} + Y^{(1)} + Y^{(2)} +$ &c. We may cause $Y^{(0)}$ to disappear, by comprising it in the arbitrary constant which we have taken for the unit; and $Y^{(1)}$ by fixing the origin of the radius at the center of gravity of the entire spheroid. This radius thus becomes,

$$1 + \alpha \{Y^{(2)} + Y^{(3)} + Y^{(4)} + \&c.\}.$$

If we then observe that

$$\left(\frac{d \left\{(1 - m^2) . \left(\frac{d Y^{(i)}}{d m}\right)\right\}}{d m}\right) = - i (i + 1) Y^{(i)} - \frac{\left(\frac{d^2 Y^{(i)}}{d \varpi^2}\right)}{1 - m^2};$$

the expression of the degree of the meridian will become

$$c - \alpha c \{5 Y^{(2)} + 11 Y^{(3)} + \ldots + (i^2 + i - 1) Y^{(i)} + \&c.\}$$
$$+ \alpha c m \left\{\left(\frac{d Y^{(2)}}{d m}\right) + \left(\frac{d Y^{(3)}}{d m}\right) + \&c.\right\}$$
$$- \alpha c . \frac{\left(\frac{d^2 Y^{(2)}}{d \varpi^2}\right) + \left(\frac{d^2 Y^{(3)}}{d \varpi^2}\right) + \&c.}{1 - m^2}.$$

If we compare these expressions of the terrestrial radius with the length of the pendulum, and the magnitude of the degree of the meridian, we see that the term $\alpha Y^{(1)}$ of the expression of the radius is multiplied by $i - 1$, in the expression of the length of the pendulum, and by $i^2 + i - 1$ in that of the degree; whence it follows, that whilst $i - 1$ is considerable, this term will be more sensible in the observations of the length of the pendulum than in that of the horizontal parallax of the moon which is proportional to the terrestrial radius; it will be still more sensible in the measures of degrees than in the lengths of the pendulum. The reason of it is, that the terms of the expression of the terrestrial radius undergo two variations in the expression of the degree of the meridian; and each differentiation multiplies these terms by the corresponding exponent of m, and this renders them the more considerable. In the expression of the variation of two consecutive degrees of the meridian, the terms of the terrestrial radius undergo three consecutive differentiations; those which disturb the figure of the earth from that of an ellipsoid, may thence become very sensible, and the ellipticity obtained by this variation may be very different from that which the observed lengths of the pendulum give. These three expressions have the advantage of being independent of the interior constitution of the earth, that is to say, of the figure and density of its shells; so that if we are going to determine the functions $Y^{(2)}$, $Y^{(3)}$, &c. by measures of degrees of meridians and parallaxes, we shall have immediately the length of the pendulum; we may therefore thus ascertain whether the law of universal gravity accords with the figure of the earth, and with the observed variations of gravity at its surface. These remarkable relations between the expressions of the degrees of the meridian and of the lengths of the pendulum may also serve to verify the hypotheses proper to represent the measures of degrees of this meridian: this will be perceptible from the application we now proceed to make to the hypothesis proposed by Bouguer, to represent the degrees measured northward in France and at the equator.

Suppose that the expression of the terrestrial radius is $1 + \alpha Y^{(2)} + \alpha Y^{(4)}$, and that we have

$$Y^{(2)} = -A\left(m^2 - \tfrac{1}{3}\right); \quad Y^{(4)} = -B\left(m^4 - \tfrac{6}{7}m^2 + \frac{3}{35}\right);$$

it is easy to see that these functions of m satisfy the equations of partial differences which $Y^{(2)}$ and $Y^{(4)}$ ought to satisfy. The variation of the degrees of the meridian will be, by what precedes,

$$\alpha c\left\{3A - \frac{102}{7}B\right\}\ {}^{2} + 15\,\alpha c B\, m^4.$$

Bouguer supposes this variation proportional to the fourth power of the sine of the latitude, or, which nearly comes to the same, to m^4; the term multiplied by m^2, therefore, being made to disappear from the preceding function, we shall have

$$B = \frac{7}{34} . A;$$

thus in this case the radius drawn from the center of gravity of the earth at its surface, will be in taking that of the equator for unity,

$$1 - \frac{7 \alpha A}{34} (4 m^2 + m^4).$$

The expression of the length l of the pendulum, will become, denoting by L, its value at the equator,

$$L + \tfrac{5}{2} \alpha \varphi . L m^2 - \frac{\alpha . A L}{34} (16 m^2 + 21 m^4).$$

Finally, the expression of the degree of the meridian, will be, calling c its length at the equator,

$$c + \frac{105}{34} . \alpha A . c . m^4.$$

We shall observe here, that agreeably to what we have just said, the term multiplied by m^4 is three times more sensible in the expression of the length of the pendulum than in that of the terrestrial radius, and five times more sensible in the expression of the length of a degree, than in that of the length of the pendulum; finally, upon the mean parallel it would be four times more sensible in the expression of the variation of consecutive degrees, than in that of the same degree. According to Bouguer, the difference of the degrees at the pole and equator is $\frac{959}{56753}$; it is the ratio which, on his hypothesis, the measures of degrees at Pello, Paris and the equator, require. This ratio is equal to $\frac{105}{34} . \alpha A$; we have therefore

$$\alpha A = 0.0054717.$$

Taking for unity the length of the pendulum at the equator, the variation of this length, in any place whatever, will be

$$- \frac{0.0054717}{34} . \{16 m^2 + 21 m^4\} + \tfrac{5}{2} \alpha \varphi . m^2.$$

By No. 563, we have $\alpha \varphi = 0.00345113$, which gives $\frac{5}{2} \alpha \varphi = 0.0086278$, and the preceding formula becomes

$$0.0060529 . m^2 - 0.0033796 . m^4.$$

At Pello, where m = sin. 74°. 22′, this formula gives 0.0027016 for the variation of the length of the pendulum. According to the observations, this variation is 0.0044625, and consequently much greater; thus, since the hypothesis of Bouguer cannot be reconciled with the observations made on the length of the pendulum, it is inadmissible.

578. Let us apply the general results which we have just found, to the case where the spheroid is not sollicited by any extraneous forces, and where it is composed of elliptic shells, whose center is at the center of gravity of the spheroid. We have seen that this case is that of the earth supposed to be originally fluid: it is also that of the earth in the hypothesis where the figures of the shells are similar. In fact, the equation (2) of No. 573 becomes at the surface where a = 1,

$$0 = Y^{(i)} \int \rho\, a^2 d a - \frac{1}{2i+1} \int \rho\, d\,(a^{i+3} Y^{(i)}) - \frac{Z^{(i)}}{4\pi}.$$

The shells being supposed similar, the value of $Y^{(i)}$ is, for each of them, the same as at the surface; it is consequently independent of a, and we have

$$Y^{(i)}. \int \rho\, a^2 d a \left\{1 - \frac{i+3}{2i+1} . a^i\right\} = \frac{Z^{(i)}}{4\pi}.$$

When i is equal to or greater than 3, $Z^{(i)}$ is nothing relatively to the earth; besides the factor $1 - \frac{i+3}{2i+1} . a^i$ is always positive; therefore $Y^{(i)}$ is then nothing. $Y^{(1)}$ is also nothing by No. 575, when we fix the origin of the radii at the center of gravity of the spheroid. Finally, by No. 577, we have $Z^{(2)}$ equal to

$$-\frac{\varphi}{2} (m^2 - \tfrac{1}{3})\, 4\pi \int \rho . a^2 d a; \quad Y^{(\,)} = -\frac{\varphi}{2} . (\mu^2 - \tfrac{1}{3}) \int \rho\, a^2 d a;$$

we have therefore

$$Y^{(2)} = -\frac{\frac{\varphi}{2} (m^2 - \frac{1}{3}) \int \rho\, a^2\, d a}{\int \rho\, a^2\, d a\, (1 - a^2)}.$$

Thus the earth is then an ellipsoid of revolution. Let us consider therefore generally the case where the figure of the earth is elliptic and of revolution.

In this case, by fixing the origin of terrestrial radii at the center of gravity of the earth, we have

$$Y^{(1)} = 0; \quad Y^{(3)} = 0; \quad Y^{(4)} = 0; \ \&c.$$
$$Y^{(2)} = -h\,(m^2 - \tfrac{1}{3}),$$

h being a function of a; moreover we have

$$Z^{(1)} = 0; \quad Z^{(3)} = 0; \quad Z^{(4)} = 0; \ \&c.$$

$$\alpha Z^{(2)} = -\frac{\alpha \varphi}{2} (m^2 - \tfrac{1}{3}) . \frac{4\pi}{3} \int \rho \, d . a^3;$$

the equation (2) of No. 573 will therefore give at the surface

$$0 = 6. \int \rho \, d (a^5 h) + 5. (\varphi - 2 h) \int \rho \, d . a^3 \quad . \quad . \quad . \quad (1)$$

This equation contains the law which ought to exist to sustain the equilibrium between the densities of the shells of the spheroid and their ellipticities; for the radius of a shell being a $\{1 + \alpha Y^{(0)} - \alpha h (\mu^2 - \frac{1}{3})\}$; if we suppose, as we may, that $Y^{(0)} = -\frac{1}{3} h$, this radius becomes a $(1 - \alpha h . \mu^2)$, and α h is the ellipticity of the shell.

At the surface, the radius is $1 - \alpha h . \mu^2$; whence we see that the decrements of the radii, from the equator to the poles, are proportional to μ^2, and consequently to the square of the sines of the latitude.

The increment of the degrees of the meridian from the equator to the poles is, by the preceding No., equal to $3 \alpha h c . \mu^2$, c being the degree of the equator; it is therefore also proportional to the square of the sine of the latitude.

The equation (1) shows us that the densities being supposed to decrease from the center to the surface, the ellipticity of the spheroid is less than in the case of homogeneity, at least whilst the ellipticities do not increase from the surface to the center in a greater ratio than the inverse ratio of the square of the distances to this center. In fact, if we suppose $h = \frac{u}{a^2}$, we shall have

$$\int \rho \, d (a^5 h) = \int \rho \, d (a^3 u) = u \int \rho \, d . a^3 + \int (d u . \int a^3 d \rho).$$

If the ellipticities increase in a less ratio than $\frac{1}{a^2}$, u increases from the center to the surface, and consequently d u is positive; besides, $d \rho$ is negative by the supposition that the densities decrease from the center to the surface; thus, $\int (d u \int a^3 d \rho)$ is a negative quantity, and making at the surface

$$\int \rho \, d (a^5 h) = (h - f) \int \rho \, d . a^3,$$

f will be a positive quantity. Hence equation (I) will give

$$h = \frac{5 \varphi - 6 f}{4};$$

α h will therefore be less than $\frac{5 \alpha \varphi}{4}$, and consequently it will be less than in the case of homogeneity, where d ϱ being equal to nothing f is also equal to zero.

Hence it follows, that in the most probable hypotheses, the flattening of the spheroid is less than $\frac{5\,\alpha\,\varphi}{4}$; for it is natural to suppose that the shells of the spheroid are denser towards the center, and that the ellipticities increase from the surface to the center in a less ratio than $\frac{1}{a^2}$, this ratio giving an infinite radius for shells infinitely near to the center, which is absurd. These suppositions are the more probable, inasmuch as they become necessary in the case where the fluid is originally fluid; then the denser shells are, as we have seen, the nearer to the center, and the ellipticities so far from increasing from the surface to the center, on the contrary, decrease.

If we suppose that the spheroid is an ellipsoid of revolution, covered with a homogeneous fluid mass of any depth whatever, by calling a′ the semi-minor axis of the solid ellipsoid, and α h′ its ellipticity, we shall have at the surface of the fluid,

$$\int \varrho\, d\, (a^5 h) = h - a'^5 h' + \int \varrho\, d\, (a^5 h)_{\prime};$$

the integral of the second member of this equation being taken relatively to the interior ellipsoid, from its center to its surface, and the density of the fluid which covers it being taken for unity. The equation (1) will give for the expression of the ellipticity α h, of the terrestrial spheroid,

$$\alpha h = \frac{5\,\alpha\,\varphi\,\{1 - a'^3 + \int \varrho\, d\, a^3\} - 6\,\alpha\, h' .\, a'^5 + 6\,\alpha \int \varrho\, d\, (a^5 h)}{4 - 10\, a'^3 + 10 . \int \varrho\, d\,.\, a^3};$$

the integrals being taken from a = 0 to a = a′.

Let us now consider the law of gravity, or which comes to the same, that of the length of the pendulum at the elliptic surface in equilibrium. The value of l, found in the preceding No., becomes in this case

$$l = L + \alpha\, L\, \{\tfrac{5}{2}\, \varphi - h\}\, (m^2 - \tfrac{1}{3});$$

making, therefore, $L' = L - \frac{1}{3}\,\alpha\, L\, (\frac{5}{2}\,\varphi - h)$, we shall have, in neglecting quantities of the order α^2,

$$l = L' + \alpha\, L'\, (\tfrac{5}{2}\, \varphi - h)\, \mu^2;$$

an equation from which it results that L′ is the length of the seconds' pendulum at the equator, and that this length increases from the equator to the poles, proportionally to the square of the sine of the latitude.

If we call $\alpha\,\epsilon$ the excess of the length of the pendulum at the pole above its length at the equator, divided by the latter, we shall have

$$\alpha\,\epsilon = \alpha\, (\tfrac{5}{2}\, \varphi - h);$$

and consequently

$$\alpha \epsilon + \alpha h = \tfrac{5}{2} \alpha \varphi;$$

a remarkable equation between the ellipticity of the earth and the variation of the length of the pendulum from the equator to the poles. In the case of homogeneity $\alpha h = \frac{5}{4} \alpha \varphi$; hence in this case $\alpha \epsilon = \alpha h$; but *if the spheroid is heterogeneous, as much as* αh *is above or below* $\frac{5}{4} \alpha \varphi$, *so much is* $\alpha \epsilon$ *above or below the same quantity.*

579. The planets being supposed covered with a fluid in equilibrium, it is necessary, in the estimate of their attractions, to know the attraction of spheroids whose surface is fluid and in equilibrium: we may express it very simply in this way. Resume the equation (5) of No. 558; the signs of integration may be made to disappear by means of equation (2) of No. 573, which gives at the surface of the spheroid,

$$\frac{4\pi}{2i+1} \int \varrho \, d\, (a^{i+3} \cdot Y^{(i)}) = \frac{4\pi}{3} Y^{(i)} \int \varrho \, d \, . \, a^3 - Z^{(i)};$$

thus fixing the origin of the radii r at the center of gravity of the spheroid which makes $Y^{(1)}$ disappear; then observing that $Z^{(1)}$ is nothing, and that $Y^{(0)}$ being arbitrary, we may suppose $\frac{4\pi}{3} Y^{(0)} - Z^{(0)} = 0$, the equation (5) of 558, will give

$$V = \frac{4\pi}{3r} \int \varrho \, d \, . \, a^3 + \frac{4\alpha\pi}{3r^3} \left\{ Y^{(2)} + \frac{Y^{(3)}}{r} + \frac{Y^{(4)}}{r^2} + \&c. \right\} \int \varrho \, d \, . \, a^3$$
$$- \frac{\alpha}{r^3} \left\{ Z^{(2)} + \frac{Z^{(3)}}{r} + \frac{Z^{(4)}}{r^2} + \&c. \right\};$$

an expression in which we ought to observe that $\frac{4\pi}{3} \int \varrho \, d \, . \, a^3$ expresses the mass of the spheroid, since, in the case of r being infinite, the value of V is equal to the mass of the spheroid divided by r. Hence the attraction of the spheroid parallel to r will be $-\left(\frac{dV}{dr}\right)$; the attraction perpendicular to this radius, in the plane of the meridian will be $-\frac{\sqrt{1-m^2}}{r} \left(\frac{dV}{dm}\right)$; finally, the attraction perpendicular to this same radius in the direction of the parallel will be

$$-\frac{\left(\frac{dV}{d\varpi}\right)}{r\sqrt{1-m^2}}.$$

The expression of V, relatively to the earth supposed elliptic, becomes

$$V = \frac{M}{r} + \frac{\frac{1}{2}\alpha\varphi - \alpha h}{r^3} M (m^2 - \tfrac{1}{3});$$

M being the mass of the earth.

580. Although the law of attraction in the inverse ratio of the square of the distance is the only one that interests us, yet equation (I) of 554 affords a determination so simple of the gravity at the surface of homogeneons spheroids in equilibrium, whatever is the exponent of the power of the distance to which the attraction is proportional, that we cannot here omit it. The attraction being as any power n of the distance, if we denote by d m a molecule of the spheroid, and by f its distance from the point attracted, the action of d m upon this point multiplied by the element — d f of its direction, will be $-d\mu\, f^n.\, d f$. The integral of this quantity, taken relatively to f, is $-\frac{d\mu\, f^{n+1}}{n+1}$, and the sum of these integrals extended to the entire spheroid is $-\frac{V}{n+1}$; supposing, as in 554, that $V = \int f^{n+1} d\mu$.

If the spheroid be fluid, homogeneous, and endowed with rotatory motion, and not sollicited by any extraneous force, we shall have at the surface, in the case of equilibrium, by No. 567,

$$\text{const.} = -\frac{V}{n+1} + \tfrac{1}{2}\, g\, r^2 (1 - m^2),$$

r being the radius drawn from the center of gravity of the spheroid at its surface, and g the centrifugal force at the distance 1 from the axis of rotation.

The gravity p at the surface of the spheroid is equal to the differential of the second member of this equation taken relatively to r, and divided by — d r, which gives

$$p = \frac{1}{n+1} \cdot \left(\frac{d V}{d r}\right) - g\, r (1 - m^2).$$

Let us now resume equation (1) of 554, which is relative to the surface,

$$\left(\frac{d V}{d r}\right) = A' - \frac{(n+1) A}{2 a} + \frac{(n+1) V}{2 a};$$

this equation, combined with the preceding ones, gives

$$p = \text{const.} + \left\{\frac{(n+1) r}{4 a} - 1\right\} g\, r (1 - m^2).$$

At the surface, r is very nearly equal to a; by making them entirely so, for the sake of simplicity, we shall have

$$p = \text{const.} + \frac{n-3}{4}\, g (1 - m^2)$$

Let P be the gravity at the equator of the spheroid, and $\alpha\ \varphi$

the ratio of the centrifugal force to gravity at the equator; we shall have

$$p = P\left\{1 + \frac{3 - n}{4} \alpha \varphi . m^2\right\};$$

whence it follows that, from the equator to the poles, gravity varies as the square of the sine of the latitude. In the case of nature, where n = —2, we have

$$p = P\{1 + \tfrac{5}{4} \alpha\varphi . m^2\};$$

which agrees with what we have before found.

But it is remarkable that if n = 3, we have p = P, that is to say, that if the attraction varies as the cube of the distance, the gravity at the surface of homogeneous spheroids is every where the same, whatever may be the motion of rotation.

581. We have only retained, in the research of the figure of the celestial bodies, quantities of the order α; but it is easy, by the preceding analysis, to extend the approximations to quantities of the order α^2, and to superior orders. For that purpose, consider the figure of a homogeneous fluid mass in equilibrium, covering a spheroid differing but little from a sphere, and endowed with a rotatory motion; which is the case of the earth and planets. The condition of equilibrium at the surface gives, by No. 557, the equation

$$\text{const.} = V - \frac{g}{2} r^2 (m^2 - \tfrac{1}{3}).$$

The value of V is composed, 1st, of the attraction of the spheroid covered by the fluid upon the molecule of the surface, determined by the coordinates r, θ, and ϖ; 2dly, of the attraction of the fluid mass upon this molecule. But the sum of these two attractions is the same as the sum of the attractions, 1st, of a spheroid supposing the density of each of its shells diminished by the density of the fluid; 2dly, of a spheroid of the same density as the fluid, and whose exterior surface is the same as that of the fluid. Let V′ be the first of these attractions and V″ the second, so that V=V′+V″; we shall have, supposing g of the order α and equal to α g′,

$$\text{const.} = V' + V'' - \frac{\alpha g'}{2} . r^2 . (m^2 - \tfrac{1}{3}).$$

We have seen in 553 that V′ may be developed into a series of the form

$$\frac{U^{(0)}}{r} + \frac{U^{(1)}}{r^2} + \frac{U^{(2)}}{r^3} + \&c.$$

$U^{(i)}$ being subject to the equation of partial differences,

$$0 = \left(\frac{d\left\{(1 - m^2)\left(\frac{dU^{(i)}}{dm}\right)\right\}}{dm}\right) + \frac{\left(\frac{d^2 U^{(i)}}{d\varpi^2}\right)}{1 - m^2} + i(i+1) U^{(i)}.$$

and by the analysis of 561, we may determine $U^{(i)}$, with all the accuracy that may be wished for, when the figure of the spheroid is known.

In like manner V″ may be developed into a series of the form

$$\frac{U_{,}^{(0)}}{r} + \frac{U_{,}^{(1)}}{r^2} + \frac{U_{,}^{(2)}}{r^3} + \&c.$$

$U_{,}^{(i)}$ being subject to the same equation of partial differences as $U^{(i)}$. If we take for the unit of density that of the fluid, we have, by 561,

$$U_{,}^{(i)} = \frac{4\pi}{(i+3)(2i+1)} \cdot Z^{(i)};$$

r^{i+3} being supposed developed into the series

$$Z^{(0)} + Z^{(1)} + Z^{(2)} + \&c.$$

in which $Z^{(i)}$ is subject to the same equation of partial differences, as $U^{(i)}$. The equation of equilibrium will therefore become

$$\text{const.} = \frac{U^{(0)}}{r} + \frac{U_{,}^{(0)}}{r} + \Sigma \cdot \frac{1}{r^{i+1}}\left\{U^{(i)} + \frac{4\pi}{(i+3)(2i+1)} Z^{(i)}\right\}$$
$$- \alpha g' r^2 (m^2 - \tfrac{1}{3});$$

i being equal to greater than unity.

If the distance r from the molecule attracted to the center of the spheroid were infinite, V would be equal to the sum of the masses of the spheroid and fluid divided by r; calling, therefore, m this mass, we have $U^{(0)} + U_{,}^{(0)} = m$. Carrying the approximation only to quantities of the order α^2, we may suppose

$$r = 1 + \alpha y + \alpha^2 y';$$

which gives

$$r^{i+3} = 1 + (i+3)\alpha y + \frac{(i+2)(i+3)}{1.2}\alpha^2 . y^2 + (i+3)\alpha^2 . y'.$$

Suppose

$$y = Y^{(1)} + Y^{(2)} + Y^{(3)} + \&c.$$
$$y' = Y'^{(1)} + Y'^{(2)} + Y'^{(3)} + \&c.$$
$$y'' = M^{(0)} + M^{(1)} + M^{(2)} + \&c.$$

$Y^{(i)}$, $Y'^{(i)}$, and $M^{(i)}$ being subject to the same equation of partial differences as $U^{(i)}$; we shall have

$$Z^{(i)} = (i+3)\alpha Y^{(i)} + \frac{(i+2)(i+3)}{1.2}\alpha^2 M^{(i)} + (i+3)\alpha^2 Y'^{(i)}.$$

Then observe that $U^{(i)}$ is a quantity of the order α, since it would be nothing if the spheroid were a sphere; thus carrying the approximation only to terms of the order α^2, $U^{(i)}$ will be of this form $\alpha U'^{(i)} + \alpha^2 U''^{(i)}$. Substituting therefore these values in the preceding equation of equilibrium, and there changing r into $1 + \alpha y + \alpha^2 y'$, we shall have to quantities of the order α^3,

$$\text{const.} = \mu\,\{1 - \alpha y + \alpha^2 y^2 - \alpha^2 y'\}$$

$$+ \Sigma \left\{ \begin{array}{l} \alpha\, U'^{(i)} + \alpha^2\, U''^{(i)} - (i+1)\,\alpha^2 y\, U'^{(i)} \\ + \dfrac{4\,\alpha\,\pi}{2i+1} Y^{(i)} - \dfrac{4\,\alpha^2\,\pi\,(i+1)}{2i+1} y\, Y^{(i)} + \dfrac{4\,\alpha^2\,\pi}{2i+1}\, Y^{(i)} \\ + \dfrac{4\,\alpha^2\,\pi\,(i+2)}{2\,(2i+1)}\, M^{(i)} \end{array} \right\}$$

$$- \frac{\alpha\, g'\,(1 + 2\,\alpha y)}{2}\,(m^2 - \tfrac{1}{3}).$$

Equating separately to zero the terms of the order α, and those of the order α^2, we shall have the two equations,

$$\Sigma \left(\mu - \frac{4\,\pi}{2i+1}\right) Y^{(i)} = \Sigma\, U'^{(i)} - \frac{g'}{2}(m^2 - \tfrac{1}{3});$$

$$\Sigma \left(\mu - \frac{4\,\pi}{2i+1}\right) Y'^{(i)} = C' + \Sigma \left\{ \begin{array}{l} U''^{(i)} - (i+1)\, y\, U'^{(i)} - \dfrac{4\pi\,(i+1)}{2i+1}\, y\, Y^{(i)} \\ + \left\{ \mu + \dfrac{4\,\pi\,(i+2)}{2\,(2i+1)} \right\} M^{(i)} \end{array} \right\}$$

$$- g'\, y\,(m^2 - \tfrac{1}{3});$$

C' being an arbitrary constant. The first of these equations detects $Y^{(i)}$ and consequently the value of y. Substituting in the second member of the second equation, we shall develope by the method of No. 560. in a series of the form

$$N^{(0)} + N^{(1)} + N^{(2)} + \&c.$$

$N^{(i)}$ being subject to the same equation of partial differences as $U^{(i)}$, and we shall determine the constant C' in such a manner that $N^{(0)}$ is nothing; thus we shall have

$$Y'^{(i)} = \frac{N^{(i)}}{\mu - \dfrac{4\,\pi}{2i+1}}$$

and consequently

$$Y' = \frac{N^{(1)}}{\mu - \frac{4}{3}\pi} + \frac{N^{(2)}}{\mu - \frac{4}{5}\pi} + \frac{N^{(3)}}{\mu - \frac{4}{7}\pi} + \&c.$$

The expression of the radius r of the surface of the fluid will thus be determined to quantities of the order α^3, and we may, by the same process, carry the approximation as far as we wish. We shall not dwell any longer upon this object, which has no other difficulty than the length of calculations; but we shall derive from the preceding analysis this important conclusion, namely, that we may affirm that the equilibrium is rigorously possible, although we cannot assign the rigorous figure which satisfies it; for we may find a series of figures, which, being substituted in the equation of equilibrium, leave remainders successively smaller and smaller, and which become less than any given quantity.

COMPARISON OF THE PRECEDING THEORY WITH OBSERVATIONS.

582. To compare with observations the theory we have above laid down, we must know the curve of the terrestrial meridians, and those which we trace by a series of geodesic operations. If through the axis of rotation of the earth, and through the zenith of a plane at its surface we imagine a plane to pass produced to the heavens; this plane will trace a great circle which will be the meridian of the plane: all points of the surface of the earth which have their zenith upon this circumference, will lie under the same celestial meridian, and they will form, upon this surface, a curve which will be the corresponding terrestrial meridian.

To determine this curve, represent by $u = 0$ the equation of the surface of the earth; u being a function of three rectangular coordinates x, y, z. Let x', y', z', be the three coordinates of the vertical which passes through the place on the earth's surface determined by the coordinates x, y, z; we shall have by the theory of curved surfaces, the two following equations,

$$0 = \left(\frac{d\,u}{d\,x}\right) d\,y' - \left(\frac{d\,u}{d\,y}\right) d\,x';$$

$$0 = \left(\frac{d\,u}{d\,x}\right) d\,z' - \left(\frac{d\,u}{d\,z}\right) d\,x'.$$

Adding the first multiplied by the indeterminate λ to the second, we get

$$d\,z' = \frac{\left(\frac{d\,u}{d\,z}\right) + \lambda \left(\frac{d\,u}{d\,y}\right)}{\left(\frac{d\,u}{d\,x}\right)} \,.\, d\,x' - \lambda\, d\,y'.$$

This equation is that of any plane parallel to the said vertical: this vertical produced to infinity coinciding with the celestial meridian, whilst its foot is only distant by a finite quantity from the plane of this meridian, may be deemed parallel to that plane. The differential equation of this plane may therefore be made to coincide with the preceding one by suitably determining the indeterminate λ.

Let

$$d\,z' = a\,d\,x' + b\,d\,y',$$

be the equation of the plane of the celestial meridian; comparing it with the preceding one, we shall get

$$\left(\frac{d\,u}{d\,z}\right) - a \left(\frac{d\,u}{d\,x}\right) - b \left(\frac{d\,u}{d\,y}\right) = 0; \quad . \quad . \quad . \quad . \quad . \quad (a)$$

To get the constants a, b, we shall suppose known the coordinates of the foot of the vertical parallel to the axes of rotation of the earth and that of a given place on its surface. Substituting successively these coordinates in the preceding equation, we shall have two equations, by means of which we shall determine a and b. The preceding equation combined with that of the surface $u = 0$, will give the curve of the terrestrial meridian which passes through the given plane.

If the earth were any ellipsoid whatever, u would be a rational and entire function of the second degree in x, y, z; the equation (a) would therefore then be that of a plane whose intersection with the surface of the earth, would form the terrestrial meridian: in the general case, this meridian is a curve of double curvature.

In this case the line determined by geodesic measures, is not that of the terrestrial meridian. To trace this line, we form a first horizontal triangle of which one of the angles has its summit at the origin of this curve, and whose two other summits are any visible objects. We determine the direction of the first side of the curve, relatively to two sides of the triangle, and to its length from the point where it meets the side which joins the two objects. We then form a second horizontal triangle with these objects, and a third one still farther from the origin of the curve. This second triangle is not in the plane of the first; it has nothing in common with the former, but the side formed by the two first objects; thus the first side of the curve being produced, lies above the plane of this second triangle; but we bend it down upon the plane so as always to form the same angles with the side common to the two triangles, and it is easy to see that for this purpose it must be bent along a vertical to this plane. Such is therefore the characteristic property of the curve traced by geodesic operations. Its first side, of which the direction may be supposed any whatever, touches the earth's surface; its second side is this tangent produced and bent vertically; its third is the tangent of the second side bent vertically, and so on.

If through the point where the two sides meet, we draw in the tangent plane at the surface of the spheroid, a line perpendicular to one of the sides, it is clear that it will be perpendicular to the other; whence it follows, that the sum of the sides is the shortest line which can be drawn upon the surface between their extreme points. Thus the lines traced by geodesic operations, have the property of being the shortest we can draw upon the surface of the spheroid between any two of their points; and p.294, Vol. I. they would be described by a body moving uniformly in this surface.

Let x, y, z be the rectangular coordinates of any part whatever of the curve; x + d x, y + d y, z + d z will be those of points infinitely near to it. Call d s the element of the curve, and suppose this element produced by a quantity equal to d s; x + 2 d x, y + 2 d y, z + 2 d z will be the coordinates of extremity of the curve thus produced. By bending it vertically, the coordinates of this extremity will become $x + 2\,dx + d^2x$, $y + 2\,dy + d^2y$, $z + 2\,dz + d^2z$; thus $-d^2x$, $-d^2y$, $-d^2z$ will be the coordinates of the vertical, taken from its foot; we shall therefore have by the nature of the vertical, and by supposing that $u = 0$ is the equation of the earth's surface,

$$0 = \left(\frac{du}{dx}\right) d^2y - \left(\frac{du}{dy}\right) d^2x;$$

$$0 = \left(\frac{du}{dx}\right) d^2z - \left(\frac{du}{dz}\right) d^2x;$$

equations which are different from those of the terrestrial meridian. In these equations d s must be constant; for it is clear that the production of d s meets the foot of the vertical at an infinitely small quantity of the fourth order nearly.

Let us see what light is thrown upon the subject of the figure of the earth by geodesic measures, whether made in the directions of the meridians, or in directions perpendicular to the meridians. We may always conceive an ellipsoid touching the terrestrial surface at every point of it, and upon which, the geodesic measures of the longitudes and latitudes from the point of contact, for a small extent, would be the same as at the surface itself. If the entire surface were that of an ellipsoid, the tangent ellipsoid would every where be the same; but if, as it is reasonable to suppose, the figure of the meridians is not elliptic, then the tangent ellipsoid varies from one country to another, and can only be determined by geodesic measures, made in different directions. It would be very interesting to know the osculating ellipsoids at a great number of places on the earth's surface.

Let $u = x^2 + y^2 + z^2 - 1 - 2\alpha u'$, be the equation to the surface of the spheroid, which we shall suppose very little different from a sphere whose radius is unity, so that α is a very small quantity whose square may be neglected. We may always consider u′ as a function of two variables x, y; for by supposing it a function of x, y, z, we may eliminate z by means of the equation $z = \sqrt{1 - x^2 - y^2}$. Hence, the three equations found above, relatively to the shortest line upon the earth's surface, become

$$\left.\begin{aligned} x\,d^2y - y\,d^2x &= \alpha\left(\frac{d\,u'}{d\,x}\right)d^2y - \alpha\left(\frac{d\,u'}{d\,y}\right)d^2x;\\ x\,d^2z - z\,d^2x &= \alpha\left(\frac{d\,u'}{d\,x}\right)d^2z;\\ y\,d^2z - z\,d^2y &= \alpha\left(\frac{d\,u'}{d\,y}\right).\,d^2z. \end{aligned}\right\} \;.\;.\; (O)$$

This line we shall call the *Geodesic line.*

Call r the radius drawn from the center of the earth to its surface, θ the angle which this radius makes with the axis of rotation, which we shall suppose to be that of z, and φ the angle which the plane formed by this axis and by r makes with the plane of x, y; we shall have

$$x = r\sin.\theta.\cos.\varphi;\; y = r\sin.\theta\sin.\varphi;\; z = r\cos.\theta;$$

whence we derive

$$r^2\sin.^2\theta.\,d\,\varphi = x\,d\,y - y\,d\,x;$$

$$-r^2\,d\,\theta = (x\,d\,z - z\,d\,x)\cos.\varphi + (y\,d\,z - z\,d\,y)\sin.\varphi$$

$$d\,s^2 = d\,x^2 + d\,y^2 + d\,z^2 = d\,r^2 + r^2\,d\,\theta^2 + r^2\,d\,\varphi^2\sin.^2\theta.$$

Considering then u', as a function of x, y, and designating by ψ the latitude; we may suppose in this function r = 1, and $\psi = 100^\circ - \theta$, which gives

$$x = \cos.\psi\cos.\varphi;\; y = \cos.\psi\sin.\varphi;$$

thus we shall have

$$\left(\frac{d\,u'}{d\,x}\right)d\,x + \left(\frac{d\,u'}{d\,y}\right)d\,y = \left(\frac{d\,u'}{d\,\psi}\right)d\,\psi + \left(\frac{d\,u'}{d\,\varphi}\right)d\,\varphi;$$

but we have

$$x^2 + y^2 = \cos.^2\psi;\; \frac{y}{x} = \tan.\varphi;$$

whence we derive

$$d\,\psi = -\frac{x\,d\,x + y\,d\,y}{\sin.\psi\cos.\psi};\; d\,\varphi = \frac{x\,d\,y - y\,d\,x}{x^2}.\cos.^2\varphi.$$

Substituting these values of $d\,\psi$ and of $d\,\varphi$ in the preceding differential equation in u', and comparing separately the coefficients of d x and d y; we shall have

$$\left(\frac{d\,u'}{d\,x}\right) = -\frac{\cos.\varphi}{\sin.\psi}.\left(\frac{d\,u'}{d\,\psi}\right) - \frac{\sin.\varphi}{\cos.\psi}.\left(\frac{d\,u'}{d}\right);$$

$$\left(\frac{d\,u'}{d\,y}\right) = -\frac{\sin.\varphi}{\sin.\psi}.\left(\frac{d\,u'}{d\,\psi}\right) + \frac{\cos.\varphi}{\cos.\psi}.\left(\frac{d\,u'}{d\,\varphi}\right);$$

which give

$$\left(\frac{d\,u'}{d\,x}\right)d^2y - \left(\frac{d\,u'}{d\,y}\right)d^2x = -\frac{\left(\frac{d\,u'}{d\,\psi}\right)}{\sin.\psi\cos.\psi}.(x\,d^2y - y\,d^2x)$$

$$-\frac{\left(\frac{d\,u'}{d\,\varphi}\right)}{\cos.^2\psi}(x\,d^2x + y\,d^2y).$$

But neglecting quantities of the order α, we have $x\, d^2 y - y\, d^2 x = 0$; and the two equations

$$x\, d^2 z - z\, d^2 x = 0, \quad y\, d^2 z - z\, d^2 y = 0,$$

give

$$z\, d^2 z = \frac{z^2 (x\, d^2 x + y\, d^2 y)}{x^2 + y^2}$$

and

$$x^2 + y^2 + z^2 = 1$$

gives

$$x\, d^2 x + y\, d^2 y + z\, d^2 z + d\, s^2 = 0;$$

substituting for $z\, d^2 z$ its preceding value, we shall have

$$x\, d^2 x + y\, d^2 y = -(x^2 + y^2)\, d\, s^2 = -d\, s^2 \cos.^2 \psi;$$

wherefore

$$\left(\frac{d\, u'}{d\, x}\right) d^2 y - \left(\frac{d\, u'}{d\, y}\right) . d^2 x = \left(\frac{d\, u'}{d\, \varphi}\right) . d\, s^2.$$

The first of equations (O), will thus give by integration,

$$r^2\, d\, \varphi \sin.^2 \theta = c\, d\, s + \alpha\, d\, s \int d\, s \left(\frac{d\, u'}{d\, \varphi}\right); \quad . \; . \; . \; . \; . \quad (p)$$

c being the arbitrary constant.

The second of equations (O) gives

$$d . (x\, d\, z - z\, d\, x) = \alpha \left(\frac{d\, u'}{d\, x}\right) d^2 z;$$

but it is easy to see by what precedes, that we have

$$d^2 z = -d\, s^2 . \sin. \psi;$$

we have therefore

$$d\, (x\, d\, z - z\, d\, x) = -\alpha\, d\, s^2 \left(\frac{d\, u'}{d\, x}\right) \sin. \psi;$$

in like manner we have

$$d\, (y\, d\, z - z\, d\, y) = -\alpha\, d\, s^2 \left(\frac{d\, u'}{d\, y}\right) \text{siu.}\, \psi;$$

we shall therefore have

$$r^2\, d\, \theta = c'\, d\, s \sin. \varphi + c''\, d\, s \cos. \varphi$$
$$- \alpha\, d\, s \cos. \varphi \int d\, s \left\{ \left(\frac{d\, u'}{d\, \psi}\right) \cos. \varphi + \left(\frac{d\, u'}{d\, \varphi}\right) \sin. \varphi \tan. \psi \right\}$$
$$- \alpha\, d\, s \sin. \varphi \int d\, s \left\{ \left(\frac{d\, u'}{d\, \psi}\right) \sin. \varphi - \left(\frac{d\, u'}{d\, \varphi}\right) \cos. \varphi \tan. \psi \right\}; \quad . \quad (q)$$

First consider the case in which the first side of the Geodesic line is parallel to the corresponding plane of the celestial meridian. In this case $d\, \varphi$ is of the order α, as also $d\, r$; we have, therefore, neglecting quantities of the order α^2, $d\, s = -r\, d\, \theta$, the arc s being supposed to increase from the equator to the poles. ψ expressing the latitude, it is easy to see that we have $\theta = 100^\circ - \psi - \left(\frac{d\,r}{d\,\psi}\right)$, which gives

$$d\,\theta = -d\,\psi - \alpha\,d\,\psi\left(\frac{d^2\,u'}{d\,\psi^2}\right);$$

we have therefore

$$d\,s = d\,\psi.\left\{1 + \alpha\,u' + \alpha\left(\frac{d^2\,u'}{d\,\psi^2}\right)\right\}.$$

Thus naming ε the difference in latitude of the two extreme points of the arc s, we shall have

$$s = \varepsilon + \alpha\,\varepsilon\left\{u_{\prime}' + \left(\frac{d^2\,u_{\prime}'}{d\,\psi^2}\right)\right\} + \frac{\alpha\,\varepsilon^2}{1.2}.\left\{\left(\frac{d\,u_{\prime}'}{d\,\psi}\right) + \left(\frac{d^3\,u_{\prime}'}{d\,\psi^3}\right)\right\} + \&c.;$$

$u_{\prime}'$ being here the value of u' at the origin of s.

If the earth were a solid of revolution, the geodesic line would be always in the plane of the same meridian; it departs from it if the parallels are not circles; the observations of this deflection may therefore clear up this important point of the theory of the earth. Resume the equation (p) and observe that in the present case, d φ and the constant c of this equation are of the order α, and that we may there suppose r = 1, d s = d ψ, $\theta = 100^\circ - \psi$; we shall thus get

$$d\,\varphi\cos.^2\psi = c\,d\,\psi + \alpha\,d\,\psi\int d\,\psi\left(\frac{d\,u'}{d\,\varphi}\right).$$

However, if we call V the angle which the plane of the celestial meridian makes with that of x, y, whence we compute the origin of the angle φ; we shall have d x' = tan. V = d y'; x', y', z' being the coordinates of that meridian whose differential equation, as we have seen in the preceding No., is

$$d\,z' = a\,d\,x' + b\,d\,y'.$$

Comparing it with the preceding one, we see that a, b are infinite and such that $-\frac{a}{b}$ = tan. V, the equation (a) of the preceding No. thus gives

$$0 = \left(\frac{d\,u}{d\,x}\right).\tan.V - \left(\frac{d\,u}{d\,y}\right),$$

whence we derive

$$0 = x\tan.V - y - \alpha\left(\frac{d\,u'}{d\,x}\right)\tan.V + \alpha\left(\frac{d\,u'}{d\,y}\right).$$

We may suppose V = φ, in the terms multiplied by α; moreover $\frac{y}{x}$ = tan. φ; w have therefore

$$\cos.\,\psi\,\cos.\,\varphi\,\{\tan.\,\varphi - \tan.\,V\} = \frac{\alpha\left(\frac{d\,u'}{d\,\varphi}\right)}{\cos.\,\psi\,\cos.\,\varphi},$$

which gives

$$\varphi - V = \frac{\alpha\left(\frac{d\,u'}{d\,\varphi}\right)}{\cos.^2\,\psi}.$$

The first side of the *Geodesic line*, being supposed parallel to the plane of the celestial meridian, the differentials of the angle V, and of the distance $(\varphi - V)\cos.\,\psi$ from the origin of the curve to the plane of the celestial meridian ought to be nothing at this origin; we have therefore at this point

$$\frac{d\,\varphi}{d\,\psi} = (\varphi - V)\tan.\,\psi = \frac{\alpha\left(\frac{d\,u'}{d\,\varphi}\right)\tan.\,\psi}{\cos.^2\,\psi};$$

and consequently, the equation (p)' gives

$$c = \alpha\left(\frac{d\,u_{\prime}'}{d\,\varphi}\right)\tan.\,\psi_{\prime},$$

$u_{\prime}$ and $\psi_{\prime}$ being referred to the origin of the arc s.

At the extremity of the measured arc, the side of the curve makes with the plane of the corresponding celestial meridian an angle very nearly equal to the differential of $(\varphi - V)\cos.\,\psi$, divided by $d\,\psi$, V being supposed constant in the differentiation; by denoting therefore this angle by ϖ, we shall have

$$\varpi = \frac{d\,\varphi}{d\,\psi}\cos.\,\psi - (\varphi - V)\sin.\,\psi.$$

If we substitute for $\frac{d\,\varphi}{d\,\psi}$ its value obtained from the equation (p), and for $\varphi - V$, its preceding value, we shall have

$$\varpi = \frac{\alpha}{\cos.\,\varphi}\cdot\left\{\left(\frac{d\,u_{\prime}'}{d\,\varphi}\right)\tan.\,\psi_{\prime} - \left(\frac{d\,u'}{d\,\varphi}\right)\tan.\,\psi + \int d\,\psi\left(\frac{d\,u'}{d\,\varphi}\right)\right\};$$

the integral being taken from the origin of the measured arc, to its extremity. Call ϵ the difference in latitude of its two extreme points; ϵ being supposed sufficiently small for ϵ^2 to be rejected, we shall have

$$\varpi = -\frac{\alpha\,\epsilon\,\tan.\,\psi}{\cos.\,\psi}\left\{\left(\frac{d\,u'}{d\,\varphi}\right)\tan.\,\psi + \left(\frac{d^2\,u'}{d\,\varphi\,d\,\psi}\right)\right\};$$

in which the values of ψ, $\left(\frac{d\,u'}{d\,\varphi}\right)$, and $\left(\frac{d^2\,u'}{d\,\varphi\,d\,\psi}\right)$ must be referred, for the greater exactness, to the middle of the measured arc. The angle ϖ must be supposed positive, when it quits the meridian, in the direction of the increments of φ.

To obtain the difference in longitude of the two meridians corresponding to the extremities of the arc, we shall observe, that $u'_{\prime}$, $V_{\prime}$, $\psi_{\prime}$, and $\varphi_{\prime}$, being the values of u', V, ψ, and φ, at the first extremity, we have

$$\varphi_{\prime} - V_{\prime} = \frac{\alpha\left(\frac{d\,u'_{\prime}}{d\,\varphi}\right)}{\cos.^2 \psi_{\prime}}; \quad \varphi - V = \frac{\alpha\left(\frac{d\,u'}{d\,\varphi}\right)}{\cos.^2 \psi};$$

but we have very nearly, neglecting the square of ε,

$$\varphi - \varphi_{\prime} = \frac{c\,\varepsilon}{\cos.^2 \psi_{\prime}}; \quad c = \alpha\left(\frac{d\,u'_{\prime}}{d\,\varphi}\right) \tan.\,\psi_{\prime};$$

we shall have, therefore,

$$V - V = -\frac{\alpha\,\varepsilon}{\cos.^2 \psi_{\prime}} \cdot \left\{ \left(\frac{d\,u'_{\prime}}{d\,\varphi}\right) \tan.\,\psi_{\prime} + \left(\frac{d^2\,u'_{\prime}}{d\,\varphi\,d\,\psi}\right) \right\};$$

whence results this very simple equation,

$$(V - V_{\prime}) \sin.\,\psi_{\prime} = \varpi;$$

thus we may, by observation alone, and independently of the knowledge of the figure, determine the difference in longitude of the meridians corresponding to the extremities of the measured arc; and if the value of the angle ϖ is such that we cannot attribute it to errors of observations, we shall be certain that the earth is not a spheroid of revolution.

Let us now consider the case where the first side of the *Geodesic line* is perpendicular to the corresponding plane of the celestial meridian. If we take this plane for that of x, y, the cosine of the angle formed by this side upon the plane, will be $\frac{\sqrt{d\,x^2 + d\,z^2}}{d\,s}$; thus this cosine being nothing at the origin, we have $d\,x = 0$, $d\,z = 0$, which gives

$$d.\,r \sin.\,\theta \cos.\,\varphi = 0; \quad d.\,r \cos.\,\theta = 0;$$

and consequently

$$r\,d\,\theta = r\,d\,\varphi \sin.\,\theta.\cos.\,\theta.\tan.\,\varphi;$$

but we have, to quantities of the order α^2, $d\,s = r\,d\,\varphi \sin.\,\theta$; we shall have, therefore, at the origin,

$$\frac{d\,\theta}{d\,s} = \frac{\tan.\,\varphi.\cos.\,\theta}{r}.$$

The constant c'', of the equation (q), is equal to the value of $x\,d\,z - z\,d\,x$, at the origin; it is therefore nothing, and the equation (q) gives at the origin,

$$\frac{d\,\theta}{d\,s} = \frac{c'}{r^2} \sin.\,\varphi;$$

we have, therefore, observing that φ is here of the order α, and that thus neglecting quantities of the order α^2, we have sin. φ = tan. φ,

$$c' = r_{,} \cos. \theta_{,},$$

the quantities $r_{,}$ and $\theta_{,}$ being relative to the origin; therefore, if we consider that at this origin the angle φ is what we have before called it, $\varphi_{,} - V_{,}$, and whose value we have found equal to $\frac{\alpha\left(\frac{d u_{,}'}{d \varphi}\right)}{\cos.^2 \psi_{,}}$; we shall have at this point

$$\frac{d \theta_{,}}{d s} = \alpha\left(\frac{d u_{,}'}{d \varphi}\right) \frac{\sin. \psi_{,}}{\cos.^2 \psi_{,}}.$$

The equation (q) then gives

$$\frac{d^2 \theta_{,}}{d s^2} = \frac{\cos. \theta_{,}}{r_{,}} \cdot \frac{d \varphi_{,}}{d s} - \alpha . \left(\frac{d u_{,}'}{d \psi}\right);$$

but we have

$$\frac{d \varphi_{,}}{d s} = \frac{1}{r_{,} \sin. \theta_{,}}; \quad r_{,} = 1 + \alpha u_{,}'; \quad \theta_{,} = 100^{\circ} - \psi_{,} - \alpha\left(\frac{d u_{,}'}{d \psi}\right);$$

we shall get therefore

$$\frac{d^2 \theta_{,}}{d s^2} = (1 - 2 \alpha u_{,}') \tan. \psi_{,} + \alpha\left(\frac{d u_{,}'}{d \psi}\right) \tan.^2 \psi_{,}.$$

Observing that at the origin,

$$\frac{d \varphi_{,}}{d s} = \frac{1}{r_{,} \sin. \theta_{,}} = \frac{1}{\cos. \psi_{,}}\left\{1 - \alpha u_{,}' + \alpha\left(\frac{d u_{,}'}{d \psi}\right) \tan. \psi_{,}\right\},$$

the equation (p) gives

$$c = r_{,} \sin. \theta_{,};$$

whence we get

$$\frac{d^2 \varphi_{,}}{d s^2} = - \frac{2 \alpha . \frac{d u_{,}'}{d s}}{r_{,} \sin. \theta_{,}} - \frac{2 . \frac{d \theta_{,}}{d s} \cos. \theta_{,}}{r_{,} \sin.^2 \theta_{,}} + \frac{\alpha\left(\frac{d u_{,}'}{d \varphi_{,}}\right)}{\cos.^2 \psi_{,}},$$

and consequently

$$\frac{d^2 \varphi_{,}}{d s^2} = - \alpha\left(\frac{d u_{,}'}{d \varphi}\right) . \frac{2 - \cos.^2 \psi_{,}}{\cos.^4 \psi_{,}}.$$

The equation

$$\theta = 100^{\circ} - \psi - \alpha\left(\frac{d u'}{d \psi}\right),$$

gives, by retaining amongst the terms of the order s^2, only those which are independent of α,

$$\psi - \psi_{,} = - s . \frac{d \theta_{,}}{d s} - \tfrac{1}{2} s^2 . \frac{d^2 \theta_{,}}{d s^2} - \frac{\alpha s}{\cos. \psi_{,}} . \left(\frac{d^2 u_{,}'}{d \varphi d \psi}\right);$$

wherefore

$$\psi - \psi_{,} = -\frac{\alpha\, s}{\cos.\psi_{,}}\cdot\left\{\left(\frac{d\,u_{,}'}{d\,\varphi}\right)\tan.\psi_{,} + \alpha\left(\frac{d^2\,u_{,}'}{d\,\varphi\,d\,\psi}\right) - \tfrac{1}{2}\,s^2\tan.\psi_{,}\right\}.$$

The difference of latitudes at the two extremities of the measured arc, will therefore give

$$-\frac{\alpha\, s}{\cos.\psi_{,}}\cdot\left\{\left(\frac{d\,u_{,}'}{d\,\varphi}\right)\tan.\psi_{,} + \alpha\left(\frac{d^2\,u_{,}'}{d\,\varphi\,d\,\psi}\right)\right\};$$

It is remarkable, that for the same arc, measured in the direction of the meridian, this function, by what precedes, is equal to $\frac{\varpi}{\tan.\psi_{,}}$; it may thus be determined in two ways, and we shall be able to judge whether the values thus found of the difference of latitudes, or of the azimuthal angle ϖ, are due to the errors of observations, or to the excentricity of the terrestrial parallels.

Retaining only the first power of s, we have

$$\varphi - \varphi_{,} = s\cdot\frac{d\,\varphi_{,}}{d\,s} = \frac{s}{\cos.\psi_{,}}\cdot\left\{1 - \alpha\,u_{,}' + \alpha\left(\frac{d\,u_{,}'}{d\,\psi}\right)\tan.\psi_{,}\right\}.$$

$\varphi - \varphi_{,}$ is not the difference in longitude of the two extremities of the arc s; this difference is equal to $V - V_{,}$; but we have, by what precedes,

$$\varphi - V = \frac{\alpha\left(\frac{d\,u'}{d\,\varphi}\right)}{\cos.^2\psi};$$

which gives

$$\varphi - V - (\varphi_{,} - V_{,}) = \frac{\alpha\, s\left(\frac{d^2\,u_{,}'}{d\,\varphi\,.\,d\,s}\right)}{\cos.^2\psi_{,}} = \frac{\alpha\, s\left(\frac{d^2\,u_{,}'}{d\,\varphi^2}\right)}{\cos.^3\psi_{,}};$$

wherefore

$$V - V_{,} = \frac{s}{\cos.\psi_{,}}\cdot\left\{1 - \alpha\,u_{,}' + \alpha\left(\frac{d\,u_{,}'}{d\,\psi}\right)\tan.\psi_{,} - \frac{\alpha\left(\frac{d^2\,u_{,}'}{d\,\varphi^2}\right)}{\cos.^2\psi_{,}}\right\}.$$

For greater exactness, we must add to this value of $V - V_{,}$ the term depending on s^3, and independent of α, which we obtain in the hypothesis of the earth being a sphere. This term is equal to $-\tfrac{1}{3}\,s^3\cdot\frac{\tan.^2\psi_{,}}{\cos.\psi_{,}}$; thus we have

$$V - V_{,} = \frac{s}{\cos.\psi_{,}}\cdot\left\{1 - \alpha.\,u_{,}' + \alpha.\left(\frac{d\,u_{,}'}{d\,\psi}\right)\tan.\psi_{,} - \frac{\alpha\left(\frac{d^2\,u_{,}'}{d\,\varphi^2}\right)}{\cos.^2\psi_{,}} - \tfrac{1}{3}\,s^2\tan.^2\psi'\right\}.$$

It remains to determine the azimuthal angle at the extremity of the arc s. For that purpose, call x′, and y′, the coordinates x, y, referred to the meridian of the last extremity of the arc s; it is easy to see that the cosine of the azimuthal angle is equal to $\frac{\sqrt{d x'^2 + d z^2}}{d s}$. If we refer the coordinates x, y, to the plane of the meridian corresponding to the first extremity of the arc; its first side being supposed perpendicular to the plane of this meridian, we shall have

$$\frac{d x_{\prime}}{d s} = 0; \ \frac{d z_{\prime}}{d s} = 0; \ \frac{d y_{\prime}}{d s} = 1;$$

wherefore, retaining only the first power of s,

$$\frac{d x}{d s} = s \cdot \frac{d^2 x_{\prime}}{d s^2}; \ \frac{d z}{d s} = s \cdot \frac{d^2 z_{\prime}}{d s^2};$$

but we have

$$x' = x \cos.(V - V_{\prime}) + y \sin.(V - V_{\prime});$$

thus $V - V_{\prime}$ being, by what precedes, of the order α, we shall have

$$\frac{d x'}{d s} = s \cdot \frac{d^2 x_{\prime}}{d s^2} + (V - V_{\prime}) \frac{d y_{\prime}}{d s}.$$

Again, we have

$$x = r \sin.\theta \cos.\varphi; \quad z = r \cos.\theta;$$

we therefore shall obtain, rejecting quantities of the order α^2, and observing that $\varphi_{\prime}$, $\frac{d^2 \varphi_{\prime}}{d s^2}$, and $\frac{d \theta_{\prime}}{d s}$ are quantities of the order α,

$$\frac{d^2 x_{\prime}}{d s^2} = \alpha \cdot \frac{d^2 u_{\prime}'}{d s^2} \sin.\theta_{\prime} + r_{\prime} \cdot \frac{d^2 \theta_{\prime}}{d s^2} \cos.\theta_{\prime} - r_{\prime} \sin.\theta_{\prime} \cdot \frac{d \varphi_{\prime}^2}{d s^2}.$$

Thence we have

$$\alpha \cdot \frac{d^2 u_{\prime}'}{d s^2} = \alpha \left(\frac{d^2 u_{\prime}'}{d \varphi^2}\right) \frac{d \varphi_{\prime}^2}{d s^2} - \alpha \left(\frac{d u_{\prime}'}{d \psi}\right) \cdot \frac{d^2 \theta_{\prime}}{d s^2} = \frac{\alpha \left(\frac{d^2 u_{\prime}'}{d \varphi^2}\right)}{\cos.^2 \psi_{\prime}} - \alpha \left(\frac{d u_{\prime}'}{d \psi}\right) \tan.\psi_{\prime};$$

moreover, $d s = r_{\prime} \sin.\theta_{\prime} . d \varphi_{\prime}$; we shall, therefore, have by substituting for $r_{\prime}$, $\theta_{\prime}$, $\frac{d \varphi_{\prime}}{d s}$, and $\frac{d^2 \theta}{d s^2}$, their preceding values,

$$\frac{d^2 x_{\prime}}{d s^2} = (1 - \alpha u_{\prime}') \cdot \frac{\sin.^2 \psi_{\prime}}{\cos.\psi_{\prime}} + \alpha \left(\frac{d u_{\prime}'}{d \psi}\right) \tan.^2 \psi_{\prime} \sin.\psi_{\prime}$$

$$- \frac{1}{\cos.\psi_{\prime}} \cdot \left\{1 - \alpha u_{\prime} + \alpha \left(\frac{d u_{\prime}'}{d \psi}\right) \tan.\psi_{\prime}\right\} + \frac{\alpha \left(\frac{d^2 u_{\prime}'}{d \varphi^2}\right)}{\cos.\psi_{\prime}}$$

Neglecting the superior powers of s, we have, as we have seen,

$$V - V_{\prime} = \frac{s}{\cos.\psi_{\prime}} \cdot \left\{1 - \alpha u_{\prime}' + \alpha \left(\frac{d u_{\prime}'}{d \psi}\right) \tan.\psi_{\prime} - \frac{\alpha \left(\frac{d^2 u_{\prime}'}{d \varphi^2}\right)}{\cos.^2 \psi_{\prime}}\right\},$$

and $\frac{d y_{\prime}}{d s} = 1$; we therefore have

$$\frac{d x_{,}}{d s} = s(1 - \alpha u_{,}') \frac{\sin.^2 \psi_{,}}{\cos. \psi_{,}} + \alpha s \left(\frac{d u_{,}'}{d \psi}\right) \tan.^2 \psi_{,} . \sin. \psi_{,} - \alpha s \left(\frac{d^2 u_{,}'}{d \varphi^2}\right) \frac{\sin.^2 \psi_{,}}{\cos.^3 \psi_{,}};$$

in like manner we shall find

$$\frac{d z}{d s} = -s(1 - \alpha u_{,}') \sin. \psi_{,} - \alpha s \left(\frac{d u_{,}'}{d \psi}\right) \tan.^2 \psi_{,} \cos. \psi_{,} + \alpha s \left(\frac{d^2 u_{,}'}{d \varphi^2}\right) \frac{\sin. \psi_{,}}{\cos.^2 \psi_{,}};$$

the cosine of the azimuthal angle, at the extremity of the arc s, will thus be

$$s \tan. \psi_{,} \left\{ 1 - \alpha u_{,}' + \alpha \left(\frac{d u_{,}'}{d \psi}\right) \tan. \psi_{,} - \frac{\alpha \left(\frac{d^2 u_{,}'}{d \varphi^2}\right)}{\cos.^2 \psi_{,}} \right\}.$$

This cosine being very small, it may be taken for the complement of the azimuthal angle, which consequently is equal to

$$100^\circ - s \tan. \psi_{,} \left\{ 1 - \alpha u_{,}' + \alpha \left(\frac{d u_{,}'}{d \psi}\right) \tan. \psi_{,} - \frac{\alpha \left(\frac{d^2 u_{,}'}{d \varphi^2}\right)}{\cos.^2 \psi_{,}} \right\}.$$

For the greater exactness, we must add to this angle that part depending on s^3, and independent of α, which we obtain in the hypothesis of the earth's sphericity. This part is equal to $\frac{1}{3} s^3 (\frac{1}{2} + \tan.^2 \psi_{,}) \tan. \psi_{,}$, Thus the azimuthal angle at the extremity of the arc s is equal to

$$100^\circ - s \tan. \psi_{,} \left\{ 1 - \alpha u_{,}' + \alpha \left(\frac{d u_{,}'}{d \psi}\right) \tan. \psi_{,} - \frac{\alpha \left(\frac{d^2 u_{,}'}{d \varphi^2}\right)}{\cos.^2 \psi_{,}} - \frac{1}{3} s^2 (\tfrac{1}{2} \tan.^2 \psi_{,}) \right\}.$$

The radius of curvature of the Geodesic line, forming any angle whatever with the plane of the meridian, is equal to

$$\frac{d s^2}{\sqrt{(d^2 x)^2 + (d^2 y)^2 + (d^2 z)^2}},$$

d s being supposed constant; let R be this radius. The equation $x^2 + y^2 + z^2 = 1 + 2 \alpha u'$ gives

$$x d^2 x + y d^2 y + z d^2 z = - d s^2 + \alpha d^2 u';$$

if we add the square of this equation to the squares of equations (O), we shall have, rejecting terms of the order α^2,

$$(x^2 + y^2 + z^2) \{(d^2 x)^2 + (d^2 y)^2 + (d^2 z)^2\} = d s^4 - 2 \alpha d s^2 d^2 u'$$

whence we derive

$$R = 1 + \alpha u' + \alpha \frac{d^2 u'}{d s^2}.$$

In the direction of the meridian, we have

$$\alpha . \frac{d^2 s u'}{d \ ^2} = \alpha \left(\frac{d^2 u'}{d \psi^2}\right);$$

wherefore

$$R = 1 + \alpha u' + \alpha \left(\frac{d^2 u'}{d \psi^2}\right).$$

In the direction perpendicular to the meridian, we have by what precedes,

$$\alpha \cdot \frac{d^2 u'_{,}}{d s^2} = \frac{\alpha \left(\frac{d^2 u'_{,}}{d \varphi^2}\right)}{\cos.^2 \psi_{,}} - \alpha \left(\frac{d u'_{,}}{d \varphi}\right) \tan. \psi_{,};$$

wherefore

$$R = 1 + \alpha u'_{,} - \alpha \left(\frac{d u'_{,}}{d \psi}\right) \tan. \psi_{,} + \frac{\alpha \left(\frac{d^2 u'_{,}}{d \varphi^2}\right)}{\cos.^2 \psi_{,}}.$$

If in the preceding expression of $V - V_{,}$, we make $\frac{s}{R} = s'$, it takes this very simple form relative to a sphere of the radius R,

$$V - V_{,} = \frac{s'}{\cos. \psi_{,}} \cdot \left\{1 - \frac{1}{3} s'^2 . \tan.^2 \psi_{,}\right\}.$$

The expression of the azimuthal angle becomes

$$100^{\circ} - s' \tan. \psi_{,} \{1 - \tfrac{1}{3} s'^2 (\tfrac{1}{2} + \tan.^2 \psi_{,})\}.$$

Call λ, the angle which the first side of the *Geodesic line* forms with the plane corresponding to the celestial meridian, we shall have

$$\frac{d^2 u'}{d s^2} = \left(\frac{d u'}{d \varphi}\right) \frac{d^2 \varphi}{d s^2} + \left(\frac{d u'}{d \psi}\right) \frac{d^2 \psi}{d s^2} + \left(\frac{d^2 u'}{d \varphi^2}\right) \frac{d \varphi^2}{d s^2} + 2 \left(\frac{d^2 u'}{d \varphi d \psi}\right) \frac{d \varphi}{d s} \frac{d \psi}{d s} + \left(\frac{d^2 u'}{d \psi^2}\right) \frac{d \psi^2}{d s^2}.$$

But supposing the earth a sphere, we have

$$\frac{d \varphi_{,}}{d s} = \frac{\sin. \lambda}{\cos. \psi_{,}}; \quad \frac{d^2 \varphi_{,}}{d s^2} = \frac{2 \sin. \lambda \cos. \lambda}{\cos \psi_{,}} \tan. \psi_{,};$$

$$\frac{d \psi_{,}}{d s} = \cos. \lambda; \quad \frac{d^2 \psi_{,}}{d s^2} = - \sin.^2 \lambda \tan. \psi_{,};$$

wherefore,

$$\frac{d^2 u'_{,}}{d s^2} = 2 \frac{\sin. \lambda \text{ eos. } \lambda}{\cos. \psi_{,}} \left\{ \left(\frac{d u'_{,}}{d \varphi}\right) \tan. \psi_{,} + \left(\frac{d^2 u'_{,}}{d \varphi d \psi}\right) \right\} - \sin.^2 \lambda \tan. \psi_{,} \left(\frac{d u'_{,}}{d \psi}\right)$$
$$+ \left(\frac{d^2 u'_{,}}{d \varphi^2}\right) \frac{\sin.^2 \lambda}{\cos.^2 \psi_{,}} + \left(\frac{d^2 u'_{,}}{d \psi^2}\right) \cos.^2 \lambda;$$

the radius of curvature R, in the direction of this *Geodesic line*, is therefore

$$1 + \alpha u'_{,} + 2 \alpha \frac{\sin. \lambda \cos. \lambda}{\cos. \psi_{,}} \left\{ \left(\frac{d u'_{,}}{d \varphi}\right) \tan. \psi_{,} + \left(\frac{d^2 u'_{,}}{d \varphi d \psi}\right) \right\} - \alpha \sin.^2 \lambda \tan. \psi_{,} \left(\frac{d u'_{,}}{d \psi}\right)$$
$$+ \alpha \left(\frac{d^2 u'_{,}}{d \varphi^2}\right) \cdot \frac{\sin.^2 \lambda}{\cos.^2 \psi_{,}} + \alpha \left(\frac{d^2 u'_{,}}{d \psi^2}\right) \cos.^2 \lambda.$$

To abridge this, let

$$K = 1 + \alpha u'_{,} - \tfrac{1}{2} \alpha \tan. \psi_{,} \left(\frac{d u'_{,}}{d \psi}\right) + \tfrac{1}{2} \cdot \frac{\alpha \left(\frac{d^2 u'_{,}}{d \varphi^2}\right)}{\cos.^2 \psi_{,}} + \tfrac{1}{2} \alpha \left(\frac{d^2 u'_{,}}{d \psi^2}\right);$$

$$A = \frac{\alpha}{\cos. \psi_{,}} \cdot \left\{ \left(\frac{d\, u_{,}'}{d\, \varphi}\right) \tan. \psi_{,} + \left(\frac{d^2\, u_{,}'}{d\, \varphi\, d\, \psi_{,}}\right) \right\};$$

$$B = \frac{\alpha}{2} \tan. \psi_{,} \left(\frac{d\, u_{,}'}{d\, \psi_{,}}\right) - \frac{\frac{\alpha}{2}\left(\frac{d^2\, u_{,}'}{d\, \varphi^2}\right)}{\cos.^2 \psi_{,}} + \frac{\alpha}{2}\left(\frac{d^2\, u_{,}'}{d\, \psi_{,}^2}\right);$$

we shall have

$$R = K + A \sin. 2\, \lambda + B \cos. 2\, \lambda.$$

The observations of azimuthal angles, and of the difference of the latitudes at the extremities of the two *geodesic lines*, one measured in the direction of the meridian, and the other in the direction perpendicular to the meridian, will give, by what precedes, the values of A, B and K; for the observations give the radii of curvature in these two directions. Let R, and R′ be these radii; we shall have

$$K = \frac{R' + R''}{2};$$

$$B = \frac{R' - R''}{2};$$

and the value of A will be determined, either by the azimuth of the extremity of the arc measured in the direction of the meridian, or by the difference in latitude of the two extremities of the arc measured in a direction perpendicular to the meridian. We shall thus get the radius of curvature of the *geodesic line*, whose first side forms any angle whatever with the meridian.

If we call 2 E, an angle whose tangent is $\frac{A}{B}$, we shall have

$$R = K + \sqrt{A^2 + B^2} . \cos. (2\, \lambda - 2\, E);$$

the greatest radius of curvature corresponds with $\lambda = E$; the corresponding geodesic line forms therefore the angle E, with the plane of the meridian. The least radius of curvature corresponds with $\lambda = 100^{\circ} + E$; let r be the least radius, and r′ the greatest, we shall have

$$R = r + (r' - r) \cos.^2 (\lambda - E),$$

$\lambda - E$ being the angle which the *geodesic line* corresponding to R, forms with that which corresponds with r′.

We have already observed, that at each point of the earth's surface, we may conceive an osculatory ellipsoid upon which the degrees, in all directions, are sensibly the same to a small extent around the point of osculation. Express the radius of this ellipsoid by the function

$$1 - \alpha \sin.^2 \psi \{1 + h \cos. 2\, (\varphi + \beta)\},$$

the longitudes φ being reckoned from a given meridian. The expression of the terrestrial meridian measured in the direction of the meridian, will be, by what precedes,

$$\epsilon - \frac{\alpha\,\epsilon}{2} \cdot \{1 + h \cos. 2\,(\varphi + \beta)\} \cdot \{1 + 3 \cos. 2\,\psi - 3\,\epsilon \sin. 2\,\psi\}.$$

If the measured arc is considerable, and if we have observed, as in France, the latitudes of some points intermediate between the extremity; we shall have by these measures, both the length of the radius taken for unity, and the value of $\alpha\ \{1 + h \cos. 2\,(\varphi + \beta)\}$. We then have, by what precedes,

$$\varpi = -2\,\alpha\, h \,.\, \epsilon \,.\, \frac{\tan.^2 \psi\,(1 + \cos.^2 \psi)}{\cos. \psi} \,.\, \sin. 2\,(\varphi + \beta);$$

the observation of the azimuthal angles at the two extremities of the arc will give $\alpha\, h \sin. 2\,(\varphi + \beta)$. Finally, the degree measured in the direction perpendicular to the meridian, is

$$1^{\circ} + 1^{\circ} . \alpha\,\{1 + h \cos. 2\,(\varphi + \beta)\} \sin.^2 \psi + 4^{\circ} . \alpha\, b \tan.^2 \psi \cos. 2\,(\varphi + \beta);$$

the measure of this degree will therefore give the value of $\alpha\, h \sin. 2\,(\varphi+\beta)$. Thus the osculatory ellipsoid will be determined by these several measures: it would be necessary for an arc so great, to retain the square of ϵ in the expression of the angle ϖ; and the more so, if, as it has been observed in France, the azimuthal angle does not vary proportionally to the measured arc: at the same time we must add a term of the form $\alpha\, k \sin. \psi \cos. \psi \sin. (\varphi + \beta')$, to get the most general expression of this radius.

583. The elliptic figure is the most simple after that of the sphere: we have seen above that this ought to be the figure of the earth and planets, on the supposition of their being originally fluid, if besides they have retained their primitive figure. It was natural therefore to compare with this figure the measured degrees of the meridian; but this comparison has given for the figure of the meridians different ellipses, and which disagree too much with observations to be admissible. However, before we renounce entirely the elliptic, we must determine that in which the greatest *defect* of the measured degrees, is smaller than in every other elliptic figure, and see whether it be within the limits of the errors of observations. We arrive at this by the following method.

Let $a^{(1)}$, $a^{(2)}$, $a^{(3)}$, &c. be the measured degrees of the meridians; $p^{(1)}$, $p^{(2)}$, $p^{(3)}$, &c. the squares of the sines of the corresponding latitudes: suppose that in the ellipse required, the degree of the meridian is expressed by the formula $z + p\,y$; calling $x^{(1)}$, $x^{(2)}$, $x^{(3)}$, &c. the errors of observation, we shall have the following equations, in which we shall suppose that $p^{(1)}$, $p^{(2)}$, $p^{(3)}$, &c. form an increasing progression,

$$a^{(1)} - z - p^{(1)} y = x^{(1)}$$
$$a^{(2)} - z - p^{(2)} y = x^{(2)} \quad \ldots\ldots\ldots \quad (A)$$
$$- - - - - - - - - - - - -$$
$$a^{(n)} - z - p^{(n)} y = x^{(n)}$$

n being the number of measured degrees.

We shall eliminate from these equations the unknown quantities z and y, and we shall have n — 2 equations of condition, between the n errors $x^{(1)}, x^{(2)}, \ldots\ldots x^{(n)}$. We must, however, determine that system of errors, in which the greatest, abstraction being made of the signs, is less than in every other system.

First suppose that we have only one equation of condition, which may be represented by

$$a = m\,x^{(1)} + n\,x^{(2)} + p\,x^{(3)} + \&c.$$

a being positive. We shall have the system of the values of $x^{(1)}$, $x^{(2)}$, &c. which gives, not regarding signs, the least value to the greatest of them; supposing them all nearly equal, and to the quotient of *a* divided by the sum of the coefficients, m, n, p, &c. taken positively. As to the sign which each quantity ought to have, it must be the same as that of its coefficient in the proposed equation.

If we have two equations of condition between the errors, the system which will give the smallest value possible to the greatest of them will be such that, signs being abstracted, all the errors will be equal to one another, with the exception of one only which will be smaller than the rest, or at least not greater. Supposing therefore that $x^{(1)}$ is this error, we shall determine it in function $x^{(2)}$, $x^{(3)}$, &c. by means of one of the proposed equations of condition; then substituting this value of $x^{(1)}$ in the other equation of condition, we shall form one between $x^{(2)}$, $x^{(3)}$, &c.; which represent by the following

$$a = m\,x^{(2)} + n\,x^{(3)} + \&c.$$

a being positive; we shall have, as above, the values of $x^{(2)}$, $x^{(3)}$, &c. by dividing a by the sum of the coefficients m, n, &c. taken positively, and by giving successively to the quotient the signs of m, n, &c. These values substituted in the expression of $x^{(1)}$ in terms of $x^{(2)}$, $x^{(3)}$, &c. will give the value of $x^{(1)}$; and if this value, abstracting signs, is not greater than that of $x^{(2)}$, this system of values will be that which we must adopt; but if greater, then the supposition that $x^{(1)}$ is the least error, is not legitimate, and we must successively make the same supposition as to $x^{(2)}$, $x^{(3)}$, &c. until we arrive at an error which is in this respect satisfactory.

If we have three equations of condition between the errors; the system which will give the smallest value possible to the greatest of them, will be such, that, abstracting signs, all the errors will be equal, with exception of two, which will be less than the others.

Supposing therefore that $x^{(1)}$, $x^{(2)}$ are these two errors, we shall climinate them from the third of the equations of condition by means of the other two, and we shall have an equation between the errors $x^{(3)}$, $x^{(4)}$, &c.: represent it by

$$a = m\,x^{(3)} + n\,x^{(4)} + \&c.$$

a being positive. We shall have the values of $x^{(3)}$, $x^{(4)}$, &c. by dividing a by the sum of the coefficients m, n, &c. taken positively, and by giving successively to the quotient, the signs of m, n, &c. These values substituted in the expressions of $x^{(1)}$, and of $x^{(2)}$ in terms of $x^{(3)}$, $x^{(4)}$, &c. will give the values of $x^{(1)}$, and of $x^{(2)}$, and if these last values, abstracting signs, do not surpass $x^{(3)}$, we shall have the system of errors, which we ought to adopt; but if one of these values exceed $x^{(3)}$, the supposition that $x^{(1)}$, and $x^{(2)}$ are the smallest errors is not legitimate, and we must use the same supposition upon another combination of errors $x^{(1)}$, $x^{(2)}$, &c. taken two and two, until we arrive at a combination in which this suppositiou is legitimate. It is easy to extend this method to the case where we should have four or more equations of condition, between the errors $x^{(1)}$, $x^{(2)}$, &c. These errors being thus known, it will be easy to obtain the values of z and y.

The method just exposed, applies to all questions of the same nature; thus, having the number n of observations upon a comet, we may by this means determine that parabolic orbit, in which the greatest error is, abstracting signs, less than in any other parabolic orbit, and thence recognise whether the parabolic hypothesis can represent these observations. But when the number of observations is considerable, this method becomes too tedious, and we may in the present problem, easily arrive at the required system of errors, by the following method.

Conceive that $x^{(i)}$, abstracting signs, is the greatest of the errors $x^{(1)}$, $x^{(2)}$, &c.; we shall first observe, that therein must exist another error $x^{(i')}$, equal, and having a contrary sign to $x^{(i)}$; otherwise we might, by making z to vary properly in the equation

$$a^{(i)} - z - p^{(i)}.\,y = x^{(i)},$$

diminish the error $x^{(i)}$, retaining to it the property of being the extreme error, which is against the hypothesis. Next we shall observe that $x^{(i)}$ and $x^{(i')}$ being the two extreme errors, one positive, and the others negative, and equal to one another, there ought to exist a third error $x^{(i'')}$, equal, abstracting signs, to $x^{(i)}$. In fact, if we take the equation corresponding to $x^{(i)}$, from the equation corresponding to $x^{(i')}$, we shall have

$$a^{(i')} - a^{(i)} - \{p^{(i')} - p^{(i)}\}.\, y = x^{(i')} - x^{(i)}.$$

The second member of this equation is, abstracting signs, the sum of the extreme errors, and it is clear, that in varying y suitably, we may diminish it, preserving to it the property of being the greatest of the sums which we can obtain by adding or subtracting the errors $x^{(1)}$, $x^{(2)}$, &c. taken two and two; provided there is no third error $x^{(i'')}$ equal, abstracting signs, to $x^{(i)}$; but the sum of the extreme errors being diminished, and these errors being made equal, by means of the value of z, each of these errors will be diminished, which is contrary to the hypothesis. There exists therefore three errors $x^{(i)}$, $x^{(i')}$, $x^{(i'')}$ equal to one another, abstracting signs, and of different signs the one from the other two.

Suppose that this one is $x^{(i')}$; then the number i′ will fall between the two numbers i and i″. To show this, let us imagine that it is not the case, and that i′ is below or above both the numbers i, i″. Taking the equation corresponding to i′, successively from the two equations corresponding to i and to i″, we shall have

$$a^{(i)} - a^{(i')} - (p^{(i)} - p^{(i')})\, y = x^{(i)} - x^{(i')};$$
$$a^{(i'')} - a^{(i')} - (p^{(i'')} - p^{(i')})\, y = x^{(i'')} - x^{(i')}.$$

The second members are equal and have the same sign; these are also, abstracting signs, the sum of the extreme errors; but it is evident, that varying y suitably, we may diminish each of these sums, since the coefficient of y, has the same sign in the two first members: moreover, we may, by varying z properly, preserve to $x^{(i')}$ the same value; $x^{(i)}$ and $x^{(i'')}$ will therefore then be, abstracting signs, less than $x^{(i')}$ which will become the greatest of the errors without having an equal; and in this case, we may, as we have seen, diminish the extreme error; which is contrary to the hypothesis. Thus the number i′ ought to fall between i and i″.

Let us now determine which of the errors $x^{(1)}$, $x^{(2)}$, &c. are the extreme errors. For that purpose, take the first of the equations (A) successively from the following ones, and we shall have this series of equations,

$$a^{(2)} - a^{(1)} - (p^{(2)} - p^{(1)})\, y = x^{(2)} - x^{(1)},$$
$$a^{(3)} - a^{(1)} - (p^{(3)} - p^{(1)})\, y = x^{(3)} - x^{(1)}; \quad \ldots \quad (B)$$
&c.

Suppose y infinite; the first members of these equations will be negative, and then the value of $x^{(1)}$ will be greater than $x^{(2)}$, $x^{(3)}$, &c.: diminishing y continually, we shall at length arrive at a value that will render positive one of the first members, which, before arriving at this state, will be nothing. To know which of these members first becomes equal to zero, we shall form the quantities,

$$\frac{a^{(2)} - a^{(1)}}{p^{(2)} - p^{(1)}};\ \frac{a^{(3)} - a^{(1)}}{p^{(3)} - p^{(1)}};\ \frac{a^{(4)} - a^{(1)}}{p^{(4)} - p^{(1)}};\ \&c.$$

Call $\beta^{(1)}$ the greatest of these quantities, and suppose it to be $\frac{a^{(r)} - a^{(1)}}{p^{(r)} - p^{(1)}}$: if there are many values equal to $\beta^{(1)}$, we shall consider that which corresponds to the number r the greatest, substituting $\beta^{(1)}$ for y, in the $(r - 1)^{th}$ of the equations (B), $x^{(r)}$ will be equal to $x^{(1)}$, and diminishing y, it will be equal to $x^{(1)}$, the first member of this equation then becoming positive. By the successive diminutions of y, this member will increase more rapidly than the first members of the equations which precede it; thus, since it becomes nothing when the preceding ones are still negative, it is clear that, in the successive diminutions of y, it will always be the greatest which proves that $x^{(r)}$ will be constantly greater than $x^{(1)}$, $x^{(2)}, \ldots x^{(r-1)}$, when y is less than $\beta^{(1)}$.

The first members of the equations (B) which follow the $(r - 1)^{th}$ will be at first negative, and whilst that is the case, $x^{(r+1)}$, $x^{(r+2)}$, &c. will be less than $x^{(1)}$, and consequently less than $x^{(r)}$, which becomes the greatest of all the errors $x^{(1)}, x^{(2)}, \ldots x^{(n)}$, when y begins to be less than $\beta^{(1)}$. But continuing to diminish y, we shall get a value of it, such that some of the errors $x^{(r+1)}$, $x^{(r+2)}$, &c. begin to *exceed* $x^{(r)}$.

To determine this value of y, we shall take the rth of equations (A) successively from the following ones, and we shall have

$$a^{(r+1)} - a^{(r)} - \{p^{(r+1)} - p^{(r)}\}\, y = x^{(r+1)} - x^{(r)};$$
$$a^{(r+2)} - a^{(r)} - \{p^{(r+2)} - p^{(r)}\}\, y = x^{(r+2)} - x^{(r)}.$$

Then we shall form the quantities

$$\frac{a^{(r+1)} - a^{(r)}}{p^{(r+1)} - p^{(r)}};\ \frac{a^{(r+2)} - a^{(r)}}{p^{(r+2)} - p^{(r)}};\ \&c.$$

Call $\beta^{(2)}$, the greatest of these quantities, and suppose that it is $\frac{a^{(r')} - a^{(r)}}{p^{(r')} - p^{(r)}}$: if many of these quantities are equal to $\beta^{(2)}$, we shall suppose that r′ is the greatest of the numbers to which they correspond. Then $x^{(r)}$ will be the greatest of the errors $x^{(1)}$, $x^{(2)}$, &c. . . . $x^{(n)}$ so long as y is comprised between $\beta^{(1)}$, and $\beta^{(2)}$; but when by diminishing y, we shall arrive at $\beta^{(2)}$; then $x^{(r')}$ will begin to exceed $x^{(r)}$, and to become the greatest of the errors.

To determine within what limits we shall form the quantities

$$\frac{a^{(r'+1)} - a^{(r')}}{p^{(r'+1)} - p^{(r')}};\ \frac{a^{(r'+2)} - a^{(r')}}{p^{(r'+2)} - p^{(r')}};\ \&c.$$

Let $\beta^{(3)}$ be the greatest of these quantities, and suppose that it is

$\frac{a^{(r''+1)} - a^{(r'')}}{p^{(r''+1)} - p^{(r'')}}$: if several of the quantities are equal to $\beta^{(3)}$, we shall suppose that r'' is the greatest of the numbers to which they correspond, $x^{(r)}$ will be the greatest of all the errors from $y = \beta^{(2)}$, to $y = \beta^{(3)}$. When $y = \beta^{(3)}$, then $x^{(r'')}$ begins to be this greatest error. Thus proceding, we shall form the two series,

$$x^{(1)};\ x^{(r)};\ x^{(r')};\ x^{(r'')};\ \ldots\ x^{(n)}$$

$$\infty;\ \beta^{(1)};\ \beta^{(2)};\ \beta^{(3)};\ \ldots\ \beta^{(q)}; -\infty;\ \ldots\ldots \quad (C)$$

The first indicates the errors $x^{(1)}$, $x^{(r)}$, $x^{(r')}$, &c. which become successively the greatest: the second series formed of decreasing quantities, indicates the limits of y, between which these errors are the greatest; thus, $x^{(1)}$ is the greatest error from $y = \infty$, to $y = \beta^{(1)}$; $x^{(r)}$ is the greatest error from $y = \beta^{(1)}$, to $y = \beta^{(2)}$; $x^{(r')}$ is the greatest error from $y = \beta^{(2)}$, to $y = \beta^{(3)}$, and so on.

Resume now the equations (B) and suppose y negative and infinite. The first members of these equations will be positive, $x^{(1)}$ will therefore then be the least of the errors $x^{(1)}$, $x^{(2)}$, &c.: augmenting y continually, some of these members will become negative, and then $x^{(1)}$ will cease to be the least of the errors. If we apply here the reasoning just used in the case of the greatest errors, we shall see that if we call $\lambda^{(1)}$ the least of the quantities

$$\frac{a^{(2)} - a^{(1)}}{p^{(2)} - p^{(1)}};\ \frac{a^{(3)} - a^{(1)}}{p^{(3)} - p^{(1)}};\ \frac{a^{(4)} - a^{(1)}}{p^{(4)} - p^{(1)}};\ \&c.$$

and if we suppose that it is $\frac{a^{(s)} - a^{(1)}}{p^{(s)} - p^{(1)}}$, s being the greatest of the numbers to which $\lambda^{(1)}$ corresponds, if several of these quantities are equal to $\lambda^{(1)}$, $x^{(1)}$ will be the least of the errors from $y = -\infty$, to $y = \lambda^{(1)}$. In like manner if we call $\lambda^{(2)}$ the least of the quantities

$$\frac{a^{(s+1)} - a^{(s)}}{p^{(s+1)} - p^{(s)}};\ \frac{a^{(s+2)} - a^{(s)}}{p^{(s+2)} - p^{(s)}};\ \&c.$$

and suppose it to be $\frac{a^{(s')} - a^{(s)}}{p^{(s')} - p^{(s)}}$, s' being the greatest of the numbers to which $\lambda^{(2)}$ corresponds, if several of these quantities are equal to $\lambda^{(2)}$; $x^{(s)}$ will be the smallest of the errors from $y = \lambda^{(1)}$, to $y = \lambda^{(2)}$; and so forth. In this manner we shall form the two series

$$x^{(1)};\ x^{(s)};\ x^{(s')};\ x^{(s'')};\ \ldots\ x^{(p)}$$

$$-\infty;\ \lambda^{(1)};\ \lambda^{(2)};\ \lambda^{(3)};\ \ldots\ \lambda^{(q')};\ \infty;\ \ldots\ldots \quad (D)$$

The first indicates the errors $x^{(1)}$, $x^{(s)}$, $x^{(s')}$, &c. which are successively the least as we augment y: the second series formed of increasing terms, indicates the limits of the values of y between which each of these errors is the least; thus $x^{(1)}$ is the least of the errors from $y = -\infty$, to $y = \lambda^{(1)}$ $x^{(s)}$ is the least of the errors, from $y = \lambda^{(1)}$, to $y = \lambda^{(2)}$, and thus of the rest.

Hence the value of y which, to the required ellipse, will be one of the quantities $\beta^{(1)}$, $\beta^{(2)}$, $\beta^{(3)}$; &c. $\lambda^{(1)}$, $\lambda^{(2)}$, &c.; it will be in the first series, if the two extreme errors of the same sign are positive. In fact, these two errors being then the greatest, they are in the series $x^{(1)}$, $x^{(r)}$, $x^{(r')}$, &c.; and since one and the same value of y renders them equal they ought to be consecutive, and the value of y which suits them, can only be one of the quantities $\beta^{(1)}$, $\beta^{(2)}$, &c.; because two of these errors cannot at the same time be made equal and the greatest, except by one only of these quantities. Here, however, is a method of determining that of the quantities $\beta^{(1)}$, $\beta^{(2)}$, &c. which ought to be taken for y.

Conceive, for example, that $\beta^{(3)}$ is this value; then there ought to be found by what precedes between $x^{(r')}$, and $x^{(r'')}$, an error which will be the *minimum* of all the errors, since $x^{(r')}$, and $x^{(r'')}$ will be the *maxima* of these errors; thus in the series $x^{(1)}$, $x^{(s)}$, $x^{(s')}$, &c. some one of the numbers s, s', &c. will be comprised between r and r'. Suppose it to be s. That $x^{(s)}$ may be the last of the value of y, it ought to be comprised between $\lambda^{(1)}$ and $\lambda^{(2)}$; therefore if $\beta^{(3)}$ is comprised by these limits, it will be the value sought of y, and it will be useless to seek others. In fact, suppose we take that of the equations (A), which answers to $x^{(s)}$ successively from the two equations which respond to $x^{(r')}$ and to $x^{(r'')}$; we shall have

$$a^{(r')} - a^{(s)} - \{p^{(r')} - p^{(s)}\}\, y = x^{(r')} - x^{(s)};$$
$$a^{(r'')} - a^{(s)} - \{p^{(r'')} - p^{(s)}\}\, y = x^{(r'')} - x^{(s)}.$$

All the members of these equations being positive, by supposing $y = \beta^{(3)}$, it is clear, that if we augment y, the quantity $x^{(r')} - x^{(s)}$ will increase; the sum of the extreme errors, taken positively, will be therefore augmented. If we diminish y, the quantity $x^{(r'')} - x^{(s)}$ will be augmented, and consequently also the sum of their extremes; $\beta^{(3)}$ is therefore the value of y, which gives the least of these sums; whence it follows that it is the only one which satisfies the problem.

We shall try in this way the values of $\beta^{(1)}$, $\beta^{(2)}$, $\beta^{(3)}$, &c., which is easily done by inspection; and if we arrive at a value which fulfils the preceding conditions, we shall be assured of the value required of y.

If any of these values of β does not fulfil these conditions, then this value of y will be some one of the terms of the series $\lambda^{(1)}$, $\lambda^{(2)}$, &c. Conceive, for example, that it is $\lambda^{(2)}$, the two extreme errors $x^{(s)}$ and $x^{(s')}$ will then be negative, and it will have, by what precedes, an intermediate error, which will be a *maximum*, and which will fall consequently in the series $x^{(1)}$, $x^{(r)}$, $x^{(r')}$, &c. Suppose that this is $x^{(r)}$, r being then necessarily comprised between s and s'; $\lambda^{(2)}$ ought, therefore, to be comprised between $\beta^{(1)}$ and $\beta^{(2)}$. If that is the case, this will be a proof that $\lambda^{(2)}$ is the value required of y. We shall try thus all the terms of the series $\lambda^{(2)}$, $\lambda^{(3)}$, $\lambda^{(4)}$, &c. up to that which fulfils the preceding conditions.

When we shall have thus determined the value of y, we shall easily obtain that of z. For this, suppose that $\beta^{(2)}$ is the value of y, and that the three extreme errors are $x^{(r)}$, $x^{(r')}$, $x^{(s)}$; we shall have $x^{(s)} = -x^{(r)}$, and consequently

$$a^{(r)} - z - p^{(r)} . y = x^{(r)};$$

$$a^{(s)} - z - p^{(s)} . y = -x^{(r)};$$

whence we get

$$z = \frac{a^{(r)} + a^{(s)}}{2} - \frac{p^{(r)} + p^{(s)}}{2} . y;$$

then we shall have the greatest error $x^{(r)}$, by means of the equation

$$x^{(r)} = \frac{a^{(r)} - a^{(s)}}{2} + \frac{p^{(s)} - p^{(r)}}{2} y.$$

584. The ellipse determined in the preceding No. serves to recognise whether the hypothesis of an elliptic figure is in the limits of the errors of observations; but it is not that which the measured degrees indicate with the greatest probability. This last ellipse, it seems, should fulfil the following conditions, viz. 1st, that the sum of the errors committed in the measures of the entire measured arcs be nothing: 2dly, that the sum of these errors, all taken positively, may be a *minimum*. Thus considering the entire ones instead of the degrees which have thence been deduced, we give to each of the degrees by so much the more influence upon the ellipticity which thence results for the earth, as the corresponding arc is considerable, as it ought to be. The following is a very simple method of determining the ellipse which satisfies these two conditions.

Resume the equations (A) of 589, and multiply them respectively by the numbers which express how many degrees the measured arcs contain, and which we will denote by $i^{(1)}$, $i^{(2)}$, $i^{(3)}$, &c. Let A be the sum of the quantities $i^{(1)} . a^{(1)}$, $i^{(2)} . a^{(2)}$, &c. divided by the sum of the numbers $i^{(1)}$, $i^{(2)}$, &c.; let, in like manner, P denote the sum of the quantities $i^{(1)} . p^{(1)}$, $i^{(2)} . p^{(2)}$, &c. divided by the sum of the numbers $i^{(1)}$, $i^{(2)}$, &c.; the condition that the sum of the errors $i^{(1)} . x^{(1)}$, $i^{(2)} . x^{(2)}$, &c. is nothing, gives

$$0 = A - z - P . y.$$

If we take this equation from each of the equations A of the preceding No., we shall have equations of the following form:

$$\left.\begin{array}{l} b^{(1)} - q^{(1)}.\ y = x^{(1)} \\ b^{(2)} - q^{(2)}.\ y = x^{(2)} \\ b^{(3)} - q^{(3)}.\ y = x^{(3)} \\ \&c. \end{array}\right\}; \quad \ldots \quad (O)$$

Form the series of quotients $\frac{b^{(1)}}{q^{(1)}}$, $\frac{b^{(2)}}{q^{(2)}}$, &c. and dispose them according to their order of magnitude, beginning with the greatest; then multiply the equations O, to which they respond, by the corresponding numbers $i^{(1)}$, $i^{(2)}$, &c.; finally, dispose these thus multiplied in the same order as the quotients.

The first members of the equations disposed in this way, will form a series of terms of the form

$$h^{(1)}\,y - c^{(1)};\quad h^{(2)}\,y - c^{(2)};\quad h^{(3)}\,y - c^{(3)};\ \&c. \quad \ldots \quad (P)$$

in which we shall suppose $h^{(1)}$, $h^{(2)}$ positive, by changing the sign of the terms when y has a negative coefficient. These terms are the errors of the measured arcs, taken positively or negatively.

Then it is evident, that in making y infinite, each term of this series becomes infinite; but they decrease as we diminish y, and end by being negative—at first, the first, then the second, and so on. Diminishing y continually, the terms once become negative continue to be so, and decrease without ceasing. To get the value y, which renders the sum of these terms all taken positively a *minimum*, we shall add the quantities $h^{(1)}$, $h^{(2)}$, &c. as far as when their sum begins to surpass the semi-sum of all these quantities; thus calling F this sum, we shall determine r such that

$$h^{(1)} + h^{(2)} + h^{(3)} + \ldots + h^{(r)} > \tfrac{1}{2}\,F;$$
$$h^{(1)} + h^{(2)} + h^{(3)} + \ldots + h^{(r-1)} < \tfrac{1}{2}\,F.$$

We shall then have $y = \frac{c^{(r)}}{h^{(r)}}$, so that the error will be nothing relatively to the same degree which corresponds to that of the equations (O), of which the first member equated to zero, gives this value of y.

To show this, suppose that we augment y by the quantity $\delta\,y$, so that $\frac{c^{(r)}}{h^{(r)}} + \delta\,y$ may be comprised between $\frac{c^{(r-1)}}{h^{(r-1)}}$ and $\frac{c^{(r)}}{h^{(r)}}$. The (r — 1) first terms of the series (P) will be negative, as in the case of $y = \frac{c^{(r)}}{h^{(r)}}$; but in taking them with the sign +, their sum will decrease by the quantity

$$\{h^{(1)} + h^{(2)} \ldots\ldots h^{(r-1)}\}\ \delta\,y.$$

The first term of this series, which is nothing when $y = \frac{c^{(r)}}{h^{(r)}}$, will become positive and equal to $h^{(r)}\,\delta y$; the sum of this term and the following, which are positive, will increase by the quantity

$$\{h^{(r)} + h^{(r+1)} + \&c.\}\,\delta y;$$

but by supposition we have

$$h^{(1)} + h^{(2)} \ldots . h^{(r-1)} < h^{(r)} + h^{(r+1)} + \&c.;$$

the entire sum of the terms of the series (P), all taken positively, will therefore be augmented, and as it is equal to the sum of the errors $i^{(1)}.\,x^{(1)} + i^{(2)}.\,x^{(2)}$, &c. of the entire measured arcs, all taken with the sign $+$, this last sum will be augmented by the supposition of $y = \frac{c^{(r)}}{h^{(r)}} + \delta y$. It is easy to prove, in the same way, that by augmenting y, so as to be comprised between $\frac{c^{(r-1)}}{h^{(r-1)}}$ and $\frac{c^{(r-2)}}{h^{(r-2)}}$, or between $\frac{c^{(r-2)}}{h^{(r-2)}}$ and $\frac{c^{(r-3)}}{h^{(r-3)}}$, &c. the sum of the errors taken with the sign $+$ will be greater than when $y = \frac{c^{(r)}}{h^{(r)}}$.

Now diminish y by the quantity δy so that $\frac{c^{(r)}}{h^{(r)}} - \delta y$ may be comprised between $\frac{c^{(r)}}{h^{(r)}}$ and $\frac{c^{(r+1)}}{h^{(r+1)}}$, the sum of the negative terms of the series (P) will increase, in changing their sign, by the quantity

$$\{h^{(1)} + h^{(2)} + \ldots . h^{(r)}\}\,\delta y;$$

and the sum of the positive terms of the same series will decrease by the quantity

$$\{h^{(r+1)} + h^{(r+2)} + \&c.\}\,\delta y;$$

and since we have

$$h^{(1)} + h^{(2)} + \ldots . h^{(r)} > h^{(r+1)} + h^{(r+2)} + \&c.,$$

it is clear that the entire sum of the errors, taken with the sign $+$, will be augmented. In the same manner we shall see that, by diminishing y, so that it should be between $\frac{c^{(r+1)}}{h^{(r+1)}}$ and $\frac{c^{(r+2)}}{h^{(r+2)}}$, or between $\frac{c^{(r+2)}}{h^{(r+2)}}$ and $\frac{c^{(r+3)}}{h^{(r+3)}}$, &c. the sum of the errors taken with the sign $+$ is greater than when $y = \frac{c^{(r)}}{h^{(r)}}$; this value of y is therefore that which renders this sum a *minimum*.

The value of y gives that of z by means of the equation

$$z = A - P . y.$$

The preceding analysis being founded on the variation of the degrees from the equator to the poles, being proportional to the square of the sine of the latitude, and this law of variation subsisting equally for gravity, it is clear that it applies also to observations upon the length of the seconds' pendulum.

The practical application of the preceding theory will fully establish its soundness and utility. For this purpose, ample scope is afforded by the actual admeasurements of arcs on the earth's surface, which have been made at different times and in different countries. Tabulated below you have such results as are most to be depended on for care in the observations, and for accuracy in the calculations.

Latitudes.	Lengths of Degrees.	Where made.	By whom made.
0°.0000	25538ᴿ.85	Peru.	Bouguer.
37 .0093	25666 .65	Cape of Good Hope	La Caille.
43 .5556	25599 .60	Pennsylvania.	Mason & Dixon.
47 .7963	25640 .55	Italy.	Boscovich & le Maire.
51 .3327	25658 .28	France.	Delambre & Mechain.
53 .0926	25683 .30	Austria.	Liesganig.
73 .7037	25832 .25	Laponia.	Clairaut, &c.

SUPPLEMENT

TO

BOOK III.

FIGURE OF THE EARTH.

585. If a fluid body had no motion about its axis, and all its parts were at rest, it would put on the form of a sphere; for the pressures on all the columns of fluid upon the central particle would not be equal unless they were of the same length. If the earth be supposed to be a fluid body, and to revolve round its axis, each particle, besides its gravity, will be urged by a centrifugal force, by which it will have a tendency to recede from the axis. On this account, Sir Isaac Newton concluded that the earth must put on a spheroidical form, the polar diameter being the

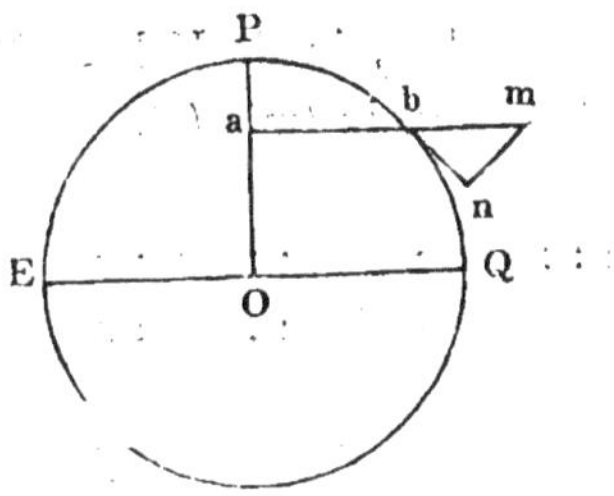

shortest. Let P E Q represent a section of the earth, P p the axis, E Q the equator, (b m) the centrifugal force of a part revolving at (b). This force is resolved into (b n), (n m), of which (b n) draws fluid from (b) to Q, and therefore tends to diminish P O, and increases E Q.

It must first be considered what will be the form of the curve P E p, and then the ratio of P O : G O may be obtained.

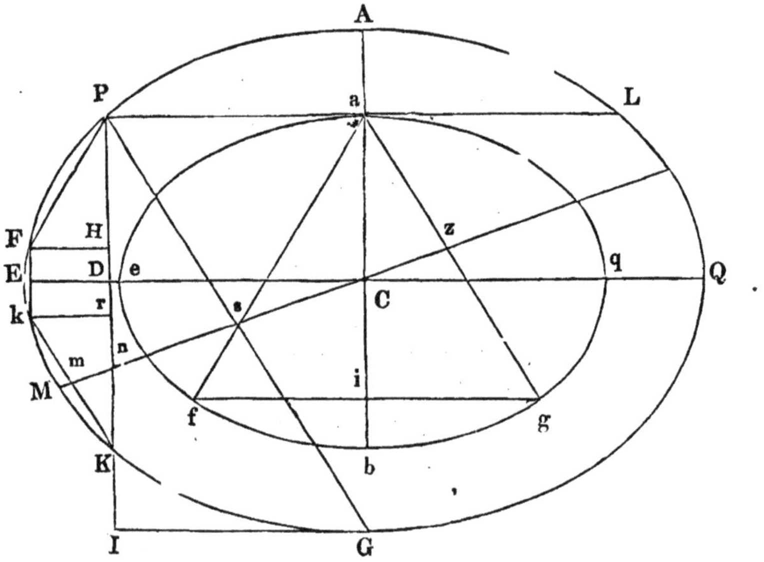

586. Lemma, Let E A Q, e a q, be similar and concentric ellipses, of which the interior is touched at the extremity of the minor axis by P a L; draw a f, a g, making any equal angle with a C; draw P F and P G respectively parallel to a f, a g; then will $PF + PG = af + ag$. For draw P K, F k perpendicular to E Q, and F H, k r perpendicular to P K, $\therefore FE = EK$, $\therefore HD = Dr$ and $PD = DK$, $\therefore PH = Kr$; also $FH = Kr$, $\therefore$ if K k be joined, $Kk = PF$; draw the diameter M C z bisecting K k, G P, a g, in (m), (s), (z).

Then

$$Km : Kn :: Ps : Pn :: az : aC :: ag : ab.$$

$$\therefore Km + Ps : Kn + Pn :: ag : ab$$

but

$$Kn + nP = KP = 2\,PD = 2\,aC = ab \;\therefore\; Km + Ps = ag.$$

$$\therefore 2\,Km + 2\,Ps = 2\,ag, \text{ or } PF + PG = ag + af.$$

Cor. $PH + PI = 2\,ai$. For

$$PF : PH :: PG : PI :: ag : ai.$$

$$\therefore PF + PG : PH + PI :: ag : ai :: 2\,ag : 2\,ai.$$

but

$$PF + PG = 2\,ag, \;\therefore PH + PI = 2\,ai.$$

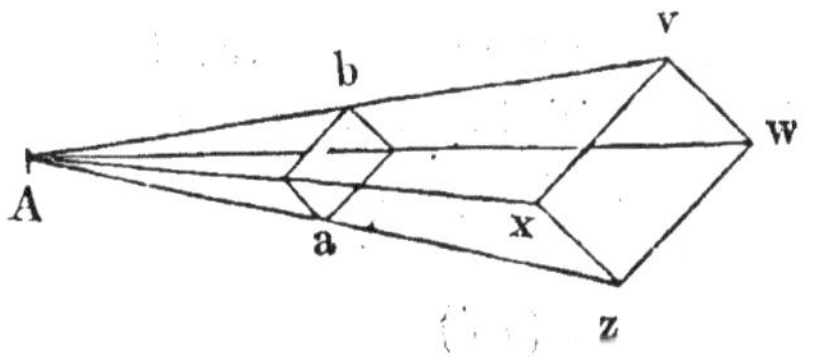

587. The attraction of a particle A towards any pyramid, the area of whose base is indefinitely small, ∝ length, the angle A being given, and the attraction to each particle varying as $\frac{1}{\text{distance}^2}$.

For let

$$a = \text{area } (v\,x\,z\,w)$$
$$m = (A\,z)$$
$$x = (A\,a)$$

Then section $a\,b = \frac{\text{section } v\,x\,z\,w\,.\,(A\,a)^2}{(A\,z)^2} = \frac{a\,x^2}{m^2}$

$$\therefore \varphi'\,.\,\text{attraction} = \frac{a\,x^2\,x'}{m^2\,x^2} = \frac{a\,x'}{m^2}$$

$$\therefore \text{attraction} = \frac{a\,x}{m^2}.$$

∴ attractions of particles at vertices of similar pyramids ∝ lengths.

588. If two particles be similarly situated in respect to two similar solids, the attraction to the solids ∝ lengths of solids.

For if the two solids be divided into similar pyramids, having the particles in the vertices, the attractions to all the corresponding pyramids ∝ their lengths ∝ lengths of solids, since the pyramids being similarly situated in the two similar solids, their lengths must be as the lengths of the solids: ∴ whole attractions ∝ lengths of the solids, or as any two lines similarly situated in them.

Cor. 1. Attraction of (a) to the spheroid a q f: attraction of A to A Q F : : a C : A C.

Cor. 2. The gravitation of two particles P and p in one diameter P C are proportional to their distances from the center. For the gravitation of (p) is the same as if all the matter between the surfaces A Q E, a q e, were taken away (Sect. XIII. Prop. XCI. Cor. 3.) ∴ P and p are similarly situated in similar solids, ∴ attractions on P and p are proportional to P C and p C, lines similarly situated in similar solids.

589. All particles equally distant from E Q gravitate towards E Q with equal forces.

For P G and P F may be considered as the axes of two very slender pyramids, contained between the plane of the figure and another plane, making a very small angle with it. In the same manner we may conceive of (a f) and (a g). Now the gravity of P to these pyramids is as P F + P G; and in the direction P d is as P H + P I. Again, the gravity of (a) to the pyramids (a f), (a g) is as (a f + a g), or in the direction (a i) as 2 a i; but P H + P I = 2 a i : ∴ gravity of P in the direction P d = gravity of (a) in the same direction.

It is evident, by carrying the ordinate (f g) along the diameter from (b) to (a); the lines (a f), (a g) will diverge from (a b), and the pyramids of which these lines are the axes, will compose the whole surface of the interior ellipse. The pyramids, of which P F, P G are the axes, will, in like manner, compose the surface of the exterior ellipse, and this is true for every section of the spheroid passing through P m. Hence the attraction of P to the spheroid P A Q in the direction P d equals the attraction of (a) to the spheroid (p a q) in the same direction.

590. Attraction of P in the direction P D : attraction of A in the same direction : : P D : A C.

For the attraction of (a) in the direction P D : attraction of A in the same direction : : P D : A C, and the attraction of (a) = attraction of P. ∴ attraction of P : attraction of A : : P D : A C.

Similarly, the attraction of P in the direction E C : attraction of A in the direction E C : : P a : E C.

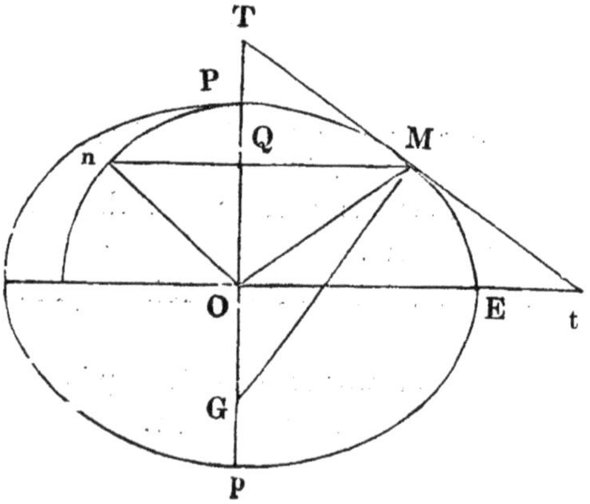

591. Draw M G perpendicular to the ellipse at M, and with the radius O P describe the arc P n.

Then Q G : Q M : : Q M : Q T

$$\therefore Q G = \frac{Q M^2}{Q T}.$$

And

$$O Q : O P :: O P : O T$$

$$\therefore O Q = \frac{O P^2}{O T}.$$

$$\therefore Q G : Q O :: \frac{Q M^2}{Q T} : \frac{O P^2}{O T} :: \frac{Q M^2 . O T}{Q T} : O P$$

but

$$O T : O Q :: O P^2 : Q O^2$$

$$\therefore O T : T Q :: O P^2 : O P^2 - O Q^2$$

$$:: O P^2 : n Q^2$$

$$:: O P^2 : P Q . Q p :: O E^2 : Q M^2$$

$$\therefore \frac{O T}{T Q} = \frac{O E^2}{Q M^2}.$$

$$\therefore Q G : Q O :: O E^2 : O P^2$$

$$\therefore Q G = \frac{O E^2}{O P^2} . Q O.$$

592. A fluid body will preserve its figure if the direction of its gravity, at every point, be perpendicular to its surface; for then gravity cannot put its surface in motion.

593. If the particles of a homogeneous fluid attract each other with forces varying as $\frac{1}{\text{distance}^2}$, and it revolve round an axis, it will put on the form of a spheroid.

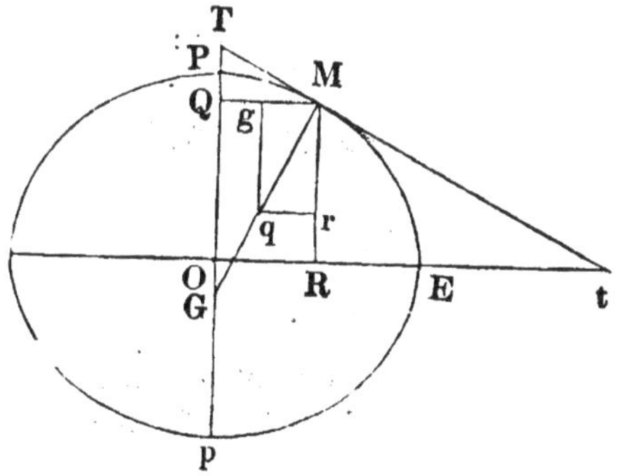

For if P E p P be a fluid, P p the axis round which it revolves, then may the spheroid revolve in such a time that the centrifugal force of any particle M combined with its gravity, may make this whole force act perpendicularly to the surface. For let E = attraction at the equator, P = attraction at the pole, F = centrifugal force at the equator.

Then (590),

attraction of M in the direction M R : P : : Q O : P O

$$\therefore \text{attraction of M in the direction M R} = \frac{P \cdot QO}{PO}.$$

Similarly, the attraction of M in the direction $MQ = \frac{E \cdot OR}{OE}$.

But the centrifugal force of bodies revolving in equal times $\propto$ radii.

$$F \propto \frac{V^2}{r} \propto \frac{c^2}{r \cdot P^2}$$

$$\propto \frac{4\pi^2 r}{P^2}$$

(and P being given) $\propto$ r

$$\therefore \text{centrifugal force of M} = \frac{F \cdot OR}{OE}.$$

$$\therefore \text{whole force of M in the direction MO} = \frac{(E - F) \cdot OR}{OE}.$$

Take $Mr = \frac{P \cdot QO}{PO}$, $Mg = \frac{(E - F) \cdot OR}{OE}$, complete the parallelogram, and M q will be the compound force; O E and O P $\therefore$ must have such a ratio to each other that M q may be always perpendicular to the curve. Suppose M q perpendicular to the curve, then, by similar triangles, q g or M r : M g : : Q G : Q M.

$$:: \frac{P \cdot QO}{PO} : \frac{(E-F)\,OR}{OE} :: \frac{OE^2}{OP^2} \cdot QO : OR$$

$$\therefore \frac{P \cdot QO \cdot OR}{PO} = (E - F) \cdot \frac{OR}{OE} \cdot \frac{OE^2}{OP^2} \cdot QO$$

$$\therefore P = (E - F) \cdot \frac{OE}{OP}$$

$$\therefore P : E - F :: OE : OP,$$

in which no lines are concerned except the two axes; $\therefore$ to a spheroid having two axes in such a ratio, the whole force will, at every point, be perpendicular to the surface, and $\therefore$ the fluid will be at rest.

594. The attraction of any point M in the direction $MR = \frac{P \cdot MR}{PO}$; $\therefore$ if P be represented by P O, M R will represent the attraction of M in the direction M R, and M v will represent the whole attraction acting perpendicularly to the surface.

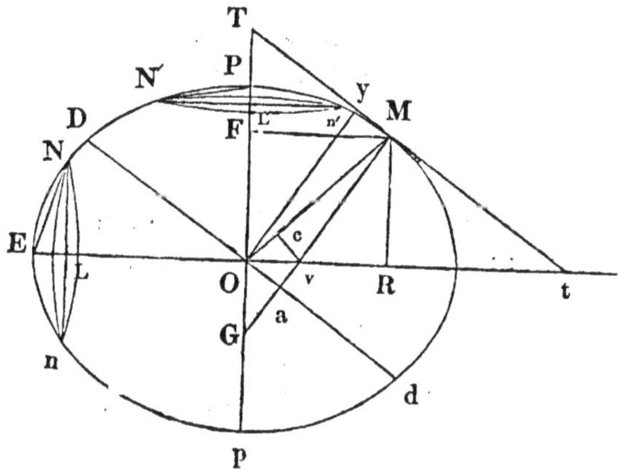

Draw (v c) perpendicular to M O.

Then

M O : M a : : M v : M c : : attraction in the direction M v : M O.

$$\therefore \text{attraction in the direction } MO = \frac{Mv \cdot Ma}{MO} = \frac{OP^2}{MO} \propto \frac{1}{MO}.$$

By similar triangles T O y, M v R, (the angle T O y being equal to the angle v M R.)

$$TO : Oy :: vM : MR$$

$$\therefore TO \cdot MR = Oy \cdot vM = Ma \cdot Mv = TO \cdot OF = OP^2.$$

595. Required the attraction of an oblong spheroid on a particle placed at the extremity of the major axis, the excentricity being very small.

Let axis major : axis minor : : 1 : 1 — n. Attraction of the circle N n (Prop XC.)

$$\propto 1 - \frac{EL}{EN} \propto 1 - \frac{x}{\sqrt{n^2 + (1-n)^2 (2n - n^2)}}$$

$$\propto 1 - x\{2x - n \cdot (4x - 2n^2)\}^{-\frac{1}{2}}$$

$$\propto 1 - x\{(2x)^{-\frac{1}{2}} + \frac{1}{2}(2x)^{-\frac{3}{2}} n \cdot (4n - 2n^2)\}$$

$$\propto 1 - \frac{\sqrt{x}}{\sqrt{2}} - \frac{n}{4\sqrt{2}} \cdot (4\sqrt{x} - 2x^{\frac{1}{2}}).$$

$$\therefore A' \propto x' - \frac{x^{\frac{1}{2}} x'}{\sqrt{2}} - \frac{n}{4\sqrt{2}} \cdot (4x^{\frac{1}{2}} x' - 2x^{\frac{3}{2}} x')$$

$$\therefore A \propto x - \frac{\sqrt{2}}{3} \cdot x^{\frac{3}{2}} - \frac{n}{4\sqrt{2}} \cdot \left(\frac{8x^{\frac{3}{2}}}{3} - \frac{4x^{\frac{5}{2}}}{5}\right)$$

Let x = 2 E O = 2,

$$\therefore A \propto 2 - \frac{4}{3} - \frac{n}{4\sqrt{2}} \cdot \left(\frac{16\sqrt{2}}{3} - \frac{16\sqrt{2}}{5}\right)$$

$$\propto \frac{2}{3} - \frac{8n}{15} \propto 1 - \frac{4n}{5}.$$

∴ attraction of the oblong spheroid on E : attraction of a circumscribed sphere on E : : (since in the sphere n = 0.)

$$: : 1 - \frac{4n}{5} : 1.$$

596. Required the attraction of an oblate spheroid on a particle placed at the extremity of the minor axis.

Let axis minor : axis major : : 1 : 1 + n.

$$\therefore A' \propto x' \left\{1 - \frac{x}{\sqrt{x^2 + (1+n)^2 \cdot (2x - x^2)}}\right\}$$

$$\propto x' \left\{1 - \frac{x}{\sqrt{2x + 4nx - 2nx^2}}\right\}$$

$$\propto x' \left\{1 - x\left((2x)^{-\frac{1}{2}} - \frac{1}{2}(2x)^{-\frac{3}{2}} \cdot \overline{4nx - 2nx^2}\right)\right\}$$

$$\propto x' - \frac{x^{\frac{1}{2}} x'}{\sqrt{2}} + \frac{n x^{\frac{1}{2}} x'}{\sqrt{2}} - \frac{n x^{\frac{3}{2}} x'}{2\sqrt{2}}$$

$$\therefore A \propto x - \frac{\sqrt{2} \cdot x^{\frac{3}{2}}}{3} + \frac{\sqrt{2} \cdot n x^{\frac{3}{2}}}{3} - \frac{n x^{\frac{5}{2}}}{5\sqrt{2}}$$

∴ whole attraction

$$\propto 2 - \frac{4}{3} + \frac{4n}{3} - \frac{4n}{5} \propto \frac{2}{3} + \frac{8n}{15} \propto 1 + \frac{4n}{5}$$

∴ attraction of the oblate sphere on P : attraction of the sphere inscribed on P : : $1 + \frac{4n}{5}$: 1.

Since these spheroids, by hypothesis, approximate to spheres, they may, without sensible error, be assumed for spheres, and their attractions will be nearly proportional to their quantities of matter. But oblong sphere : oblate : : oblate : circumscribed sphere. ∴ A of oblong sphere on E : A′ of oblate on E : : A′ : A″ of circumscribed sphere on E.

$$\therefore A' : A'' : : A : A' : : \sqrt{A} : \sqrt{A''} : : \sqrt{1 - \frac{4n}{5}} : 1 : : 1 - \frac{2n}{5} : 1$$

Also

att^{n}. of oblate sph. on P : att^{n}. of insc^{d}. sph. on P : : $1 + \frac{4n}{5} : 1$

att^{n}. of insc^{d}. sph. on P : att^{n}. of circumsc^{d}. sph. on E : : $1 : 1 + n$

att^{n}. of circumsc^{d}. sph. on E : attr^{n}. of oblate sph. on E : : $1 : 1 - \frac{2n}{5}$

∴ attraction of the oblate sphere on P : attraction of the oblate sphere on E : : $1 + \frac{4n}{5} : \overline{1+n} \cdot \overline{1 - \frac{2n}{5}}$

$$: : 1 + \frac{4n}{5} : 1 + \frac{3n}{5} : : 1 + \frac{n}{5} : 1 \text{ nearly.}$$

$$1 + \frac{3n}{5}\Big) 1 + \frac{4n}{5} \Big(1 + \frac{n}{5}$$

$$\begin{array}{l} 1 + \frac{3n}{5} \\ \hline \frac{n}{5} \\ \frac{n}{5} + \frac{3n^2}{25} \\ \hline -\frac{3n^2}{25} \end{array}$$

$$\therefore P : E : : 1 + \frac{n}{5} : 1;$$

but (593), $P : E - F : : OE : OP$

$$: : 1 + n : 1 : : P + F : E \text{ nearly}$$

$$\overline{1+n} \cdot \overline{E - F} = P$$

$$\therefore \overline{1+n} \cdot E - F - nF = P$$

$$\therefore \overline{1+n} \cdot E - nF = P + F$$

and since (n) is very small, as also F compared with E,

$$\therefore \overline{1+n} \cdot E = P + F$$

$$\therefore 1 + n : 1 : : P + F : E$$

$$\therefore P = E + \frac{nE}{5}$$

$$\therefore E + \frac{nE}{5} + F : E : : 1 + n : 1$$

$$\therefore E + \frac{nE}{5} + F = E + nE$$

$$\therefore F = \frac{4nE}{5}$$

$$\therefore n = \frac{5F}{4E}$$

$$\therefore 4E : 5F : : 1 : n$$

or "four times the primitive gravity at the equator : five times the centrifugal force at the equator : : one half polar axis : excentricity."

597. The centrifugal force opposed to gravity $\propto \cos.^2$ latitude.

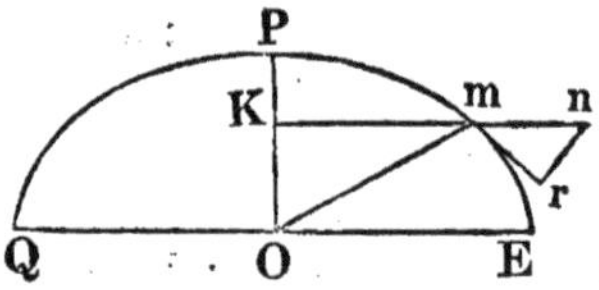

Let (m n) = centrifugal force at (m), F = centrifugal force at E.

∴ (n r) is that part of the centrifugal force at (m) which is opposed to gravity.

Now

$$\left.\begin{array}{l} \mathrm{F : m\,n :: O\,E : K\,m} \\ \text{and} \\ \mathrm{m\,n : n\,r :: O\,m : K\,m} \end{array}\right\} \therefore \mathrm{F : n\,r :: O\,m^2 : K\,m^2} \quad :: r^2 : \cos.^2 \text{lat.}$$

$$\therefore \mathrm{m\,r} \propto \cos.^2 \text{lat.}$$

598. From the equator to the pole, the increase of the length of a degree of the meridian $\propto \sin.^2$ lat.

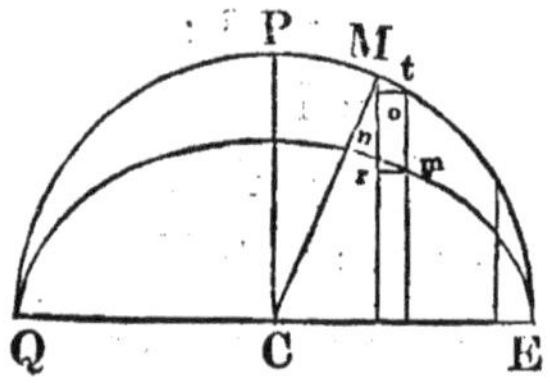

$$\mathrm{n\,r : M\,s :: n\,G : M\,G :: C\,P : C\,R :: 1 - n : 1.}$$

$$\therefore \mathrm{n\,r} = \overline{1 - \mathrm{n}} \,.\, \mathrm{M\,S} = \overline{1 - \mathrm{n}} \,.\, \varphi' \sin.\theta = 1 - \mathrm{n} \,.\, \cos.\theta \,.\, \theta'$$

$$\mathrm{m\,r} = \mathrm{s\,t} = \varphi' . \cos.\theta = -\sin.\theta . \theta' .$$

$$\therefore \mathrm{m\,r}^2 = \sin.^2\theta . \theta^2$$

$$\therefore \mathrm{m\,n}^2 = \mathrm{n\,r}^2 + \mathrm{m\,r}^2 = \theta'^2 . \sin.^2\theta + (1 - \mathrm{n})^2 . \cos.^2\theta . \theta'^2$$

$$= \theta'^2 . (\sin.^2\theta + \overline{1 - 2\,\mathrm{n}} \,.\, \cos.^2\theta)$$

$$= \theta'^2 (\sin.^2\theta + \cos.^2\theta - 2\,\mathrm{n} \,.\, \cos.^2\theta)$$

$$= \theta'^2 . (1 - 2\,\mathrm{n} \,.\, \cos.^2\theta)$$

$$\therefore \mathrm{m\,n} = \theta' . (1 - \mathrm{n} \,.\, \cos.^2\theta)$$

∴ at the equator, since

$$\theta = 0; \; \mathrm{m'\,n'} = \theta' (1 - \mathrm{n})$$

$$\therefore \text{ increase } = \theta' (1 - n.\cos.^2 \theta - 1 + n) = \theta'.n(1 - \cos.^2 \theta)$$
$$= \theta'.n \sin.^2 \theta,$$

$$\therefore \text{ increase } \propto n\,\theta'.\sin.^2 \theta$$
$$\propto \sin.^2 \theta, \propto \sin.^2 \text{latitude}.$$

599. Given the lengths of a degree at two given latitudes, required the ratio between the polar and equatorial diameters.

Let P and p be the lengths of a degree at the pole and equator, m and n the lengths in latitudes whose sines are S and s, and cosines C and c. Then as length of a degree ∞ radius of curvature, (for the arc of the meridian intercepted between an angle of one degree, which is called the length of a degree, may be supposed to coincide with the circle of curvature for that degree, and will $\therefore \propto$ radius of curvature.)

$$\propto \frac{CD^2}{PF}.$$

Now at the pole CD^2 becomes AC^2, and P F becomes B C

$$\therefore \text{ length of a degree } \propto \frac{AC^2}{BC}, \propto \frac{a^2}{b};$$

similarly the length of a degree at the equator

$$\propto \frac{BC^2}{AC}, \propto \frac{b^2}{a},$$

$$P : p :: \frac{a^2}{b} : \frac{b^2}{a} :: a^3 : b^3 :: 1 : (1 - n')^3.$$

Now

$$m - p : n - p \,(598) :: S^2 : s^2,$$
$$\therefore m - n : n - p \quad :: S^2 - s^2 : S^2,$$
$$\therefore n - p = \frac{\overline{m-n}.S^2}{S^2 - s^2},$$

but

$$P - p : n - p :: 1^2 : s^2 :: P - p : \frac{\overline{m-n}.S^2}{S^2 - s^2},$$

$$\therefore P - p = \frac{m - n}{S^2 - s^2},$$

$$\therefore P = p + \frac{m - n}{S^2 - s^2},$$

$$\frac{\overline{m-n}.s^2}{S^2 - s^2} = n - p,$$

$$\therefore P = n - \frac{\overline{m-n}.S^2}{S^2 - s^2},$$

$$= \frac{n S^2 - n s^2 - m s^2 + n s^2}{S^2 - s^2} = \frac{n S^2 - m s^2}{S^2 - s^2},$$

$$\therefore P = p + \frac{m - n}{S^2 - s^2} = \frac{n S^2 - m s^2 + m - n}{S^2 - s^2},$$

$$= \frac{m.(1 - s^2) - n.(1 - S^2)}{S^2 - s^2},$$

$$= \frac{m c^2 - n C^2}{S^2 - s^2},$$

$$\therefore P : p :: \frac{m c^2 - n C^2}{S^2 - s^2} : \frac{n S^2 - m s^2}{S^2 - s^2},$$

$$:: m c^2 - n C^2 : n S^2 - m s^2 :: 1 : (1 - n')^3$$

$$\therefore (m c^2 - n C^2)^{\frac{1}{3}} : (n S^2 - m s^2)^{\frac{1}{3}} :: 1 : 1 - n'.$$

600. The variation in the length of a pendulum $\propto$ sin.² latitude.

Let l = length of a pendulum vibrating seconds at the equator.

L = length of one vibrating seconds at latitude θ.

The force of gravity at the pole = 1, ∴ the force of gravity at the equator = 1 — F, and the force of gravity in latitude θ (603) = $1 - F. \cos.^2 \theta$,

$$\therefore L : l :: 1 - F. \cos.^2 \theta : 1 - F \text{ (since } \delta \propto \delta F)$$

$$\therefore L - l : l :: F. (1 - \cos.^2 \theta) : 1 - F :: F. \sin.^2 \theta : 1 - F,$$

$$\therefore L - l = \frac{l F. \sin^2 \theta}{1 - F} \propto \sin.^2 \theta.$$

From the poles to the equator, the decrease of the length of a pendulum always vibrating in the same time, $\propto$ cos.² latitude.

Let L' = length of a pendulum vibrating seconds at the pole,

$$\therefore L' : L :: 1 : 1 - F. \cos^2 \theta,$$

$$\therefore L' : L' - L :: 1 : F. \cos^2 \theta,$$

$$\therefore L' - L \propto \cos.^2 \theta.$$

601. The increase of attraction from the equator to the pole $\propto$ sin.² lat.

Let

$$O E : O P :: 1 : 1 - n.$$

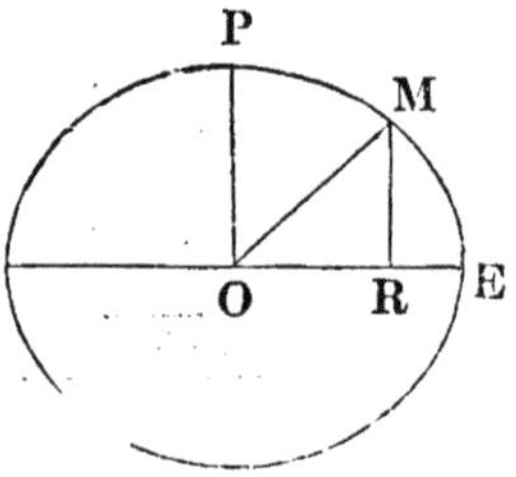

Let

$M O = a$, the angle $M O E = \theta$,

$$\therefore M R^2 = \frac{P O^2}{O E^2}. \{O E^2 - O R^2\},$$

or

$$a^2. \sin.^2 \theta = (1 - n)^2. (1 - a^2 \cos.^2 \theta)$$

$$= \overline{1 - 2 n}. (1 - a^2 \cos.^2 \theta)$$

$$\therefore a^2. \{ \sin.^2 \theta + \overline{1 - 2 n}. \cos.^2 \theta\} = 1 - 2 n$$

$$\therefore a^2 = \frac{1 - 2n}{\sin.^2\theta + \cos.^2\theta - 2n.\cos.^2\theta} = \frac{1 - 2n}{1 - 2n.\cos.^2\theta},$$

$$\therefore a = \frac{1 - n}{1 - n.\cos.^2\theta} = \overline{1 - n}.\frac{1 + n.\cos.^2\theta}{1^2 - n^2.\cos.^4\theta} = \overline{1 - n}.(1 + n\cos.^2\theta),$$

$$= 1 - n(1 - \cos.^2\theta) = 1 - n.\sin.^2\theta, \therefore$$

$$\therefore \frac{1}{a} = \frac{1}{1 - n.\sin.^2\theta} = 1 + n.\sin.^2\theta = \frac{1}{MO},$$

but (594) the attraction in the direction M O $\propto \frac{1}{MO}$,

$\therefore$ attraction in the direction M O (A) : attraction at E (A′)

$$:: 1 + n.\sin.^2\theta : 1,$$

$$\therefore A - A' : A' :: n.\sin.^2\theta : 1,$$

$$\therefore A - A' = A'.n.\sin.^2\theta,$$

$\therefore$ increase of attraction $\propto \sin.^2\theta \propto \sin.^2$ latitude.

602. Given the lengths of two pendulums vibrating seconds in two known latitudes; find the lengths of pendulums that will vibrate seconds at the equator and pole.

Let L, l be the lengths of pendulums vibrating seconds at the equator and pole.

L′, l′ be the lengths in given latitudes whose sines are S, s, cosines C, c.

$$\therefore L' - L : l' - L :: S^2 : s^2$$

$$\therefore L's^2 - Ls^2 = l'S^2 - Ls^2$$

$$\therefore L.(S^2 - s^2) = l'S^2 - L's^2,$$

$$\therefore L = \frac{l'S^2 - L's^2}{S^2 - s^2}.$$

Again

$$L' - L : l - L :: S^2 : 1,$$

$$\therefore L' - L = l S^2 - L S^2,$$

$$\therefore l = \frac{L' - L.(1 - S^2)}{S^2},$$

$$= \frac{L'}{S^2} - \frac{(l'S^2 - L's^2)(1 - S^2)}{S^2.(S^2 - s^2)},$$

$$= \frac{L'S^2 - L's^2 - l'S^2 + l'S^4 + L's^2 - L'S^2s^2}{S^2.(S^2 - s^2)},$$

$$= \frac{L'S^2 - l'.S^2 + l'.S^4 - L'.S^2s^2}{S^2.(S^2 - s^2)},$$

$$= \frac{L'.(1 - s^2) - l'.(1 - S^2)}{S^2 - s^2} = \frac{L'c^2 - l'C^2}{S^2 - s^2}.$$

603. Given the lengths of two pendulums vibrating seconds in two known latitudes; required the ratio between the equatorial and polar diameters.

Since the lengths $\propto$ forces, the times being the same,

$$\therefore L : l :: \text{force at the equator} : \text{force at the pole}$$

$$:: (10)\ \frac{1}{1} : \frac{1}{1-n} :: 1 - n : 1 :: OP : OE,$$

$$\therefore OP : OE :: \text{polar diameter} : \text{equatorial diameter}$$

$$:: L : l :: l' S^2 - L' s^2 : L' c^2 - l' C^2.$$

604. To compare the space described in one second by the force of gravity in any given latitude, with that which would be described in the same time, if the earth did not revolve round its axis.

The space which would be described by a body, if the rotatory motion of the earth were to cease, equals the space actually described by a body at the pole in the same time; and if the force at the pole equal 1, the force at the latitude θ (597) equal $1 - F . \cos.^2 \theta$, and since $S = m F T^2$, and T is the same, $\therefore S \propto F$.

$\therefore$ space actually described when the earth revolves : space which would be described if the earth were at rest :: $1 - F. \cos.^2 \theta : 1$.

605. Let the earth be supposed a sphere of a given magnitude, and to revolve round its axis in a given time; to compare the weight of a body at the equator, with its weight in a given latitude.

The centrifugal force $= \frac{V^2}{r} = \frac{4 \pi^2 . r}{P^2} = F$ equal a given quantity, since (r) and P are known. Now the force at the equator $= 1 - F$, and the force at latitude $\theta = 1 - F . \cos.^2 \theta$, and the weight $\propto$ attractive force

$$\therefore W : W' :: 1 - F : 1 - F . \cos.^2 \theta.$$

606. Find the ratio of the times of oscillation of a pendulum at the equator and at the pole, supposing the earth to be a sphere, and to revolve round its axis in a given time.

$L \propto F T^2$ but L is constant, $\therefore T^2 \propto \frac{1}{F}$,

$\therefore$ T. oscillation at the pole : T. oscillation at the equator

$$:: \sqrt{\text{force at the equator}} : \sqrt{\text{force at the pole}}$$

$$:: \sqrt{1 - F} : 1.$$

ON THE THEORY OF THE TIDES.

607. If a spherical body at rest be acted upon by some other body, it may put on the form of a spheroid.

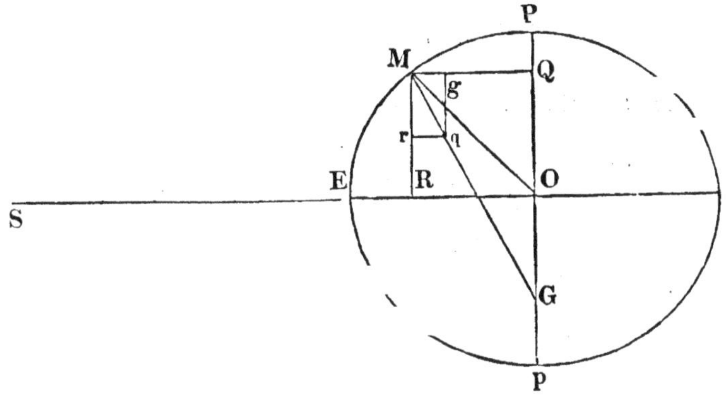

Let P E p be the earth at rest; (S) a body acting upon it; (O) its center; (M) a particle on its surface.

Let P = polar, E = equatorial, } attraction on the earth.

Then the attraction on M is parallel to $M Q = \frac{E . O R}{O E}$.

Similarly the attraction on M is parallel to $M R = \frac{P . O Q}{O P}$.

Let (m) = mean addititious force of S on P.

(n) = mean addititious force of S on E.

Now since the addititious force (Sect. XI.) $\propto$ distance,

$\therefore$ the whole addititious force of S on $M = \frac{m . M O}{P O}$,

and

$\frac{m . M O}{P O}$: addititious force in the direction M R :: M O : M R,

$\therefore$ addititious force in the direction $M R = \frac{m . M R}{P O} = \frac{m . O Q}{P O}$.

Again, since

$m : n :: P O : E O$,

$\therefore \frac{m}{P O} = \frac{n}{E O}$,

$\therefore$ whole addititious force of S on $M = \frac{n . M O}{E O}$,

∴ addititious force in the direction M Q $= \frac{n \cdot MQ}{EO} = \frac{n \cdot OR}{EO}$,

∴ whole disturbing force of S on M in the direction M Q = twice the addititious force in that direction, and is negative $= -\frac{2n \cdot OR}{OE}$.

∴ whole attraction of M in the direction M Q $= \{E - 2n\} \cdot \frac{OR}{OE}$,

and the whole attraction of M in the direction M R $= \{P + m\} \cdot \frac{OQ}{OP}$.

$$\text{Take } Mg = \{E - 2n\} \cdot \frac{OR}{OE} \Big\}$$
$$Mr = \{P + m\} \cdot \frac{OQ}{OP} \Big\}$$

complete the parallelogram (m q), and produce M q to meet P p in G.

Now if the surface be at rest, M G will be perpendicular to the surface.

$$\therefore Mr : Mg :: gq : gM :: GQ : QM,$$

or

$$\{P + m\} \cdot \frac{OQ}{OP} : \{E - 2n\} \frac{OR}{OE} :: \frac{OE^2}{OP^2} OQ \cdot OR.$$

$$\therefore P + m : E - 2n :: OE : OP,$$

∴ figure may be an ellipse.

608. Suppose the Moon to move in the equator; to find the greatest elevation of tide.

Let A B C D be the undisturbed sphere; M P m K a spheroid formed by the attraction of the Moon; M the place to which the Moon is vertical.

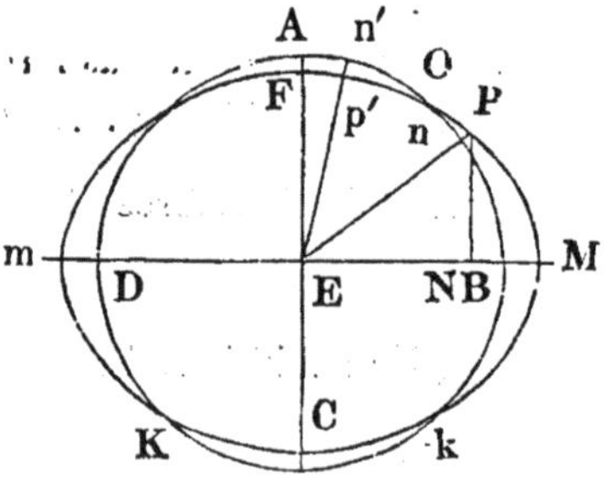

Let

$$\left\{\begin{array}{l} AE = 1. \\ EM = 1 + \alpha \\ EF = 1 - \beta \end{array}\right\}$$

Then since the sphere and spheroid have the same solid content,

$$\therefore \frac{4\pi \cdot (AE)^3}{3} = \frac{4\pi \cdot EM \cdot (FE)^2}{3},$$

$$\therefore \frac{4\pi}{3} = \frac{4\pi \cdot (1+\alpha) \cdot (1-\beta)^2}{3} = \frac{4\pi \cdot (1+\alpha)(1 - 2\beta + \beta^2)}{3},$$

$$\therefore 1 = 1 + \alpha - 2\beta - 2\alpha\beta + \beta^2 + \alpha\beta^2$$
$$= 1 + \alpha - 2\beta \text{ nearly, } (\alpha) \text{ and } (\beta) \text{ being very small,}$$
$$\therefore \alpha = 2\beta \text{ or greatest elevation} = 2 \times \text{greatest depression.}$$

614. To find the greatest height of the tide at any place, as (n).

Let

$$E P = \theta - \angle P E M = \theta - \alpha + \beta = \frac{3\alpha}{2} = E M - E F = M,$$

$$\therefore P N^2 = \varrho^2 . \sin.^2 \theta = \frac{E F^2}{E M^2} . \{E M^2 - E N^2\}$$
$$= \left(\frac{1 - \beta}{1 + \alpha}\right)^2 . \{(1 + \alpha)^2 - \varrho^2 \cos.^2 \theta\}.$$

Now $\frac{(1 - \beta)^2}{(1 + \alpha)^2}$ by actual division (all the terms of two or more dimensions being neglected) $= 1 - 2 . (\alpha + \beta) = 1 - 2 M$,

$$\therefore P N^2 = \varrho^2 . \sin.^2 \theta = (1 - 2 M) . \{1 + 2\alpha - \varrho^2 . \cos.^2 \theta\}$$
$$(\text{since } 2\alpha = \frac{3\alpha}{2} . \frac{4}{3} = \frac{4}{3} M) = (1 - 2 M) \{1 + \frac{4 M}{3} - \varrho^2 . \cos.^2 \theta\}.$$
$$\therefore \varrho^2 . \{\sin.^2 \theta + (1 - 2 M) . \cos.^2 \theta\} = (1 - 2 M) . \left(1 + \frac{4 M}{3}\right)$$
$$\therefore \varrho^2 = \frac{1 - \frac{2 M}{3}}{\sin.^2 \theta + \cos.^2 \theta - 2 M . \cos.^2 \theta} = \frac{1 - \frac{2 M}{3}}{1 - 2 M . \cos.^2 \theta},$$
$$\therefore \varrho = \frac{1 - \frac{M}{3}}{1 - M . \cos.^2 \theta} = (1 + M . \cos.^2 \theta) . \left(1 - \frac{M}{3}\right) \text{ nearly,}$$
$$= 1 + M . \cos.^2 \theta - \frac{M}{3},$$
$$\therefore \varrho - 1 = M . \cos.^2 \theta - \frac{M}{3} = E P - E n = P n = \text{elevation required.}$$

615. Similarly if the angle $M E p = \theta'$, $\therefore E p = 1 + M \cos.^2 \theta' - \frac{M}{3}$,

$$\therefore 1 - E p = p n' = \text{depression} = \frac{M}{3} - M . \cos.^2 \theta$$
$$= M - M . \cos.^2 \theta' - \frac{2 M}{3} = M \sin.^2 \theta' - \frac{2 M}{3}.$$

616. $B M = a = \frac{2 M}{3}$,

$$\therefore B M - P n = \frac{2 M}{3} + M . \sin.^2 \theta - \frac{2 M}{3}$$
$$= M . \sin.^2 \theta \propto \sin.^2 \theta,$$

∴ greatest elevation $\propto$ sin.^{2} horizontal angle from the time of high tide.

617. At (O) P n = 0,

$$\therefore M \,.\, \cos.^2 \theta - \frac{M}{3} = 0,$$

$$\therefore M \,.\, \cos.^2 \theta = \frac{M}{3},$$

$$\therefore \cos.\, \theta = \frac{1}{\sqrt{3}},$$

$$\therefore \theta = 54^\circ \;,\; 44'.$$

Hitherto we have considered the moon only as acting on the spheroid. Now let the sun also act, and let the elevation be considered as that produced by the joint action of the sun and moon in their different positions.

Let us suppose a spheroid to be formed by the action of the sun, whose semi-axis major = (1 + a), axis minor = (1 — b).

618. Let (a + b) = S, (φ) = the angular distance of any place from the point to which the sun is vertical. It may be shown in the same manner as was proved in the case of the moon, that

$$S \,.\, \cos.^2 \varphi - \frac{S}{3} = \text{elevation due to the sun,}$$

and

$$S \,.\, \sin.^2 \varphi' - \frac{2\,S}{3} = \text{depression due to the sun,}$$

(φ') being the angular distance of the place of low water from the point to which the sun is vertical,

$$\therefore M \,.\, \cos.^2 \theta + S \,.\, \cos.^2 \varphi - \frac{M + S}{3} = \text{compound elevation.}$$

Similarly $M \,.\, \sin.^2 \theta' + S \,.\, \sin.^2 \varphi' - \frac{2}{3}\,\overline{M + S}$ = compound depression.

619. Let the sun and moon be both vertical to the same place,

$$\therefore \theta = \varphi = 0,$$

$$\therefore M + S - \frac{M + S}{3} = \frac{2}{3}\,\overline{M + S} = \text{compound elevation,}$$

and

$$\theta' = \varphi' = 90^\circ,$$

$$\therefore M + S - \tfrac{2}{3} \,.\, \overline{M + S} = \tfrac{1}{3}\,\overline{M + S} = \text{compound depression,}$$

∴ compound elevation + compound depression = $\overline{M + S}$ = height of spring tide.

620. Let the moon be in the quadratures with the sun, then at a place under the moon,

$$(\theta) = 0, \text{ and } (\varphi) = 90^\circ,$$

$$\therefore \text{ compound elevation } = M - \frac{M+S}{3},$$

also $(\theta') = 90$, and $(\varphi') = 0$,

$$\therefore \text{ compound depression } = M - \tfrac{2}{3}.\overline{M+S},$$

$\therefore$ height of the tide at the place under the moon $= 2M - \overline{M+S}$

$= M + S =$ height of neap tide.

Similarly at a place under the sun, height of tide $= S - M$.

621. Given the elongation of the sun and moon, to find the place of compound high tide.

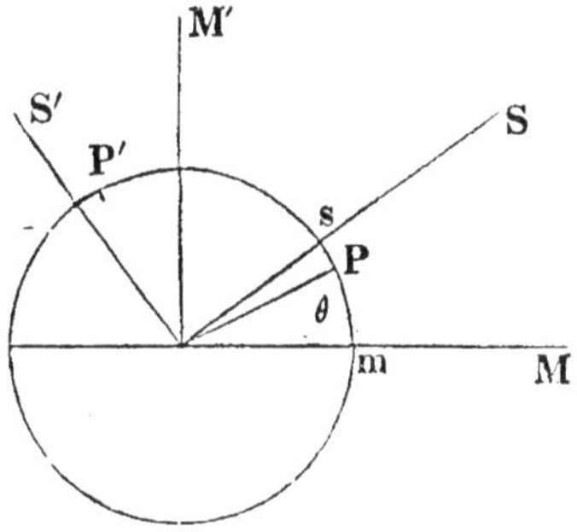

Compound elevation $= M \cos.^2 \theta + S \cos.^2 \varphi - \frac{M+S}{3} =$ maximum at high water.

$\therefore - 2 M \cos. \theta \sin. \theta\, \theta' - 2 S \cos. \varphi \sin. \varphi\, \varphi' = 0$,

but

$(\theta + \varphi) =$ elongation $= E$

$=$ constant quantity,

$\therefore \theta' + \varphi' = 0$

$\therefore \theta' = - \varphi'$,

$\therefore 2 M \cos. \theta \sin. \theta = 2 S \cos. \varphi \sin. \varphi'$,

$\therefore M \sin. 2\theta = S \sin. 2\varphi$,

$\therefore M : S :: \sin. 2\varphi : \sin. 2\theta$,

$\therefore M + S : M - S :: \sin. 2\varphi + \sin. 2\theta : \sin. 2\varphi - \sin. 2\theta$,

$:: \tan. (\varphi + \theta) : \tan. (\varphi - \theta)$,

and since $(\varphi + \theta)$ is known, $\therefore (\varphi - \theta)$ is obtained, and $\therefore (\varphi)$ and (θ) are found, i. e. the distance of the sun and moon from the place of compound high tide is determined.

622. Let P be the place of high tide,

P' the place of low water, 90° distant from P,

$P m = \theta - P m 1 = 90 + \theta = \theta' - P s = \varphi - P' s$

$= 90 - \varphi = \varphi'$.

Now the greatest depression $= M \sin.^2 \theta' + S \sin.^2 \varphi' - \tfrac{2}{3}\, \overline{M+S}$,

but

$\sin.^2 \theta' = \sin.^2 (90 + \theta) = \sin.^2$ supplemental angle $(90 - \theta) = \cos.^2 \theta$,

and

$\sin.^2 \varphi' = \sin.^2 (90 - \varphi) = \cos.^2 \varphi$,

$\therefore$ the greatest depression $= M \cos.^2 \theta + S \cos.^2 \varphi - \tfrac{2}{3}\, \overline{M+S}$,

and the greatest elevation $= M \cos.^2 \theta + S \cos.^2 \varphi - \tfrac{1}{3}\, \overline{M+S}$,

$\therefore$ the greatest whole tide $=$ the greatest elevation $+$ greatest depression

$$= 2\,M\cos.^2\theta + 2\,S\cos.^2\varphi - \overline{M + S},$$
$$= M\,\{2\cos.^2\theta - 1\} + S\,(2\cos.^2\varphi - 1)$$
$$= M\cos.\,2\,\theta + S\cos.\,2\,\varphi.$$

623. Hence Robison's construction.

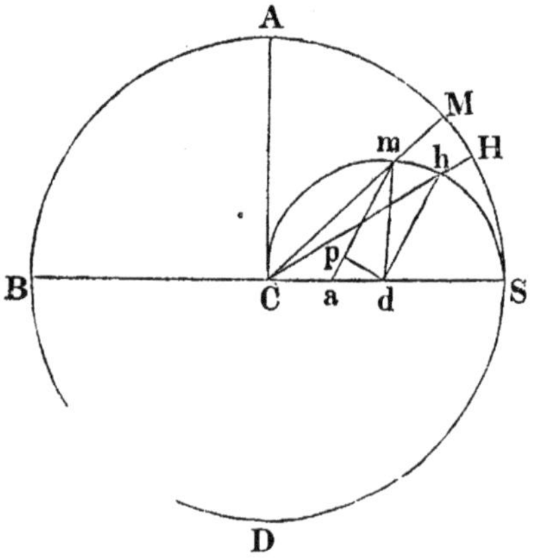

Let A B D S be a great circle, S and M the places to which the sun and moon are vertical; on S C, as diameter, describe a circle, bisect S C in (d); and take S d : d a :: M : S. Take the angle S C M $= (\varphi + \theta)$, and let C M cut the inner circle in (m), join (m a) and draw (h d) parallel to it; through (h) draw C h H meeting the outer circle in H; then will H be the place of high water.

For draw (d p) perpendicular to (m a) and join (m d).

Let the angle S C H $= \varphi$, and the angle M C H $= \theta$.

$$\text{Since } M : S :: Sd : da$$
$$\therefore M + S : M - S :: Sd + da : Sd - da$$
$$:: dm + da : dm - da$$
$$:: \tan.\frac{dam + dma}{2} : \tan.\frac{dam - dma}{2}$$
$$:: \tan.\frac{Sdm}{2} : \tan.\frac{dam - dma}{2}$$
$$:: \tan.\,SCM : \tan.\frac{Sdh - mdh}{2}$$
$$:: \tan.\,SCM : \tan.\,(SCH - HCM)$$
$$:: \tan.\,(\varphi + \theta) : \tan.\,(\varphi - \theta)$$

$\therefore$ H is the place of high water 621.

Also (m a) equals the height of the whole tide. For (a p) $=$ a d. cos. p a d
$$= S.\cos.\,Sdh = S.\cos.\,2\,\varphi$$

and

$$(pm) = md.\cos.\,pmd = M.\cos.\,mdh = M.\cos.\,2\,\theta$$

$\therefore$ a m = a p + p m = M. cos. 2 θ + S. cos. 2 φ = height of the tide.

At new moon, $\theta = \varphi = 0$ } $\therefore$ tide = M + S = spring tide.
At full moon, $\theta = 0$, $\varphi = 180^{\circ}$ }

When the moon is in quadratures, (m a) coincides with C A,

$\therefore \theta = 0, \varphi = 90^{\circ}$,

$\therefore$ tide = M — S = neap tide.

624. The fluxion of the tide, i. e. the increase or decrease in the height of the tide $\propto \varphi'$. (m a) $\propto \varphi'$. {M. cos. 2 θ + S. cos. 2 φ}. But the sun for any place is considered as constant,

$\therefore \varphi'$. (m a) $\propto$ — M. sin. 2 θ. 2 θ',

$\therefore \varphi'$. (m a) is a maximum at the octants of the tide with the moon

$\propto$ — M. sin. 2 θ

since at the octants, 2 $\theta = 90^{\circ}$.

The fluxion of the tide is represented in the figure by (d p).

For let (m u) be a given arc of the moon's synodical motion, draw (n v) perpendicular on (m a), $\therefore$ (m v) is the difference of the tides.

Now m u : m v : : m d : d p and m u and m d are constant, $\therefore$ m v $\propto$ d p and d p is a maximum, when it coincides with (d a), i. e. when the tide is in octants; for then 2 (m a d) = 90°.

625. At the new and full moon, it is high water when the sun and

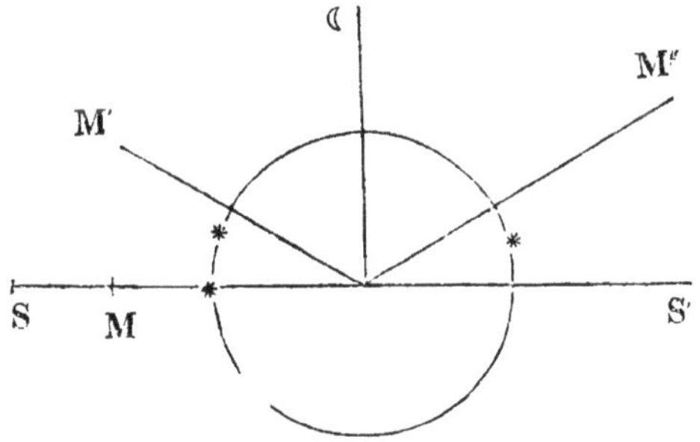

moon are on the meridian; i. e. at noon and midnight. At the quadratures of the moon, it is high water when the moon is on the meridian, because then (m) coincides with C.

For let M. cos.$^2 \theta$ + S. cos.$^2 \varphi$ — $\frac{M + S}{3}$ = maximum; then since in quadratures $(\varphi + \theta) = 90^{\circ}$, $\therefore \varphi = 90^{\circ} - \theta$,

$\therefore$ M. cos.$^2 \theta$ + S. sin.$^2 \theta$ — $\frac{1}{3}\overline{M + S}$ = maximum,

$\therefore$ 2 M. cos. θ. sin. θ. θ' = 2 S. sin. θ. cos. θ. θ',

$\therefore \overline{M - S}$. 2. sin. θ. cos. θ = $\overline{M - S}$. sin. 2 θ = 0, $\therefore$ sin. 2 θ = 0,

$\therefore \theta = 0$, that is, the moon is on the meridian.

626. From the new moon to the quadratures, the place of high tide follows the moon, i. e. is westward of it; since the moon moves from west to east, from the quadratures to the full moon, the place of high tide is before the moon. There is therefore some place at which its distance from the moon (θ) equals a maximum.

Now (621) $M : S :: \sin. 2\varphi : \sin. 2\theta$

$$\therefore M . \sin. 2\theta = S . \sin. 2\varphi$$

$$\therefore M . 2\theta' . \cos. 2\theta = S . 2\varphi' . \cos. 2\varphi = 0$$

$$\therefore \cos. 2\varphi = 0, \therefore \varphi = 45^{\circ}.$$

627. By (621) $M . \sin. 2\theta = S . \sin. 2\varphi$

$$\therefore \theta' . M . \cos. 2\theta = \varphi' . S . \cos. 2\varphi$$

but

$$\varphi + \theta = e, \therefore \varphi' + \theta' = e', \therefore \theta' = e' - \varphi'$$

$$\therefore (e' - \varphi') M . \cos. 2\theta = \varphi' . S . \cos. 2\varphi$$

$$\therefore e' . M . \cos. 2\theta = \varphi' . \{S . \cos. 2\varphi + M . \cos. 2\theta\}$$

$$\therefore \varphi = \frac{e' . M . \cos. 2\theta}{M . \cos. 2\theta + S . \cos. 2\varphi}.$$

Next, considering the moon to be out of the equator, its action on the tides will be affected by its declination, and the action of the sun will not be considered.

By Art. (614) the elevation $= M \cos.^2 \theta - \frac{M}{3}$

$$\therefore \text{elevation above low water mark} = M . \cos.^2 \theta - \frac{M}{3} + b$$

now

$$b = \frac{a}{2} = \frac{M}{3}$$

$$\therefore \text{elevation above low water} = M . \cos.^2 \theta$$

$$= \text{magnitude of the tide.}$$

Let the angle Z P M which measures the time from the moon's passing the meridian equal t.

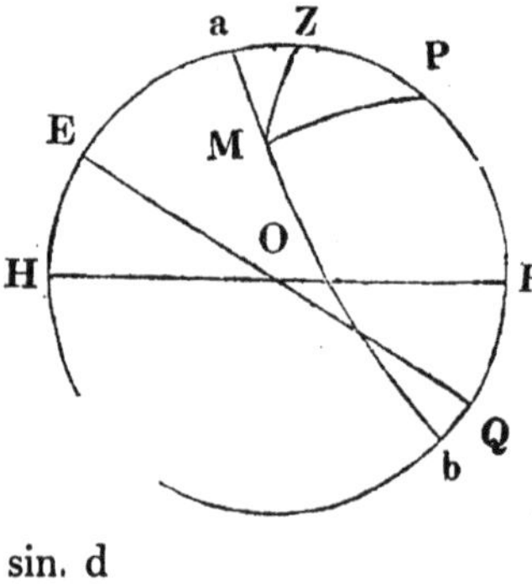

Let the latitude of the place $= 90^{\circ} - PZ = l$

Let the declination $= 90^{\circ} - PM = d$

$$\cos. ZPM = \frac{\cos. ZM - \cos. ZP \cos. PM}{\sin. ZP \sin. ZM}$$

or

$$\cos. t = \frac{\cos. \theta - \sin. l \sin. d}{\cos. l \cos. d}$$

$$\therefore \cos. \theta = \cos. t \cos. l \cos. d + \sin. l \sin. d$$

$\therefore$ magnitude of the tide $= M. \{\cos. t \cos. l \cos d + \sin. l \sin. d\}^2$
$\therefore$ for the same place and the same declination of the moon, the magnitude of the tide depends upon the value of ($\cos. t$). Now the greatest and least values of ($\cos. t$) are ($+ 1$) and ($- 1$), and since the moon only acts, it is high water when the moon is on the meridian, and the mean tide $= \frac{\text{greatest} + \text{least}}{2}$,

$$\text{greatest} = M. \{\sin. l \sin. d + \cos. l \cos. d\}^2$$
$$\text{least} = M. \{\sin. l \sin. d - \cos. l \cos. d\}^2$$
$$\therefore \frac{\text{greatest} + \text{least}}{2} = M. \{\sin.^2 l \sin.^2 d + \cos.^2 l \cos.^2 d\}$$
$$2 \sin.^2 l = 1 - \cos. 2 l$$
$$2 \sin.^2 d = 1 - \cos. 2 d$$
$$\therefore 4. \sin.^2 l \sin.^2 d = 1 - \{\cos. 2 l + \cos. 2 d\} + \cos. 2 l \cos. 2 d$$
$$2. \cos.^2 l = \cos. 2 l + 1$$
$$2. \cos.^2 d = \cos. 2 d + 1$$
$$\therefore 4. \cos.^2 l \cos.^2 d = 1 + (\cos. 2 l + \cos. 2 d) + \cos. 2 l \cos. 2 d$$
$$\therefore 4. \{\sin.^2 l \sin.^2 d + \cos.^2 l \cos.^2 d\} = 2 + 2. \cos. 2 l \cos. 2 d$$
$$\therefore \text{mean tide} = M. \sin.^2 l \sin.^2 d + \cos.^2 l \cos.^2 d$$
$$= M. \frac{1 + \cos. 2 l \cos. 2 d}{2}$$

It is low water at that place from whose meridian the moon is distant 90°, $\therefore \cos. \theta = 0$, $\therefore$ for low water

$$\cos. t = - \frac{\sin. l \sin. d}{\cos. l \cos. d} = - \tan. l \tan. d.$$

When $(l + d) = 90^\circ$, $\therefore \tan. l \tan. d = \tan. l \tan. (90^\circ - l)$

$$= \tan. l \cot. l = \frac{\tan. l}{\tan. l} = 1$$

$\therefore \cos. t = - 1$, $\therefore t = 180^\circ$, $\therefore$ time from the moon's passing the meridian in this case equals twelve hours, $\therefore$ under these circumstances there is but one tide in twenty-four hours.

When $l = d$, $\therefore \cos. t = - \tan.^2 l$.
and the greatest elevation $= M \{\cos. t \cos. l \cos. d + \sin. l \sin. d\}^2$

$$(\text{since } \cos. t = 1) = M. \{\cos.^2 l + \sin.^2 l\} = M.$$

When $d = 0$, $\therefore$ greatest elevation $= M \cos.^2 l$.

When $l = 0$, $\therefore$ greatest elevation $= M \cos.^2 d$.

At high water $t = 0$, $\therefore$ greatest elevation when the moon is in the meridian above the horizon, or, the superior tide $= M \{\cos. l \cos. d + \sin. l \sin. d\}^2 = M \cos.^2 (l - d) = T$.

For the inferior tide $t = 180^\circ$, $\therefore \cos. t = - 1$,

$\therefore$ inferior tide $= M\,\{\sin. l \sin. d - \cos. l \cos. d\}^2$

$= M\,\{-1\,(\cos. l \cos. d - \sin. l \sin. d)\}^2$

$= M \cos.^2 (l + d) = T'$.

Hence Robison's construction.

With C P = M, as a radius, describe a circle P Q p E representing

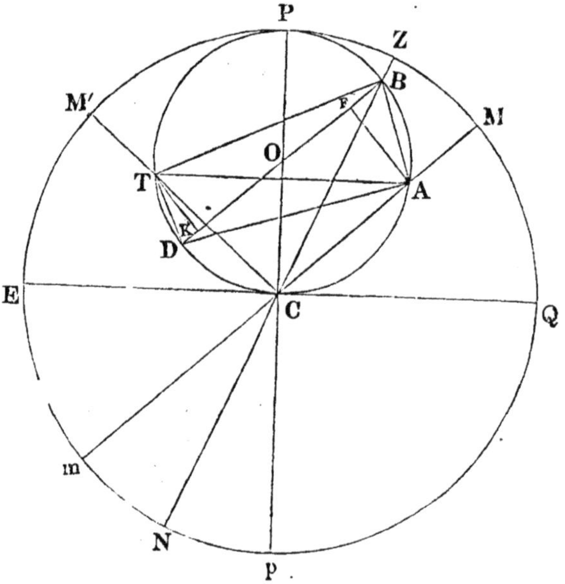

a terrestrial meridian; P, p, the poles of the earth; E Q the equator; (Z) the zenith; (N) the nadir of a place on this meridian; M the place of the moon. Then

$$\left.\begin{array}{r} \text{Z Q latitude of the place} = l \\ \text{M Q declination} = d \end{array}\right\} \therefore \text{Z M the zenith distance} = l - d.$$

Join C M, cutting the inner circle in A; draw A T parallel to E Q. Join C T and produce it to M′; then M′ is the place of the moon after half a revolution, ∴ M′ N = nadir distance

$= \text{M E} + \text{E N} = \text{M Q} + \text{Z Q} = l + d.$

Join C Z cutting the inner circle in B; join B with the center O and produce it to D; join A D, B T, A B, D T; and draw T K, A F perpendiculars on B D.

$$\left.\begin{array}{l} \angle \text{A D B} = \angle \text{B C A} = \text{Z Q} - \text{M Q} = l - d \\ \angle \text{T D B} = 180^\circ - \angle \text{T C B} = \angle \text{M C N} = l + d \end{array}\right\} \text{and the angles B A D,}$$

B T Z are right angles

$$\text{B D} : \text{D A} :: \text{D A} : \text{D F} = \frac{\text{D A}^2}{\text{B D}} = \frac{\text{B D}^2 . \cos. \text{B D A}}{\text{B D}} = \text{BD}.\cos.^2 (l - d)$$

$= M \cos.^2 (l - d) =$ height of the sup^r. tide.

Again

$$B D : D T :: D T : D K = \frac{D T^2}{B D} = \frac{B D^2 . \cos. B D T}{B D} = B D \cos. \overline{l + d}$$

$$= M \cos. \overline{l + d} = \text{point of the inferior tide.}$$

If the moon be in the zenith, the superior tide equals the maximum.

For then $l - d = 0$, $\therefore \cos. \overline{l - d} = \text{maximum}$, and $B D = D F$.

If the moon be in the equator, $d = 0$, $\therefore D F = D K$.

The superior tide $= M \cos.^2 (l - d) = T$

The inferior tide $= M \cos.^2 (l + d) = T$.

Now $T > T'$, if (d) be positive, i. e. if the moon and place be both on *the same* side of the equator.

$T < T'$ if (d) be negative, i. e. if the moon and place be on *different* sides of the equator.

If (d) $= 90^\circ - l$, $\therefore D K = M \cos.^2 (l + 90^\circ - l) = M \cos.^2 90^\circ = 0$.

If (d) $= 90^\circ + l$, and in this case (d) be positive, and (l) negative,

$\therefore D F = \cos.^2 (d - l). M = M \cos.^2 (90^\circ + l - l) = M \cos.^2 90^\circ = 0$.

PROBLEMS

FOR

VOLUME III.

PROB. I. *The altitude* P R *of the pole is equal to the latitude of the place.*

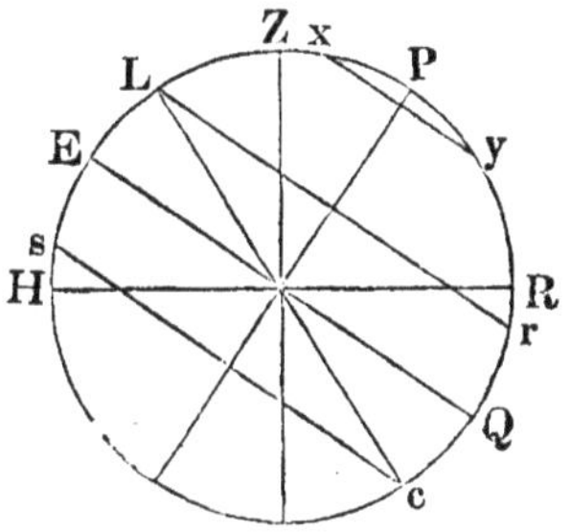

For Z E measures the latitude.

= P R by taking Z P from E P and Z R.

PROB. 2. *One half the sum of the greatest and least altitudes of a circumpolar star is equal to the altitude of the pole.*

The greatest and least altitudes are at x, y on the meridian.

Also

$$xR + yR = xy + 2yR = 2(Py + Ry) = 2 \text{. altitude of the pole.}$$

PROB. 3. *One half the difference of the sun's greatest and least meridian altitudes is equal to the inclination of the ecliptic to the equator.*

The sun's declination is greatest at L, at which time it describes the parallel L r.

$\therefore$ L H is the greatest altitude,

The sun's declination is least at C, when it describes the parallel s C.

$\therefore$ s H is the least altitude,

and

$$\tfrac{1}{2} . (LH - sH) = \tfrac{1}{2} Ls = LE.$$

PROB. 4. *One half the sum of the sun's greatest and least meridian altitudes is equal to the colatitude of the place.*

$$\tfrac{1}{2}(LH + sH) = \tfrac{1}{2}(HE + EL + HE - Es)$$
$$= \tfrac{1}{2}(2\,HE) = HE.$$

PROB. 5. *The angle which the equator makes with the horizon is equal to the colatitude* = E H.

PROB. 6. When the sun describes b a in twelve hours, he will describe c a in six; if on the meridian at a it be noon, at c it will be six o'clock. Also at d he will be due east. He travels 15° in one hour. The angle a P x, measured by the number of degrees contained in a x (supposing x equals the sun's place), converted into the time at the rate of 15° for one hour, *gives the time from apparent noon*, or from the sun's arrival at a.

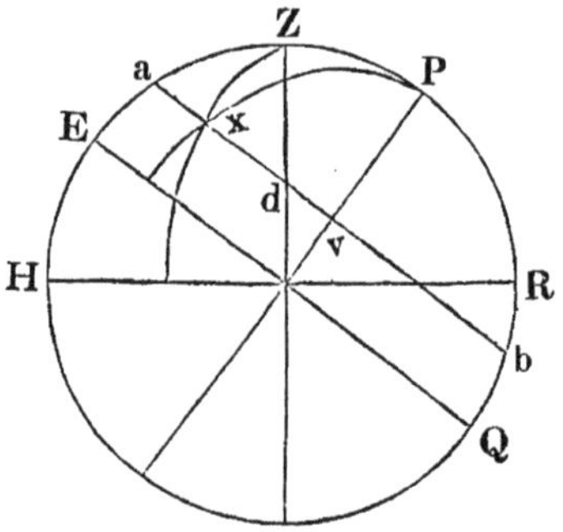

PROB. 7. *Given the sun's declination, and latitude of the place ; find the time of rising, and azimuth at that time.*

Given Z E, ∴ Z P = colat. given.
Given b c, ∴ P b = codec. given.
Given b Z = 90°.

Required the angle Z P b, measuring a b, which measures the time from sun rise to noon.

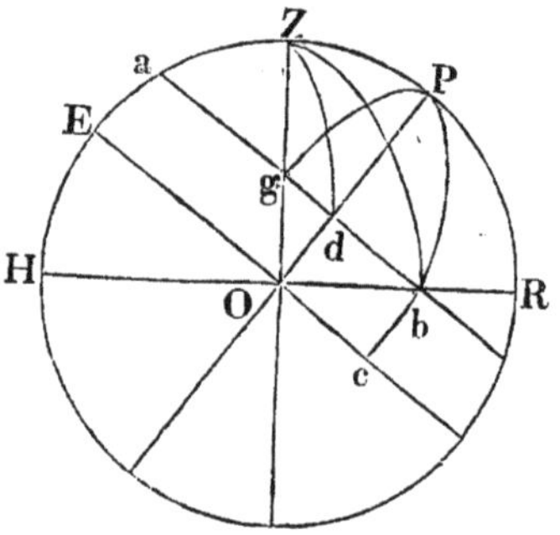

Take the angles adjacent to the side 90°, and complements of the other three parts, for the circular parts.

∴ r . cos. Z P b = cot. Z P cot. P b

or

r . cos. hour ∠ = tan. lat. tan. dec.

∴ log. tan. lat. + log. tan. dec. — 10 = log. cos. hour ∠ required.

Also the angle P Z b measures b R, the azimuth referred to the north, and

r . cos. P b = cos. P Z . cos. Z

$$\therefore \cos. Z = \frac{r . \cos. p}{\sin. L}.$$

PROB. 7. (a) r. cos. hour ∠ = tan. latitude tan. declination, *for sun rise.*

Hence *the length of the day* = 2 . cos. hour ∠ = $\frac{2 . \text{tan. lat. tan. dec.}}{}$

h may be found thus, from △ Z P b cos. $h = \frac{\text{cos. Z b} - \text{Z cos.P. cos.P b}}{\text{sin. Z P . sin. P b}}$

$= (\text{since Z b} = 90^0,) - \frac{\text{sin. L}}{\text{cos. L}} \cdot \frac{\text{cos. p}}{\text{sin. p}}$, or since h > 90⁰,

$-$ cos. h $=$ $-$ tan. L . cot. p, or cos. h $=$ tan. L . cot. p.

and the angle P Z b may be similarly found,

$$\text{r . cos. Z} = \frac{\text{cos. P b} - \text{cos. Z P . cos. Z b}}{\text{sin. Z P . sin. Z b}}$$

$$= \frac{\text{cos. p}}{\text{cos. L}}.$$

Prob. 8. *Find the sun's altitude at six o'clock in terms of the latitude and declination.*

The sun is at d at six o'clock. The angle Z P d = right angle.

Z p = colat. P d = codec. Required Z d (= coalt.)

r . cos. Z d = cos. Z P . cos. d P

or

r . sin. altitude = sin. latitude × sin. declination.

Prob. 9. *Find the time when the sun comes to the prime vertical (that vertical whose plane is perpendicular to the meridian as well as to the horizon), and his altitude at that time, in terms of the latitude and declination.*

Z P = colatitude. P g = codeclination. The angle P Z g = right angle. Required the angle Z P g.

∴ r . cos. Z P g = tan. Z P . cot. P g.

= cot. latitude tan. declination.

Also required Z g equal to the coaltitude,

r . cos. P g = cos. P Z . cos. Z g.

$$\therefore \frac{\text{r . sin. declination}}{\text{sin. latitude}} = \text{sin. altitude.}$$

Prob. 10. *Given the latitude, declination, and altitude of the sun; find the hour and azimuth.*

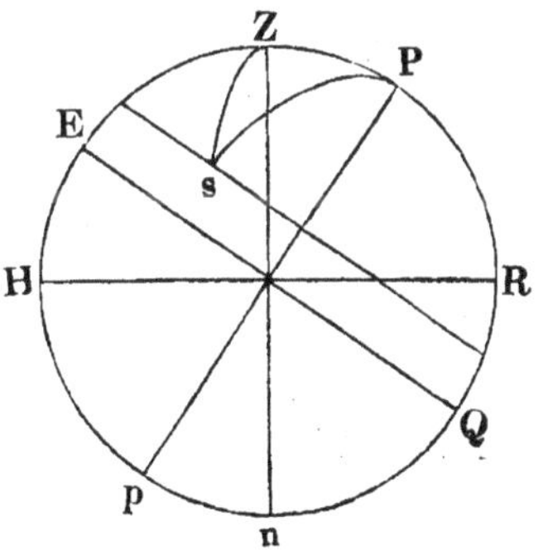

Let s be the place.

Given Z P, Z s, P s. Find the angle Z P s.

Let Z P, Z s, P s = a, b, c, be given, to find B.

$$\text{sin. B} = \frac{2\,r}{\text{sin. a . sin. c}} \times$$

$$\sqrt{\text{s . (s} - \text{a) . (s} - \text{b) . (s} - \text{c)}}$$

where $s = \frac{a + b + c}{2}$.

Also find C. $\sqrt{\qquad\qquad}$. (Or by Nap. 1st and 2d Anal.)

$$\sin. C = \frac{2 r}{\sin. a . \sin. b}.$$

$$\text{Similarly, } \sin. A = \sin. \angle \text{ of position} = \frac{2 r}{\sin. b . \sin. c} \sqrt{\qquad}.$$

Prob. 11. *Given the error in the altitude. Find the error in the time in terms of latitude and azimuth.*

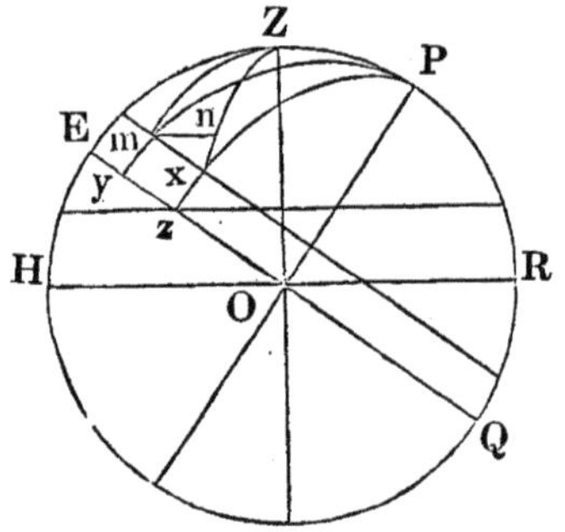

Let m n be parallel to H, and n x be the error in the altitude.

$\therefore \angle$ m P x = error in the time = y z.

y z : m x : : rad. : cos. m y

m x : x n : : rad. : sin. n m x

$\therefore$ y z : x n : : r^2 : cos. my. sin. n m x

or

$$y z = \frac{r^2 . n x}{\cos. m y . \sin. n m x}$$

$$= \frac{r^2 . n x}{\cos. m y . \sin. Z x P};$$

but

$$\frac{\sin. Z x P}{\sin. x Z P} = \frac{\sin. Z P}{\sin. P x}$$

$$\therefore \sin. Z x P = \frac{\sin. P Z . \sin. x Z P}{\cos. m y}$$

$$\therefore y z = \frac{r^2 . n x}{\cos. L. \sin. \text{azimuth}}.$$

Cor. Sin. of the azimuth is greatest when a z = 90°, or when the sun is on the prime vertical, $\therefore$ y z is then least.

Also, the perpendicular ascent of a body is quickest on the prime vertical, for if y z and the latitude be given, n x $\propto$ azimuth, which is the greatest.

Prob. 12. *Given the latitude and declination. Find the time when twilight begins.*

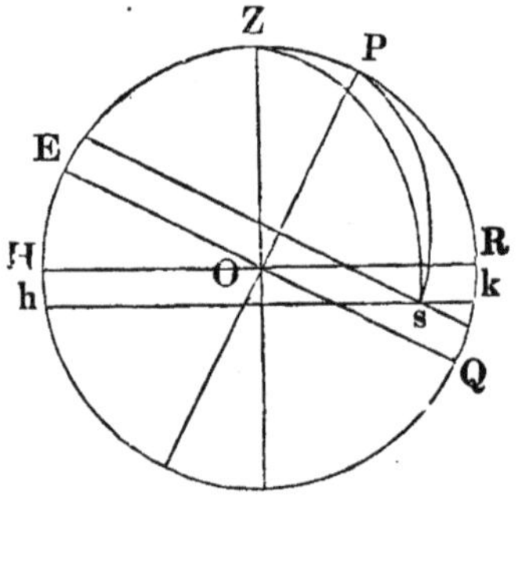

(Twilight begins when the sun is 18° below the horizon.)

h k is parallel to H R and 18° below H R.

$\therefore$ Twilight begins when the sun is in h k.

$\therefore$ Z s = 90° + 18°, P s = D, Z P = colat.
Find the angle Z P s.

PROB. 13. *Find the time when the apparent diurnal motion of a fixed star is perpendicular to the horizon in terms of the latitude and declination.*

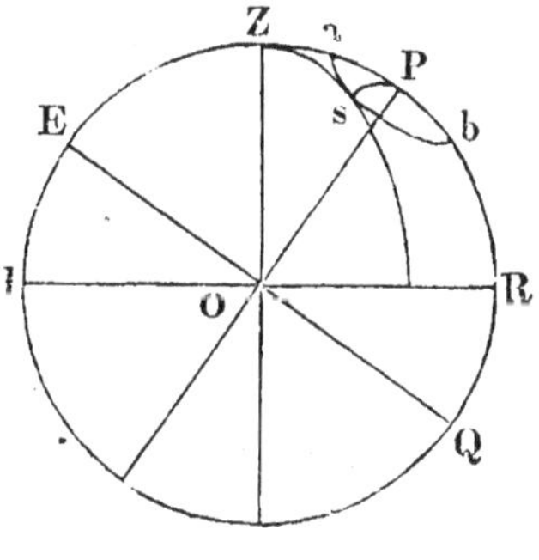

Let a b be the parallel described by the star.

Draw a vertical circle touching it at s.

∴ s is the place where the motion appears perpendicular to H R.

∴ Z P, P s, and ∠ Z S P $= 90^0$ is given.
Find Z P s.

PROB. 14. *Find the time of the shortest twilight, in terms of the latitude and declination.*

a b is parallel to H R 18^0 below H R.

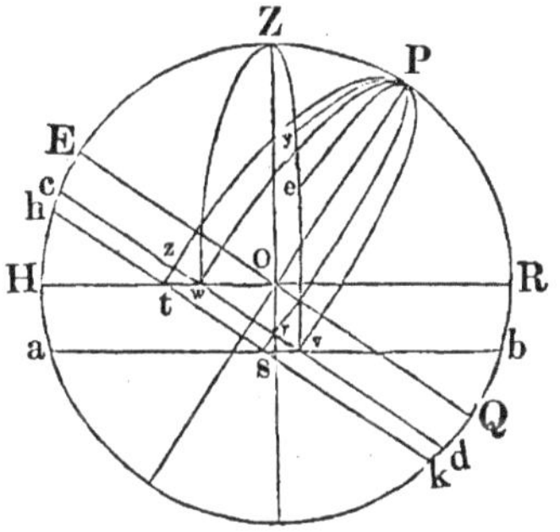

The parallels of declination c d, h k, are indefinitely near each other.

The angles v P w, s P t, measure the durations of twilight for c d, h k.

Since twilight is shortest, the increment of duration is nothing.

∴ v P w = s P t

∴ v r = w z

and r s = t z

and the angle v r s = right angle

= w z t.

∴ ∠ r v s = z w t, and ∠ Z w c $= 90^0 -$ z w t $= 90^0 -$ Z w P.

∴ ∠ z w t = Z w P.

Similarly,

∠ r v s = Z v P

∴ Z w P = Z a P.

Take v e $= 90^0$. Join P e. Draw P y perpendicular to Z c.

In the triangles Z P w, P v e, Z w = e v, P w = P v, and the angles contained are equal, ∴ Z P = P e.

∴ In the triangles Z P y, P e y, Z P = P e, P y com; and the angles at y are right angles.

∴ Z e is bisected in y.

r . cos. P v = cos. P y . cos. v y

r . cos. P e = cos. P y . cos. y e.

$\therefore$ cos. P v : cos. P e :: cos. v y : cos. y e

(but v y is greater than 90°, $\therefore$ therefore cos. v y is negative.)

:: — cos. (— compl. y e) : cos. y e

:: sin. y e : cos. y e

:: tan. y e : r.

$$\therefore \cos. p = \frac{\sin. L. \tan. y e}{r} = \frac{\sin. L. \tan. \frac{18^0}{2}}{r} = \frac{\sin. L. \tan. 9^o}{r}$$

P Z is never greater than 90°, Z y is equal to 9, $\therefore$ P y is never greater than 90°, $\therefore$ cos. P y is always positive; v y is always greater than 90°, $\therefore$ cos. v y is always negative, $\therefore$ cos. P v is negative, $\therefore$ the sun's declination is south.

Also, if instead of R b = 18°, we take it equal to 2 s equal the sun's diameter, we get from the expression $\sin. D = \frac{\sin. L. \tan. s}{r}$ the time when the sun is the shortest time in bringing his body over the horizon.

PROB. 15. *Find the duration of the shortest twilight.*

$\angle$ w P Z = v P e, $\therefore \angle$ Z P e = v P w.

$\therefore$ 2 Z P e is equal to the duration of the shortest twilight.

r . sin. Z y = sin. Z P . sin. Z P y

or

$$\sin. Z P y = \frac{\sin. 90^0 . r}{\cos. L.},$$

which doubled is equal to the duration required.

PROB. (A). Given the sun's azimuth at six, and also the time when due east. Find the latitude.

From the triangle Z P c,

r . cos. L = tan. P c . cot. P Z c.

From the triangle Z P d,

r . cos. h = cot. L . cot. P d.

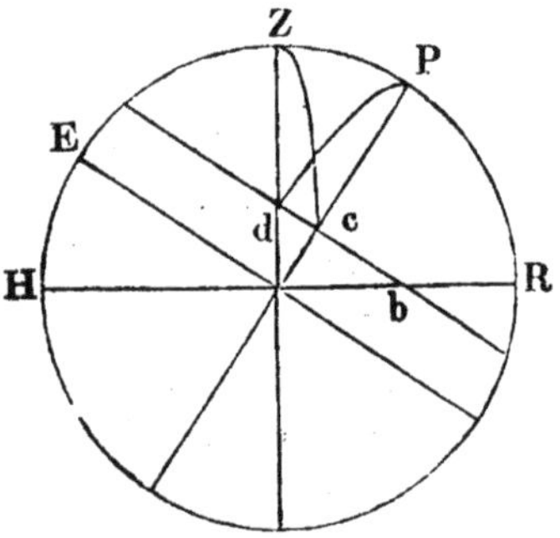

$$\therefore \tan. P c = \frac{\cos. L}{\cot. Z}$$

$$\cot. P d = \frac{\cos. h}{\cot. L}.$$

$$\therefore \tan. P d = \frac{\cot. L}{\cos. h}$$

$$\therefore \frac{\cos. L}{\cot. Z} = \frac{\cot. L}{\cos. h}$$

$$\therefore \sin. L = \frac{\cot. Z}{\cos. h}.$$

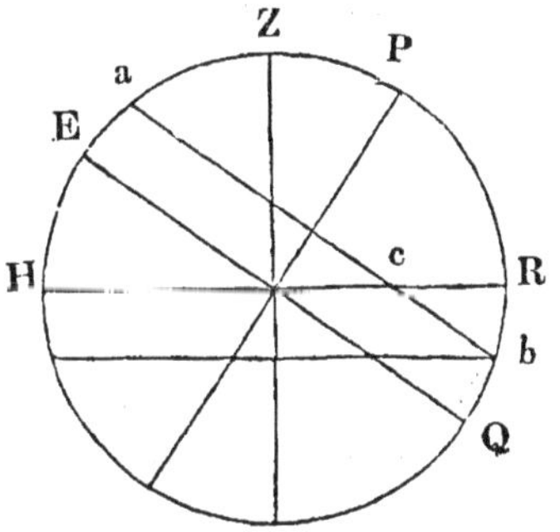

Prob. 16. *Find the declination when it is just twilight all night.*

Dec. $bQ = QR - bR$

$= \text{colat.} - 18^0$

$= 90^0 - L - 18^0$

$= 72^0 - L$

Prob. 17. Given the declination, find the latitude, the sun being due east, when one half the time has elapsed between his rising and noon.

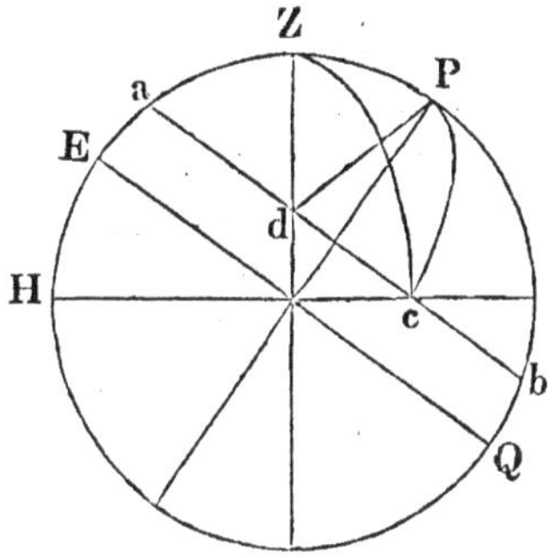

Given $\angle ZPc$, and $ZPd = \frac{1}{2} ZPc$.

Given also $Pd = p$,

and $\angle PZd$ right angle.

$\because$ by Nap.

$r \cdot \cos. h = \tan. ZP \cdot \cot. p$

$\because \cot. L = \dfrac{r \cdot \cos. h}{\cot. p}$.

If the angle ZPc be not given.

From the triangle ZPd,

$. \cos. ZPd = \tan. ZP \cdot \cot. p.$

From the triangle ZPc,

$r \cdot \cos. ZPc = \cot. ZP \cdot \cot. p$,

or $\left.\begin{array}{l}\cos. h = \cot. \lambda. \cot. p \\ \cos. 2h = \tan. \lambda. \cot. p\end{array}\right\}$

$$= 2 \cos.^2 h - 1 = 2 \cot.^2 \lambda. \cot.^2 p - 1 = \frac{2 \cot.^2 p - 1}{\tan.^2 \lambda}.$$

$$\therefore \tan.^3 \lambda. \cot. p = 2 \cot.^2 p - \tan.^2 \lambda.$$

$$\therefore \tan.^3 \lambda + \frac{\tan.^2 \lambda}{\cot. p} - 2 \cot. p = 0,$$

from the solution of which cubic equation, $\tan. \lambda$ is found.

Prob. 18. Given the angle between two and three o'clock in the horizontal dial equal to a. Find the longitude.

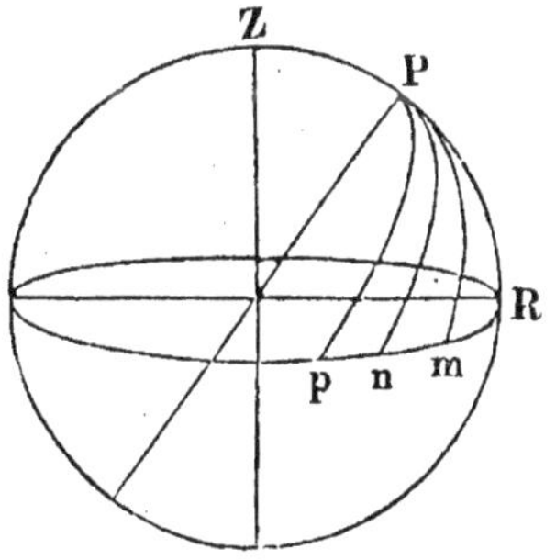

From the triangle PRn,

$r . \sin. PR = \tan. Rn . \cot. 30$

$= \tan. Rn . \sqrt{3}.$

From the triangle PRp,

$r . \sin. PR = \tan. Rp . \cot. 45$

$= \tan Rp.$

$$\therefore \tan. np = \tan. a = \tan. Rp - Rn$$
$$= \frac{\tan. Rp - \tan. Rn}{1 + \tan. Rp \cdot \tan. Rn}$$
$$= \frac{\sin. \lambda \left(1 - \frac{1}{\surd 3}\right)}{1 + \frac{\sin.^2 \lambda}{\surd 3}} = \frac{\sin. \lambda . (\surd 3 - 1)}{\surd 3 + \sin.^2 \lambda}.$$

PROB. 19. In what longitude is the angle between the hour lines of twelve and one on the horizontal dial equal to twice the angle between the same hour lines of the vertical sun dial?

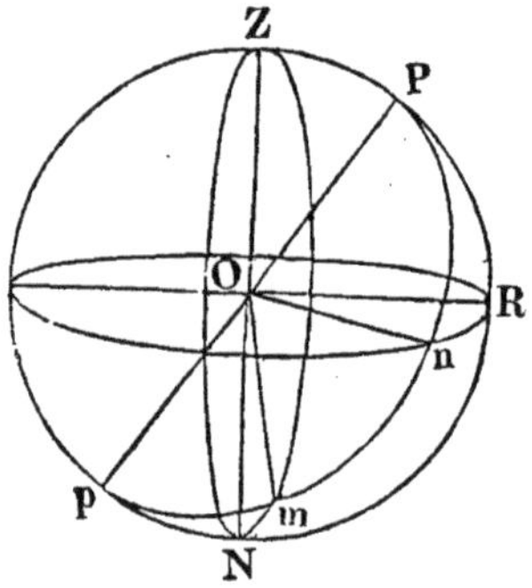

From the triangle P R n,

$$\sin. \lambda = \cot. 15 . \tan. Rn$$

From the triangle p N m,

$$\sin. pM = \cot. 15 . \tan. Nm$$
$$= \cos. \lambda = \cot. 15 \cdot \tan. \frac{Rn}{2}.$$
$$\therefore \frac{\sin. \lambda}{\cos. \lambda} = \frac{\tan. Rn}{\tan. \frac{Rn}{2}} = \tan. \lambda$$
$$= \frac{\frac{\tan. \frac{Rn}{2} + \tan. \frac{Rn}{2}}{1 - \tan.^2 \frac{Rn}{2}}}{\tan. \frac{Rn}{2}} = \frac{2}{1 - \tan.^2 \frac{Rn}{2}}.$$

PROB. 20. *Given the altitude, latitude, and declination of the sun, find the time.*

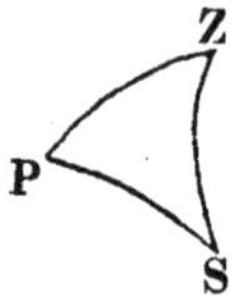

$$\cos. h = \frac{\cos. ZS - \cos. ZP . \cos. PS}{\sin. ZP . \sin. PS}$$
$$= \frac{\sin. A - \sin. L . \cos. p}{\cos. L . \sin. p}$$
$$\therefore 1 + \cos. h = \frac{\cos. L . \sin. p + \sin. A - \sin. L . \cos. p}{\cos. L . \sin. p}$$
$$= \frac{\sin. (p - L) + \sin. A}{\cos. L . \sin. p}$$

or

$$2 \cos.^2 \frac{h}{2} = \frac{2 . \cos \left(\frac{A + p - L}{2}\right) . \sin. \left(\frac{A + L - p}{2}\right)}{\cos. L . \sin. p}$$

$\therefore \cos.^2 \frac{h}{2} = \left(\frac{\cos.\,(\text{——}).\,\sin.\,(\text{——})}{\text{———}}\right)$ the form adapted to the Logarithmic computation, or, see Prob. (18).

Prob. 21. Given a star's right ascension and declination. Find the latitude and longitude of the star.

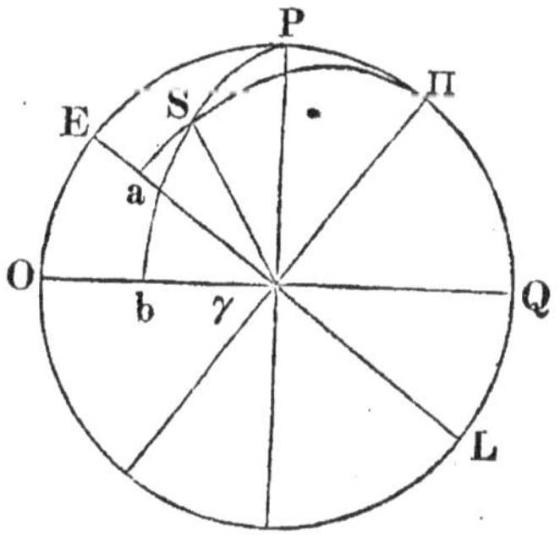

Given

γ b, b S, ∠ S b γ right angle

∴ find ∠ S γ b and S γ.

∴ ∠ S γ a = S γ b — Obl.

∴ S γ is known, ∠ S γ a is known

and S a γ is a right angle,

∴ find S a = latitude

γ a = longitude.

Given the sun's right ascension and declination. Find the obliquity of the ecliptic.

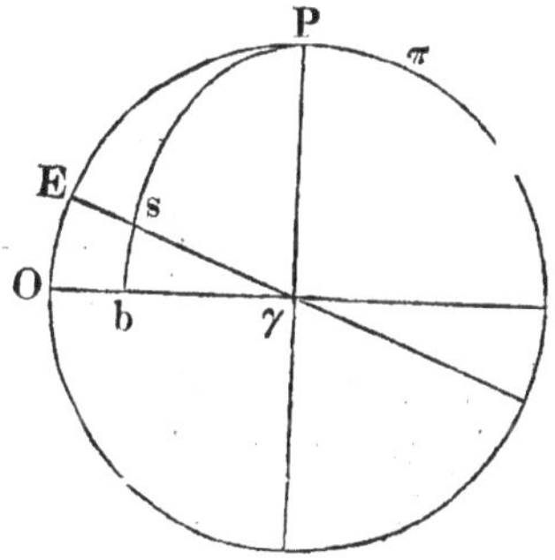

P S being known P γ = 90°, ∠ S P γ = R A,

∴ in the △ S P γ, ∠ S γ P is known.

∴ obliquity = 90° — S γ P is known.

Prob. 22. In what latitude does the twilight last all night? Declination given.

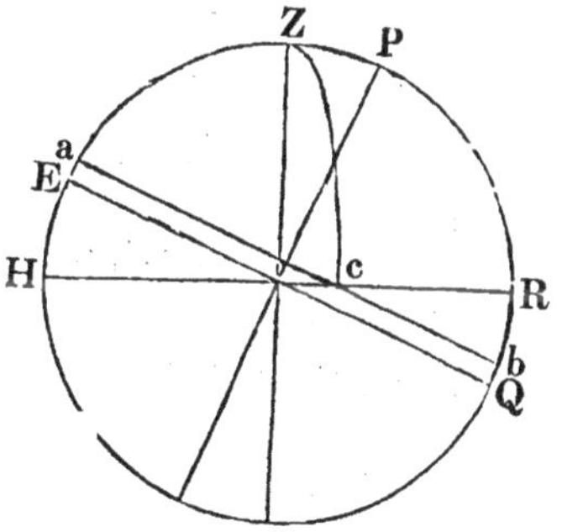

(Twilight begins when the sun is 18° below the horizon in his ascent, and ends when he is there in his descent, lasting in each case as long as he is in travelling 18°.)

R Q = H E = colat. = b Q + b R

= D + 18°.

∴ 90° — 18 — D = L

= 72° — D.

(See Prob. 16.)

Find the general equation for the hour at which the twilight begins.

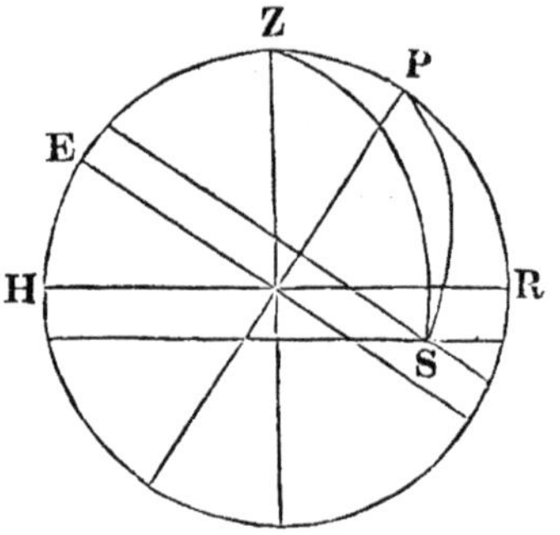

Let the sides P Z, P S, Z S, be a b c.

$$\text{Then } \sin.^2 \frac{C}{2} = \frac{\sin.\left(\frac{a+b+c}{2} - a\right)\sin.\left(\frac{a+b+c}{2} - b\right)}{\sin.\ a\ .\ \sin.\ b}.$$

or

$$\sin.^2 \frac{h}{2} = \frac{\left\{\begin{array}{l}\sin.\left(\frac{\text{colat.} + p + 108^0}{2} - \text{colat.}\right) \\ \times\left(\frac{\text{sin. cotan.} + p + 108^0}{2} - p\right)\end{array}\right\}}{\cos.\ L\ .\ \sin.\ p}.$$

Prob. 24. Given the difference between the times of rising of the stars, and their declinations: required the latitude of the place.

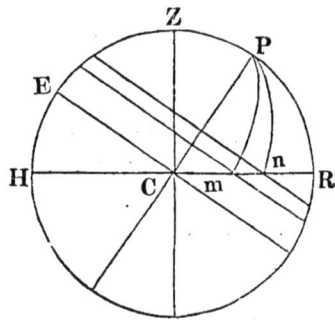

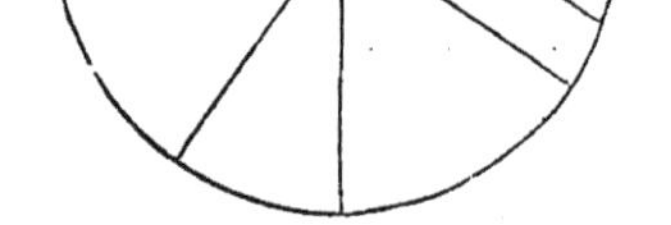

Given P m, P n, and the ∠ m P n included.

From Napier's first and second analogies, the ∠ P m n is known,

∴ P m C = complement of P m n is known,

∴ P C = 90°, P m is given, and the ∠ P m C is found,

∴ P R = latitude is known.

Prob. 25. Given the sun in the equator, also latitude and altitude: find the time.

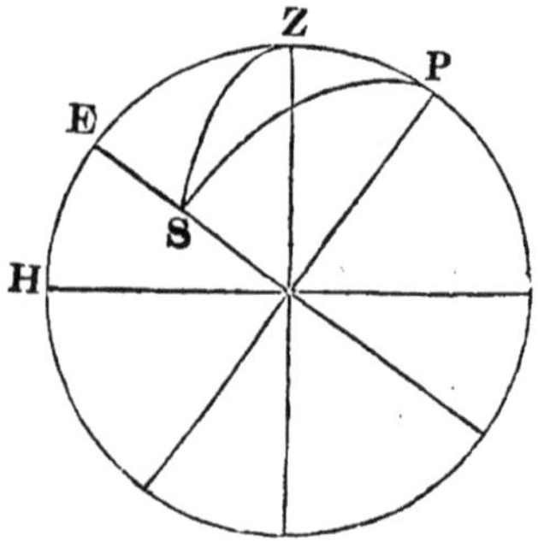

Given

Z P, Z S, P S = 90° find the ∠ Z P S.

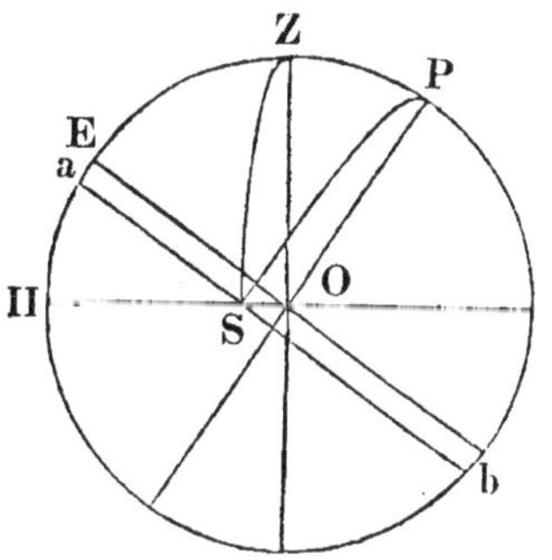

PROB. 26. The sun's declination $= 8^0$ south, required the latitude, when he rises in the south-east point of the horizon, and also the time of rising.

$P S = 90^0 + 8^0$, $Z S = 90^0$, $\angle S Z P = 45^0 + 90^0$.

Find Z P, and the $\angle$ Z P S.

PROB. 27. Determine a point in E Q, that the sum of the arcs drawn from it to two given places on the earth's surface shall be minimum.

Let A, B, be the spectator's situations, whereof the latitude and longitude are known.

Let E Q be the equator, p the point required; a b = difference of the longitudes is known. Let a p = x.

$\therefore$ p b = a — x. Let L, L′ be the latitudes.

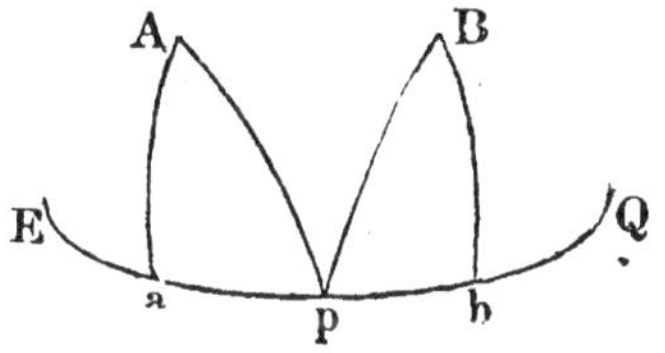

In $\triangle$ A a p, r. cos. A p = cos. L′. cos. x.

In $\triangle$ B b p, r. cos. B p = cos. L′. cos. $\overline{a - x}$,

$\therefore$ cos. L . cos. x + cos. L′. cos. (a—x) = max.

$\therefore$ cos. L . (— sin. x) . d x + cos. L′. $\times$ sin. (a — x). (— d x) = 0,

$\therefore$ — cos. L . sin. x = cos. L′. sin. a. cos. x — cos. L′. cos. a. sin. x.

Let sin. x = y

$$\therefore -\cos. L . y = \cos. L'. \sin. a. \sqrt{1 - y^2} - \cos. L'. \cos. a. y$$

$\therefore$ transposing and squaring

$$\cos.^2 L. y^2 - 2. \cos. L. \cos. L'. \cos.^2 y^2 + \cos.^2 L'. \cos.^2 a. y^2$$
$$= \cos.^2 L'. \sin.^2 a - \cos.^2 L'. \sin.^2 a\, y^2,$$

$$\therefore y^2 = \&c. = n. \text{ and } y = \sqrt{n}.$$

PROB. 28. To a spectator situated within the tropics, the sun's azimuth will admit of a maximum twice every day, from the time of his leaving the solstice till his declination equal the latitude of the place. Required proof.

a b the parallel of declination passing through capricorn.

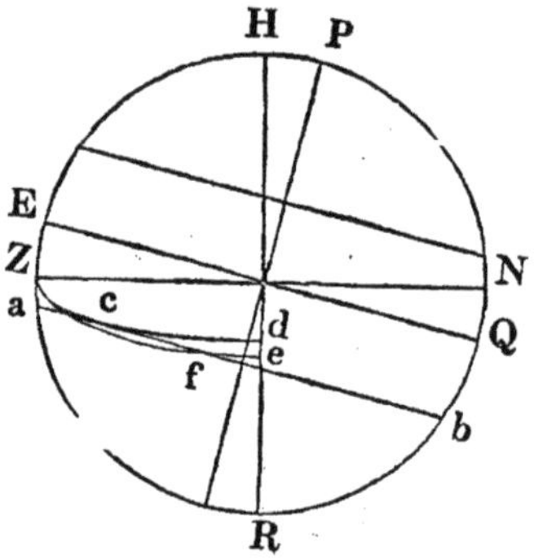

From Z a circle may be drawn touching the parallel of the declination till this parallel coincides with Z. ∴ every day till that time the sun will have a maximum azimuth twice a day, and at that time he will have it only once at Z.

(Also the sun will have the same azimuth twice a day, i. e. he will be twice at f.)

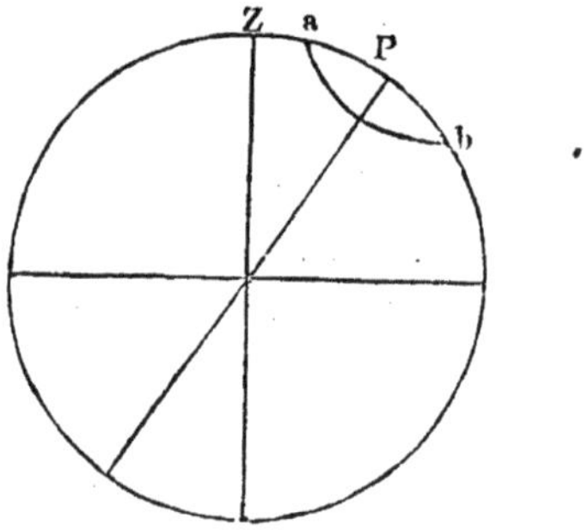

PROB. 29. The true zenith distance of the polar star when it first passes the meridian is equal to m, and at the second passage is equal to n. Required the latitude.

Given b Z = m, a Z = n,

$$Z P = \text{colat.} = \tfrac{1}{2}.\ \overline{m + n}.$$

PROB. 30. If the sun's declination E e, is greater than E Z, draw the circle Z m touching the parallel of the declination,

∴ R m is the greatest azimuth that day

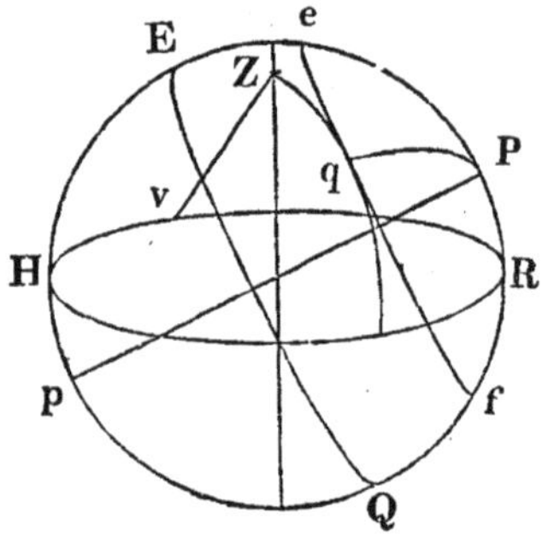

If Z v be a straight line drawn perpendicular to the horizon, the shadow of this line being always opposite the sun, will, in the morning as the sun rises from f, recede from the south point H, till the sun reaches his greatest azimuth, and then will approach H; also twice in the day the shadow will be upon every particular point, because the sun has the same azimuth twice a day, in this situation. ∴ shadow will go backwards upon the horizon.

But if we consider P p a straight line or the earth's axis produced, the sun will revolve about it, ∴ the shadow will not go backwards.

$$r.\ \text{cot.}\ Z P q = \text{tan.}\ P q.\ \text{cot.}\ P Z,$$

or

cot. (time of the greatest azimuth) = tan. p. tan. L.

All the bodies in our system are elevated by refraction 33′, and depressed by parallax.

∴ at their rise they will be distant from Z, $90° + 33'$ — horizontal parallax.

A fix d star has no parallax, ∴ distance from Z = $90° + 33'$.

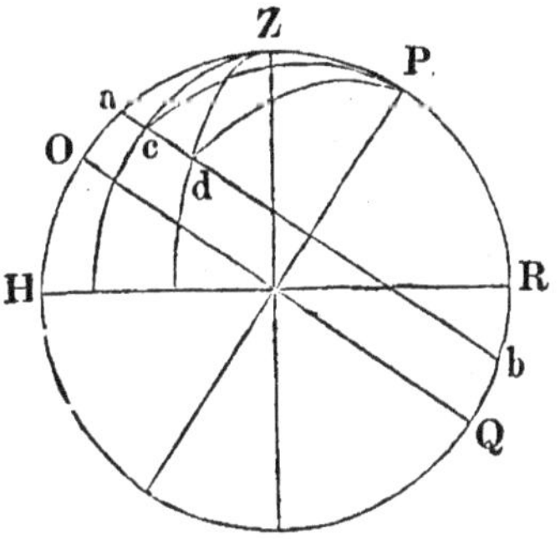

Prob. 31. Given two altitudes and the time between them, and the declination. Find the latitude of the place.

Given Z c, Z d, P c, P d, ∠ c P d.

From △ c P d, find c d, and ∠ P d c.

From △ Z c d, find ∠ Z d c,

∴ Z d p = c d P — c d Z,

∴ From △ Z P d, find Z P = colat.

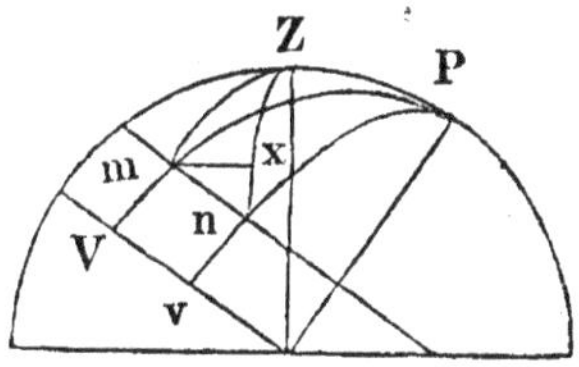

Prob. 32. To find the time in which the sun passes the meridian or the horizontal wire of a telescope.

Let m n equal the diameter of the sun equal d″ in space.

V v : m n : : r : cosine declination,

$$\therefore \text{V v} = \frac{\text{m n}}{\text{cosine declination}} \text{ radius } 1,$$

= d″. second declination in seconds of space,

∴ 15″ in space : 1″ in time

$$: : \text{d}'' \text{ second dec.} : \frac{\text{d}'' \text{ second dec.}}{15''}$$

= time in seconds of passing the merid

Hence the sun's diameter in R A = V v = d″. second declination.

(n x = d″ = sun's diameter)

V v : m n : : r : sin. P n

m n : n x : : r : sin. x n P

V v : n x : : r^2 : sin. P n . sin. Z n P,

$$\therefore \text{V v} = \frac{r^2 . \text{n x}}{\text{sin. P n . sin. Z n P}} = \frac{r^2 \text{n x}}{\text{sin. Z P . sin. P Z n}}$$

$$= \frac{r^2 . \text{d}''}{\cos. \lambda . \text{ sin. azimuth}}$$

$$\therefore \text{time of describing V v} = \frac{r^2 . \text{d}''}{15'' . \cos. \lambda . \text{ sin. azimuth}}$$

which also gives the time of the sun's rising above the horizon.

Prob. 33. *Flamstead's method of determining the right ascension of a star.*

Lemma. The right ascension of stars passing the meridian at different times, differs as the difference of the times of their passing.

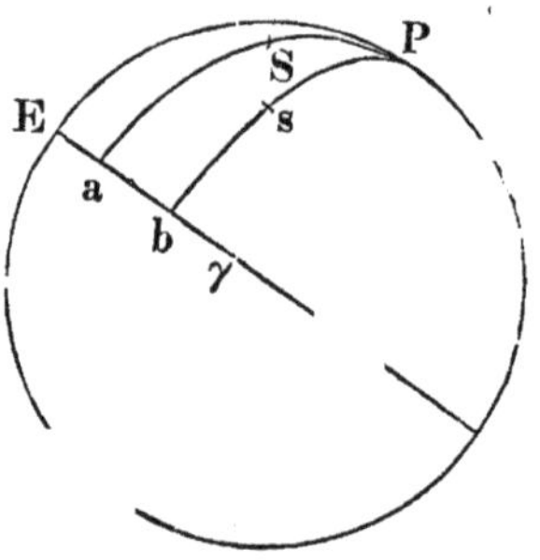

For the angle a P b measures the difference of the times of passing, which is measured by a b = a γ — b γ.

Hence, as the interval of the times of the succeeding passages of any fixed star : 360 (the difference of its right ascensions between those times) : : the interval between the passages of any two fixed stars : to the difference of their right ascensions.

Let A G c be the equator, A B C the ecliptic, S the place of a star, S m a secondary to the equator. Let the sun be near the equinox at P, when on the meridian.

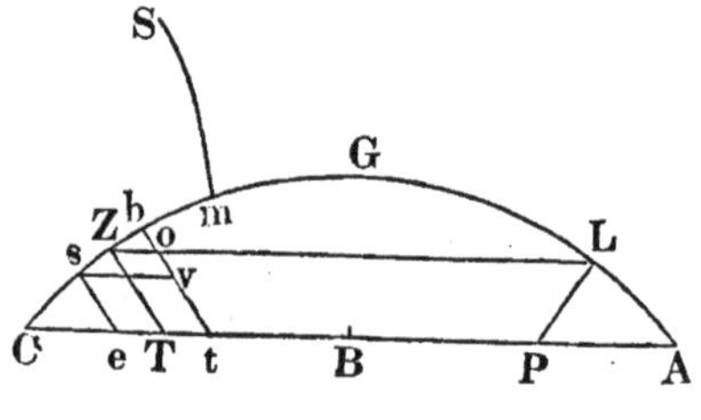

Take C T = P A, ∴ the sun's declination at T = that at P. Draw P L, T Z, perpendicular to A G c.

∴ Z L parallel to A C.

Observe the meridian altitude of the sun at P, and the time of the passage of his center over the meridian.

Observe what time the star passes over the meridian, thence find the apparent difference of their right ascensions.

When the sun approaches T, observe his meridian altitude on one day, when he is close to T, and the next day when he has passed through T, so that at t it may be greater, and at e less than the meridian altitude at P. Draw t b, and e s, perpendiculars.

Observe on the two days before mentioned, the differences b m, s m, of the sun's right ascension, and that of the star.

Draw s v parallel to A C.

Considering the variation of the right ascension and declination to be uniform for a short time, v b (change of the meridian altitudes in one day) : o b difference of the declinations) : : s b (= s m — b m) : Z b., Whence Z b. Add or substract Z b to or from T m. Whence Z m. Add, or take the difference of, (according to circumstances), Z m, L m, whence Z L, $\therefore \frac{180 - Z L}{2}$ gives A L, the sun's right ascension at the time of the first observation.

$\therefore$ A L + L m = the star's right ascension. Whence the right ascension of all the stars.

PROB. 34. Given the altitudes of two known stars. Find λ.

Right ascensions being known, $\therefore$ â b = the difference of right ascensions, is known,

$\therefore \angle$ a P b is known.

$\therefore$ From $\triangle$ s P σ, $\angle$ s σ P is known, and σ s,

From $\triangle$ Z s σ, $\angle$ s σ Z is known,

$\therefore \angle$ Z σ P is known,

$\therefore$ from $\triangle$ Z σ P, Z P is known.

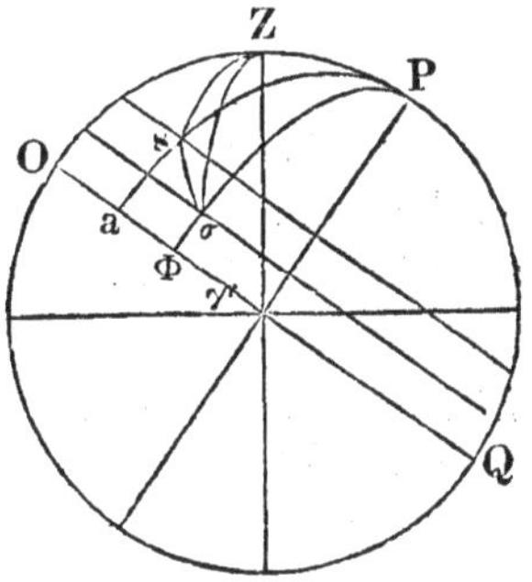

PROB. 35. Given the apparent diameter of a planet, at the nearest and most distant points of the earth's orbit. Required the radius of the planet's orbit.

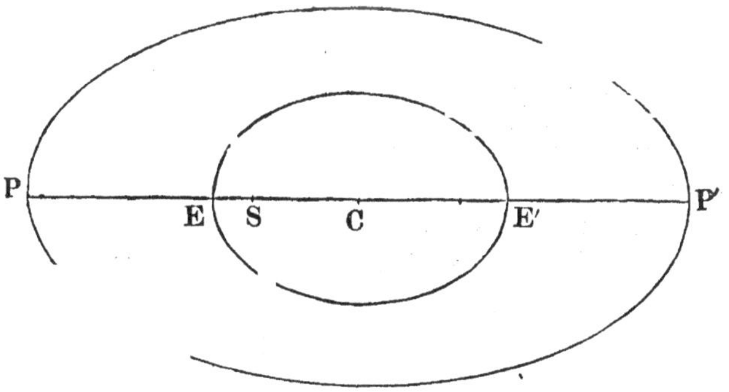

$D \propto \frac{1}{\text{distance}}$; D greatest, D′ nearest diameter.

$$\therefore D : D' :: E P : E' P$$
$$:: E P - E C : C P + E C,$$
$$\therefore D\, C P + D\, E C = D'\, C P - D'\, E C,$$
$$\therefore C P = E C\, \frac{D + D'}{D' - D}.$$

PROB. 36. Given the sun's greatest apparent diameter, and least, as 101 and 100. Find the excentricity of the earth's orbit.

$D \propto \frac{1}{\text{rad.}}$; the sun at S, P P′ the earth's orbit.

$$100 : 101 :: S P : S P' :: C P - C S : C P + C S$$

$$\therefore 100\, C P + 100\, C S = 101\, C P - 101\, C S$$

$$\therefore 201\, C S = C P$$

$$\therefore C S = \frac{C P}{201},$$ on the same scale of notation.

PROB. 37. Two places are on the same meridian. Find the hour on a given day, when the sun will have the same altitude at each place.

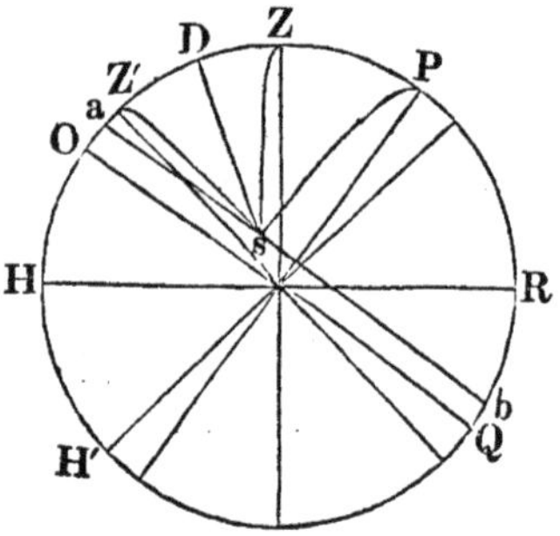

Z Z′, two zeniths of places, ∴ Z Z′ is known, S the place of the sun in the parallel a b, Z S = S Z′.

From S draw perpendicular S D,

$$\therefore Z D = Z' D,$$

$$\therefore P Z + \frac{Z Z'}{2} = P D,$$ is known,

P S is known, ∠ S D P right ∠,

∴ ∠ D P S = hour is known.

PROB. 38. Find the time in which the sun passes the vertical wire of a telescope.

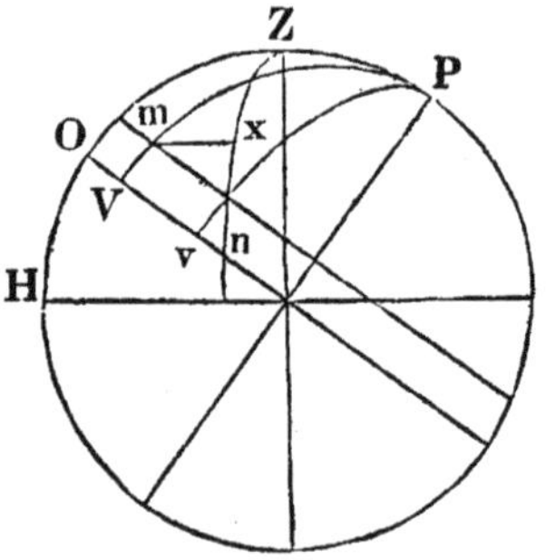

Meridian = the vertical wire,

∴ the time of passing the meridian = the time of passing the vertical wire.

Take m n = the sun's diameter = d.

V v : m n : r : cos. declination,

$$\therefore V v = \frac{d\, r}{\text{cos. dec.}},$$

∴ V v converted into the time at the rate of 15′ for 1° = the time required.

PROB. 40. If a man be in the arctic circle, the longest day = 24 hours, the shortest = 0.

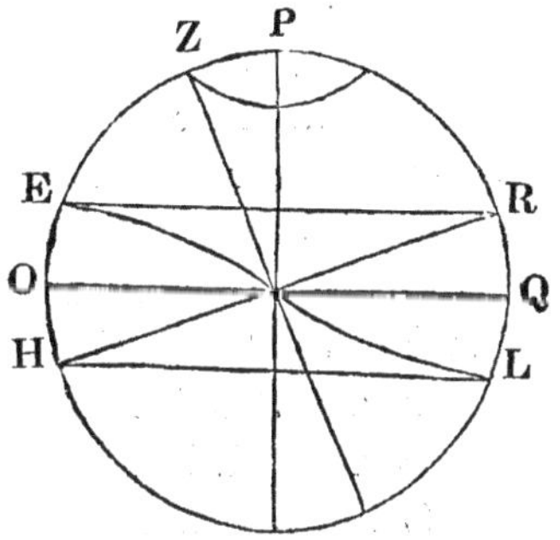

P Z = obliquity = Q R,

$\therefore$ Z R = P Q = 90

Z H = P Q = 90^0

$\therefore$ H R is the horizon, and the nearest parallel touches at R,

$\therefore$ the day = 24 hours, and the farthest parallel touches at H,

$\therefore$ the day = 0 hours.

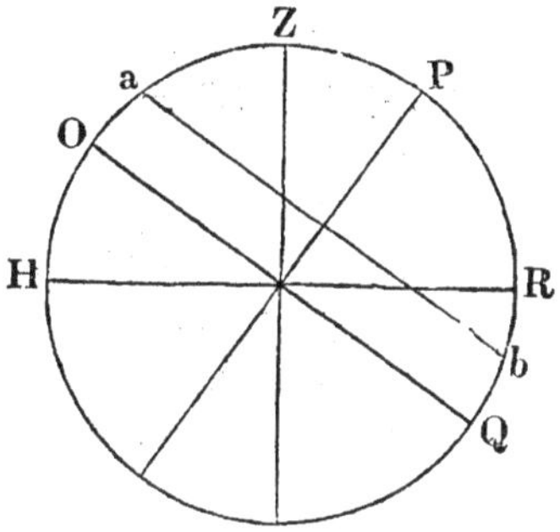

Prob. 41. Given the sun's meridian altitude = 62^0, midnight depression = 22^0. Find the longitude and declination.

Q a = b Q

or H a — H Q = R Q — R b

= H Q — R b,

$\therefore \frac{\text{H a} + \text{R b}}{2} = \text{H Q} = \cos. \lambda$

$= 42^0, \therefore \lambda = 48^0,$

$\therefore$ D = 62 — 42 = 20.

Prob. 42. Given the sun's declination, apparent diameter, altitude, and longitude. Find the time of passing the horizontal wire of a telescope.

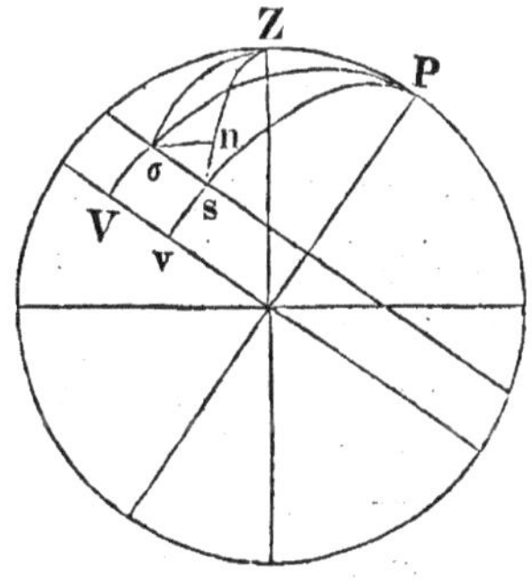

s = the place of the sun.

Take s n in a vertical circle = the sun's diameter = d.

Draw n σ parallel to the horizon,

V v : σ s : : r : cos. dec.

σ s : n s : : r : sin. n s P,

$\therefore$ V v : d : : r^2 : sin. P s sin. n s P

: : r^2 : sin. Z P sin. P Z s,

$\therefore \text{V v} = \frac{\text{d } r^2}{\text{sin. cos. } \lambda \text{ sin. azimuth,}}$ converted into the time, gives the time required

Prob. 43. Given the longitude, right ascension, and declination of two stars; find the time when both are on the same azimuthal circle, and also of the azimuth.

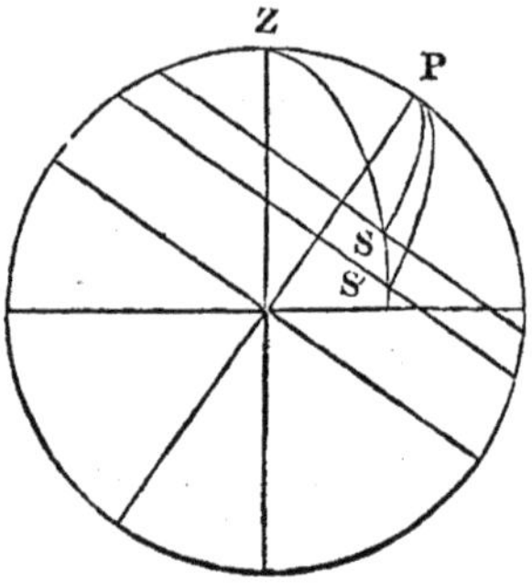

Given P S, P S′, and $\angle$ S P S′ = difference of right ascension.

$\therefore \angle$ P S S′ is known,

$\therefore \angle$ P S Z is known,

and Z P given, and P S given,

$\therefore \angle$ P Z S, is known = azimuth, and Z P S = time for the first star, or (Z P S + S P S′) = time for the second star.

Prob. 44. Given the longitude and declination. Find the time when the sun ascends perpendicular to H R.

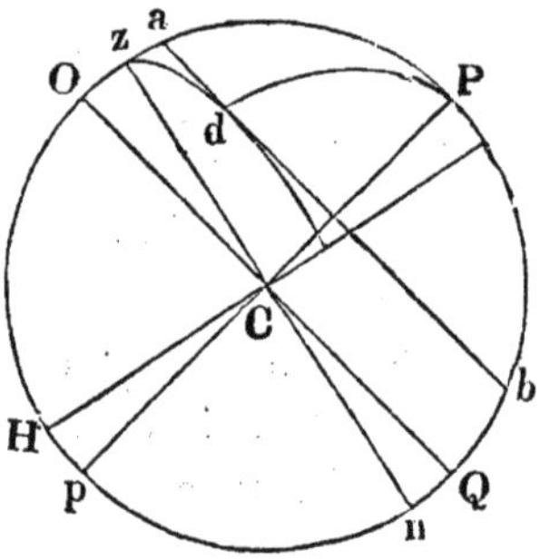

D must be greater than λ, or a Q greater than Z Q.

Draw the vertical circle tangent to the parallel of declination, at d.

P Z given, P d given, $\angle$ P d Z is a right $\angle$,

$\therefore \angle$ Z P d is known.

Prob. 45. Find the length of the longest day in longitude = 45^0.

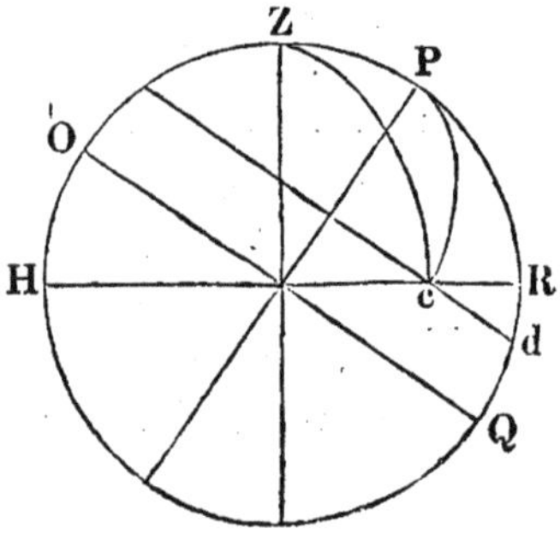

Q d = obliquity,

$\therefore$ P d = 90 — obliquity = P c,

Z P = 45,

Z c = 90,

$\therefore$ 2 hours is known.

PROB. 46. Find the right ascension and declination of a star, when in a line with two known stars, and also in another line with two other known stars.

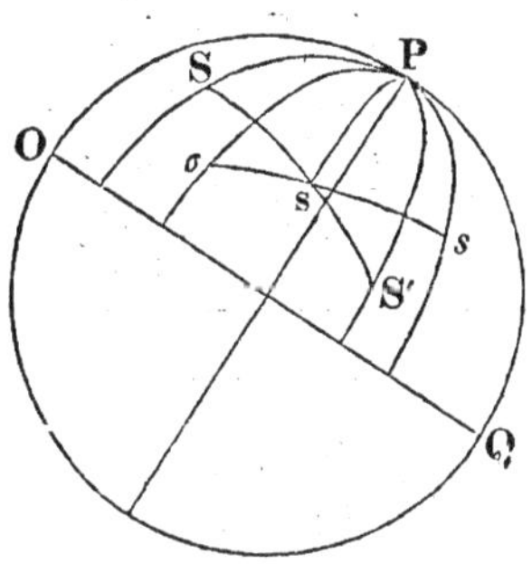

The star is in the same line with S, S′, and in the same line with *s*, *σ*,

∴ in the intersection s.

PROB. 47. The least error in the time due to the given error in altitude $= b''$. Find the longitude.

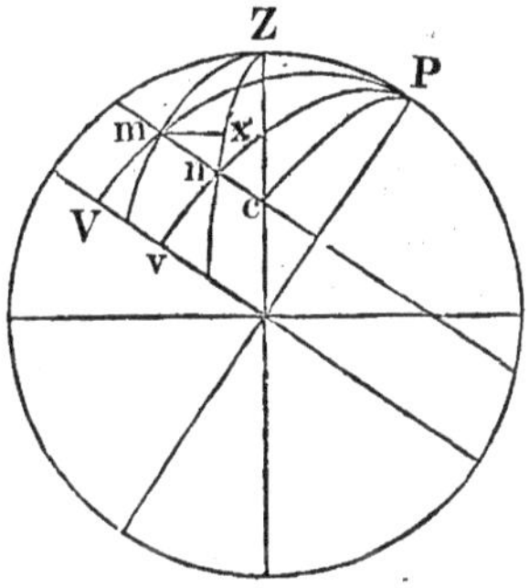

n x is the given error in altitude,

V v : m n : : r : cos. declination,

m n : n x : : r : sin. x n P.

V v : n x : : r^2 : sin. P n sin. Z n P,

$$V v = \frac{n x r^2}{\sin. P n \sin. Z n P}$$

$$= \frac{n x r^2}{\sin. Z P \sin. P Z n}$$

$$= \frac{n x r^2}{\cos. \lambda \sin. \text{azimuth}},$$

∴ V v is least when the sin. azimuth is greatest, or the azimuth = 90°, i. e. the prime vertical.

$$\therefore b = \frac{n x r^2}{\cos. \lambda},$$

$$\therefore \cos. \lambda = \frac{n x r^2}{b}.$$

PROB. 48. Given two altitudes, and two azimuths of the sun. Find the longitude.

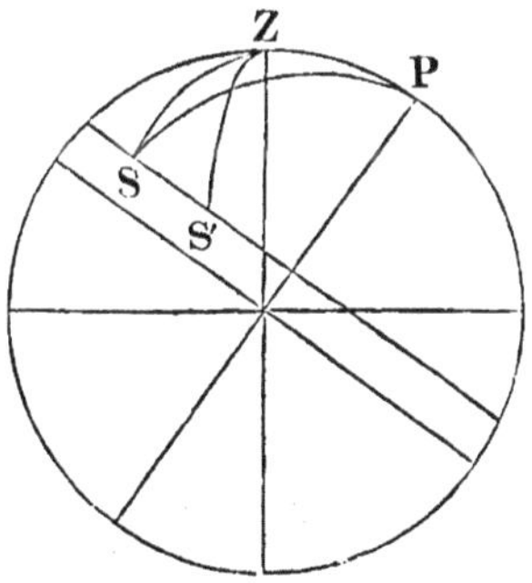

Z S is known, Z S′ is known, ∠ S Z S′ = difference of the azimuth,

∴ ∠ Z S S′ is known,

∴ ∠ Z S P = Z S S′ — 90° is known,

∴ Z S P, Z S, S Z P, known,

find Z P.

PROB. 49. Near the solstice, the declination $\propto$ longitude, nearly.

$$r \sin. D = \sin. L \sin. \gamma,$$

$$\therefore r\, d\,(D) \cos. D = \sin. \gamma\, d\,(L) \cos. L$$

$$\therefore r \frac{d\,(D)}{d\,(L)} = \frac{\sin. \gamma \cos. L}{\cos. D}$$

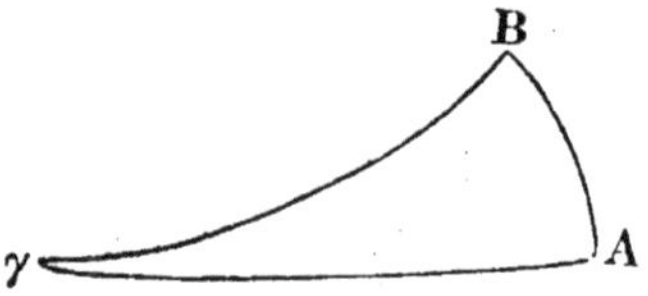

$$= \frac{\sin. \gamma \cos. 90 - d\,(L)}{\cos. \gamma}, \text{ since } D$$

may be considered the measure of γ,

$$= \tan. \gamma \sin. d\,(L)$$

$$= \tan. \gamma\, d\,(L), \text{ since } d\,(L) \text{ small,}$$

$$\therefore \frac{d(D)}{d\,(L)^2} = \frac{\tan. \gamma}{r} = \text{constant quantity,}$$

$$\therefore d\,(D) \propto d\,(L) \text{ nearly.}$$

PROB. 50. Given the apparent time T of the revolution of a spot on the sun's surface, find the real time.

Considering the spot as the inferior planet in inferior conjunction, $T = \frac{P\,p}{P - p}$ where P equals the earth's periodic time, p equals the planet's,

$$\therefore T\,P - T\,p = P\,p,$$

$$\therefore p = \frac{T\,P}{T + P}.$$

PROB. 51. The sun's declination equal 8° south, find the latitude of the place where he rises in the south east point, and also the time of his rising.

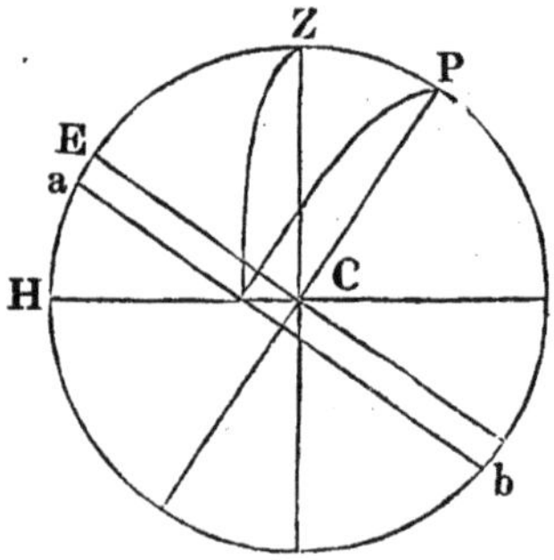

$$Z\,c = 90^0,\ P\,c = 98^0,\ \angle\, c\,Z\,S = 135^0,$$

whence Z P, and $\angle$ h

Prob. 52. How high must a man be raised to see the sun at midnight?

Z P = R Q. Take P d = Q b

$\therefore$ b d = 90°.

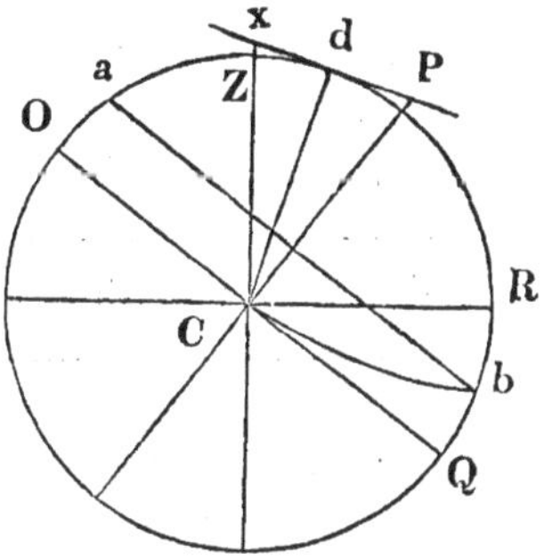

Draw x d to the tangent at d,
$\therefore$ if the person be raised to Z x, he will see the sun at b,

$\angle$ d C b = 90° = x C R,

$\therefore$ x C d = R C b measured by R b given.

$\therefore$ in the rectilinear $\triangle$ x d C, $\angle$ x d C = right angle,

$\angle$ x C d being known from the dec.

C d = radius of the earth.

$\therefore$ C x being known,

$\therefore$ C x — 90°, or Z x is known.

Prob. 53. Given the latitudes and longitudes of two places, find the straight line which joins them. They lie in the same declination of the circle.

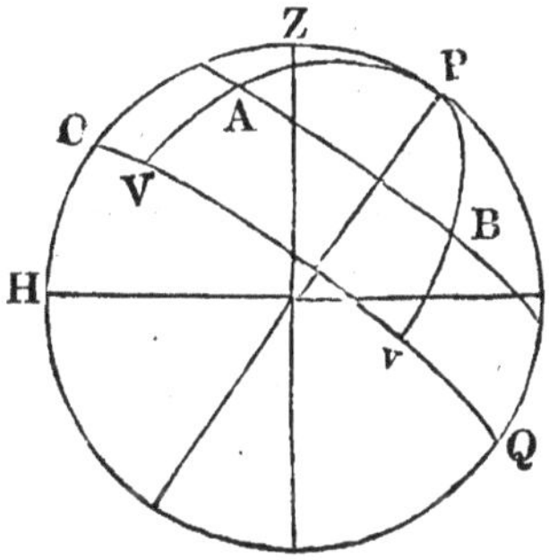

V v : A B : : 1 : cosine declination,

$\therefore$ A B is known,

and the straight line joining A, B, is the chord of A, B, $= 2 \sin. \frac{A B}{2}$.

Prob. 54. A clock being properly adjusted to keep the sidereal time, required to find when γ is on the meridian.

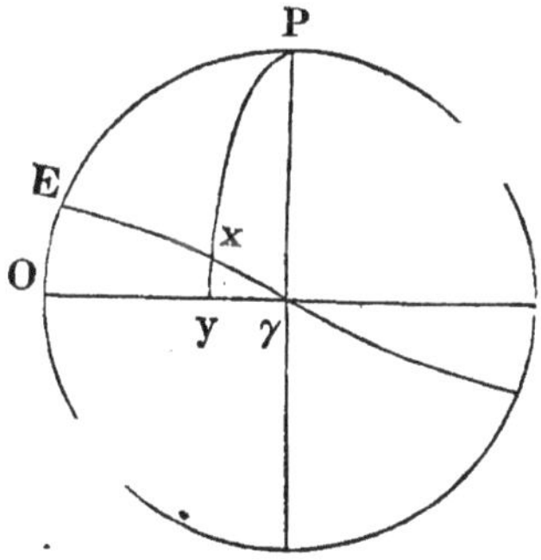

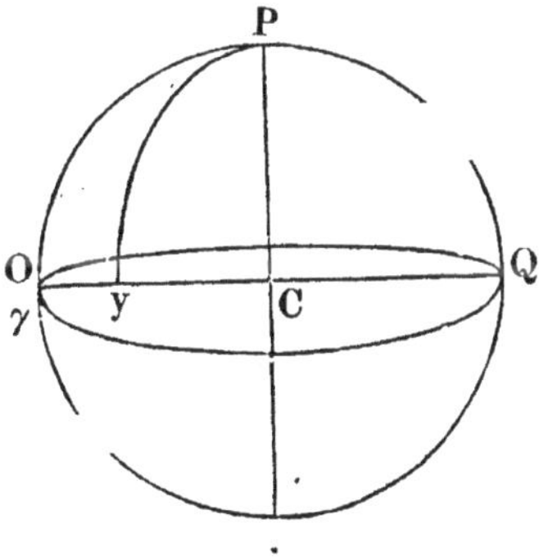

Observe the sun's center on the meridian, when the declination = x y, is known,

$\angle$ x y γ = right angle
x γ y = I, being known,
x y is known.

Whence y γ = time from noon to γ being on the meridian, or from γ being on the meridian to noon, whence two values of γ y are found.

If the declination north and before solstice the > value gives the time,
——————————— after——— < ——————————
If the declination south and before———12+< ——————————
——————————— after———12+> ——————————

(Thelwall.)

PROB. 55. Given the sun's declination and longitude, find his right aseension, his oblique ascension, his azimuth and amplitude, and the time of his rising, and the length of the day.

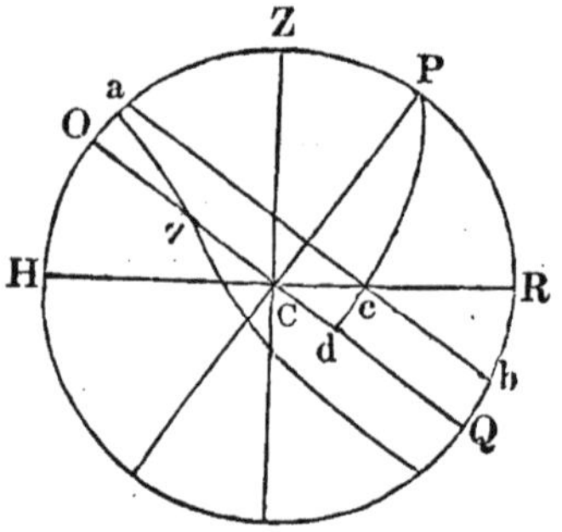

γ C is given, from $\triangle$ c C d, c d is given; I. and right angle, find c d.
$\therefore$ C γ = R A, C d = oblique asc^{n}.
and C d measures $\angle$ C P c,
.. 90^0 + C P c = time of rising,
2 (90^0 + C P c) = length of the day.

NOTES.

To show that (see p. 16.)

$$\Sigma\,.\,\mu \times \Sigma\,.\,\mu \frac{x\,dy - y\,dx}{dt} + \Sigma\,.\,\mu y \times \Sigma\,.\,\mu \frac{dx}{dt} - \Sigma\,.\,\mu x \times \Sigma\,\mu \frac{dy}{dt}$$

$$= \Sigma\,.\,\mu\mu' \left\{ \frac{(x'-x)(dy'-dy) - (y'-y)(dx'-dx)}{dt} \right\}.$$

Not considering the common factor $\frac{1}{dt}$, we have

$$\Sigma\,.\,\mu \times \Sigma\,.\,\mu\,(x\,dy - y\,dx)$$
$$= (\mu + \mu' + \mu'' + \ldots)\,\{\mu\,(x\,dy - y\,dx)$$
$$+ \mu'\,(x'\,dy' - y'\,dx') + \mu''\,(x''\,dy'' - y''\,dx'') + \&c.\}$$
$$= \mu^2\,(x\,dy - y\,dx) + \mu'^2\,(x'\,dy' - y'\,dx')$$
$$+ \mu''^2\,(x''\,dy'' - y''\,dx'') + \&c.$$
$$+ \mu\mu'\,(x\,dy - y\,dx + x'\,dy' - y'\,dx')$$
$$+ \mu\mu''\,(x\,dy - y\,dx + x''\,dy'' - y''\,dx'') + \&c.$$
$$+ \mu'\mu''\,(x'\,dy' - y'\,dx') + (x''\,dy'' - y''\,dx'')$$
$$+ \mu'\mu'''\,(x'\,dy' - y'\,dx' + x'''\,dy''' - y'''\,dx''') + \&c$$
$$+ \mu''\mu'''\,(x''\,dy'' - y''\,dx'' + x'''\,dy''' - y'''\,dx''') + \&c.$$
$$\&c.$$

the law of which is evident

Again,

$$\Sigma\,.\,\mu y \times \Sigma\,.\,\mu\,dx - \Sigma\,.\,\mu x \times \Sigma\,.\,\mu\,dy$$
$$= (\mu y + \mu' y' + \mu'' y'' + \ldots)\,(\mu\,dx + \mu'\,dx' + \mu''\,dx'' + \ldots\,\&c.)$$
$$- (\mu x + \mu' x' + \mu'' x'' + \ldots)\,(\mu\,dy + \mu'\,dy' + \mu''\,dy'' + \ldots)$$
$$= -\mu^2\,(x\,dy - y\,dx) - \mu'^2\,(x'\,dy' - y'\,dx') - \&c.$$
$$+ \mu\mu'\,(y\,dx' - x\,dy' + y'\,dx - x'\,dy)$$
$$+ \mu\mu''\,(y\,dx'' - x\,dy'' + y''\,dx - x''\,dy) + \&c.$$
$$+ \mu'\mu''\,(y'\,dx'' - x'\,dy'' + y''\,dx' - x''\,dy')$$
$$+ \mu'\mu'''\,(y'\,dx''' - x'\,dy''' + y'''\,dx' - x'''\,dy') + \&c.$$
$$+ \&c.$$

Hence by adding together these results the aggregate is

$$\mu\mu'(x\,dy - y\,dx + x'\,dy' - y'\,dx' + y\,dx' - x\,dy' + y'\,dx - x'\,dy) + \mu\mu''(x\,dy - y\,dx + \&c.) + \&c.$$

$$\mu'\mu''(x'\,dy' - y'\,dx' + x''\,dy'' - y''\,dx'' + y'\,dx'' - x'\,dy'' + y''\,dx' - x''\,dy') + \&c.$$

&c.

But

$$\begin{aligned} x\,dy - y\,dx + x'\,dy' - y'\,dx' + y\,dx' - x\,dy' + y'\,dx - x'\,dy &= dy\,(x - x') + dx\,(y' - y) + dy'\,(x' - x) + dx'\,(y - y') \\ &= (x' - x)\,(dy' - dy) - (y' - y).(dx' - dx); \end{aligned}$$

and in like manner the coefficients of $\mu\mu''$, $\mu\mu'''$ ..... $\mu'\mu''$, $\mu'\mu'''$, ... &c. are found to be respectively

$$(x'' - x)\,(dy'' - dy) - (y'' - y)\,(dx'' - dx),$$
$$(x''' - x)\,(dy''' - dy) - (y''' - y)\,(dx''' - dx),$$

$$(x'' - x')\,dy'' - dy') - (y'' - y')\,(dx'' - dx'),$$
$$(x''' - x')\,(dy''' - dy') - (y''' - y')\,(dx''' - dx')$$

&c.

Hence then the sum of all the terms in $\mu\mu'$, $\mu\mu''$ ..... $\mu'\mu''$, $\mu'\mu''$ ..... $\mu''\mu'''$, $\mu'\mu''''$ ..... is briefly expressed by

$$\Sigma\,.\,\mu\mu'\,\{(x' - x)\,(dy' - dy) - (y' - y)\,(dx' - dx)\}$$

and the suppressed coefficient $\frac{1}{dt}$ being restored, the only difficulty of p. 16 will be fully explained.

That $\Sigma\,.\left(\frac{d\lambda}{dx}\right) = 0$, &c. has been shown.

2. To show that $\int(2\,\Sigma\,.\,\mu\,dx \times \Sigma\,.\,\mu\,d^2x) = (\Sigma\,.\,\mu\,dx)^2$.......... page 17.

$$\begin{aligned} \Sigma\,.\,\mu\,d^2x &= \mu\,d^2x + \mu'\,d^2x' + \&c. \\ &= d\,.\,\mu\,dx + d\,.\,\mu'\,dx' + \&c. \\ &= d\,(\mu\,dx + \mu'\,dx' + \&c.) \\ &= d\,.\,\Sigma\,.\,\mu\,dx. \end{aligned}$$

Henee

$$\begin{aligned} \int(2\,\Sigma\,.\,\mu\,dx \times \Sigma\,.\,\mu\,d^2x) &= \int 2\,.\,\Sigma\,\mu\,dx \times d\,.\,\Sigma\,.\,\mu\,dx \\ &= (\Sigma\,.\,\mu\,dx)^2 \end{aligned}$$

being of the form $\int 2\,n\,du = u^2$.

3. To show that (page 17).

$$\Sigma.\mu \times \Sigma.\mu\,(d\,x^2 + d\,y^2 + d\,z^2)$$
$$- \{(\Sigma.\mu\,d\,x)^2 + (\Sigma.\mu\,d\,y)^2 + (\Sigma.\mu\,d\,z)^2\}$$
$$= \Sigma.\mu\mu'\,\{(d\,x' - d\,x)^2 + (d\,y' - d\,y)^2 + (d\,z' - d\,z)^2\}.$$

Since the quantities are similarly involved, for brevity, let us find the value of $\Sigma.\mu \times \Sigma.\mu\,d\,x^2 - (\Sigma.\mu\,d\,x)^2$.

$$\text{It} = (\mu + \mu' + \mu'' + \ldots.)\,(\mu\,d\,x^2 + \mu'\,d\,x'^2 + \mu''\,d\,x''^2 + \ldots.)$$
$$- (\mu\,d\,x + \mu'\,d\,x' + \mu''\,d\,x'' + \ldots.)^2;$$

Consequently when the expression is developed, the terms $\mu^2\,d\,x^2$, $\mu'^2\,d\,x'^2$, $\mu''^2\,d\,x''^2$, &c. will be destroyed, and the remaining ones will be

$$\mu\mu'\,(d\,x^2 + d\,x'^2 - 2\,d\,x\,d\,x') = \mu\mu'\,(d\,x' - d\,x)^2$$
$$\mu\mu''\,(d\,x^2 + d\,x''^2 - 2\,d\,x\,d\,x'') = \mu\mu''\,(d\,x'' - d\,x)^2$$

$$\mu'\mu''\,(d\,x'^2 + d\,x''^2 - 2\,d\,x'\,d\,x'') = \mu'\mu''\,(d\,x'' - d\,x')^2$$
$$\mu'\mu'''\,(d\,x'^2 + d\,x'''^2 - 2\,d\,x'\,d\,x''') = \mu'\mu'''\,(d\,x''' - d\,x')^2$$

$$\mu''\mu'''\,(d\,x''^2 + d\,x'''^2 - 2\,d\,x''\,d\,x''') = \mu''\mu'''\,(d\,x''' - d\,x'')^2$$
&c.

Hence, of the partial expression

$$\Sigma.\mu \times \Sigma.\mu\,d\,x^2 - (\Sigma.\mu\,d\,x)^2 = \Sigma.\mu\mu'\,(d\,x' - d\,x)^2.$$

In like manner

$$\Sigma.\mu \times \Sigma.\mu\,d\,y^2 - (\Sigma.\mu\,d\,y)^2 = \Sigma.\mu\mu'\,(d\,y' - d\,y)^2$$
$$\Sigma.\mu \times \Sigma.\mu\,d\,z^2 - (\Sigma.\mu\,d\,z)^2 = \Sigma.\mu\mu'\,(d\,z' - d\,z)^2$$

and the aggregate of these three, whose first members amount to the proposed form, is

$$\Sigma.\mu\mu'\,\{(d\,x' - d\,x)^2 + (d\,y' - d\,y)^2 + (d\,z' - d\,z)^2\}$$

4. To show that (p. 19.)

$$\frac{\Sigma.\frac{\mu\,x}{\varrho^3}}{\Sigma.\mu} = \frac{3\,x'}{(\varrho')^3}$$

nearly.

It is shown already in page 19 that

$$\frac{x}{\varrho^3} = \frac{x}{(\varrho')^3} - \frac{3x'}{(\varrho')^5}(x'x_, + y'y_, + z'z_,).$$

But since $x_, = x - x'$, $y_, = y - y'$, $z_, = z - z'$, by substitution and multiplying both members by μ, we get

$$\frac{\mu x}{\varrho^3} = \frac{\mu x}{(\varrho')^3} - \frac{3x'}{(\varrho')^5}(x' . \mu x + y' . \mu y + z' . \mu z) + \frac{3x'}{(\varrho')^3} . \mu$$

nearly.

Similarly

$$\frac{\mu' x'}{\varrho'^3} = \frac{\mu' x'}{(\varrho')^3} - \frac{3x'}{(\varrho')^5}(x' . \mu' x' + y' . \mu' y' + z' . \mu' z') + \frac{3x'}{(\varrho')^3} . \mu'$$

nearly.

&c.

Hence

$$\Sigma . \frac{\mu x}{\varrho^3} = \frac{\Sigma . \mu x}{(\varrho')^3} - \frac{3x'}{(\varrho')^5}(x' \Sigma . \mu x + y' \Sigma . \mu y + z' \Sigma . \mu z) + \frac{3x'}{(\varrho')^3} \Sigma \mu.$$

But by the property of the center of gravity,

$$\Sigma . \mu x = 0, \ \Sigma . \mu y = 0, \ \Sigma . \mu z = 0.$$

Hence

$$\Sigma . \frac{\mu x}{\varrho^3} = \frac{3x'}{(\varrho')^3} \times \Sigma \mu \text{ nearly.}$$

5. To show that (p. 22.)

$$\frac{x}{\varrho} d^2 x + \frac{y}{\varrho} d^2 y + \frac{z}{\varrho} d^2 z = d^2 \varrho - \varrho d v^2 \cos.^2 \theta - \varrho d \theta^2$$

and that

$$\frac{x}{\varrho}\left(\frac{dQ}{dx}\right) + \frac{y}{\varrho}\left(\frac{dQ}{dy}\right) + \frac{z}{\varrho}\left(\frac{dQ}{dz}\right) = \left(\frac{dQ}{d\varrho}\right).$$

First, we have

$$x d^2 x + y d^2 y + z d^2 z$$
$$= d(x dx + y dy + z dz) - (dx^2 + dy^2 + dz^2).$$

But

$$x^2 + y^2 + z^2 = \varrho^2,$$
$$x dx + y dy + z dz = \varrho . d \varrho$$

and because

$$x = \varrho \cos. \theta \times \cos. v$$
$$y = \varrho \cos. \theta \times \sin. v$$
$$z = \varrho \sin. \theta.$$

$$\therefore dx^2 + dy^2 = \{d(\varrho \cos.\theta) . \cos. v - \varrho \cos.\theta \times dv \sin. v\}^2$$
$$+ \{(\varrho \cos.\theta) \sin. v + \varrho \cos.\theta\, dv \cos. v\}^2$$
$$= (d . \varrho \cos.\theta)^2 + \varrho^2 dv^2 \cos.^2\theta,$$
$$\therefore dx^2 + dy^2 + dz^2 = (d . \varrho \sin.\theta)^2 + (d . \varrho \cos.\theta)^2 + \varrho^2 dv^2 \cos.^2\theta$$
$$= d\varrho^2 + \varrho^2 d\theta^2 + \varrho^2 dv^2 \cos.^2\theta.$$

Hence, since also

$$d . \varrho\, d\varrho = d\varrho^2 + \varrho\, d^2\varrho,$$
$$\frac{x\, d^2x + y\, d^2y + z\, d^2z}{\varrho} = d^2\varrho - \varrho\, dv^2 \cos.^2\theta - \varrho\, d\theta^2.$$

Secondly, since ϱ is evidently independent of the angles θ and v, the three equations (1), give us

$$\left(\frac{dx}{d\varrho}\right) = \cos.\theta \cos. v = \frac{x}{\varrho},$$
$$\left(\frac{dy}{d\varrho}\right) = \cos.\theta \sin. v = \frac{y}{\varrho},$$
$$\left(\frac{dz}{d\varrho}\right) = \sin.\theta \quad = \frac{z}{\varrho}$$

Hence

$$\frac{x}{\varrho}\left(\frac{dQ}{dx}\right) + \frac{y}{\varrho}\left(\frac{dQ}{dy}\right) + \frac{z}{\varrho}\left(\frac{dQ}{dz}\right)$$
$$= \left(\frac{dQ}{dx}\right)\left(\frac{dx}{d\varrho}\right) + \left(\frac{dQ}{dy}\right)\left(\frac{dy}{d\varrho}\right) + \left(\frac{dQ}{dz}\right)\left(\frac{dz}{d\varrho}\right).$$

But since Q is a function of ϱ (observe the equations 1), and ϱ is a function of x, y, z, viz. $\sqrt{x^2 + y^2 + z^2}$,

$$dQ = \left(\frac{dQ}{d\varrho}\right) d\varrho = \left(\frac{dQ}{d\varrho}\right) \times \left\{dx\left(\frac{d\varrho}{dx}\right) + dy\left(\frac{d\varrho}{dy}\right) + dz\left(\frac{d\varrho}{dz}\right)\right\}$$
$$= dx . \left(\frac{dQ}{d\varrho}\right)\left(\frac{d\varrho}{dx}\right) + dy\left(\frac{dQ}{d\varrho}\right)\left(\frac{d\varrho}{dy}\right) + dz\left(\frac{dQ}{d\varrho}\right)\left(\frac{d\varrho}{dz}\right).$$

But

$$dx\left(\frac{dQ}{d\varrho}\right)\left(\frac{d\varrho}{dx}\right) = dx . \left(\frac{dQ}{dx}\right) = d\varrho\left(\frac{dx}{d\varrho}\right)\left(\frac{dQ}{dx}\right)$$

and like transformations may be effected in the other two terms. Consequently we have

$$dQ = d\varrho\left(\frac{dx}{d\varrho}\right)\left(\frac{dQ}{dx}\right) + d\varrho\left(\frac{dy}{d\varrho}\right)\left(\frac{dQ}{dy}\right) + d\varrho\left(\frac{dz}{d\varrho}\right)\left(\frac{dQ}{dz}\right).$$

Hence and from what was before proved, we get

$$\frac{d^2\varrho}{dt^2} - \frac{\varrho\, d v^2}{d t^2}\cos.^2\theta - \frac{\varrho\, d\theta^2}{d t^2} = \frac{x}{\varrho}\left(\frac{dQ}{dx}\right) + \frac{y}{\varrho}\left(\frac{dQ}{dy}\right) + \frac{y}{\varrho}\left(\frac{dQ}{dz}\right)$$

$$= \left(\frac{dQ}{d\varrho}\right).$$

6. To show that $x\,d^2y - y\,d^2x = d\,(\varrho^2\,dv\cos.^2\theta)$, and that $x\left(\frac{dQ}{dy}\right) - y\left(\frac{dQ}{dx}\right) = \left(\frac{dQ}{dv}\right)$, see p. 22.

First, since

$$x\,d^2y = d\,.\,x\,dy - dx\,dy,$$
$$y\,d^2x = d\,.\,y\,dx - dx\,dy,$$
$$\therefore x\,d^2y - y\,d^2x = d\,.(x\,dy - y\,dx).$$

But from equations (1), p. 22,

$$dy = \sin.v\,.\,d\,(\varrho\cos.\theta) + \varrho\cos.\theta\,.\cos.v\,dv$$
$$dx = \sin.v\,.\,d\,(\varrho\cos.\theta) - \varrho\cos.\theta\,.\sin.v\,dv,$$

$$\left.\begin{array}{l}\therefore x\,dy = \sin.v\cos.v\,.\,\dfrac{d\,(\varrho^2\cos.^2\theta)}{z} + \varrho^2\cos.^2\theta\cos.^2v\,dv \\ y\,dx = \sin.v\cos.v\,.\,\dfrac{d\,(\varrho^2\cos.^2\theta)}{z} - \varrho^2\cos.^2\theta\sin.^2v\,dv\end{array}\right\}$$

the difference of which is

$$\varrho^2\cos.^2\theta \times dv.$$

Consequently

$$x\,d^2y - y\,d^2x = d\,.(\varrho^2\,dv\cos.^2\theta).$$

Secondly by equations (1) p. 22, we have

$$\left(\frac{dy}{dv}\right) = \varrho\cos.\theta\cos.v = x$$

$$\left(\frac{dx}{dv}\right) = -\varrho\cos.\theta\sin.v = -y,$$

$$\therefore x\left(\frac{dQ}{dy}\right) - y\left(\frac{dQ}{dx}\right) = \left(\frac{dy}{dv}\right)\left(\frac{dQ}{dy}\right) + \left(\frac{dx}{dv}\right)\left(\frac{dQ}{dx}\right).$$

But since, dividing the two first of the equations (1) p. 22, we have $\frac{y}{x} = \tan.v$, v is a function of x, y only. Consequently, as in the note preceding this it may be shown that

$$\left(\frac{dQ}{dv}\right) = \left(\frac{dy}{dv}\right)\left(\frac{dQ}{dy}\right) + \left(\frac{dx}{dv}\right)\left(\frac{dQ}{dx}\right)$$
$$= x\left(\frac{dQ}{dy}\right) - y\left(\frac{dQ}{dx}\right).$$

Hence

$$\frac{d.\left(\varrho^2 \frac{dv}{dt} \cos.^2 \theta\right)}{dt} = \left(\frac{dQ}{dv}\right).$$

7. To find the value of $\left(\frac{dQ}{d\theta}\right)$ in terms of ϱ, v, θ, (see the last line but two of p. 22).

Since θ is a function of x, y, z, we have

$$\left(\frac{dQ}{d\theta}\right) = \left(\frac{dQ}{dx}\right)\left(\frac{dx}{d\theta}\right) + \left(\frac{dQ}{dy}\right)\left(\frac{dy}{d\theta}\right) + \left(\frac{dQ}{dz}\right)\left(\frac{dz}{d\theta}\right).$$

But from equations (1) p. 22, we get

$$\left(\frac{dx}{d\theta}\right) = -\varrho \sin. \theta \cos. v$$
$$\left(\frac{dy}{d\theta}\right) = -\varrho \sin. \theta \sin. v$$
$$\left(\frac{dz}{d\theta}\right) = \varrho \cos. \theta.$$

Hence multiplying the values of

$$\left(\frac{dQ}{dx}\right), \quad \left(\frac{dQ}{dy}\right), \quad \left(\frac{dQ}{dz}\right),$$

viz.

$$\frac{d^2 x}{dt^2}, \quad \frac{d^2 y}{dt^2}, \quad \frac{d^2 z}{dt^2}$$ (see p. 22),

by the partial differences we get

$$\left(\frac{dQ}{d\theta}\right) = \frac{1}{dt^2}\{d^2 z\, \varrho \cos. \theta - d^2 y . \varrho \sin. \theta \sin. v - d^2 x\, \varrho \sin. \theta \cos. v\}$$

Now the first term gives

$$\varrho \cos. \theta . d^2 z = \varrho\, \{d^2 \varrho \sin. \theta \cos. \theta + 2\, d\varrho\, d\theta \cos.^2 \theta$$
$$+ \varrho \cos.^2 \theta\, d^2 \theta - \varrho\, d\theta^2 \sin. \theta \cos. \theta\},$$

and the two other terms gives when added, by means of the equations (1) p. 22,

$$-\frac{\sin. \theta}{\cos. \theta}(y\, d^2 y + x\, d^2 x) = -\frac{\sin. \theta}{\cos. \theta}\{d\,(y\, dy + x\, dx) - (dx^2 + dy^2)\}$$

But

$$d(y\,d\,y + x\,d\,x) = \tfrac{1}{2} d\,.(d\,.\overline{x^2 + y^2}) = \tfrac{1}{2} d^2 (\varrho^2 \cos.^2 \theta)$$
$$= d\,\{\varrho \cos.\theta\, d\,(\varrho \cos.\theta)\}$$
$$= (d\,.\,\varrho \cos.\theta)^2 + \varrho \cos.\theta\, d^2 (\varrho \cos.\theta)$$

and

$$d\,x^2 + d\,y^2 = (d\,.\,\varrho \cos.\theta)^2 + \varrho^2 \cos.^2 \theta\,.\,d\,v^2.$$

Hence

$$-\frac{\sin.\theta}{\cos.\theta}(y\,d^2 y + x\,d^2 x)$$
$$= -\frac{\sin.\theta}{\cos.\theta}\{\varrho \cos.\theta\, d^2 (\varrho \cos.\theta) - \varrho^2 \cos.^2\theta\, d\,v^2\}$$
$$= -\varrho \sin.\theta\,\{d^2 (\varrho \cos.\theta) - \varrho \cos.\theta\, d\,v^2\}$$
$$= -\varrho \sin.\theta\,\{d^2 \varrho \cos.\theta - 2\,d\,\varrho\,d\,\theta \sin.\theta - d^2\theta\,\varrho \sin.\theta$$
$$- \overline{d\,\theta^2 + d\,v^2}\,\varrho \cos.\theta\}.$$

Adding this value to the preceding one of the first term, we have

$$\left(\frac{d\,Q}{d\,\theta}\right) = \frac{\varrho}{d\,t^2} \times \{\varrho\, d^2\theta + 2\,d\,\varrho\,d\,\theta + \varrho\,d\,v^2 \sin.\theta \cos.\theta\}$$
$$= \varrho^2 . \frac{d^2\theta}{d\,t^2} + \varrho^2 \frac{d\,v^2}{d\,t^2} \sin.\theta \cos.\theta + \frac{2\,\varrho\,d\,\varrho\,d\,\theta}{d\,t^2},$$

the value required.

8. To develope $\frac{1}{1 + e \cos.\theta}$ in terms of the cosines of θ and of its multiples, see p. 25.

If c be the member whose hyperbolic logarithm is unity, we know that

$$\cos.\theta = \frac{c^{\theta\sqrt{-1}} + c^{-\theta\sqrt{-1}}}{2}$$

which value of cos. θ being substituted in the proposed expression, we have

$$\frac{1}{1 + e \cos.\theta} = \frac{2\,c^{\theta\sqrt{-1}}}{e\,c^{2\theta\sqrt{-1}} + 2\,c^{\theta\sqrt{-1}} + e}$$
$$= \frac{2\,c^{\theta\sqrt{-1}}}{e} \times \frac{1}{c^{2\theta\sqrt{-1}} + \frac{2}{e} c^{\theta\sqrt{-1}} + 1}.$$

But since

$$c^{2\theta\sqrt{-1}} + \frac{2}{e} c^{\theta\sqrt{-1}} + 1 = 0$$

gives

$$c^{\theta\sqrt{-1}} = -\frac{1}{e} \pm \sqrt{\frac{1}{e^2} - 1} = -\left(\frac{1}{e} \mp \frac{\sqrt{1-e^2}}{e}\right)$$

and since, if we make

$$\frac{1}{e}\left(1 - \sqrt{1-e^2}\right) = \lambda \text{ which also } = \frac{e}{1+\sqrt{1-e^2}},$$

we also have

$$\frac{1}{e}\left(1 + \sqrt{1-e^2}\right) = \frac{1}{\lambda};$$

the expression proposed becomes

$$\frac{1}{1 + e\cos.\theta} = \frac{2\,c^{\theta\sqrt{-1}}}{e} \times \frac{1}{\left(c^{\theta\sqrt{-1}} + \lambda\right)\left(c^{\theta\sqrt{-1}} + \frac{1}{\lambda}\right)}$$

$$= \frac{2\lambda}{e} \times \frac{1}{\left(1 + \lambda c^{\theta\sqrt{-1}}\right)\left(1 + \lambda c^{-\theta\sqrt{-1}}\right)}$$

$$= \frac{2\lambda}{e\,(1-\lambda^2)} \times \left(\frac{1}{1+\lambda c^{\theta\sqrt{-1}}} - \frac{\lambda c^{-\theta\sqrt{-1}}}{1+\lambda c^{-\theta\sqrt{-1}}}\right)$$

But

$$\frac{\lambda}{e} = \frac{1}{1+\sqrt{1-e^2}},$$

and

$$1 - \lambda^2 = \frac{2\sqrt{1-e^2}}{1+\sqrt{1-e^2}},$$

$$\therefore \frac{1}{1+e\cos.\theta} = \frac{1}{\sqrt{(1-e^2)}} \times \left(\frac{1}{1+\lambda c^{\theta\sqrt{-1}}} - \frac{\lambda c^{-\theta\sqrt{-1}}}{1+\lambda c^{-\theta\sqrt{-1}}}\right),$$

which when $\theta = \nu - \varpi$ is the same expression as that in page 25.

Again by division

$$\frac{1}{1+\lambda c^{\theta\sqrt{-1}}} = 1 - \lambda c^{\theta\sqrt{-1}} + \lambda^2 c^{2\theta\sqrt{-1}} - \&c.$$

and

$$\frac{\lambda c^{-\theta\sqrt{-1}}}{1+\lambda c^{-\theta\sqrt{-1}}} = \lambda c^{-\theta\sqrt{-1}} - \lambda^2 c^{-2\theta\sqrt{-1}} + \&c.$$

Taking the latter from the former, we get

$$\frac{\sqrt{1-e^2}}{1+e\cos.\theta} = 1 - \lambda\left(c^{\theta\sqrt{-1}} + c^{-\theta\sqrt{-1}}\right) + \lambda^2\left(c^{2\theta\sqrt{-1}} + c^{-2\theta\sqrt{-1}}\right)$$

$$= 1 - 2\lambda\cos.\theta + 2\lambda^2\cos.2\theta - 2\lambda^3\cos.3\theta + \&c.$$

and substituting for θ the value $v - \varpi$, we get the expression in page 25.

9. To demonstrate the Theorem of page 28.

Let us take the case of three variables x, y, z. Then our system of differential equations is

$$0 = H x + G \frac{d x}{d t} + F \frac{d^2 x}{d t^2};$$

$$0 = H y + G \frac{d y}{d t} + F \frac{d^2 y}{d t^2};$$

$$0 = H z + G \frac{d z}{d t} + F \frac{d^2 z}{d t^2};$$

in which F, G, H, are symmetrical functions of x, y, z; that is such as would not be altered by substituting x for y, and y for x; and so on for the other variables taken in pairs; for instance, functions of this kind

$$\sqrt{x^2 + y^2 + z^2 + t^2}, \ (x y + x z + x t + y z + y t + z t)^{\frac{p}{q}},$$

$$(x y z + x z t + y z t),$$

log. (x y z t) and so on.

Multiply the first of the equations by the arbitrary α, the second by β, and the third by γ and add them together; the result is

$$0 = H (\alpha x + \beta y + \gamma z) + G \left(\alpha \frac{d x}{d t} + \beta \frac{d y}{d t} + \gamma \frac{d z}{d t}\right)$$

$$+ F \left(\alpha \frac{d^2 x}{d t^2} + \beta \frac{d^2 y}{d t^2} + \gamma \frac{d^2 z}{d t^2}\right).$$

Now since α, β, γ, are arbitrary, we may assume

$$\alpha x + \beta y + \gamma z = 0,$$

which gives

$$\alpha \frac{d x}{d t} + \beta \frac{d y}{d t} + \gamma \frac{d z}{d t} = 0,$$

$$\alpha \frac{d^2 x}{d t^2} + \beta \frac{d^2 y}{d t^2} + \gamma \frac{d^2 z}{d t^2} = 0.$$

and substituting for x, $\frac{d x}{d t}$, $\frac{d^2 x}{d t^2}$, their values hence derived in the first of the proposed equations, we have

$$-\frac{H}{\alpha}(\beta y+\gamma z)-\frac{G}{\alpha}\left(\beta\frac{dy}{dt}+\gamma\frac{dz}{dt}\right)-\frac{F}{\alpha}\left(\beta\frac{dy^2}{dt^2}+\gamma\frac{d^2z}{dt^2}\right)=$$

$$-\frac{\beta}{\alpha}\left(Hy+G\frac{dy}{dt}+F\frac{d^2y}{dt^2}\right)-\frac{\gamma}{\alpha}\left(Hz+G\frac{dz}{dt}+F\frac{d^2z}{dt^2}\right)=$$

$$-\frac{\beta}{\alpha}\times 0-\frac{\gamma}{\alpha}\times 0=0.$$

In the same it will appear that

$$\alpha x+\beta y+\gamma z=0$$

verifies each of the other two equations. It is therefore the integral of each of them, and may be put under the form

$$z=ax+by$$

in valuing only two arbitraries a and b, which are sufficient, two arbitraries only being required to complete the integral of an equation of the second order.

In the equations (θ) p. 27.

$$\frac{m}{\varrho^3}=H,\quad G=0\quad\text{and } F=1$$

and ϱ^3 being $=(x^2+y^2+z^2)^{\frac{3}{2}}$ is symmetrical with regard to x, y, z. Hence the theorem here applies and gives for the integral of any of the equations θ

$$z=ax+by,$$

see page 28.

Again, let us now take four variables x, y, z, u; then the theorem proposes the integration of

$$0=Hx+G\frac{dx}{dt}+F\frac{d^2x}{dt^2}$$

$$0=Hy+G\frac{dy}{dt}+F\frac{d^2y}{dt^2}$$

$$0=Hz+G\frac{dz}{dt}+F\frac{d^2z}{dt^2}$$

$$0=Hu+G\frac{du}{dt}+F\frac{d^2u}{dt^2}.$$

Multiplying these by the arbitraries α, β, γ, δ and adding them we get, as before

$$\begin{aligned}0=H&(\alpha x+\beta y+\gamma z+\delta u)\\&+G\left(\alpha\frac{dx}{dt}+\beta\frac{dy}{dt}+\gamma\frac{dz}{dt}+\delta\frac{du}{dt}\right)\\&+F\left(\alpha\frac{d^2x}{dt^2}+\beta\frac{d^2y}{dt^2}+\gamma\frac{d^2z}{dt^2}+\delta\frac{d^2u}{dt^2}\right)\end{aligned}$$

Assume

$$\alpha x + \beta y + \gamma z + \delta u = 0.$$

and upon trial it will be found as before, that this equation satisfies each of the four proposed equations, or it is their integrals supposing them to subsist simultaneously. As before, however, there are more arbitraries than are necessary for the integral of each, two only being required.

Hence the integral of each will be of the form

$$x + y + \gamma z + \delta u = 0.$$

This form might have been obtained at once, by adding the two last of the proposed equations multiplied by γ and δ to the two first of them, and assuming the coefficient of $H = 0$, as before.

In the same manner if we have (n) differential equations of the i-th order, the order involving the n variables $x^{(1)}, x^{(2)} \ldots x^{(n)}$, and of the general form

$$0 = H x^{(s)} + G \frac{d x^{(s)}}{d t} + F \frac{d^2 x^{(s)}}{d t^2} + \ldots A \frac{d^{i-1} x^{(s)}}{d t^{i-1}} + \frac{d^i x^{(s)}}{d t^i}.$$

we shall find by multiplying i of them (for instance the i wherein first $s = 1, 2 \ldots i$) by the arbitraries $\alpha^{(1)}, \alpha^{(2)}, \ldots \alpha^{(i)}$; adding these results together and their aggregate to the sum of the other equations; and assuming the coefficient of $H = 0$, that

$$\alpha^{(1)} x^{(1)} + \alpha^{(2)} x^{(2)} + \ldots \alpha^{(i)} x^{(i)} + x^{i+1} + x^{i+2} + \ldots x^n = 0$$

will satisfy each of the proposed differential equations subsisting simultaneously; and since it has an arbitrary for every integration, it must be the complete integral of any one of them.

This result is the same in substance as that enunciated in the theorem of p. 28, inasmuch as it is obtained by adding together the equations whose first members are $x^{(1)}, x^{(2)}$, &c. and making such arrangements as are permitted by a change of the arbitraries. In short if we had multiplied the i last equations instead of the i first by the arbitraries, and added the results to the n — i first equations, our assumption would have been

$$x^{(1)} + x^{(2)} + \ldots x^{(n-i)} + \alpha^{(1)} x^{(n-i+1)} + \alpha^{(2)} x^{(n-i+2)} + \ldots \alpha^i x^n = 0 \ldots (\alpha)$$

which is derived at once by adding together the n — i equations in page 28.

If we wish to obtain these $n - i$ equations from the equation (a), it may be effected by making assumptions of the required form, provided by so doing we do not destroy the arbitrary nature $\alpha^{(1)}, \alpha^{(2)}, \ldots\ldots \alpha^{(i)}$. The necessary assumptions do, however, evidently still leave them arbitrary. Those assumptions are therefore legitimate, and will give the forms of Laplace.

END OF VOLUME SECOND.

Lightning Source UK Ltd.
Milton Keynes UK
UKHW010830051218
333501UK00011B/478/P

9 781330 165928